aplia™

Get a Be[tter] Grade in Statistics!

Log in now to Aplia™ for Statistics. Score better on exams, practice homework problems, access your eBook, and more!

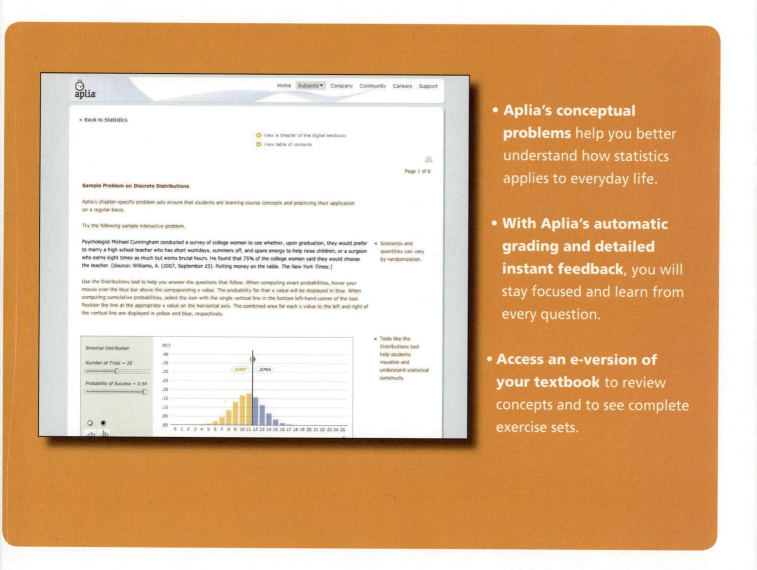

- **Aplia's conceptual problems** help you better understand how statistics applies to everyday life.

- **With Aplia's automatic grading and detailed instant feedback,** you will stay focused and learn from every question.

- **Access an e-version of your textbook** to review concepts and to see complete exercise sets.

To get started, use the access code that is packaged with this Hybrid Edition. Check with your instructor to verify that Aplia for Statistics is required for your course before purchasing.

www.aplia.com/statistics

HYBRID EDITION

Understanding
Basic Statistics

Charles Henry Brase
Regis University

Corrinne Pellillo Brase
Arapahoe Community College

BROOKS/COLE
CENGAGE Learning·

Australia • Brazil • Japan • Korea • Mexico • Singapore • Spain • United Kingdom • United States

Understanding Basic Statistics,
Hybrid Edition
Charles Henry Brase and
Corrinne Pellillo Brase

Senior Sponsoring Editor: Molly Taylor

Senior Development Editor: Jay Campbell

Assistant Editor: Shaylin Walsh Hogan

Associate Media Editor: Andrew Coppola

Senior Marketing Manager: Barb Bartoszek

Marketing Director: Mandee Eckersley

Marketing Coordinator: Michael Ledesma

Marketing Communications Manager:
Mary Anne Payumo

Content Project Manager: Jessica Rasile

Art Director: Linda Helcher

Manufacturing Planner: Doug Bertke

Rights Acquisition Specialist:
Shalice Shah-Caldwell

Production Service/Compositor:
Graphic World Inc.

Chapter Opener Photo: Francois
van Heerden/Shutterstock.com

TI-83 Plus and TI-84 Plus are registered
trademarks of Texas Instruments, Inc.
SPSS is a registered trademark of SPSS, Inc.
Minitab is a registered trademark of
Minitab, Inc.
Microsoft Excel Screen Shots reprinted by
permission from Microsoft Corporation.
Excel, Microsoft, and Windows are either
registered trademarks or trademarks of
Microsoft Corporation in the United States
and/or other countries.

For product information and technology assistance, contact us at
Cengage Learning Customer & Sales Support, 1-800-354-9706
For permission to use material from this text or product,
submit all requests online at **www.cengage.com/permissions**.
Further permissions questions can be e-mailed to
permissionrequest@cengage.com

Library of Congress Control Number: 2011943906

ISBN-13: 978-1-133-11414-7

ISBN-10: 1-133-11414-8

Brooks/Cole
20 Channel Center Street
Boston, MA 02210

Cengage Learning is a leading provider of customized learning solutions with office locations around the globe, including Singapore, the
United Kingdom, Australia, Mexico, Brazil, and Japan. Locate your local
office at **international.cengage.com/region.**

Cengage Learning products are represented in Canada by Nelson
Education, Ltd.

To learn more about Brooks/Cole, visit **www.cengage.com/brookscole.**

Purchase any of our products at your local college store or at our preferred online store **www.cengagebrain.com.**

Instructors: Please visit **login.cengage.com** and log in to access
instructor-specific resources.

Printed in the United States of America
1 2 3 4 5 6 7 16 15 14 13 12

This book is dedicated to the memory of
a great teacher, mathematician, and friend

Burton W. Jones

Professor Emeritus, University of Colorado

Options That SAVE Your Students Money

ONLINE LEARNING WITH EBOOK

Teach more, grade less with Aplia.

Aplia™ is an online interactive learning solution that improves comprehension and outcomes by increasing student effort and engagement.

Founded by a professor to enhance his own courses, **Aplia** provides automatically graded assignments with detailed, immediate explanations on every question, and innovative teaching materials. Our easy-to-use system has been used by more than 1,000,000 students at over 1800 institutions.

www.aplia.com/statistcs

Aplia Printed Access Card
ISBN: 9781133726005
Aplia can be bundled with the text. Contact your Cengage Learning representative to find out more about bundling options.

BOOK RENTALS UP TO 60% OFF

CENGAGE**brain**.com

Students have the CHOICE to **purchase the eBook** or **rent the text** at CengageBrain.com

OR

an **eTextbook in PDF format** is also available for instant access for your students at **www.coursesmart.com**.

CourseSmart

Students can rent Brase/Brase's *Understanding Basic Statistics*, 6th Edition, for up to 60% off list price.

CUSTOM SOLUTIONS

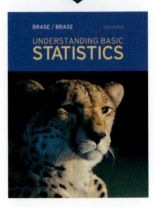

Custom Solutions to Fit Every Need

Contact your Cengage Learning representative to learn more about what custom solutions are available to meet your course needs.

- Adapt existing Cengage Learning content by adding or removing chapters

- Incorporate your own materials

CONTENTS

Critical Thinking

Students need to develop critical thinking skills in order to understand and evaluate the limitations of statistical methods. *Understanding Basic Statistics* makes students aware of method appropriateness, assumptions, biases, and justifiable conclusions.

CRITICAL THINKING

Sampling frame

We call the list of individuals from which a sample is actually selected the *sampling frame*. Ideally, the sampling frame is the entire population. However, from a practical perspective, not all members of a population may be accessible. For instance, using a telephone directory as the sample frame for residential telephone contacts would not include unlisted numbers.

Undercoverage

When the sample frame does not match the population, we have what is called *undercoverage*. In demographic studies, undercoverage could result if the homeless, fugitives from the law, and so forth, are not included in the study.

A **sampling frame** is a list of individuals from which a sample is actually selected.

Undercoverage results from omitting population members from the sample frame.

Sampling error

In general, even when the sampling frame and the population match, a sample is not a perfect representation of a population. Therefore, information drawn from a sample may not exactly match corresponding information from the population. To the extent that sample information does not match the corresponding population information, we have an error, called a *sampling error*.

◄ **Critical Thinking**

Critical thinking is an important skill for students to develop in order to avoid reaching misleading conclusions. The Critical Thinking feature provides additional clarification on specific concepts as a safeguard against incorrect evaluation of information.

Interpretation ▶

Increasingly, calculators and computers are used to generate the numeric results of a statistical process. However, the student still needs to correctly interpret those results in the context of a particular application. The Interpretation feature calls attention to this important step. Interpretation is stressed in examples, guided exercises, and in the problem sets.

The probability is 0.9988 that $\bar{x}$ is between 2350 and 2650.

(c) *Interpretation* At the end of each day, the inspector must decide to accept or reject the accumulated milk that has been held in cold storage awaiting shipment. Suppose the 42 samples taken by the inspector have a mean bacteria count $\bar{x}$ that is *not* between 2350 and 2650. If you were the inspector, what would be your comment on this situation?

SOLUTION: The probability that $\bar{x}$ is between 2350 and 2650 for milk that is not contaminated is very high. If the inspector finds that the average bacteria count for the 42 samples is not between 2350 and 2650, then it is reasonable to conclude that there is something wrong with the milk. If $\bar{x}$ is less than 2350, you might suspect someone added chemicals to the milk to artificially reduce the bacteria count. If $\bar{x}$ is above 2650, you might suspect some other kind of biological contamination.

10. | *Critical Thinking* Consider two data sets.

Set A: $n = 5$; $\bar{x} = 10$ Set B: $n = 50$; $\bar{x} = 10$

(a) Suppose the number 20 is included as an additional data value in Set A. Compute $\bar{x}$ for the new data set. *Hint:* $\Sigma x = n\bar{x}$. To compute $\bar{x}$ for the new data set, add 20 to Σx for the original data set and divide by 6.
(b) Suppose the number 20 is included as an additional data value in Set B. Compute $\bar{x}$ for the new data set.
(c) Why does the addition of the number 20 to each data set change the mean for Set A more than it does for Set B?

11. | *Interpretation* A job-performance evaluation form has these categories:

1 = excellent; 2 = good; 3 = satisfactory; 4 = poor; 5 = unacceptable

Based on 15 client reviews, one employee had

median rating of 4; mode rating of 1.

The employee was pleased that most clients had rated her as excellent. The supervisor said improvement was needed because at least half the clients had rated the employee at the poor or unacceptable level. Comment on the different perspectives.

◄ **NEW! Critical Thinking and Interpretation Exercises**

In every section and chapter problem set, Critical Thinking problems provide students with the opportunity to test their understanding of the application of statistical methods and their interpretation of their results. Interpretation problems ask students to apply statistical results to the particular application.

Statistical Literacy

No language can be spoken without learning the vocabulary, including statistics.
Understanding Basic Statistics introduces statistical terms with deliberate care.

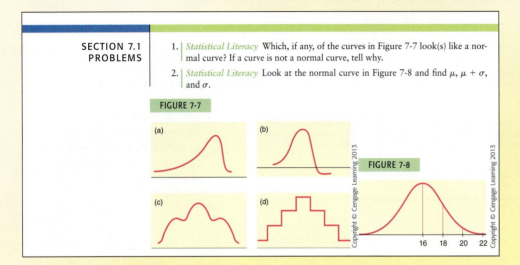

SECTION 7.1 PROBLEMS

1. *Statistical Literacy* Which, if any, of the curves in Figure 7-7 look(s) like a normal curve? If a curve is not a normal curve, tell why.

2. *Statistical Literacy* Look at the normal curve in Figure 7-8 and find μ, $\mu + \sigma$, and σ.

FIGURE 7-7

(a) (b) (c) (d)

Copyright © Cengage Learning 2013

FIGURE 7-8

16 18 20 22

Copyright © Cengage Learning 2013

◄ Statistical Literacy Problems

In every section and chapter problem set, Statistical Literacy problems test student understanding of terminology, statistical methods, and the appropriate conditions for use of the different processes.

Definition Boxes ►

Whenever important terms are introduced in text, yellow definition boxes appear within the discussions. These boxes make it easy to reference or review terms as they are used further.

Box-and-Whisker Plots

Five-number summary

The quartiles together with the low and high data values give us a very useful *five-number summary* of the data and their spread.

Five-Number Summary

Lowest value, Q_1, median, Q_3, highest value

Box-and-whisker plot

We will use these five numbers to create a graphic sketch of the data called a *box-and-whisker plot*. Box-and-whisker plots provide another useful technique from exploratory data analysis (EDA) for describing data.

IMPORTANT WORDS & SYMBOLS

Section 5.1
Probability of an event A, $P(A)$ 132
Intuition   132
Relative frequency   132
Equally likely outcomes   132
Law of large numbers   134
Statistical experiment   134
Event   134
Simple event   134
Sample space   134
Complement of event A^c 136

Section 5.2
Independent events   143
Dependent events   143

Multiplication rules of probability (for independent and dependent events)   143
More than two independent events   147
Probability of A or B 147
Mutually exclusive events   149
Addition rules (for mutually exclusive and general events)   149
More than two mutually exclusive events   151
Basic probability rules   154

Section 5.3
Multiplication rule of counting   163
Tree diagram   163
Factorial notation   166

▲ REVISED! Important Words & Symbols

The Important Words & Symbols within the Chapter Review feature at the end of each chapter summarizes the terms introduced in the Definition Boxes for student review at a glance. Page numbers for the first occurrence of the term are given for easy reference.

Linking Concepts: Writing Projects ▶

Much of statistical literacy is the ability to communicate concepts effectively. The Linking Concepts: Writing Projects feature at the end of each chapter tests both statistical literacy and critical thinking by asking the student to express their understanding in words.

LINKING CONCEPTS: WRITING PROJECTS

Discuss each of the following topics in class or review the topics on your own. Then write a brief but complete essay in which you summarize the main points. Please include formulas and graphs as appropriate.

1. An average is an attempt to summarize a collection of data into just *one* number. Discuss how the mean, median, and mode all represent averages in this context. Also discuss the differences among these averages. Why is the mean a balance point? Why is the median a midway point? Why is the mode the most common data point? List three areas of daily life in which you think the mean, median, or mode would be the best choice to describe an "average."

2. Why do we need to study the variation of a collection of data? Why isn't the average by itself adequate? We have studied three ways to measure variation. The range, the standard deviation, and, to a large extent, a box-and-whisker plot all indicate the variation within a data collection. Discuss similarities and differences among these ways to measure data variation. Why would it seem reasonable to pair the median with a box-and-whisker plot and to pair the mean with the standard deviation? What are the advantages and disadvantages of each method of describing data spread? Comment on statements such as the following: (a) The range is easy to compute, but it doesn't give much information; (b) although the standard deviation is more complicated to compute, it has some significant applications; (c) the box-and-whisker plot is fairly easy to construct, and it gives a lot of information at a glance.

---

10. *Basic Computation: Testing μ with σ Unknown* A random sample has 49 values. The sample mean is 8.5 and the sample standard deviation is 1.5. Use a level of significance of 0.01 to conduct a left-tailed test of the claim that the population mean is 9.2.
    (a) *Check Requirements* Is it appropriate to use a Student's t distribution? Explain. How many degrees of freedom do we use?
    (b) What are the hypotheses?
    (c) Compute the sample test statistic t.
    (d) Estimate the P-value for the test.
    (e) Do we reject or fail to reject H_0?
    (f) *Interpret* the results.

◀ NEW! Basic Computation Problems

These problems focus student attention on relevant formulas, requirements, and computational procedures. After practicing these skills, students are more confident as they approach real-world applications.

Expand Your Knowledge Problems ▶

Expand Your Knowledge problems present optional enrichment topics that go beyond the material introduced in a section. Vocabulary and concepts needed to solve the problems are included at point-of-use, expanding students' statistical literacy.

---

 25. *Expand Your Knowledge: Harmonic Mean* When data consist of rates of change, such as speeds, the *harmonic mean* is an appropriate measure of central tendency. For n data values,

$$\text{Harmonic mean} = \frac{n}{\sum \frac{1}{x}}, \quad \text{assuming no data value is 0}$$

Suppose you drive 60 miles per hour for 100 miles, then 75 miles per hour for 100 miles. Use the harmonic mean to find your average speed.

 26. *Expand Your Knowledge: Geometric Mean* When data consist of percentages, ratios, growth rates, or other rates of change, the *geometric mean* is a useful measure of central tendency. For n data values,

$$\text{Geometric mean} = \sqrt[n]{\text{product of the } n \text{ data values}}, \quad \text{assuming all data values are positive}$$

To find the *average growth factor* over 5 years of an investment in a mutual fund with growth rates of 10% the first year, 12% the second year, 14.8% the third year, 3.8% the fourth year, and 6% the fifth year, take the geometric mean of 1.10, 1.12, 1.148, 1.038, and 1.16. Find the average growth factor of this investment.

Note that for the same data, the harmonic mean is less than or equal to the geometric mean, which is less than or equal to the arithmetic mean (Source: *Oxford Dictionary of Statistics*).

Direction and Purpose

Real knowledge is delivered through direction, not just facts. *Understanding Basic Statistics* ensures the student knows what is being covered and why at every step along the way to statistical literacy.

Chapter Preview ▶ Questions

Preview Questions at the beginning of each chapter give the student a taste of what types of questions can be answered with an understanding of the knowledge to come.

NORMAL CURVES AND SAMPLING DISTRIBUTIONS

PREVIEW QUESTIONS

What are some characteristics of a normal distribution? What does the empirical rule tell you about data spread around the mean? (SECTION 7.1)

Can you compare apples and oranges, or maybe elephants and butterflies? In most cases, the answer is no—unless you first standardize your measurements. What are a standard normal distribution and a standard z score? (SECTION 7.2)

How do you convert any normal distribution to a standard normal distribution? How do you find probabilities of "standardized events"? (SECTION 7.3)

As humans, our experiences are finite and limited. Consequently, most of the important decisions in our lives are based on sample (incomplete) information. What is a probability sampling distribution... good

FOCUS PROBLEM

Benford's Law: The Importance of Being Number 1

Benford's Law states that in a wide variety of circumstances, numbers have "1" as their first nonzero digit disproportionately often. Benford's Law applies to such diverse topics as the drainage areas of rivers; properties of chemicals; populations of towns; figures in newspapers, magazines, and government reports; and the half-lives of radioactive atoms!

Specifically, such diverse measurements begin with "1" about 30% of the time, with "2" about 18% of time, and with "3" about 12.5% of the time. Larger digits occur less often. For example, less than 5% of the numbers in circumstances such as these begin with the digit 9. This is in dramatic contrast to a random sampling situation, in which each of the digits 1 through 9 has an equal chance of appearing.

The first nonzero digits of numbers taken from large bodies of numerical records such as tax returns, population studies, government records, and so forth, show the probabilities of occurrence as displayed in the table on the next page.

▲ Chapter Focus Problems

The Preview Questions in each chapter are followed by Focus Problems, which serve as more specific examples of what questions the student will soon be able to answer. The Focus Problems are set within appropriate applications and are incorporated into the end-of-section exercises, giving students the opportunity to test their understanding.

7. *Focus Problem: Benford's Law* Please read the Focus Problem at the beginning of this chapter. Recall that Benford's Law claims that numbers chosen from very large data files tend to have "1" as the first nonzero digit disproportionately often. In fact, research has shown that if you randomly draw a number from a very large data file, the probability of getting a number with "1" as the leading digit is about 0.301 (see the reference in this chapter's Focus Problem).

Now suppose you are an auditor for a very large corporation. The revenue report involves millions of numbers in a large computer file. Let us say you took a random sample of $n = 215$ numerical entries from the file and $r = 46$ of the entries had a first nonzero digit of 1. Let p represent the population proportion of all numbers in the corporate file that have a first nonzero digit of 1.

Direction and Purpose

Focus Points ▶

Each section opens with bulleted Focus Points describing the primary learning objectives of the section.

SECTION 3.1

Measures of Central Tendency: Mode, Median, and Mean

FOCUS POINTS
- Compute mean, median, and mode from raw data.
- Interpret what mean, median, and mode tell you.
- Explain how mean, median, and mode can be affected by extreme data values.
- Define and compute a trimmed mean.
- Compute a weighted average.

The average price of an ounce of gold is $1500. The Zippy car averages 39 miles per gallon on the highway. A survey showed the average shoe size for women is size 9.

In each of the preceding statements, *one* number is used to describe the entire sample or population. Such a number is called an *average*. There are many ways to compute averages, but we will study only three of the major ones.

The easiest average to compute is the *mode*.

Average

Mode

> The **mode** of a data set is the value that occurs most frequently.

EXAMPLE 1   MODE

Count the letters in each word of this sentence and give the mode. The numbers of letters in the words of the sentence are

5   3   7   2   4   4   2   4   8   3   4   3   4

Scanning the data, we see that 4 is the mode because more words have 4 letters than any other number. For larger data sets, it is useful to order—or sort—the data before scanning them for the mode.

LOOKING FORWARD

In our future work with inferential statistics, we will use the mean $\bar{x}$ from a random sample to estimate the population parameter μ (Chapter 8) and to make decisions regarding the value of μ (Chapter 9).

▲ **NEW!** **Looking Forward**

This feature shows students that the presented material will be used later. It helps motivate students to pay a little extra attention to key topics.

Chapter Review

SUMMARY

Organizing and presenting data are the main purposes of the branch of statistics called descriptive statistics. Graphs provide an important way to show how the data are distributed.

- Frequency tables show how the data are distributed within set classes. The classes are chosen so that they cover all data values and so that each data value falls within only one class. The number of classes and the class width determine the class limits and class boundaries. The number of data values falling within a class is the class frequency.

- A histogram is a graphical display of the information in a frequency table. Classes are shown on the horizontal axis, with corresponding frequencies on the vertical axis. Relative-frequency histograms show relative frequencies on the vertical axis. Ogives show cumulative frequencies on the vertical axis. Dotplots are like histograms, except that the classes are individual data values.

- Bar graphs, Pareto charts, and pie charts are useful to show how quantitative or qualitative data are distributed over chosen categories.

- Time-series graphs show how data change over set intervals of time.

- Stem-and-leaf displays are effective means of ordering data and showing important features of the distribution.

Graphs aren't just pretty pictures. They help reveal important properties of the data distribution, including the shape and whether or not there are any outliers.

◀ **Chapter Summaries**

The Summary within each Chapter Review feature now also appears in bulleted form, so students can see what they need to know at a glance.

Statistics is not done in a vacuum. *Understanding Basic Statistics* gives students valuable skills for the real world with technology instruction, genuine applications, actual data, and group projects.

REVISED! Tech Notes ▶

Tech Notes appearing throughout the text give students helpful hints on using TI-4 Plus and TI-*n*spire (with 84 Plus keypad) and TI-83 calculators, Microsoft Excel 2010, and Minitab to solve a problem. They include display screens to help students visualize and better understand the solution.

TECH NOTES *Stem-and-leaf display*

TI-84Plus/TI-83Plus/TI-*n*spire Does not support stem-and-leaf displays. You can sort the data by using keys Stat ➤ Edit ➤ 2:SortA.

Excel 2010 Enter your data and select the data you want to sort. On the **Home** ribbon, click the **Sort and Filter** button in the **Editing** group of the ribbon and select the desired sorting option.

Minitab Use the menu selections Graph ➤ Stem-and-Leaf and fill in the dialogue box.

Minitab Stem-and-Leaf Display (for Data in Guided Exercise 4)

```
Stem-and-Leaf of Scores      N=35
Leaf Unit=1.0

    1            8   3
    5            9   2789
   14           10   123455669
  (11)          11   01222267789
   10           12   045568
    4           13   125
    1           14   3
```

Copyright © Cengage Learning 2013

The values shown in the left column represent depth. Numbers above the value in parentheses show the cumulative number of values from the top to the stem of the middle value. Numbers below the value in parentheses show the cumulative number of values from the bottom to the stem of the middle value. The number in parentheses shows how many values are on the same line as the middle value.

USING TECHNOLOGY

Binomial Distributions

Although tables of binomial probabilities can be found in most libraries, such tables are often inadequate. Either the value of p (the probability of success on a trial) you are looking for is not in the table, or the value of n (the number of trials) you are looking for is too large for the table. In Chapter 5, we will study the normal approximation to the binomial. This approximation is a great help in many practical applications. Even so, we sometimes use the formula for the binomial probability distribution on a computer or graphing calculator to compute the probability we want.

Applications

The following percentages were obtained over many years of observation by the U.S. Weather Bureau. All data listed are for the month of December.

Location	Long-Term Mean % of Clear Days in Dec.
Juneau, Alaska	18%
Seattle, Washington	24%
Hilo, Hawaii	36%
Honolulu, Hawaii	60%
Las Vegas, Nevada	75%
Phoenix, Arizona	77%

Source: Adapted from *Local Climatological Data,* U.S. Weather Bureau publication, "Normals, Means, and Extremes" table.

In the locations listed, the month of December is a relatively stable month with respect to weather. Since weather patterns from one day to the next are more or less the same, it is reasonable to use a binomial probability model.

1. Let r be the number of clear days in December. Since December has 31 days, $0 \leq r \leq 31$. Using appropriate computer software or calculators available to you, find the probability $P(r)$ for each of the listed locations when $r = 0, 1, 2, \ldots, 31$.

You may find that using cumulative probabilities and appropriate subtraction of probabilities, rather than addition of probabilities, will make finding the solutions to Applications 3–7 easier.

3. Estimate the probability that Juneau will have at most 7 clear days in December.

4. Estimate the probability that Seattle will have from 5 to 10 (including 5 and 10) clear days in December.

5. Estimate the probability that Hilo will have at least 12 clear days in December.

6. Estimate the probability that Phoenix will have 20 or more clear days in December.

7. Estimate the probability that Las Vegas will have from 20 to 25 (including 20 and 25) clear days in December.

Technology Hints

TI-84Plus/TI-83Plus/TI-*n*spire (with TI-84 Plus keypad), Excel 2010, Minitab

The Tech Note in Section 6.2 gives specific instructions for binomial distribution functions on the TI-84Plus/TI-83Plus/TI-*n*spire (with TI-84Plus keypad) calculators, Excel 2010, and Minitab.

SPSS

In SPSS, the function **PDF.BINOM(q,n,p)** gives the probability of q successes out of n trials, where p is the probability of success on a single trial. In the data editor, name a variable r and enter values 0 through n. Name another variable Prob_r. Then use the menu choices **Transform ➤ Compute**. In the dialogue box, use Prob_r for the target variable. In the function group, select **PDF and Noncentral PDF**. In the function box, select **PDF.BINOM(q,n,p)**. Use the variable r for q and appropriate values for n and p. Note that the function **CDF.BINOM(q,n,p)**, from the **CDF and Noncentral CDF** group, gives the cumulative probability of 0 through q successes.

◀ REVISED! Using Technology

Further technology instruction is available at the end of each chapter in the Using Technology section. Problems are presented with real-world data from a variety of disciplines that can be solved by using TI-84 Plus, TI-*n*spire (with 84 Plus keypad), and TI-83 Plus calculators, Microsoft Excel 2010, and Minitab.

Real-World Skills

EXAMPLE 10   PROBABILITY REGARDING x AND $\bar{x}$

Suppose a team of biologists has been studying the Pinedale children's fishing pond. Let x represent the length of a single trout taken at random from the pond. This group of biologists has determined that x has a normal distribution with mean $\mu = 10.2$ inches and standard deviation $\sigma = 1.4$ inches.

(a) What is the probability that a *single trout* taken at random from the pond is between 8 and 12 inches long?

SOLUTION: We use the methods of Section 7.3, with $\mu = 10.2$ and $\sigma = 1.4$, to get

$$z = \frac{x - \mu}{\sigma} = \frac{x - 10.2}{1.4}$$

Therefore,

$$P(8 < x < 12) = P\left(\frac{8 - 10.2}{1.4} < z < \frac{12 - 10.2}{1.4}\right)$$

◀ **UPDATED! Applications**

Real-world applications are used from the beginning to introduce each statistical process. Rather than just crunching numbers, students come to appreciate the value of statistics through relevant examples.

Most exercises in each section ▶ are applications problems.

f years centered about the mean in which about 95% of dates) will be found.

of years centered about the mean in which almost all the data (tree-ring dates) will be found.

10. *Vending Machine: Soft Drinks* A vending machine automatically pours soft drinks into cups. The amount of soft drink dispensed into a cup is normally distributed with a mean of 7.6 ounces and standard deviation of 0.4 ounce. Examine Figure 7-3 and answer the following questions:
   (a) Estimate the probability that the machine will overflow an 8-ounce cup.
   (b) Estimate the probability that the machine will not overflow an 8-ounce cup.
   (c) The machine has just been loaded with 850 cups. How many of these do you expect will overflow when served?

11. *Pain Management: Laser Therapy* "Effect of Helium-Neon Laser Auriculotherapy on Experimental Pain Threshold" is the title of an article in the journal *Physical Therapy* (Vol. 70, No. 1, pp. 24–30). In this article, laser therapy is discussed as a useful alternative to drugs in pain management for chronically ill patients. To measure pain threshold, a machine was used that delivered low-voltage direct current to

(a) Mak

(b) *Grouped Data* Make a frequency table using four classes. Then estimate the mean and sample standard deviation using the frequency table. Compute a 75% Chebyshev interval centered about the mean.

(c) If you have a statistical calculator or computer, use it to find the actual sample mean and sample standard deviation.

11. *Performance Rating: Weighted Average* A performance evaluation for new sales representatives at Office Automation Incorporated involves several ratings done on a scale of 1 to 10, with 10 the highest rating. The activities rated include new contacts, successful contacts, total contacts, dollar volume of sales, and reports. Then an overall rating is determined by using a weighted average. The weights are 2 for new contacts, 3 for successful contacts, 3 for total contacts, 5 for dollar value of sales, and 3 for reports. What would the overall rating be for a sales representative with ratings of 5 for new contacts, 8 for successful contacts, 7 for total contacts, 9 for dollar volume of sales, and 7 for reports?

◀ **Data Highlights: Group Projects**

Using Group Projects, students gain experience working with others by discussing a topic, analyzing data, and collaborating to formulate their responses to the questions posed in the exercise.

DATA HIGHLIGHTS: GROUP PROJECTS

Old Faithful Geyser, Yellowstone National Park

Break into small groups and discuss the following topics. Organize a brief outline in which you summarize the main points of your group discussion.

1. *The Story of Old Faithful* is a short book written by George Marler and published by the Yellowstone Association. Chapter 7 of this interesting book talks about the effect of the 1959 earthquake on eruption intervals for Old Faithful Geyser. Dr. John Rinehart (a senior research scientist with the National Oceanic and Atmospheric Administration) has done extensive studies of the eruption intervals before and after the 1959 earthquake. Examine Figure 3-11. Notice the general shape. Is the graph more or less symmetrical? Does it have a single mode frequency? The mean interval between eruptions has remained steady at about 65 minutes for the past 100 years. Therefore, the 1959 earthquake did not significantly change the mean, but it did change the distribution of eruption intervals. Examine Figure 3-12. Would you say there are really two frequency modes, one shorter and the other longer? Explain. The overall mean is about the same for both graphs, but one graph has a much larger standard deviation (for eruption intervals) than the other. Do no calculations, just look at both graphs, and then explain which graph has the smaller and which has the larger standard deviation. Which distribution will have the larger coefficient of variation? In everyday terms, what would this mean if you were actually at Yellowstone waiting to see the next eruption of Old Faithful? Explain your answer.

2. Most academic advisors tell students to major in a field they really love. After all, it is true that money cannot buy happiness! Nevertheless, it is interesting to at least look at some of the higher-paying fields of study. After all, a field like

Steve Maehl, 2009/Used under license from Shutterstock.com

Making the Jump

Get to the "Aha!" moment faster. *Understanding Basic Statistics* provides the push students need to get there through guidance and example.

PROCEDURE

HOW TO TEST μ WHEN σ IS UNKNOWN

Requirements

Let x be a random variable appropriate to your application. Obtain a simple random sample (of size n) of x values from which you compute the sample mean $\bar{x}$ and the sample standard deviation s. If you can assume that x has a normal distribution or simply a mound-shaped and symmetrical distribution, then any sample size n will work. If you cannot assume this, use a sample size $n \geq 30$.

Procedure

1. In the context of the application, state the *null and alternate hypotheses* and set the *level of significance* α.
2. Use $\bar{x}$, s, and n from the sample, with μ from H_0, to compute the *sample test statistic*.

d.f. for testing μ when σ is unknown

$$t = \frac{\bar{x} - \mu}{\frac{s}{\sqrt{n}}} \text{ with degrees of freedom } d.f. = n - 1$$

3. Use the Student's t distribution and the type of test, one-tailed or two-tailed, to find (or estimate) the *P-value* corresponding to the test statistic.
4. Con...
   not r...

◀ **REVISED! Procedures**

Procedure display boxes summarize simple step-by-step strategies for carrying out statistical procedures and methods as they are introduced. Students can refer back to these boxes as they practice using the procedures.

Guided Exercises ▶

Students gain experience with new procedures and methods through Guided Exercises. Beside each problem in a Guided Exercise, a completely worked-out solution appears for immediate reinforcement.

GUIDED EXERCISE 10 | **Probability regarding $\bar{x}$**

In mountain country, major highways sometimes use tunnels instead of long, winding roads over high passes. However, too many vehicles in a tunnel at the same time can cause a hazardous situation. Traffic engineers are studying a long tunnel in Colorado. If x represents the time for a vehicle to go through the tunnel, it is known that the x distribution has mean $\mu = 12.1$ minutes and standard deviation $\sigma = 3.8$ minutes under ordinary traffic conditions. From a histogram of x values, it was found that the x distribution is mound-shaped with some symmetry about the mean.

Engineers have calculated that, *on average*, vehicles should spend from 11 to 13 minutes in the tunnel. If the time is less than 11 minutes, traffic is moving too fast for safe travel in the tunnel. If the time is more than 13 minutes, there is a problem of bad air quality (too much carbon monoxide and other pollutants).

Under ordinary conditions, there are about 50 vehicles in the tunnel at one time. What is the probability that the mean time for 50 vehicles in the tunnel will be from 11 to 13 minutes? We will answer this question in steps.

(a) Let $\bar{x}$ represent the sample mean based on samples of size 50. Describe the $\bar{x}$ distribution.

➡ From the central limit theorem, we expect the $\bar{x}$ distribution to be approximately normal, with mean and standard deviation

$$\mu_{\bar{x}} = \mu = 12.1 \quad \text{and} \quad \sigma_{\bar{x}} = \frac{\sigma}{\sqrt{n}} = \frac{3.8}{\sqrt{50}} \approx 0.54$$

(b) Find $P(11 < \bar{x} < 13)$.

Jupiter Images

➡ We convert the interval

$$11 < \bar{x} < 13$$

to a standard z interval and use the standard normal probability table to find our answer. Since

$$z = \frac{\bar{x} - \mu}{\sigma/\sqrt{n}} \approx \frac{\bar{x} - 12.1}{0.54},$$

$\bar{x} = 11$ converts to $z \approx \dfrac{11 - 12.1}{0.54} = -2.04$

and $\bar{x} = 13$ converts to $z \approx \dfrac{13 - 12.1}{0.54} = 1.67$

Therefore,

$$P(11 < \bar{x} < 13) = P(-2.04 < z < 1.67)$$
$$= 0.9525 - 0.0207$$
$$= 0.9318$$

(c) *Interpret* your answer to part (b).

➡ It seems that about 93% of the time, there should be no safety hazard for average traffic flow.

PREFACE

Welcome to the exciting world of statistics! We have written this text to make statistics accessible to everyone, including those with a limited mathematics background. Statistics affects all aspects of our lives. Whether we are testing new medical devices or determining what will entertain us, applications of statistics are so numerous that, in a sense, we are limited only by our own imagination in discovering new uses for statistics.

Overview

The hybrid edition of *Understanding Basic Statistics* continues to emphasize concepts of statistics. Statistical methods are carefully presented with a focus on understanding both the *suitability of the method* and the *meaning of the result*. Statistical methods and measurements are developed in the context of applications.

Critical thinking and interpretation are essential in understanding and evaluating information. Statistical literacy is fundamental for applying and comprehending statistical results. In this edition we have expanded and highlighted the treatment of statistical literacy, critical thinking, and interpretation.

We have retained and expanded features that made the previous editions of the text very readable. Definition boxes highlight important terms. Procedures displays summarize steps for analyzing data. Examples, exercises, and problems touch on applications appropriate to a broad range of interests.

New with the hybrid edition is CourseMate, encompassing all interactive online products and services with this text. Online homework powered by Aplia is now available through CengageBrain.com. Also available through CourseMate are more than 100 data sets (in Microsoft Excel, Minitab, SPSS, and TI-84Plus/TI-83Plus/TI-*n*spire with 84 Plus keypad ASCII file formats), lecture aids, a glossary, statistical tables, instructional videos (also available on DVDs), an Online Multimedia eBook, interactive videos (also available on DVDs), and interactive tutorials.

Understanding Basic Statistics, Hybrid Edition, is carefully designed to present *core topics* of an introductory course in statistics. Our other text, *Understandable Statistics*, Tenth Edition, is a more comprehensive text with additional topics. Both books feature the same effective pedagogy and instructional design.

Major Changes in the Hybrid Edition

With each new edition, the authors reevaluate the scope, appropriateness, and effectiveness of the text's presentation and reflect on extensive user feedback. Revisions have been made throughout the text to clarify explanations of important concepts and to update problems.

Key Features of the Hybrid Edition

Many statistics courses are evolving into lecture–lab courses or into courses for which all assignments—homework and even tests—are delivered online. Furthermore, distance learning is growing rapidly. Instructors are increasingly faced with the challenge of integrating technology and teaching the concepts and skills of statistics.

In this hybrid text, as with the hardcover text, the goal is to provide instructors and students with the tools they can use to meet this challenge.

What does hybrid mean? A hybrid text involves an integration of several products and can be used best in a course with a strong blend of lecture time and online course work.

For this hybrid edition, the end-of-section exercises have been removed from the text and are available exclusively in Aplia, an easy-to-use online homework system. This slightly smaller book will be more manageable for the student who is spending his or her homework time on a computer.

Although this text has been tailored for instructors and students who use online homework, the focus of the expositions remains unchanged.

Critical Thinking, Interpretation, and Statistical Literacy

The hybrid edition of this text continues and expands the emphasis on critical thinking, interpretation, and statistical literacy. Calculators and computers are very good at providing numerical results of statistical processes. However, numbers from a computer or calculator display are meaningless unless the user knows how to interpret the results and if the statistical process is appropriate. This text helps students determine whether or not a statistical method or process is appropriate. It helps students understand what a statistic measures. It helps students interpret the results of a confidence interval, hypothesis test, or linear regression model.

New Problems Featuring Basic Computation

Calculators and computer software automatically calculate designated statistical measurements. However, students gain an appreciation and understanding of what the measurements mean by studying and using the basic formulas. Basic computation problems focus attention on using formulas with small data sets. Students see not only how the formulas work, but also how the resulting measurements relate to the displayed data set.

There are more than 200 new and revised problems that feature basic computations, interpretation, and statistical literacy. Students are asked to check that the use of specific probability distributions and inferential methods are appropriate.

New Content

Section 2.1 now includes a discussion of cumulative frequency and ogives.

Problem sets in Sections 3.1 and 3.2 provide problems discussing the effects of data transformation (adding a constant to each data value or multiplying each data value by a constant) on the mean, median, mode, and standard deviation.

New *Expand Your Knowledge* problems present donut pie charts (Section 2.2), the geometric mean and the harmonic mean (Section 3.1), moving averages (Section 3.2), correlation of averages (Section 4.2), odds for and odds against (Section 5.1), uniform probability distribution (Section 7.1), plus four method of confidence intervals (Section 8.3), probability of two confidence intervals (Chapter 8 review), and serial correlation (Section 11.4).

Excel 2010 and TI-*n*spire Calculator (with 84 Plus keypad)

Excel 2010 instructions are included in the *Tech Notes* and *Using Technology* features.

The TI-*n*spire calculator with the 84 Plus keypad is also addressed. Instructions for using the TI-*n*spire calculator with the *n*spire keypad are included in the separate Technology Guide for the TI-84 Plus graphing calculator.

Looking Forward

A new *Looking Forward* feature briefly points out how current subject matter will be used in later chapters.

Other Changes

Examples in probability (Chapter 5) have been revised to incorporate sample spaces that are small and easy for students to visualize.

Throughout this edition, examples, guided exercises, and problem sets feature interpretation. They also address the suitability of using a specified statistical method or process for analysis.

Continuing Content

Introduction to Hypothesis Testing Using *P*-values

In keeping with the use of computer technology and standard practice in research, hypothesis testing is introduced using *P*-values. The critical region method is still supported but not given primary emphasis.

Use of Student's *t* Distribution in Confidence Intervals and Testing of Means

If the normal distribution is used in confidence intervals and testing of means, then the *population standard deviation must be known.* If the population standard deviation is not known, then under conditions described in the text, the Student's *t* distribution is used. This is the most commonly used procedure in statistical research. It is also used in statistical software packages such as Microsoft Excel, Minitab, SPSS, and TI-84Plus/TI-83Plus/TI-*n*spire calculators.

Confidence Intervals and Hypothesis Tests of Difference of Means

If the normal distribution is used, then both population standard deviations must be known. When this is not the case, the Student's *t* distribution incorporates an approximation for *t*, with a commonly used conservative choice for the degrees of freedom. Satterthwaite's approximation for the degrees of freedom as used in computer software is also discussed. The pooled standard deviation is presented for appropriate applications ($\sigma_1 \approx \sigma_2$).

Features in the Hybrid Edition

Chapter and Section Lead-ins

- *Preview Questions* at the beginning of each chapter are keyed to the sections.
- *Focus Problems* at the beginning of each chapter demonstrate types of questions students can answer once they master the concepts and skills presented in the chapter.
- *Focus Points* at the beginning of each section describe the primary learning objectives of the section.

Carefully Developed Pedagogy

- *Examples* show students how to select and use appropriate procedures.
- *Guided Exercises* within the sections give students an opportunity to work with a new concept. Completely worked-out solutions appear beside each exercise to give immediate reinforcement.
- *Definition boxes* highlight important definitions throughout the text.
- *Procedure displays* summarize appropriate conditions and key strategies for carrying out statistical procedures and methods. Conditions required for using the procedure are also stated.
- NEW! *Looking Forward* features give a brief preview of how a current topic is used later.
- *Labels* for each example or guided exercise highlight the technique, concept, or process illustrated by the example or guided exercise. In addition, labels for section and chapter problems describe the field of application and show the wide variety of subjects in which statistics is used.
- *Section and chapter problems* require the student to use all the new concepts mastered in the section or chapter. Problem sets include a variety of real-world applications with data or settings from identifiable sources. Key steps and solutions to odd-numbered problems appear at the end of the book.
- NEW! *Basic Computation problems* ask students to practice using formulas and statistical methods on very small data sets. Such practice helps students understand what a statistic measures.
- *Statistical Literacy problems* ask students to focus on correct terminology and processes of appropriate statistical methods. Such problems occur in every section and chapter problem set.
- NEW! *Interpretation problems* ask students to explain the meaning of the statistical results in the context of the application.
- *Critical Thinking problems* ask students to analyze and comment on various issues that arise in the application of statistical methods and in the interpretation of results. These problems occur in every section and chapter problem set.
- *Expand Your Knowledge problems* present enrichment topics such as dot plots, grouped data, estimation of standard deviation from a range of data values, residual plots, relationship between confidence intervals and two-tailed hypothesis tests, and more.
- *Cumulative review problem sets* occur after every third chapter and include key topics from previous chapters. Answers to *all* cumulative review problems are given at the end of the book.
- *Data Highlights and Linking Concepts* provide group projects and writing projects.
- *Viewpoints* are brief essays presenting diverse situations in which statistics is used.
- *Design and photos* are appealing and enhance readability.

Technology within the Text

- *Tech Notes* within sections provide brief point-of-use instructions for the TI-84 Plus, TI-83 Plus and TI-*n*spire (with 84 Plus keypad) calculators, Microsoft Excel 2010, and Minitab.
- *Using Technology* sections have been revised to show the use of SPSS as well as the TI-84 Plus, TI-83 Plus and TI-*n*spire (with 84 Plus keypad) calculators, Microsoft Excel, and Minitab.

Alternate Routes through the Text

Understanding Basic Statistics, Hybrid Edition, is designed to be flexible. It offers the professor a choice of teaching possibilities. In most one-semester courses, it is not practical to cover all the material in depth. However, depending on the emphasis of the course, the professor may choose to cover various topics. For help in topic selection, refer to the Table of Prerequisite Material on page 1.

- *Linear regression.* Chapter 4, Correlation and Regression, may be delayed until after Chapter 9. The descriptive topics of linear regression may then be followed immediately by the inferential topics of linear regression presented in Chapter 11.
- *Probability.* For courses requiring minimal probability, Section 5.1 (What Is Probability?) and the first part of Section 5.2 (Some Probability Rules—Compound Events) will be sufficient.

Acknowledgments

It is our pleasure to acknowledge the prepublication reviewers of this text. All of their insights and comments have been very valuable to us. Reviewers of this text include:

Reza Abbasian, Texas Lutheran University
Paul Ache, Kutztown University
Kathleen Almy, Rock Valley College
Polly Amstutz, University of Nebraska at Kearney
Delores Anderson, Truett-McConnell College
Robert J. Astalos, Feather River College
Lynda L. Ballou, Kansas State University
Mary Benson, Pensacola Junior College
Larry Bernett, Benedictine University
Kiran Bhutani, The Catholic University of America
Kristy E. Bland, Valdosta State University
John Bray, Broward Community College
Bill Burgin, Gaston College
Toni Carroll, Siena Heights University
Pinyuen Chen, Syracuse University
Emmanuel des-Bordes, James A. Rhodes State College
Jennifer M. Dollar, Grand Rapids Community College
Larry E. Dunham, Wor-Wic Community College
Andrew Ellett, Indiana University
Ruby Evans, Keiser University
Mary Fine, Moberly Area Community College
Rebecca Fouguette, Santa Rosa Junior College
Rene Garcia, Miami-Dade Community College
Larry Green, Lake Tahoe Community College
Shari Harris, John Wood Community College
Janice Hector, DeAnza College
Jane Keller, Metropolitan Community College
Raja Khoury, Collin County Community College
Diane Koenig, Rock Valley College
Charles G. Laws, Cleveland State Community College
Michael R. Lloyd, Henderson State University
Beth Long, Pellissippi State Technical and Community College
Lewis Lum, University of Portland
Darcy P. Mays, Virginia Commonwealth University
Charles C. Okeke, College of Southern Nevada, Las Vegas
Peg Pankowski, Community College of Allegheny County
Ram Polepeddi, Westwood College, Denver North Campus
Azar Raiszadeh, Chattanooga State Technical Community College
Traei Reed, St. Johns River Community College

Michael L. Russo, Suffolk County Community College
Janel Schultz, Saint Mary's University of Minnesota
Sankara Sethuraman, Augusta State University
Stephen Soltys, West Chester University of Pennsylvania
Ron Spicer, Colorado Technical University
Winson Taam, Oakland University
Jennifer L. Taggart, Rockford College
William Truman, University of North Carolina at Pembroke
Bill White, University of South Carolina Upstate
Jim Wienckowski, State University of New York at Buffalo
Stephen M. Wilkerson, Susquehanna University
Hongkai Zhang, East Central University
Shunpu Zhang, University of Alaska, Fairbanks
Cathy Zucco-Teveloff, Trinity College

We would especially like to thank George Pasles for his careful accuracy review of this text. We are especially appreciative of the excellent work by the editorial and production professionals at Brooks/Cole Cengage. In particular we thank Molly Taylor, Laura Wheel, Jessica Rasile, and Jay Campbell. Without their creative insight and attention to detail, a project of this quality and magnitude would not be possible. Finally, we acknowledge the cooperation of Minitab, Inc., SPSS, Texas Instruments, and Microsoft.

Charles Henry Brase

Corrinne Pellillo Brase

ADDITIONAL RESOURCES—GET MORE FROM YOUR TEXTBOOK!

Instructor Resources

Annotated Instructor's Edition (AIE) Answers to all exercises, teaching comments, and pedagogical suggestions appear in the margin or at the end of the text in the case of large graphs.

Solution Builder Contains complete solutions to all exercises in the text, including those in the Chapter Review and Cumulative Review Problems in online format. Solution Builder allows instructors to create customized, secure PDF printouts of solutions matched exactly to the exercises assigned for class. Available to adoptions by signing up at www.cengage.com/solutionbuilder.

ExamView® Allows instructors to create, deliver, and customize tests for class in print and online formats and features automatic grading. This electronic test bank features more than 450 questions based on the text. All test items are also provided in PDF and Microsoft® Word formats for instructors who opt not to use the software component.

PowerLecture Provides the instructor with dynamic media tools for teaching. Create, deliver, and customize tests (both print and online) in minutes with *ExamView® Computerized Testing Featuring Algorithmic Equations.* Easily build solution sets for homework or exams using Solution Builder's online solutions manual. Microsoft® PowerPoint® lecture slides and figures from the book are also included on this CD-ROM.

Student Resources

Student Solutions Manual Provides solutions to the odd-numbered section and chapter exercises and to all the Cumulative Review exercises in the student textbook.

Instructional DVDs Hosted by Dana Mosely, these text-specific DVDs cover all sections of the text and provide explanations of key concepts, examples, exercises, and applications in a lecture-based format. DVDs are close-captioned for the hearing-impaired.

Aplia Is an online interactive learning solution that helps students improve comprehension—and their grade—by integrating a variety of media and tools such as videos, tutorials, practice tests, and an interactive eBook. Created by a professor to enhance his own courses, *Aplia* provides automatically graded assignments with detailed, immediate feedback on every question, and innovative teaching materials. More than 1,000,000 students have used *Aplia* at more than 1,800 institutions.

MINITAB® and IBM SPSS Statistics CD-ROMs These statistical software packages manipulate and interpret data to produce textual, graphical, and tabular results. MINITAB and/or SPSS may be packaged with the textbook. Student versions are available.

CourseMate Brings course concepts to life with interactive learning, study, and exam preparation tools that support the printed textbook. Watch student comprehension soar as your class works with the printed textbook and the textbook-specific website. *Statistics CourseMate* goes beyond the book to deliver what you need. Find the following and more at **www.cengage.com/statistics/brase.**

- **Engagement Tracker** A first-of-its-kind tool that monitors student engagement in the course. Includes online student quizzes.

- **Technology Guides** Separate guides exist with information and examples for each of four technology tools. Guides are available for the TI-84 Plus, TI-83 Plus, and TI-*n*spire graphing calculators; Minitab software (version 14); Microsoft Excel (2010/2008/2007); and SPSS Statistics software.

- **Interactive Teaching and Learning Tools** Include glossary flashcards, online data sets (in Microsoft Excel, Minitab, SPSS, and Tl-84Plus/TI-83Plus/TI-*n*spire with 84 Plus keypad ASCII file formats), statistical tables and formulae, and more.

- **Multimedia eBook** Integrates numerous assets such as video explanations and tutorials to expand upon and reinforce concepts as they appear in the text.

CengageBrain.com Provides the freedom to purchase online homework and other materials a-la-carte—exactly what you need, when you need it.

For more information, visit **www.cengage.com/statistics/brase** or contact your local Cengage Learning sales representative.

Understanding Basic Statistics

TABLE OF PREREQUISITE MATERIAL

Chapter		Prerequisite Sections
1	Getting Started	None
2	Organizing Data	1.1, 1.2
3	Averages and Variation	1.1, 1.2, 2.1
4	Correlation and Regression	1.1, 1.2, 3.1, 3.2
5	Elementary Probability Theory	1.1, 1.2, 2.1
6	The Binomial Probability Distribution and Related Topics	1.1, 1.2, 2.1, 3.1, 3.2, 5.1, 5.2 5.3 useful but not essential
7	Normal Curves and Sampling Distributions (omit 7.6) (include 7.6)	1.1, 1.2, 2.1, 3.1, 3.2, 5.1, 5.2, 6.1 also 6.2, 6.3
8	Estimation (omit 8.3) (include 8.3)	1.1, 1.2, 2.1, 3.1, 3.2, 5.1, 5.2, 6.1, 7.1, 7.2, 7.3, 7.4, 7.5 also 6.2, 6.3, 7.6
9	Hypothesis Testing (omit 9.3) (include 9.3)	1.1, 1.2, 2.1, 3.1, 3.2, 5.1, 5.2, 6.1, 7.1, 7.2, 7.3, 7.4, 7.5 also 6.2, 6.3, 7.6
10	Inferences about Differences (omit 10.3) (include 10.3)	1.1, 1.2, 2.1, 3.1, 3.2, 5.1, 5.2, 6.1, 7.1, 7.2, 7.3, 7.4, 7.5, 8.1, 8.2, 9.1, 9.2 also 6.2, 6.3, 7.6, 9.3
11	Additional Topics Using Inference (Part I: 11.1, 11.2, 11.3) (Part II: 11.4)	1.1, 1.2, 2.1, 3.1, 3.2, 5.1, 5.2, 6.1, 7.1, 7.2, 7.3, 7.4, 7.5, 9.1 Chapter 4, 8.1, 8.2 also

1

Chance favors the prepared mind.

—LOUIS PASTEUR

Statistical techniques are tools of thought . . . not substitutes for thought.

—ABRAHAM KAPLAN

For online student resources, visit the Brase/Brase, *Understanding Basic Statistics,* 6th edition website at **http://www.cengagebrain.com/**

Louis Pasteur (1822–1895) is the founder of modern bacteriology. At age 57, Pasteur was studying cholera. He accidentally left some bacillus cultures unattended in his laboratory during the summer. In the fall, he injected laboratory animals with these bacilli. To his surprise, the animals did not die—in fact, they thrived and were resistant to cholera.

When the final results were examined, it is said that Pasteur remained silent for a minute and then exclaimed, as if he had seen a vision, "Don't you see they have been vaccinated!" Pasteur's work ultimately saved many human lives.

Most of the important decisions in life involve incomplete information. Such decisions often involve so many complicated factors that a complete analysis is not practical or even possible. We are often forced into the position of making a guess based on limited information.

As the first quote reminds us, our chances of success are greatly improved if we have a "prepared mind." The statistical methods you will learn in this book will help you achieve a prepared mind for the study of many different fields. The second quote reminds us that statistics is an important tool, but it is not a replacement for an in-depth knowledge of the field to which it is being applied.

The authors of this book want you to understand and enjoy statistics. The reading material will *tell you* about the subject. The examples will *show you* how it works. To understand, however, you must *get involved.* Guided exercises, calculator and computer applications, section and chapter problems, and writing exercises are all designed to get you involved in the subject. As you grow in your understanding of statistics, we believe you will enjoy learning a subject that has a world full of interesting applications.

GETTING STARTED

PREVIEW QUESTIONS

Why is statistics important? (SECTION 1.1)

What is the nature of data? (SECTION 1.1)

How can you draw a random sample? (SECTION 1.2)

What are other sampling techniques? (SECTION 1.2)

How can you design ways to collect data? (SECTION 1.3)

STUDIO SATO/amanaimages/Corbis

FOCUS PROBLEM

Where Have All the Fireflies Gone?

A feature article in *The Wall Street Journal* discusses the disappearance of fireflies. In the article, Professor Sara Lewis of Tufts University and other scholars express concern about the decline in the worldwide population of fireflies.

There are a number of possible explanations for the decline, including habitat reduction of woodlands, wetlands, and open fields; pesticides; and pollution. Artificial nighttime lighting might interfere with the Morse code-like mating ritual of the fireflies. Some chemical companies pay a bounty for fireflies because the insects contain two rare chemicals used in medical research and in electronic detection systems in spacecraft.

What does any of this have to do with statistics?

The truth, at this time, is that no one really knows (a) how much the world firefly population has declined or (b) how to explain the decline. The population of all fireflies is simply too large to study in its entirety.

In any study of fireflies, we must rely on incomplete information from samples. Furthermore, from these samples we must draw realistic conclusions that have statistical integrity. This is the kind of work that makes use of statistical methods to determine ways to collect, analyze, and investigate data.

Suppose you are conducting a study to compare firefly populations exposed to normal daylight–darkness conditions with firefly populations exposed to continuous light (24 hours a day). You set up two firefly colonies in a laboratory environment. The two colonies are identical except that one colony is exposed to normal daylight–darkness conditions and the other is exposed to continuous light. Each colony is populated with the same number of mature fireflies. After 72 hours, you count the number of living fireflies in each colony.

Courtesy of Corrinne and Charles Brase

Adapted from Ohio State University Firefly Files logo

After completing this chapter, you will be able to answer the following questions.

(a) Is this an experiment or an observation study? Explain.

(b) Is there a control group? Is there a treatment group?

(c) What is the variable in this study?

(d) What is the level of measurement (nominal, interval, ordinal, or ratio) of the variable?

(See Problem 11 of the Chapter 1 Review Problems.)

SECTION 1.1

What Is Statistics?

FOCUS POINTS
- Identify variables in a statistical study.
- Distinguish between quantitative and qualitative variables.
- Identify populations and samples.
- Distinguish between parameters and statistics.
- Determine the level of measurement.
- Compare descriptive and inferential statistics.

Introduction

Decision making is an important aspect of our lives. We make decisions based on the information we have, our attitudes, and our values. Statistical methods help us examine information. Moreover, statistics can be used for making decisions when we are faced with uncertainties. For instance, if we wish to estimate the proportion of people who will have a severe reaction to a flu shot without giving the shot to everyone who wants it, statistics provides appropriate methods. Statistical methods enable us to look at information from a small collection of people or items and make inferences about a larger collection of people or items.

Procedures for analyzing data, together with rules of inference, are central topics in the study of statistics.

Statistics

> **Statistics** is the study of how to collect, organize, analyze, and interpret numerical information from data.

The subject of statistics is multifaceted. The following definition of statistics is found in the *International Encyclopedia of Statistical Science*, edited by Miodrag Lovric. Professor David Hand of Imperial College London—the president of the Royal Statistical Society—presents the definition in his article "Statistics: An Overview."

> **Statistics** is both the science of uncertainty and the technology of extracting information from data.

The statistical procedures you will learn in this book should supplement your built-in system of inference—that is, the results from statistical procedures and good sense should dovetail. Of course, statistical methods themselves have no power to work miracles. These methods can help us make some decisions, but not all conceivable decisions. Remember, even a properly applied statistical procedure is no more accurate than the data, or facts, on which it is based. Finally, statistical results should be interpreted by one who understands not only the methods, but also the subject matter to which they have been applied.

The general prerequisite for statistical decision making is the gathering of data. First, we need to identify the individuals or objects to be included in the study and the characteristics or features of the individuals that are of interest.

Individuals

Variable

> **Individuals** are the people or objects included in the study.
>
> A **variable** is a characteristic of the individual to be measured or observed.

For instance, if we want to do a study about the people who have climbed Mt. Everest, then the individuals in the study are all people who have actually made it to the summit. One variable might be the height of such individuals. Other variables might be age, weight, gender, nationality, income, and so on. Regardless of the variables we use, we would not include measurements or observations from people who have not climbed the mountain.

The variables in a study may be *quantitative* or *qualitative* in nature.

Quantitative variable

Qualitative variable

> A **quantitative variable** has a value or numerical measurement for which operations such as addition or averaging make sense.
>
> A **qualitative variable** describes an individual by placing the individual into a category or group, such as male or female.

For the Mt. Everest climbers, variables such as height, weight, age, or income are *quantitative* variables. *Qualitative variables* involve non-numerical observations such as gender or nationality. Sometimes qualitative variables are referred to as *categorical variables.*

Another important issue regarding data is their source. Do the data comprise information from *all* individuals of interest, or from just *some* of the individuals?

Population data

Sample data

> In **population data,** the data are from *every* individual of interest.
>
> In **sample data,** the data are from *only some* of the individuals of interest.

It is important to know whether the data are population data or sample data. Data from a specific population are fixed and complete. Data from a sample may vary from sample to sample and are *not* complete.

Population parameter

Sample statistic

> A **population parameter** is a numerical measure that describes an aspect of a population.
>
> A **sample statistic** is a numerical measure that describes an aspect of a sample.

For instance, if we have data from *all* the individuals who have climbed Mt. Everest, then we have population data. The proportion of males in the *population* of all climbers who have conquered Mt. Everest is an example of a *parameter.*

On the other hand, if our data come from just some of the climbers, we have sample data. The proportion of male climbers in the *sample* is an example of a *statistic.* Note that different samples may have different values for the proportion of male climbers. One of the important features of sample statistics is that they can vary from sample to sample, whereas population parameters are fixed for a given population.

LOOKING FORWARD

In later chapters we will use information based on a sample and sample statistics to estimate population parameters (Chapter 8) or make decisions about the value of population parameters (Chapters 9 and 10).

EXAMPLE 1   USING BASIC TERMINOLOGY

The Hawaii Department of Tropical Agriculture is conducting a study of ready-to-harvest pineapples in an experimental field.

(a) The pineapples are the *objects* (individuals) of the study. If the researchers are interested in the individual weights of pineapples in the field, then the *variable* consists of weights. At this point, it is important to specify units of measurement and degrees of accuracy of measurement. The weights could be measured to the nearest ounce or gram. Weight is a *quantitative* variable because it is a numerical measure. If weights of *all* the ready-to-harvest pineapples in the field are included in the data, then we have a *population*. The average weight of all ready-to-harvest pineapples in the field is a *parameter*.

(b) Suppose the researchers also want data on taste. A panel of tasters rates the pineapples according to the categories "poor," "acceptable," and "good." Only some of the pineapples are included in the taste test. In this case, the *variable* is taste. This is a *qualitative* or *categorical* variable. Because only some of the pineapples in the field are included in the study, we have a *sample*. The proportion of pineapples in the sample with a taste rating of "good" is a *statistic*.

Joe Solem/Riser/Getty Images

Throughout this text, you will encounter *Guided Exercises* embedded in the reading material. These exercises are included to give you an opportunity to work immediately with new ideas. The questions guide you through appropriate analysis. Cover the answers on the right side (an index card will fit this purpose). After you have thought about or written down *your own response*, check the answers. If there are several parts to an exercise, check each part before you continue. You should be able to answer most of these exercise questions, but don't skip them—they are important.

GUIDED EXERCISE 1 | Using basic terminology

Television station QUE wants to know the proportion of TV owners in Virginia who watch the station's new program at least once a week. The station asks a group of 1000 TV owners in Virginia if they watch the program at least once a week.

(a) Identify the individuals of the study and the variable.

➡ The individuals are the 1000 TV owners surveyed. The variable is the response does, or does not, watch the new program at least once a week.

(b) Do the data comprise a sample? If so, what is the underlying population?

➡ The data comprise a sample of the population of responses from all TV owners in Virginia.

(c) Is the variable qualitative or quantitative?

➡ Qualitative—the categories are the two possible responses, does or does not watch the program.

(d) Identify a quantitative variable that might be of interest.

➡ Age or income might be of interest.

(e) Is the proportion of viewers in the sample who watch the new program at least once a week a statistic or a parameter?

➡ Statistic—the proportion is computed from sample data.

Levels of Measurement: Nominal, Ordinal, Interval, Ratio

We have categorized data as either qualitative or quantitative. Another way to classify data is according to one of the four *levels of measurement*. These levels indicate the type of arithmetic that is appropriate for the data, such as ordering, taking differences, or taking ratios.

Levels of measurement

Nominal level

Ordinal level

Interval level

Ratio level

Levels of Measurement

The **nominal level of measurement** applies to data that consist of names, labels, or categories. There are no implied criteria by which the data can be ordered from smallest to largest.

The **ordinal level of measurement** applies to data that can be arranged in order. However, differences between data values either cannot be determined or are meaningless.

The **interval level of measurement** applies to data that can be arranged in order. In addition, differences between data values are meaningful.

The **ratio level of measurement** applies to data that can be arranged in order. In addition, both differences between data values and ratios of data values are meaningful. Data at the ratio level have a true zero.

EXAMPLE 2   LEVELS OF MEASUREMENT

Identify the type of data.

Michelle Dulieu, 2009/Used under license from Shutterstock.com

(a) Taos, Acoma, Zuni, and Cochiti are the names of four Native American pueblos from the population of names of all Native American pueblos in Arizona and New Mexico.

SOLUTION: These data are at the *nominal* level. Notice that these data values are simply names. By looking at the name alone, we cannot determine if one name is "greater than or less than" another. Any ordering of the names would be numerically meaningless.

(b) In a high-school graduating class of 319 students, Jim ranked 25th, June ranked 19th, Walter ranked 10th, and Julia ranked 4th, where 1 is the highest rank.

SOLUTION: These data are at the *ordinal* level. Ordering the data clearly makes sense. Walter ranked higher than June. Jim had the lowest rank, and Julia the highest. However, numerical differences in ranks do not have meaning. The difference between June's and Jim's ranks is 6, and this is the same difference that exists between Walter's and Julia's ranks. However, this difference doesn't really mean anything significant. For instance, if you looked at grade point average, Walter and Julia may have had a large gap between their grade point averages, whereas June and Jim may have had closer grade point averages. In any ranking system, it is only the relative standing that matters. Differences between ranks are meaningless.

Korban Schwab/iStockphoto.com

(c) Body temperatures (in degrees Celsius) of trout in the Yellowstone River.

SOLUTION: These data are at the *interval* level. We can certainly order the data, and we can compute meaningful differences. However, for Celsius-scale temperatures, there is not an inherent starting point. The value 0°C may seem to be a starting point, but this value does not indicate the state of "no heat." Furthermore, it is not correct to say that 20°C is twice as hot as 10°C.

(d) Length of trout swimming in the Yellowstone River.

> **SOLUTION:** These data are at the *ratio* level. An 18-inch trout is three times as long as a 6-inch trout. Observe that we can divide 6 into 18 to determine a meaningful *ratio* of trout lengths.

In summary, there are four levels of measurement. The nominal level is considered the lowest, and in ascending order we have the ordinal, interval, and ratio levels. In general, calculations based on a particular level of measurement may not be appropriate for a lower level.

PROCEDURE

HOW TO DETERMINE THE LEVEL OF MEASUREMENT

The levels of measurement, listed from lowest to highest, are nominal, ordinal, interval, and ratio. To determine the level of measurement of data, state the *highest level* that can be justified for the entire collection of data. Consider which calculations are suitable for the data.

Level of Measurement	Suitable Calculation
Nominal	We can put the data into categories.
Ordinal	We can order the data from smallest to largest or "worst" to "best." Each data value can be *compared* with another data value.
Interval	We can order the data and also take the differences between data values. At this level, it makes sense to compare the differences between data values. For instance, we can say that one data value is 5 more than or 12 less than another data value.
Ratio	We can order the data, take differences, and also find the ratio between data values. For instance, it makes sense to say that one data value is twice as large as another.

GUIDED EXERCISE 2 | Levels of measurement

The following entries describe different data associated with a state senator. For each data entry, indicate the corresponding *level of measurement*.

(a) The senator's name is Sam Wilson. ➡ Nominal level.

(b) The senator is 58 years old. ➡ Ratio level. Notice that age has a meaningful zero. It makes sense to give age ratios. For instance, Sam is twice as old as someone who is 29.

(c) The years in which the senator was elected to the Senate are 1998, 2004, and 2010. ➡ Interval level. Dates can be ordered, and the difference between dates has meaning. For instance, 2004 is six years later than 1998. However, ratios do not make sense. The year 2000 is not twice as large as the year 1000. In addition, the year 0 does not mean "no time."

Continued

GUIDED EXERCISE 2   *continued*

(d) The senator's total taxable income last year was $878,314.

➡ Ratio level. It makes sense to say that the senator's income is 10 times that of someone earning $87,831.40.

(e) The senator surveyed his constituents regarding his proposed water-protection bill. The choices for response were strongly support, support, neutral, against, or strongly against.

➡ Ordinal level. The choices can be ordered, but there is no meaningful numerical difference between two choices.

(f) The senator's marital status is "married."

➡ Nominal level.

(g) A leading news magazine claims the senator is ranked seventh for his voting record on bills regarding public education.

➡ Ordinal level. Ranks can be ordered, but differences between ranks may vary in meaning.

CRITICAL THINKING

"Data! Data! Data!" he cried impatiently. "I can't make bricks without clay." Sherlock Holmes said these words in "The Adventure of the Copper Beeches" by Sir Arthur Conan Doyle.

Reliable statistical conclusions require reliable data. This section has provided some of the vocabulary used in discussing data. As you read a statistical study or conduct one, pay attention to the nature of the data and the ways they were collected.

When you select a variable to measure, be sure to specify the process and requirements for measurement. For example, if the variable is the weight of ready-to-harvest pineapples, specify the unit of weight, the accuracy of measurement, and maybe even the particular scale to be used. If some weights are in ounces and others in grams, the data are fairly useless.

Another concern is whether or not your measurement instrument truly measures the variable. Just asking people if they know the geographic location of the island nation of Fiji may not provide accurate results. The answers may reflect the fact that the respondents want you to think they are knowledgeable. Asking people to locate Fiji on a map may give more reliable results.

The level of measurement is also an issue. You can put numbers into a calculator or computer and do all kinds of arithmetic. However, you need to judge whether the operations are meaningful. For ordinal data such as restaurant rankings, you can't conclude that a 4-star restaurant is "twice as good" as a 2-star restaurant, even though the number 4 is twice 2.

Are the data from a sample, or do they comprise the entire population? Sample data can vary from one sample to another! This means that if you are studying the same statistic from two different samples of the same size, the data values may be different. In fact, the ways in which sample statistics vary among different samples of the same size will be the focus of our study from Section 7.4 on.

Interpretation When you work with sample data, carefully consider the population from which they are drawn. Observations and analysis of the sample are applicable to only the population from which the sample is drawn.

LOOKING FORWARD

The purpose of collecting and analyzing data is to obtain information. Statistical methods provide us tools to obtain information from data. These methods break into two branches.

Descriptive statistics

Descriptive statistics involves methods of organizing, picturing, and summarizing information from samples or populations (Chapters 2, 3, and 4).

Inferential statistics

Inferential statistics involves methods of using information from a sample to draw conclusions regarding the population (Chapters 5–11).

VIEWPOINT | The First Measured Century

The 20th century saw measurements of aspects of American life that had never been systematically studied before. Social conditions involving crime, sex, food, fun, religion, and work were numerically investigated. The measurements and survey responses taken over the entire century reveal unsuspected statistical trends. The First Measured Century is a book by Caplow, Hicks, and Wattenberg. It is also a PBS documentary available on DVD. For more information, visit the Brase/Brase statistics site at **http://www.cengagebrain.com/** *and find the link to the PBS First Measured Century documentary.*

SECTION 1.1 PROBLEMS

aplia™   Access end-of-section exercises at www.aplia.com

SECTION 1.2

Random Samples

FOCUS POINTS
- Explain the importance of random samples.
- Construct a simple random sample using random numbers.
- Simulate a random process.
- Describe stratified sampling, cluster sampling, systematic sampling, multistage sampling, and convenience sampling.

Simple Random Samples

Eat lamb—20,000 coyotes can't be wrong!

This slogan is sometimes found on bumper stickers in the western United States. The slogan indicates the trouble that ranchers have experienced in protecting their flocks from predators. Based on their experience with this sample of the coyote population, the ranchers concluded that *all* coyotes are dangerous to their flocks and should be eliminated! The ranchers used a special poison bait to get rid of the coyotes. Not only was this poison distributed on ranch land but with government cooperation it also was distributed widely on public lands.

The ranchers found that the results of the widespread poisoning were not very beneficial. The sheep-eating coyotes continued to thrive while the general population of coyotes and other predators declined. What was the problem? The sheep-eating coyotes that the ranchers had observed were not a representative sample of all coyotes. Modern methods of predator control, however, target the sheep-eating coyotes. To a certain extent, the new methods have come about through a closer examination of the sampling techniques used.

In this section, we will examine several widely used sampling techniques. One of the most important sampling techniques is a *simple random sample*.

Simple random sample

A **simple random sample** of n measurements from a population is a subset of the population selected in such a manner that every sample of size n from the population has an equal chance of being selected.

In a simple random sample, not only does every sample of the specified size have an equal chance of being selected, but every individual of the population also has an equal chance of being selected. However, the fact that each individual has an equal chance of being selected does not necessarily imply a simple random sample. Remember, for a simple random sample, every sample of the given size must also have an equal chance of being selected.

GUIDED EXERCISE 3 | **Simple random sample**

Is open space around metropolitan areas important? Players of the Colorado Lottery might think so, since some of the proceeds of the game go to fund open space and outdoor recreational space. To play the game, you pay $1 and choose any six different numbers from the group of numbers 1 through 42. If your group of six numbers matches the winning group of six numbers selected by simple random sampling, then you are a winner of a grand prize of at least $1.5 million.

(a) Is the number 25 as likely to be selected in the winning group of six numbers as the number 5?

➡ Yes. Because the winning numbers constitute a simple random sample, each number from 1 through 42 has an equal chance of being selected.

(b) Could all the winning numbers be even?

➡ Yes, since six even numbers is one of the possible groups of six numbers.

(c) Your friend always plays the numbers

1 2 3 4 5 6

Could she ever win?

➡ Yes. In a simple random sample, the listed group of six numbers is *as likely as any* of the 5,245,786 possible groups of six numbers to be selected as the winner. (See Section 5.3 to learn how to compute the number of possible groups of six numbers that can be selected from 42 numbers.)

How do we get random samples? Suppose you need to know if the emission systems of the latest shipment of Toyotas satisfy pollution-control standards. You want to pick a random sample of 30 cars from this shipment of 500 cars and test them. One way to pick a random sample is to number the cars 1 through 500. Write these numbers on cards, mix up the cards, and then draw 30 numbers. The sample will consist of the cars with the chosen numbers. If you mix the cards sufficiently, this procedure produces a random sample.

Random-number table

An easier way to select the numbers is to use a *random-number table*. You can make one yourself by writing the digits 0 through 9 on separate cards and mixing up these cards in a hat. Then draw a card, record the digit, return the card, and mix up the cards again. Draw another card, record the digit, and so on. Table 1 in the Appendix is a ready-made random-number table (adapted from the Rand Corporation, *A Million Random Digits with 100,000 Normal Deviates*). Let's see how to pick our random sample of 30 Toyotas by using this random-number table.

EXAMPLE 3   RANDOM-NUMBER TABLE

Use a random-number table to pick a random sample of 30 cars from a population of 500 cars.

SOLUTION: Again, we assign each car a different number between 1 and 500, inclusive. Then we use the random-number table to choose the sample. Table 1 in the Appendix has 50 rows and 10 blocks of five digits each; it can be thought of as a solid mass of digits that has been broken up into rows and blocks for user convenience.

You read the digits by beginning anywhere in the table. We dropped a pin on the table, and the head of the pin landed in row 15, block 5. We'll begin there and list all the digits in that row. If we need more digits, we'll move on to row 16, and so on. The digits we begin with are

| 99281 | 59640 | 15221 | 96079 | 09961 | 05371 |

Since the highest number assigned to a car is 500, and this number has three digits, we regroup our digits into blocks of 3:

| 992 | 815 | 964 | 015 | 221 | 960 | 790 | 996 | 105 | 371 |

To construct our random sample, we use the first 30 car numbers we encounter in the random-number table when we start at row 15, block 5. We skip the first three groups—992, 815, and 964—because these numbers are all too large. The next group of three digits is 015, which corresponds to 15. Car number 15 is the first car included in our sample, and the next is car number 221. We skip the next three groups and then include car numbers 105 and 371. To get the rest of the cars in the sample, we continue to the next line and use the random-number table in the same fashion. If we encounter a number we've used before, we skip it.

COMMENT   When we use the term *(simple) random sample,* we have very specific criteria in mind for selecting the sample. One proper method for selecting a simple random sample is to use a computer- or calculator-based random-number generator or a table of random numbers as we have done in the example. The term *random* should not be confused with *haphazard*!

LOOKING FORWARD

The runs test for randomness found in *Understandable Statistics,* 10th edition, by Brase and Brase, shows how to determine if two symbols are randomly mixed in an ordered list of symbols.

PROCEDURE

HOW TO DRAW A RANDOM SAMPLE

1. Number all members of the population sequentially.
2. Use a table, calculator, or computer to select random numbers from the numbers assigned to the population members.
3. Create the sample by using population members with numbers corresponding to those randomly selected.

LOOKING FORWARD

Simple random samples are key components in methods of inferential statistics that we will study in Chapters 8–11. In fact, in order to draw conclusions about a population, the methods we will study *require* that we have simple random samples from the populations of interest.

Another important use of random-number tables is in *simulation*. We use the word *simulation* to refer to the process of providing numerical imitations of "real" phenomena. Simulation methods have been productive in studying a diverse array of subjects, such as nuclear reactors, cloud formation, cardiology (and medical science in general), highway design, production control, shipbuilding, airplane design, war games, economics, and electronics. A complete list would probably include something from every aspect of modern life. In Guided Exercise 4 we'll perform a brief simulation.

Simulation

> A **simulation** is a numerical facsimile or representation of a real-world phenomenon.

GUIDED EXERCISE 4 | **Simulation**

Use a random-number table to simulate the outcomes of tossing a balanced (that is, fair) penny 10 times.

(a) How many outcomes are possible when you toss a coin once?

⟹   Two—heads or tails.

(b) There are several ways to assign numbers to the two outcomes. Because we assume a fair coin, we can assign an even digit to the outcome "heads" and an odd digit to the outcome "tails." Then, starting at block 3 of row 2 of Table 1 in the Appendix, list the first 10 single digits.

⟹   7 1 5 4 9 4 4 8 4 3

(c) What are the outcomes associated with the 10 digits?

⟹   T T T H T H H H H T

(d) If you start in a different block and row of Table 1 in the Appendix, will you get the same sequence of outcomes?

⟹   It is possible, but not very likely. (In Section 5.3 you will learn how to determine that there are 1024 possible sequences of outcomes for 10 tosses of a coin.)

TECH NOTES   Most statistical software packages, spreadsheet programs, and statistical calculators generate random numbers. In general, these devices sample with replacement. *Sampling with replacement* means that although a number is selected for the sample, it is *not removed* from the population. Therefore, the same number may be selected for the sample more than once. If you need to sample without replacement, generate more items than you need for the sample. Then sort the sample and remove duplicate values. Specific procedures for generating random samples using the TI-84Plus/TI-83Plus/TI-*n*spire (with TI-84 Plus keypad) calculator, Excel 2010, Minitab, and SPSS are shown in Using Technology at the end of this chapter. More details are given in the separate *Technology Guides* for each of these technologies.

Sampling with replacement

Other Sampling Techniques

Although we will assume throughout this text that (simple) random samples are used, other methods of sampling are also widely used. Appropriate statistical techniques exist for these sampling methods, but they are beyond the scope of this text.

Stratified sampling

One of these sampling methods is called *stratified sampling*. Groups or classes inside a population that share a common characteristic are called *strata* (plural of *stratum*). For example, in the population of all undergraduate college students, some strata might be freshmen, sophomores, juniors, or seniors. Other strata might be men or women, in-state students or out-of-state students, and so on. In the method of stratified sampling, the population is divided into at least two distinct strata. Then a (simple) random sample of a certain size is drawn from each stratum, and the information obtained is carefully adjusted or weighted in all resulting calculations.

The groups or strata are often sampled in proportion to their actual percentages of occurrence in the overall population. However, other (more sophisticated) ways to determine the optimal sample size in each stratum may give the best results. In general, statistical analysis and tests based on data obtained from stratified samples are somewhat different from techniques discussed in an introductory course in statistics. Such methods for stratified sampling will not be discussed in this text.

Systematic sampling

Another popular method of sampling is called *systematic sampling*. In this method, it is assumed that the elements of the population are arranged in some natural sequential order. Then we select a (random) starting point and select every *k*th element for our sample. For example, people lining up to buy rock-concert tickets are "in order." To generate a systematic sample of these people (and ask questions regarding topics such as age, smoking habits, income level, etc.), we could include every fifth person in line. The "starting" person is selected at random from the first five.

The advantage of a systematic sample is that it is easy to get. However, there are dangers in using systematic sampling. When the population is repetitive or cyclic in nature, systematic sampling should not be used. For example, consider a fabric mill that produces dress material. Suppose the loom that produces the material makes a mistake every 17th yard, but we check only every 16th yard with an automated electronic scanner. In this case, a random starting point may or may not result in detection of fabric flaws before a large amount of fabric is produced.

Cluster sampling

Cluster sampling is a method used extensively by government agencies and certain private research organizations. In cluster sampling, we begin by dividing the demographic area into sections. Then we randomly select sections or clusters. Every member of the cluster is included in the sample. For example, in conducting a survey of schoolchildren in a large city, we could first randomly select five schools and then include all the children from each selected school.

Multistage samples

Often a population is very large or geographically spread out. In such cases, samples are constructed through a *multistage sample design* of several stages, with the final stage consisting of clusters. For instance, the government Current Population Survey interviews about 60,000 households across the United States each month by means of a multistage sample design.

For the Current Population Survey, the first stage consists of selecting samples of large geographic areas that do not cross state lines. These areas are further broken down into smaller blocks, which are stratified according to ethnic and other factors. Stratified samples of the blocks are then taken. Finally, housing units in each chosen block are broken into clusters of nearby housing units. A random sample of these clusters of housing units is selected, and each household in the final cluster is interviewed.

Convenience sampling

Convenience sampling simply uses results or data that are conveniently and readily obtained. In some cases, this may be all that is available, and in many cases it is better than no information at all. However, convenience sampling does run the risk of being severely biased. For instance, consider a newsperson who wishes to get the "opinions of the people" about a proposed seat tax to be imposed on tickets to all sporting events. The revenues from the seat tax will then be used to support the local symphony. The newsperson stands in front of a concert hall and surveys the first five people exiting after a symphony performance who will cooperate. This method of choosing a sample will produce some opinions, and perhaps some human interest stories, but it certainly has bias. It is hoped that the city council will not use these opinions as the sole basis for a decision about the proposed tax. It is good advice to be very cautious indeed when the data come from the method of convenience sampling.

Sampling Techniques

Random sampling: Use a simple random sample from the entire population.

Stratified sampling: Divide the entire population into distinct subgroups called strata. The strata are based on a specific characteristic, such as age, income, education level, and so on. All members of a stratum share the specific characteristic. Draw random samples from each stratum.

Systematic sampling: Number all members of the population sequentially. Then, from a starting point selected at random, include every *k*th member of the population in the sample.

Cluster sampling: Divide the entire population into pre-existing segments or clusters. The clusters are often geographic. Make a random selection of clusters. Include every member of each selected cluster in the sample.

Multistage sampling: Use a variety of sampling methods to create successively smaller groups at each stage. The final sample consists of clusters.

Convenience sampling: Create a sample by using data from population members that are readily available.

CRITICAL THINKING

Sampling frame

We call the list of individuals from which a sample is actually selected the *sampling frame*. Ideally, the sampling frame is the entire population. However, from a practical perspective, not all members of a population may be accessible. For instance, using a telephone directory as the sample frame for residential telephone contacts would not include unlisted numbers.

Undercoverage

When the sample frame does not match the population, we have what is called *undercoverage*. In demographic studies, undercoverage could result if the homeless, fugitives from the law, and so forth, are not included in the study.

A **sampling frame** is a list of individuals from which a sample is actually selected.

Undercoverage results from omitting population members from the sample frame.

In general, even when the sampling frame and the population match, a sample is not a perfect representation of a population. Therefore, information drawn from a sample may not exactly match corresponding information from the population. To the extent that sample information does not match the corresponding population information, we have an error, called a *sampling error*.

Sampling error

> A **sampling error** is the difference between measurements from a sample and corresponding measurements from the respective population. It is caused by the fact that the sample does not perfectly represent the population.
>
> A **nonsampling error** is the result of poor sample design, sloppy data collection, faulty measuring instruments, bias in questionnaires, and so on.

Nonsampling error

Sampling errors do not represent mistakes! They are simply the consequences of using samples instead of populations. However, be alert to nonsampling errors, which may sometimes occur inadvertently.

VIEWPOINT | Extraterrestrial Life?

Do you believe intelligent life exists on other planets? Using methods of random sampling, a Fox News opinion poll found that about 54% of all U.S. men do believe in intelligent life on other planets, whereas only 47% of U.S. women believe there is such life. How could you conduct a random survey of students on your campus regarding belief in extraterrestrial life?

SECTION 1.2 PROBLEMS

Access end-of-section exercises at www.aplia.com

SECTION 1.3

Introduction to Experimental Design

Focus points
- Discuss what it means to take a census.
- Describe simulations, observational studies, and experiments.
- Identify control groups, placebo effects, completely randomized experiments, and randomized block experiments.
- Discuss potential pitfalls that might make your data unreliable.

Planning a Statistical Study

Planning a statistical study and gathering data are essential components of obtaining reliable information. Depending on the nature of the statistical study, a great deal of expertise and resources may be required during the planning stage. In this section, we look at some of the basics of planning a statistical study.

PROCEDURE

BASIC GUIDELINES FOR PLANNING A STATISTICAL STUDY

1. First, identify the individuals or objects of interest.

2. Specify the variables as well as the protocols for taking measurements or making observations.

3. Determine whether you will use an entire population or a representative sample. If using a sample, decide on a viable sampling method.

4. In your data-collection plan, address issues of ethics, subject confidentiality, and privacy. If you are collecting data at a business, store, college, or other institution, be sure to be courteous and to obtain permission as necessary.

5. Collect the data.

6. Use appropriate descriptive statistics methods (Chapters 2–4) and make decisions using appropriate inferential statistics methods (Chapters 8–11).

7. Finally, note any concerns you might have about your data-collection methods and list any recommendations for future studies.

One issue to consider is whether to use the entire population in a study or a representative sample. If we use data from the entire population, we have a *census*.

Census

> In a **census**, measurements or observations from the *entire* population are used.

When the population is small and easily accessible, a census is very useful because it gives complete information about the population. However, obtaining a census can be both expensive and difficult. Every 10 years, the U.S. Department of Commerce's Census Bureau is required to conduct a census of the United States. However, contacting some members of the population—such as the homeless—is almost impossible. Sometimes members of the population will not respond. In such cases, statistical estimates for the missing responses are often supplied.

Overcounting—that is, counting the same person more than once—is also a problem the Census Bureau is addressing. In fact, in 2000, slightly more people were counted twice than the estimated number of people missed. For instance, a college student living on campus might be counted on a parent's census form as well as on his or her own census form.

If we use data from only part of the population of interest, we have a *sample*.

Sample

> In a **sample**, measurements or observations from *part* of the population are used.

In the previous section, we examined several sampling strategies: simple random, stratified, cluster, systematic, multistage, and convenience. In this text, we will study methods of inferential statistics based on simple random samples.

Simulation

As discussed in Section 1.2, *simulation* is a numerical facsimile of real-world phenomena. Sometimes simulation is called a "dry lab" approach, in the sense that it is a mathematical imitation of a real situation. Advantages of simulation are that numerical and statistical simulations can fit real-world problems extremely well. The researcher can also explore procedures through simulation that might be very dangerous in real life.

Experiments and Observation

When gathering data for a statistical study, we want to distinguish between observational studies and experiments.

Observational study

> In an **observational study,** observations and measurements of individuals are conducted in a way that doesn't change the response or the variable being measured.

Experiment

> In an **experiment,** a *treatment* is deliberately imposed on the individuals in order to observe a possible change in the response or variable being measured.

EXAMPLE 4   EXPERIMENT

In 1778, Captain James Cook landed in what we now call the Hawaiian Islands. He gave the islanders a present of several goats, and over the years these animals multiplied into wild herds totaling several thousand. They eat almost anything, including the famous silversword plant, which was once unique to Hawaii. At one time, the silversword grew abundantly on the island of Maui (in Haleakala, a national park on that island, the silversword can still be found), but each year there seemed to be fewer and fewer plants. Biologists suspected that the goats were partially responsible for the decline in the number of plants and conducted a statistical study that verified their theory.

(a) To test the theory, park biologists set up stations in remote areas of Haleakala. At each station, two plots of land similar in soil conditions, climate, and plant count were selected. One plot was fenced to keep out the goats, while the other was not. At regular intervals a plant count was made in each plot. This study involved an *experiment* because a *treatment* (the fence) was imposed on one plot.

(b) The experiment involved two plots at each station. The plot that was not fenced was the *control* plot. This was the plot on which a treatment was specifically not imposed, although the plot was similar to the fenced plot in every other way.

Silversword plant, Haleakala National Park

Tim Davis/Photo Researchers

Statistical experiments are commonly used to determine the effect of a treatment. However, the design of the experiment needs to *control* for other possible causes of the effect. For instance, in medical experiments, the *placebo effect* is the improvement or change that is the result of patients' just believing in the treatment, whether or not the treatment itself is effective.

Placebo effect

> The **placebo effect** occurs when a subject receives no treatment but (incorrectly) believes he or she is in fact receiving treatment and responds favorably.

To account for the placebo effect, patients are divided into two groups. One group receives the prescribed treatment. The other group, called the *control group,* receives a dummy or placebo treatment that is disguised to look like the

Treatment group

Completely randomized experiment

real treatment. Finally, after the treatment cycle, the medical condition of the patients in the *treatment group* is compared to that of the patients in the control group.

A common way to assign patients to treatment and control groups is by using a random process. This is the essence of a *completely randomized experiment*.

> A **completely randomized experiment** is one in which a random process is used to assign each individual to one of the treatments.

EXAMPLE 5   COMPLETELY RANDOMIZED EXPERIMENT

Can chest pain be relieved by drilling holes in the heart? For more than a decade, surgeons have been using a laser procedure to drill holes in the heart. Many patients report a lasting and dramatic decrease in angina (chest pain) symptoms. Is the relief due to the procedure, or is it a placebo effect? A recent research project at Lenox Hill Hospital in New York City provided some information about this issue by using a completely randomized experiment. The laser treatment was applied through a less invasive (catheter laser) process. A group of 298 volunteers with severe, untreatable chest pain were randomly assigned to get the laser or not. The patients were sedated but awake. They could hear the doctors discuss the laser process. Each patient thought he or she was receiving the treatment.

The experimental design can be pictured as

The laser patients did well. But shockingly, the placebo group showed more improvement in pain relief. The medical impacts of this study are still being investigated.

It is difficult to control all the variables that might influence the response to a treatment. One way to control some of the variables is through *blocking*.

Block

Randomized block experiment

> A **block** is a group of individuals sharing some common features that might affect the treatment.
>
> In a **randomized block experiment,** individuals are first sorted into blocks, and then a random process is used to assign each individual in the block to one of the treatments.

A randomized block design utilizing gender for blocks in the experiment involving laser holes in the heart would be

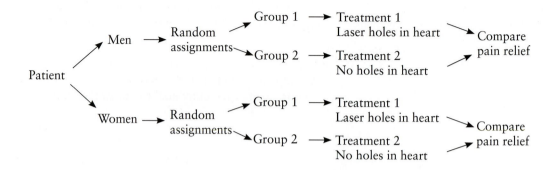

The study cited in Example 5 has many features of good experimental design.

Control group

There is a **control group.** This group receives a dummy treatment, enabling the researchers to control for the placebo effect. In general, a control group is used to account for the influence of other known or unknown variables that might be an underlying cause of a change in response in the experimental group. Such variables are called **lurking** or **confounding variables.**

Randomization

Randomization is used to assign individuals to the two treatment groups. This helps prevent bias in selecting members for each group.

Replication

Replication of the experiment on many patients reduces the possibility that the differences in pain relief for the two groups occurred by chance alone.

Double-blind experiment

Many experiments are also *double-blind*. This means that neither the individuals in the study nor the observers know which subjects are receiving the treatment. Double-blind experiments help control for subtle biases that a doctor might pass on to a patient.

LOOKING FORWARD

One-way and two-way ANOVA are analysis techniques used to study results from completely randomized experiments with several treatments or several blocks with multiple treatments. These topics can be found in *Understandable Statistics*, 10th edition, by Brase and Brase.

GUIDED EXERCISE 5 | Collecting data

Which technique for gathering data (sampling, experiment, simulation, or census) do you think might be the most appropriate for the following studies?

(a) Study of the effect of stopping the cooling process of a nuclear reactor.

    Simulation, since you probably do not want to risk a nuclear meltdown.

Continued

(b) Study of the amount of time college students taking a full course load spend watching television.

➡ Sampling and using an observational study would work well. Notice that obtaining the information from a student will probably not change the amount of time the student spends watching television.

(c) Study of the effect on bone mass of a calcium supplement given to young girls.

➡ Experimentation. A study by Tom Lloyd and others reported in *The Journal of the American Medical Association* utilized 94 young girls. Half were randomly selected and given a placebo. The other half were given calcium supplements to bring their daily calcium intake up to about 1400 milligrams per day. The group getting the experimental treatment of calcium gained 1.3% more bone mass in a year than the girls getting the placebo.

(d) Study of the credit-hour load of *each* student enrolled at your college at the end of the drop–add period this semester.

➡ Census. The registrar can obtain records for *every* student.

Spencer Grant/PhotoEdit

Likert scale

Survey pitfalls
Nonresponse

Hidden bias

Voluntary response

Surveys

Once you decide whether you are going to use sampling, census, observation, or experiments, a common means to gather data about people is to ask them questions. This process is the essence of *surveying*. Sometimes the possible responses are simply yes or no. Other times the respondents choose a number on a scale that represents their feelings from, say, strongly disagree to strongly agree. Such a scale is called a *Likert scale*. In the case of an open-ended, discussion-type response, the researcher must determine a way to convert the response to a category or number.

A number of issues can arise when using a survey.

Some Potential Pitfalls of a Survey

Nonresponse: Individuals either cannot be contacted or refuse to participate. Nonresponse can result in significant undercoverage of a population.

Truthfulness of response: Respondents may lie intentionally or inadvertently.

Faulty recall: Respondents may not accurately remember when or whether an event took place.

Hidden bias: The question may be worded in such a way as to elicit a specific response. The order of questions might lead to biased responses. Also, the number of responses on a Likert scale may force responses that do not reflect the respondent's feelings or experience.

Vague wording: Words such as *often*, *seldom*, and *occasionally* mean different things to different people.

Interviewer influence: Factors such as tone of voice, body language, dress, gender, authority, and ethnicity of the interviewer might influence responses.

Voluntary response: Individuals with strong feelings about a subject are more likely than others to respond. Such a study is interesting but not reflective of the population.

Lurking and confounding variables

Sometimes our goal is to understand the cause-and-effect relationships between two or more variables. Such studies can be complicated by *lurking variables* or *confounding variables*.

> A **lurking variable** is one for which no data have been collected but which nevertheless has influence on other variables in the study.
>
> Two variables are **confounded** when the effects of one cannot be distinguished from the effects of the other. Confounding variables may be part of the study, or they may be outside lurking variables.

For instance, consider a study involving just two variables, amount of gasoline used to commute to work and time to commute to work. Level of traffic congestion is a likely lurking variable that increases both of the study variables. In a study involving several variables such as grade point average, difficulty of courses, IQ, and available study time, some of the variables might be confounded. For instance, students with less study time might opt for easier courses.

Generalizing results

Some researchers want to *generalize* their findings to a situation of wider scope than that of the actual data setting. The true scope of a new discovery must be determined by repeated studies in various real-world settings. Statistical experiments showing that a drug had a certain effect on a collection of laboratory rats do not guarantee that the drug will have a similar effect on a herd of wild horses in Montana.

Study sponsor

The *sponsorship of a study* is another area of concern. Subtle bias may be introduced. For instance, if a pharmaceutical company is paying freelance researchers to work on a study, the researchers may dismiss rare negative findings about a drug or treatment.

GUIDED EXERCISE 6 | *Cautions about data*

Comment on the usefulness of the data collected as described.

(a) A uniformed police officer interviews a group of 20 college freshmen. She asks each one his or her name and then if he or she has used an illegal drug in the last month.

⟹ Respondents may not answer truthfully. Some may refuse to participate.

(b) Jessica saw some data showing that cities with more low-income housing have more homeless people. Does building low-income housing cause homelessness?

⟹ There may be some other confounding or lurking variables, such as the size of the city. Larger cities may have more low-income housing and more homeless people.

(c) A survey about food in the student cafeteria was conducted by placing forms for customers to pick up at the cash register. A drop box for completed forms was available outside the cafeteria.

⟹ The voluntary response likely produced more negative comments.

(d) Extensive studies on coronary problems were conducted using men over age 50 as the subjects.

⟹ Conclusions for men over age 50 may or may not generalize to other age and gender groups. These results may be useful for women or younger people, but studies specifically involving these groups may need to be performed.

Choosing Data-Collection Techniques

We've briefly discussed three common techniques for gathering data: observational studies, experiments, and surveys. Which technique is best? The answer depends on the number of variables of interest and the level of confidence needed regarding statements of relationships among the variables.

- Surveys may be the best choice for gathering information across a wide range of many variables. Many questions can be included in a survey. However, great care must be taken in the construction of the survey instrument and in the administration of the survey. Nonresponse and other issues discussed earlier can introduce bias.

- Observational studies are the next most convenient technique for gathering information on many variables. Protocols for taking measurements or recording observations need to be specified carefully.

- Experiments are the most stringent and restrictive data-gathering technique. They can be time consuming, expensive, and difficult to administer. In experiments, the goal is often to study the effects of changing only one variable at a time. Because of the requirements, the number of variables may be more limited. Experiments must be designed carefully to ensure that the resulting data are relevant to the research questions.

COMMENT   An experiment is the best technique for reaching valid conclusions. By carefully controlling for other variables, the effect of changing one variable in a treatment group and comparing it to a control group yields results carrying high confidence.

The next most effective technique for obtaining results that have high confidence is the use of observational studies. Care must be taken that the act of observation does not change the behavior being measured or observed.

The least effective technique for drawing conclusions is the survey. Surveys have many pitfalls and by their nature cannot give exceedingly precise results. A medical study utilizing a survey asking patients if they feel better after taking a specific drug gives some information, but not precise information about the drug's effects. However, surveys are widely used to gauge attitudes, gather demographic information, study social and political trends, and so on.

VIEWPOINT | Is the Placebo Effect a Myth?

Henry Beecher, former chief of anesthesiology at Massachusetts General Hospital, published a paper in The Journal of the American Medical Association (1955) in which he claimed that the placebo effect is so powerful that about 35% of patients would improve simply if they believed a dummy treatment (placebo) was real. However, two Danish medical researchers refute this widely accepted claim in The New England Journal of Medicine. They say the placebo effect is nothing more than a "regression effect," referring to a well-known statistical observation that patients who feel especially bad one day will almost always feel better the next day, no matter what is done for them. However, other respected statisticians question the findings of the Danish researchers. Regardless of the new controversy surrounding the placebo effect, medical researchers agree that placebos are still needed in clinical research. Double-blind research using placebos prevents researchers from inadvertently biasing results.

SECTION 1.3 PROBLEMS | aplia™   Access end-of-section exercises at www.aplia.com

Chapter Review

SUMMARY

In this chapter, you've seen that statistics is the study of how to collect, organize, analyze, and interpret numerical information from populations or samples. This chapter discussed some of the features of data and ways to collect data. In particular, the chapter discussed

- Individuals or subjects of a study and the variables associated with those individuals

- Data classification as qualitative or quantitative, and levels of measurement of data

- Sample and population data: Summary measurements from sample data are called statistics, and those from populations are called parameters

- Sampling strategies, including simple random, stratified, systematic, multistage, and convenience. Inferential techniques presented in this text are based on simple random samples

- Methods of obtaining data: Census, simulation, observational studies, experiments, and surveys

- Concerns: Undercoverage of a population, nonresponse, bias in data from surveys and other factors, effects of confounding or lurking variables on other variables, generalization of study results beyond the population of the study, and study sponsorship

IMPORTANT WORDS & SYMBOLS

*Indicates section of first appearance.

Is Chocolate Good for Your Heart?

A study of 7841 Harvard alumni showed that the death rate was 30% lower in those who ate candy compared with those who abstained. It turns out that candy, especially chocolate, contains antioxidants that help slow the aging process. Also, chocolate, like aspirin, reduces the activity of blood platelets that contribute to plaque and blood clotting. Furthermore, chocolate seems to raise levels of high-density lipoprotein (HDL), the good cholesterol. However, these results are all preliminary; the investigation is far from complete. A wealth of information on this topic was published in the August 2000 issue of The Journal of Nutrition. Statistical studies and reliable experimental design are indispensable in this type of research.

CHAPTER REVIEW PROBLEMS

Joe McDaniel/iStockphoto.com

1. *Critical Thinking* Sudoku is a puzzle consisting of squares arranged in 9 rows and 9 columns. The 81 squares are further divided into nine 3×3 square boxes. The object is to fill in the squares with numerals 1 through 9 so that each column, row, and box contains all nine numbers. However, there is a requirement that each number appear only once in any row, column, or box. Each puzzle already has numbers in some of the squares. Would it be appropriate to use a random-number table to select a digit for each blank square? Explain.

2. *Critical Thinking* Alisha wants to do a statistical study to determine how long it takes people to complete a sudoku puzzle (see Problem 1 for a description of the puzzle). Her plan is as follows:

    Download 10 different puzzles from the Internet.
    Find 10 friends willing to participate.
    Ask each friend to complete one of the puzzles and time him- or herself.
    Gather the completion times from each friend.

    Describe some of the problems with Alisha's plan for the study. (*Note:* Puzzles differ in difficulty, ranging from beginner to very difficult.) Are the results from Alisha's study anecdotal, or do they apply to the general population?

3. *Statistical Literacy* You are conducting a study of students doing work-study jobs on your campus. Among the questions on the survey instrument are:

    A. How many hours are you scheduled to work each week? Answer to the nearest hour.

    B. How applicable is this work experience to your future employment goals?

    Respond using the following scale: 1 = not at all, 2 = somewhat, 3 = very

    (a) Suppose you take random samples from the following groups: freshmen, sophomores, juniors, and seniors. What kind of sampling technique are you using (simple random, stratified, systematic, cluster, multistage, convenience)?
    (b) Describe the individuals of this study.
    (c) What is the variable for question A? Classify the variable as qualitative or quantitative. What is the level of the measurement?
    (d) What is the variable for question B? Classify the variable as qualitative or quantitative. What is the level of the measurement?
    (e) Is the proportion of responses "3 = very" to question B a statistic or a parameter?
    (f) Suppose only 40% of the students you selected for the sample respond. What is the nonresponse rate? Do you think the nonresponse rate might introduce bias into the study? Explain.
    (g) Would it be appropriate to generalize the results of your study to all work-study students in the nation? Explain.

4. *Radio Talk Show: Sample Bias* A radio talk-show host asked listeners to respond either yes or no to the question, "Is the candidate who spends the most on a campaign the most likely to win?" Fifteen people called in and nine said yes. What is the implied population? What is the variable? Can you detect any bias in the selection of the sample?

5. *Simulation: TV Habits* One cable station knows that approximately 30% of its viewers have TiVo and can easily skip over advertising breaks. You are to design a simulation of how a random sample of seven station viewers would respond to the question, "Do you have TiVo?" How would you assign the random digits 0 through 9 to the responses "Yes" and "No" to the TiVo question? Use your random-digit assignment and the random-number table to generate the responses from a random sample of seven station viewers.

6. *General: Type of Sampling* Categorize the type of sampling (simple random, stratified, systematic, cluster, or convenience) used in each of the following situations.
   (a) To conduct a pre-election opinion poll on a proposed amendment to the state constitution, a random sample of 10 telephone prefixes (first three digits of the phone number) was selected, and all households from the phone prefixes selected were called.
   (b) To conduct a study on depression among the elderly, a sample of 30 patients in one nursing home was used.
   (c) To maintain quality control in a brewery, every 20th bottle of beer coming off the production line was opened and tested.
   (d) Subscribers to the magazine *Sound Alive* were assigned numbers. Then a sample of 30 subscribers was selected by using a random-number table. The subscribers in the sample were invited to rate new compact-disc players for a "What the Subscribers Think" column.
   (e) To judge the appeal of a proposed TV sitcom, a random sample of 10 people from each of three different age categories was selected and those chosen were asked to rate a pilot show.

7. *General: Gathering Data* Which technique for gathering data (observational study or experiment) do you think was used in the following studies? Explain.
   (a) The U.S. Census Bureau tracks population age. In 1900, the percentage of the population that was 19 years old or younger was 44.4%. In 1930, the percentage was 38.8%; in 1970, the percentage was 37.9%; and in 2000, the percentage in that age group was down to 28.5% (Reference: *The First Measured Century*, T. Caplow, L. Hicks, and B. J. Wattenberg).
   (b) After receiving the same lessons, a class of 100 students was randomly divided into two groups of 50 each. One group was given a multiple-choice exam covering the material in the lessons. The other group was given an essay exam. The average test scores for the two groups were then compared.

8. *General: Experiment* How would you use a completely randomized experiment in each of the following settings? Is a placebo being used or not? Be specific and give details.
   (a) A charitable nonprofit organization wants to test two methods of fundraising. From a list of 1000 past donors, half will be sent literature about the successful activities of the charity and asked to make another donation. The other 500 donors will be contacted by phone and asked to make another donation. The percentage of people from each group who make a new donation will be compared.
   (b) A tooth-whitening gel is to be tested for effectiveness. A group of 85 adults have volunteered to participate in the study. Of these, 43 are to be given a gel that contains the tooth-whitening chemicals. The remaining 42 are to be given a similar-looking package of gel that does not contain the tooth-whitening chemicals. A standard method will be used to evaluate the whiteness of teeth for all participants. Then the results for the two groups will be compared. How could this experiment be designed to be double-blind?

(c) Consider the experiment described in part (a). Describe how you would use a randomized block experiment with blocks based on age. Use three blocks: donors under 30 years old, donors 30 to 59 years old, and donors 60 years old and over.

9. *Student Life: Data-Collection Project* Make a statistical profile of your own statistics class. Items of interest might be
   (a) Height; age; gender; pulse; number of siblings; marital status
   (b) Number of college credit hours completed (as of beginning of term); grade point average
   (c) Major; number of credit hours enrolled in this term
   (d) Number of scheduled work hours per week
   (e) Distance from residence to first class; time it takes to travel from residence to first class
   (f) Year, make, and color of car usually driven

   What directions would you give to people answering these questions? For instance, how accurate should the measurements be? Should age be recorded as of last birthday?

10. *Census: Website* *Census and You,* a publication of the Census Bureau, indicates that "Wherever your Web journey ends up, it should start at the Census Bureau's site." Visit the Brase/Brase statistics site at **http://www.cengagebrain.com/** and find a link to the Census Bureau's site, as well as to FedStats, another extensive site offering links to federal data. The Census Bureau site touts itself as the source of "official statistics," but it is willing to share the spotlight. It now has links to other "official" sources: other federal agencies, foreign statistical agencies, and state data centers. If you have access to the Internet, try the Census Bureau's site.

11. *Focus Problem: Fireflies* Suppose you are conducting a study to compare firefly populations exposed to normal daylight–darkness conditions with firefly populations exposed to continuous light (24 hours a day). You set up two firefly colonies in a laboratory environment. The two colonies are identical except that one colony is exposed to normal daylight–darkness conditions and the other is exposed to continuous light. Each colony is populated with the same number of mature fireflies. After 72 hours, you count the number of living fireflies in each colony.
   (a) Is this an experiment or an observation study? Explain.
   (b) Is there a control group? Is there a treatment group?
   (c) What is the variable in this study?
   (d) What is the level of measurement (nominal, interval, ordinal, or ratio) of the variable?

DATA HIGHLIGHTS: GROUP PROJECTS

1. Use a random-number table or random-number generator to simulate tossing a fair coin 10 times. Generate 20 such simulations of 10 coin tosses. Compare the simulations. Are there any strings of 10 heads? of 4 heads? Does it seem that in most of the simulations, half the outcomes are heads and half are tails? In Chapter 6, we will study the probabilities of getting from 0 to 10 heads in such a simulation.

2. Use a random-number table or random-number generator to generate a random sample of 30 distinct values from the set of integers 1 to 100. Instructions for doing this using the TI-84Plus/TI-83Plus/TI-*n*spire (with TI-84 Plus keypad), Excel 2010, Minitab, and SPSS are given in Using Technology at the end of this chapter. Generate five such samples. How many of the samples include the number 1? the number 100? Comment about the differences among the samples. How well do the samples seem to represent the numbers between 1 and 100?

LINKING CONCEPTS: WRITING PROJECTS

Discuss each of the following topics in class or review the topics on your own. Then write a brief but complete essay in which you summarize the main points. Please include formulas and graphs as appropriate.

1. What does it mean to say that we are going to use a sample to draw an inference about a population? Why is a random sample so important for this process? If we wanted a random sample of students in the cafeteria, why couldn't we just choose the students who order Diet Pepsi with their lunch? Comment on the statement "A random sample is like a miniature population, whereas samples that are not random are likely to be biased." Why would the students who order Diet Pepsi with lunch not be a random sample of students in the cafeteria?

2. In your own words, explain the differences among the following sampling techniques: simple random sample, stratified sample, systematic sample, cluster sample, multistage sample, and convenience sample. Describe situations in which each type might be useful.

USING TECHNOLOGY

General spreadsheet programs such as Microsoft Excel, specific statistical software packages such as Minitab or SPSS, and graphing calculators such as the TI-84Plus/TI-83Plus/TI-nspire all offer computing support for statistical methods. Applications in this section may be completed using software or calculators with statistical functions. Select keystroke or menu choices are shown for the TI-84Plus/TI-83Plus/TI-nspire (with TI-84Plus keypad) calculators, Minitab, Excel 2010, and SPSS in the Technology Hints portion of this section. More details can be found in the software-specific *Technology Guides* that accompany this text.

Applications

Most software packages sample *with replacement*. That is, the same number may be used more than once in the sample. If your applications require sampling without replacement, draw more items than you need. Then use sort commands in the software to put the data in order, and delete repeated data.

1. Simulate the results of tossing a fair die 18 times. Repeat the simulation. Are the results the same? Did you expect them to be the same? Why or why not? Do there appear to be equal numbers of outcomes 1 through 6 in each simulation? In Chapter 5, we will encounter the law of large numbers, which tells us that we would expect equal numbers of outcomes only when the simulation is very large.

2. A college has 5000 students, and the registrar wishes to use a random sample of 50 students to examine credit-hour enrollment for this semester. Write a brief description of how a random sample can be drawn. Draw a random sample of 50 students. Are you sampling with or without replacement?

Technology Hints: Random Numbers

TI-84Plus/TI-83Plus/TI-nspire (with TI-84Plus keypad)

The TI-nspire calculator with the TI-84Plus keypad installed works exactly like other TI-84Plus calculators. Instructions for the TI-84Plus, TI-83Plus, and the TI-nspire (with the TI-84Plus keypad installed) calculators are included directly in this text as well as in a separate *Technology Guide*. When the nspire keypad is installed, the required keystrokes and screen displays are different from those with the TI-84Plus keypad. A separate *Technology Guide* for this text provides instructions for using the nspire keypad to perform statistical operations, create statistical graphs, and apply statistical tests.

The instructions that follow apply to the TI-84Plus, TI-83Plus, and TI-nspire (with the TI-84 keypad installed) calculators.

To select a random set of integers between two specified values, press the **MATH** key and highlight **PRB** with **5:randInt** (low value, high value, sample size). Press Enter and fill in the low value, high value, and sample size. To store the sample in list L1, press the **STO➡** key and then L1. The screen display shows two random samples of size 5 drawn from the integers between 1 and 100.

```
randInt(1,100,5)
{63 89 13 46 47}
randInt(1,100,5)
{29 82 99 50 41}
```

Excel 2010

Many statistical processes that we will use throughout this text are included in the **Analysis ToolPak** add-in for Excel. This is an add-in that is included with the standard versions of Excel 2010. To see if you have the add-in installed, click the green **File** tab in the upper-left corner of the spreadsheet, press the **Excel Options** button, and select **Add-ins.** Check that **Analysis ToolPak** is in the Active Application Add-ins list. If it is not, select it for an active application.

To select a random number between two specified integer values, first select a cell in the active worksheet. Then type the command **=RANDBETWEEN(bottom number, top number)** in the formula bar, where the bottom number is the lower specified value and the top number is the higher specified value.

Alternatively, access a dialogue box for the command by clicking the **Insert function** button f_x on the ribbon. You will see an insert function dialogue box. Select the category **All,** and then scroll down until you reach **RANDBETWEEN** and click OK. Fill in the bottom and top numbers. In the display shown, the bottom number is 1 and the top number is 100.

RANDBETWEEN	▼	✕ ✓ f_x	=RANDBETWEEN(1,100)		
	A	B	C	D	E	F
1	46					
2	47					
3	49					
4	73					
5	11					

Minitab

To generate random integers between specified values, use the menu selection **Calc ➤ Random Data ➤ Integer.** Fill in the dialogue box to get five random numbers between 1 and 100.

⊞ Worksheet 2 ***	
	C1
↓	
1	8
2	35
3	33
4	9
5	15

SPSS

SPSS is a research statistical package for the social sciences. Data are entered in the data editor, which has a spreadsheet format. In the data-editor window, you have a choice of data view (default) or variable view. In variable view, you name variables, declare type (numeric for measurements, string for category), determine format, and declare measurement type. The choices for measurement type are scale (for ratio or interval data), ordinal, or nominal. Once you have entered data, you can use the menu bar at the top of the screen to select activities, graphs, or analysis appropriate to the data.

SPSS supports several random-sample activities. In particular, you can select a random sample from an existing data set or from a variety of probability distributions.

Selecting a random integer between two specified values involves several steps. First, in the data editor, enter the sample numbers in the first column. For instance, to generate five random numbers, list the values 1 through 5 in the first column. Notice that the label for the first column is now var00001. SPSS does not have a direct function for selecting a random sample of integers. However, there is a function for sampling values from the uniform distribution of all real numbers between two specified values. We will use that function and then truncate the values to obtain a random sample of integers between two specified values.

Use the menu options **Transform ➤ Compute**. In the dialogue box, type in var00002 as the target variable. In the Function group select **Random Number**. Then under Functions and Special Variables, select **Rv.Uniform**. In the Numeric Expression box, replace the two question marks with the minimum and maximum. Use 1 as the minimum and 101 as the maximum. The maximum is 101 because numbers between 100 and 101 truncate to 100.

The random numbers from the uniform distribution now appear in the second column under var00002. You can visually truncate the values to obtain random integers. However, if you want SPSS to truncate the values for you, you can again use the menu choices **Transform ➤ Compute**. In the dialogue box, enter var00003 for the target variable. In the Function Group select **Arithmetic**. Then in the Functions and Special Variables box select **Trunc(1)**. In the Numeric Expression box use var00002 in place of the question mark representing numexpr. The random integers between 1 and 100 appear in the third column under var00003.

Copyright © Cengage Learning 2013

Copyright © Cengage Learning 2013

2

Fyodor Savintsev/ITAR-TASS/Corbis News/Corbis

World History Archive/Alamy

In dwelling upon the vital importance of sound observation, it must never be lost sight of what observation is for. It is not for the sake of piling up miscellaneous information or curious facts, but for the sake of saving life and increasing health and comfort.

—FLORENCE NIGHTINGALE,
Notes on Nursing

Florence Nightingale (1820–1910) has been described as a "passionate statistician" and a "relevant statistician." She viewed statistics as a science that allowed one to transcend his or her narrow individual experience and aspire to the broader service of humanity. She was one of the first nurses to use graphic representation of statistics, illustrating with charts and diagrams how improved sanitation decreased the rate of mortality. Her statistical reports about the appalling sanitary conditions at Scutari (the main British hospital during the Crimean War) were taken very seriously by the English secretary at war, Sidney Herbert. When sanitary reforms recommended by Nightingale were instituted in military hospitals, the mortality rate dropped from an incredible 42.7% to only 2.2%.

ORGANIZING DATA

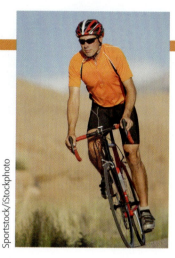

Sportstock/iStockphoto

PREVIEW QUESTIONS

What are histograms? When are they used? (SECTION 2.1)

What are common distribution shapes? (SECTION 2.1)

How can you select graphs appropriate for given data sets? (SECTION 2.2)

How can you quickly order data and, at the same time, reveal the distribution shape? (SECTION 2.3)

FOCUS PROBLEM

Say It with Pictures

Edward R. Tufte, in his book *The Visual Display of Quantitative Information*, presents a number of guidelines for producing good graphics. According to the criteria, a graphical display should

- show the data;
- induce the viewer to think about the substance of the graphic rather than about the methodology, the design, the technology, or other production devices;
- avoid distorting what the data have to say.

Vandystadt/Photo Researchers Inc

As an example of a graph that violates some of the criteria, Tufte includes a graphic that appeared in a well-known newspaper. Figure 2-1(a), on the next page, shows a figure similar to the problem graphic, whereas part (b) of the figure shows a better rendition of the data display.

After completing this chapter, you will be able to answer the following questions:

(a) Look at the graph in Figure 2-1(a). Is it essentially a bar graph? Explain. What are some of the flaws of Figure 2-1(a) as a bar graph?

(b) Examine Figure 2-1(b), which shows the same information. Is it essentially a time-series graph? Explain. In what ways does the second graph seem to display the information in a clearer manner?

(See Problem 5 of the Chapter 2 Review Problems.)

FIGURE 2-1

(a) Fuel Economy Standards for Autos

Set by Congress and supplemented by the Transportation Department
Source: Redrawn from *The Visual Display of Quantitive Information* by Edward R. Tufte, p. 58.
Copyright © 1983. Reprinted by permission of Graphic Press.

Note: Fuel-economy standards for passenger cars remained at 27.5 miles per gallon through 2010. It is anticipated that the mileage standard will increase soon.

(b) Required Fuel Economy Standards: New Cars Built from 1978 to 1985

Source: Redrawn from *The Visual Display of Quantitative Information* by Edward R. Tufte, p. 58. Copyright © 1983. Reprinted by permission of Graphic Press.

SECTION 2.1

Frequency Distributions, Histograms, and Related Topics

FOCUS POINTS

- Organize raw data using a frequency table.
- Construct histograms, relative-frequency histograms, and ogives.
- Recognize basic distribution shapes: uniform, symmetrical, skewed, and bimodal.
- Interpret graphs in the context of the data setting.

Frequency Tables

When we have a large set of quantitative data, it's useful to organize it into smaller intervals or *classes* and count how many data values fall into each class. A frequency table does just that.

Frequency distribution

A **frequency distribution** partitions data into classes or intervals and shows how many data values are in each class. The classes or intervals are constructed so that each data value falls into exactly one class.

Frequency table

A **frequency table** displays each data class along with the number (frequency) of data in that class. Other items such as the midpoint of each class may be included as well.

Constructing a frequency table and corresponding frequency distribution involves a number of steps. Example 1 demonstrates the steps.

EXAMPLE 1   FREQUENCY TABLE

A task force to encourage carpooling did a study of one-way commuting distances of workers in the downtown Dallas area. A random sample of 60 of these workers was taken. The commuting distances of the workers in the sample are given in Table 2-1. Make a frequency table for these data.

TABLE 2-1	One-Way Commuting Distances (in Miles) for 60 Workers in Downtown Dallas								
13	47	10	3	16	20	17	40	4	2
7	25	8	21	19	15	3	17	14	6
12	45	1	8	4	16	11	18	23	12
6	2	14	13	7	15	46	12	9	18
34	13	41	28	36	17	24	27	29	9
14	26	10	24	37	31	8	16	12	16

Copyright © Cengage Learning 2013

SOLUTION:

(a) First decide how many classes you want. Five to 15 classes are usually used. If you use fewer than five classes, you risk losing too much information. If you use more than 15 classes, the data may not be sufficiently summarized. Let the spread of the data and the purpose of the frequency table be your guides when selecting the number of classes. In the case of the commuting data, let's use *six* classes.

Class width

(b) Next, find the *class width* for the six classes.

PROCEDURE

HOW TO FIND THE CLASS WIDTH (INTEGER DATA)

1. Compute: $\dfrac{\text{Largest data value} - \text{smallest data value}}{\text{Desired number of classes}}$

2. Increase the computed value to the next highest whole number.

Note: To ensure that all the classes taken together cover the data, we need to increase the result of Step 1 to the *next whole number,* even if Step 1 produces a whole number. For instance, if the calculation in Step 1 produces the value 4, we make the class width 5.

To find the class width for the commuting data, we observe that the largest distance commuted is 47 miles and the smallest is 1 mile. Using six classes, the class width is 8, since

$$\text{Class width} = \frac{47 - 1}{6} \approx 7.7 \ \ (\text{increase to 8})$$

(c) Now we determine the data range for each class.

Class limits

The **lower class limit** is the lowest data value that can fit in a class. The **upper class limit** is the highest data value that can fit in a class. The **class width** is the difference between the *lower* class limit of one class and the *lower* class limit of the next class.

The smallest commuting distance in our sample is 1 mile. We use this *smallest* data value as the lower class limit of the *first* class. Since the class width is 8, we add 8 to 1 to find that the *lower* class limit for the *second* class is 9. Following this pattern, we establish *all* the lower class limits. Then we fill in the *upper class limits* so that the classes span the entire range of data. Table 2-2, on the next page, shows the upper and lower class limits for the commuting distance data.

(d) Now we are ready to tally the commuting distance data into the six classes and find the frequency for each class.

| TABLE 2-2 | Frequency Table of One-Way Commuting Distances for 60 Downtown Dallas Workers (Data in Miles) | | | |

Class Limits Lower–Upper	Class Boundaries Lower–Upper	Tally	Frequency	Class Midpoint
1–8	0.5–8.5	⅃⅃⅃⅃⅃⅃ IIII	14	4.5
9–16	8.5–16.5	⅃⅃⅃⅃⅃⅃⅃⅃⅃⅃ I	21	12.5
17–24	16.5–24.5	⅃⅃⅃⅃ I	11	20.5
25–32	24.5–32.5	⅃⅃ I	6	28.5
33–40	32.5–40.5	IIII	4	36.5
41–48	40.5–48.5	IIII	4	44.5

PROCEDURE

HOW TO TALLY DATA

Tallying data is a method of counting data values that fall into a particular class or category.

To tally data into classes of a frequency table, examine each data value. Determine which class contains the data value and make a tally mark or vertical stroke (I) beside that class. For ease of counting, each fifth tally mark of a class is placed diagonally across the prior four marks (⅃⅃⅃⅃).

Class frequency

The *class frequency* for a class is the number of tally marks corresponding to that class.

Table 2-2 shows the tally and frequency of each class.

Class midpoint or class mark

(e) The center of each class is called the *midpoint* (or *class mark*). The midpoint is often used as a representative value of the entire class. The midpoint is found by adding the lower and upper class limits of one class and dividing by 2.

$$\text{Midpoint} = \frac{\text{Lower class limit} + \text{upper class limit}}{2}$$

Table 2-2 shows the class midpoints.

Class boundaries

(f) There is a space between the upper limit of one class and the lower limit of the next class. The halfway points of these intervals are called *class boundaries*. These are shown in Table 2-2.

PROCEDURE

HOW TO FIND CLASS BOUNDARIES (INTEGER DATA)

To find **upper class boundaries,** add 0.5 unit to the upper class limits.

To find **lower class boundaries,** subtract 0.5 unit from the lower class limits.

Problems 17 and 18 show the procedure for finding the class width, class limits, and class boundaries for noninteger decimal data.

Relative frequency

Basic frequency tables show how many data values fall into each class. It's also useful to know the *relative frequency* of a class. The relative frequency of a class is the proportion of all data values that fall into that class. To find the

TABLE 2-3	Relative Frequencies of One-Way Commuting Distances	
Class	Frequency *f*	Relative Frequency *f/n*
1–8	14	$14/60 \approx 0.23$
9–16	21	$21/60 \approx 0.35$
17–24	11	$11/60 \approx 0.18$
25–32	6	$6/60 \approx 0.10$
33–40	4	$4/60 \approx 0.07$
41–48	4	$4/60 \approx 0.07$

relative frequency of a particular class, divide the class frequency *f* by the total of all frequencies *n* (sample size).

$$\text{Relative frequency} = \frac{f}{n} = \frac{\text{Class frequency}}{\text{Total of all frequencies}}$$

Table 2-3 shows the relative frequencies for the commuter data of Table 2-1. Since we already have the frequency table (Table 2-2), the relative-frequency table is obtained easily. The sample size is $n = 60$. Notice that the sample size is the total of all the frequencies. Therefore, the relative frequency for the first class (the class from 1 to 8) is

$$\text{Relative frequency} = \frac{f}{n} = \frac{14}{60} \approx 0.23$$

The symbol $\approx$ means "is approximately equal to." We use the symbol because we rounded the relative frequency. Relative frequencies for the other classes are computed in a similar way.

The total of the relative frequencies should be 1. However, rounded results may make the total slightly higher or lower than 1.

LOOKING FORWARD

Sorting data into classes can be somewhat tedious. It is an easier task if the data are first ordered from smallest to largest. Stem-and-leaf diagrams, presented in Section 2.3, provide a convenient way to order data by hand.

Let's summarize the procedure for making a frequency table that includes relative frequencies.

PROCEDURE

How to make a frequency table

1. Determine the number of classes and the corresponding class width.

2. Create the distinct classes. We use the convention that the *lower class limit* of the first class is the smallest data value. Add the class width to this number to get the *lower class limit* of the next class.

3. Fill in *upper class limits* to create distinct classes that accommodate all possible data values from the data set.

4. Tally the data into classes. Each data value should fall into exactly one class. Total the tallies to obtain each *class frequency*.

5. Compute the *midpoint* (class mark) for each class.

6. Determine the *class boundaries*.

<table>
<tr><td>PROCEDURE

Relative-frequency table</td><td>HOW TO MAKE A RELATIVE-FREQUENCY TABLE

First make a frequency table. Then, for each class, compute the relative frequency, f/n, where f is the class frequency and n is the total sample size.</td></tr>
</table>

Histograms and Relative-Frequency Histograms

Histogram
Relative-frequency histogram

Histograms and *relative-frequency histograms* provide effective visual displays of data organized into frequency tables. In these graphs, we use bars to represent each class, where the width of the bar is the class width. For histograms, the height of the bar is the class frequency, whereas for relative-frequency histograms, the height of the bar is the relative frequency of that class.

<table>
<tr><td>PROCEDURE

Problems 19, 20, and 21 introduce dotplots, which are diagrams that show the frequency with which each data value occurs.</td><td>HOW TO MAKE A HISTOGRAM OR A RELATIVE-FREQUENCY HISTOGRAM

1. Make a frequency table (including relative frequencies) with the designated number of classes.

2. Place class boundaries on the horizontal axis and frequencies or relative frequencies on the vertical axis.

3. For each class of the frequency table, draw a bar whose width extends between the corresponding class boundaries. For histograms, the height of each bar is the corresponding class frequency. For relative-frequency histograms, the height of each bar is the corresponding class relative frequency.</td></tr>
</table>

EXAMPLE 2   HISTOGRAM AND RELATIVE-FREQUENCY HISTOGRAM

Make a histogram and a relative-frequency histogram with six bars for the data in Table 2-1 showing one-way commuting distances.

SOLUTION: The first step is to make a frequency table and a relative-frequency table with six classes. We'll use Table 2-2 and Table 2-3. Figures 2-2 and 2-3

FIGURE 2-2

Histogram for Dallas Commuters:
One-Way Commuting Distances

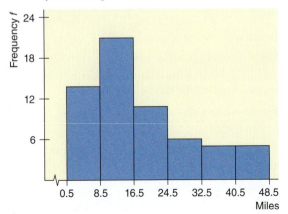

FIGURE 2-3

Relative-Frequency Histogram for Dallas Commuters:
One-Way Commuting Distances

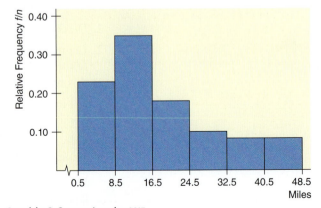

show the histogram and relative-frequency histogram. In both graphs, class boundaries are marked on the horizontal axis. For each class of the frequency table, make a corresponding bar with horizontal width extending from the lower boundary to the upper boundary of the respective class. For a histogram, the height of each bar is the corresponding class frequency. For a relative-frequency histogram, the height of each bar is the corresponding relative frequency. Notice that the basic shapes of the graphs are the same. The only difference involves the vertical axis. The vertical axis of the histogram shows frequencies, whereas that of the relative-frequency histogram shows relative frequencies.

COMMENT   The use of class boundaries in histograms assures us that the bars of the histogram touch and that no data fall on the boundaries. Both of these features are important. But a histogram displaying class boundaries may look awkward. For instance, the mileage range of 8.5 to 16.5 miles shown in Figure 2-2 isn't as natural a choice as a mileage range of 8 to 16 miles. For this reason, many magazines and newspapers do not use class boundaries as labels on a histogram. Instead, some use lower class limits as labels, with the convention that *a data value falling on the class limit is included in the next higher class (class to the right of the limit)*. Another convention is to label midpoints instead of class boundaries. Determine the default convention being used before creating frequency tables and histograms on a computer.

GUIDED EXERCISE 1 | **Histogram and relative-frequency histogram**

An irate customer called Dollar Day Mail Order Company 40 times during the last two weeks to see why his order had not arrived. Each time he called, he recorded the length of time he was put on hold before being allowed to talk to a customer service representative. See Table 2-4.

TABLE 2-4 Length of Time on Hold (in Minutes)

1	5	5	6	7	4	8	7	6	5
5	6	7	6	6	5	8	9	9	10
7	8	11	2	4	6	5	12	13	6
3	7	8	8	9	9	10	9	8	9

Copyright © Cengage Learning 2013

(a) What are the largest and smallest values in Table 2-4? If we want five classes in a frequency table, what should the class width be?

⇒ The largest value is 13; the smallest value is 1. The class width is

$$\frac{13 - 1}{5} = 2.4 \approx 3 \qquad \textit{Note: Increase the value to 3.}$$

(b) Complete the following frequency table.

TABLE 2-5 Time on Hold

Class Limits Lower–Upper	Tally	Frequency	Midpoint
1–3	____	____	____
4–____	____	____	____
____–9	____	____	____
____–____	____	____	____
____–____	____	____	____

Copyright © Cengage Learning 2013

⇒ **TABLE 2-6 Completion of Table 2-5**

Class Limits Lower–Upper	Tally	Frequency	Midpoint
1–3	|||	3	2
4–6	ℕℕ ℕℕ ℕℕ	15	5
7–9	ℕℕ ℕℕ ℕℕ ||	17	8
10–12	||||	4	11
13–15			1	14

Copyright © Cengage Learning 2013

Continued

(c) Recall that the class boundary is halfway between the upper limit of one class and the lower limit of the next. Use this fact to find the class boundaries in Table 2-7 and to complete the partial histogram in Figure 2-4.

TABLE 2-7 **Class Boundaries**

Class Limits	Class Boundaries
1–3	0.5–3.5
4–6	3.5–6.5
7–9	6.5–___
10–12	___–___
13–15	___–___

TABLE 2-8 **Completion of Table 2-7**

Class Limits	Class Boundaries
1–3	0.5–3.5
4–6	3.5–6.5
7–9	6.5–9.5
10–12	9.5–12.5
13–15	12.5–15.5

FIGURE 2-4

FIGURE 2-5 Completion of Figure 2-4

(d) Compute the relative class frequency f/n for each class in Table 2-9 and complete the partial relative-frequency histogram in Figure 2-6.

TABLE 2-9 **Relative Class Frequency**

Class	f/n
1–3	3/40 = 0.075
4–6	15/40 = 0.375
7–9	___
10–12	___
13–15	___

TABLE 2-10 **Completion of Table 2-9**

Class	f/n
1–3	0.075
4–6	0.375
7–9	0.425
10–12	0.100
13–15	0.025

FIGURE 2-6

FIGURE 2-7 Completion of Figure 2-6

LOOKING FORWARD

We will see relative-frequency distributions again when we study probability in Chapter 5. There we will see that if a random sample is large enough, then we can estimate the probability of an event by the relative frequency of the event. The relative-frequency distribution then can be interpreted as a *probability distribution*. Such distributions will form the basis of our work in inferential statistics.

Distribution Shapes

Histograms are valuable and useful tools. If the raw data came from a random sample of population values, the histogram constructed from the sample values should have a distribution shape that is reasonably similar to that of the population.

   Several terms are commonly used to describe histograms and their associated population distributions.

Mound-shaped symmetrical distribution

(a) *Mound-shaped symmetrical:* This term refers to a histogram in which both sides are (more or less) the same when the graph is folded vertically down the middle. Figure 2-8(a) shows a typical mound-shaped symmetrical histogram.

Uniform distribution

(b) *Uniform or rectangular:* These terms refer to a histogram in which every class has equal frequency. From one point of view, a uniform distribution is symmetrical with the added property that the bars are of the same height. Figure 2-8(b) illustrates a typical histogram with a uniform shape.

Skewed left
Skewed right

(c) *Skewed left or skewed right:* These terms refer to a histogram in which one tail is stretched out longer than the other. The direction of skewness is on the side of the *longer* tail. So if the longer tail is on the left, we say the histogram is skewed to the left. Figure 2-8(c) shows a typical histogram skewed to the left and another skewed to the right.

Bimodal distribution

(d) *Bimodal:* This term refers to a histogram in which the two classes with the largest frequencies are separated by at least one class. The top two frequencies of these classes may have slightly different values. This type of situation sometimes indicates that we are sampling from two different populations. Figure 2-8(d) illustrates a typical histogram with a bimodal shape.

FIGURE 2-8

Types of Histograms
Copyright © Cengage Learning 2013

(a) Typical mound-shaped symmetrical histogram

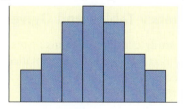

(b) Typical uniform or rectangular histogram

(c) Typical skewed histogram

Skewed left   Skewed right

(d) Typical bimodal histogram

LOOKING FORWARD

Normal and Student's *t* distributions are two important probability distributions used extensively in inferential statistics (Chapters 7–10). Both of these distributions are mound-shaped symmetrical. Another useful probability distribution is the Chi-square distribution (Chapter 11), which is skewed right.

CRITICAL THINKING

A bimodal distribution shape might indicate that the data are from two different populations. For instance, a histogram showing the heights of a random sample of adults is likely to be bimodal because two populations, male and female, were combined.

If there are gaps in the histogram between bars at either end of the graph, the data set might include *outliers*.

Outliers

> **Outliers** in a data set are the data values that are very different from other measurements in the data set.

Outliers may indicate data-recording errors. Valid outliers may be so unusual that they should be examined separately from the rest of the data. For instance, in a study of salaries of employees at one company, the CEO's salary may be so high and unique for the company that it should be considered separately from the other salaries. Decisions about outliers that are not recording errors need to be made by people familiar with both the field and the purpose of the study.

TECH NOTES

The TI-84Plus/TI-83Plus/TI-*n*spire calculators, Excel 2010, and Minitab all create histograms. However, each technology automatically selects the number of classes to use. In Using Technology at the end of this chapter, you will see instructions for specifying the number of classes yourself and for generating histograms such as those we create "by hand."

Cumulative-Frequency Tables and Ogives

Sometimes we want to study cumulative totals instead of frequencies. Cumulative frequencies tell us how many data values are smaller than an upper class boundary. Once we have a frequency table, it is a fairly straightforward matter to add a column of cumulative frequencies.

Cumulative frequency

> The **cumulative frequency** for a class is the sum of the frequencies for *that class* and *all previous classes*.

Ogive

An *ogive* (pronounced "oh-jīve") is a graph that displays cumulative frequencies.

PROCEDURE	HOW TO MAKE AN OGIVE
	1. Make a frequency table showing class boundaries and cumulative frequencies.
	2. For each class, make a dot over the *upper class boundary* at the height of the cumulative class frequency. The coordinates of the dots are (upper class boundary, cumulative class frequency). Connect these dots with line segments.
	3. By convention, an ogive begins on the horizontal axis at the lower class boundary of the first class.

EXAMPLE 3   CUMULATIVE-FREQUENCY TABLE AND OGIVE

Aspen, Colorado, is a world-famous ski area. If the daily high temperature is above 40°F, the surface of the snow tends to melt. It then freezes again at night. This can result in a snow crust that is icy. It also can increase avalanche danger.

Table 2-11 gives a summary of daily high temperatures (°F) in Aspen during the 151-day ski season.

TABLE 2-11   **High Temperatures during the Aspen Ski Season (°F)**

Class Boundaries			
Lower	Upper	Frequency	Cumulative Frequency
10.5	20.5	23	23
20.5	30.5	43	66 (sum 23 + 43)
30.5	40.5	51	117 (sum 66 + 51)
40.5	50.5	27	144 (sum 117 + 27)
50.5	60.5	7	151 (sum 144 + 7)

Copyright © Cengage Learning 2013

(a) The cumulative frequency for a class is computed by adding the frequency of that class to the frequencies of previous classes. Table 2-11 shows the cumulative frequencies.

(b) To draw the corresponding ogive, we place a dot at cumulative frequency 0 on the lower class boundary of the first class. Then we place dots over the *upper class boundaries* at the height of the cumulative class frequency for the corresponding class. Finally, we connect the dots. Figure 2-9 shows the corresponding ogive.

(c) Looking at the ogive, estimate the total number of days with a high temperature lower than or equal to 40°F.

Ph. Royer/Photo Researchers Inc.

FIGURE 2-9

Ogive for Daily High Temperatures (°F) during the Aspen Ski Season

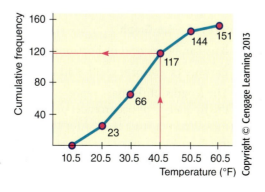

Copyright © Cengage Learning 2013

SOLUTION: Following the red lines on the ogive in Figure 2-9, we see that 117 days have had high temperatures of no more than 40°F.

VIEWPOINT | Mush, You Huskies!

In 1925, the village of Nome, Alaska, had a terrible diphtheria epidemic. Serum was available in Anchorage but had to be brought to Nome by dogsled over the 1161-mile Iditarod Trail. Since 1973, the Iditarod Dog Sled Race from Anchorage to Nome has been an annual sporting event, with a current purse of more than $600,000. Winning times range from more than 20 days to a little over 9 days.

To collect data on winning times, visit the Brase/Brase statistics site at **http://www.cengagebrain.com** *and find the link to the Iditarod. Make a frequency distribution for these times.*

SECTION 2.1 PROBLEMS

 aplia™   Access end-of-section exercises at www.aplia.com

SECTION 2.2

Bar Graphs, Circle Graphs, and Time-Series Graphs

FOCUS POINTS
- Determine types of graphs appropriate for specific data.
- Construct bar graphs, Pareto charts, circle graphs, and time-series graphs.
- Interpret information displayed in graphs.

Histograms provide a useful visual display of the distribution of data. However, the data must be quantitative. In this section, we examine other types of graphs, some of which are suitable for qualitative or category data as well.

Bar graph

Let's start with *bar graphs*. These are graphs that can be used to display quantitative or qualitative data.

> **Features of a Bar Graph**
>
> 1. Bars can be vertical or horizontal.
>
> 2. Bars are of uniform width and uniform spacing.
>
> 3. The lengths of the bars represent values of the variable being displayed, the frequency of occurrence, or the percentage of occurrence. The same measurement scale is used for the length of each bar.
>
> 4. The graph is well annotated with title, labels for each bar, and vertical or horizontal scale or actual value for the length of each bar.

EXAMPLE 4   BAR GRAPH

Figure 2-11 shows two bar graphs depicting the life expectancies for men and women born in the designated years. Let's analyze the features of these graphs.

SOLUTION: The graphs are called *clustered bar graphs* because there are two bars for each year of birth. One bar represents the life expectancy for men, and the other represents the life expectancy for women. The height of each bar represents the life expectancy (in years).

Changing scale

An important feature illustrated in Figure 2-11(b) is that of a *changing scale*. Notice that the scale between 0 and 65 is compressed. The changing scale amplifies the apparent difference between life expectancies for men and women, as well as the increase in life expectancies from those born in 1980 to those born in 2010.

> **Changing Scale**
>
> Whenever you use a change in scale in a graphic, warn the viewer by using a squiggle (‿⁄‿) on the changed axis. Sometimes, if a single bar is unusually long, the bar length is compressed with a squiggle in the bar itself.

LOOKING FORWARD

Cluster bar graphs displaying percentages are particularly useful visuals for showing how different populations fit different categories. For instance, in Example 2 in Chapter 11, we see a cluster graph showing preferences of males and females for dogs, cats, other pets, or no pets. This graph displays the sample data for an inferential test called a *test of homogeneity*.

FIGURE 2-11

Life Expectancy

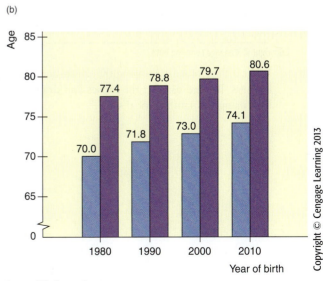

Source: U.S. Census Bureau

Pareto charts

Quality control is an important aspect of today's production and service industries. Dr. W. Edwards Deming was one of the developers of total quality management (TQM). In his book *Out of the Crisis,* he outlines many strategies for monitoring and improving service and production industries. In particular, Dr. Deming recommends the use of some statistical methods to organize and analyze data from industries so that sources of problems can be identified and then corrected. *Pareto charts* (pronounced "Pah-rāy-tō") are among the many techniques used in quality-control programs.

> A **Pareto chart** is a bar graph in which the bar height represents frequency of an event. In addition, the bars are arranged from left to right according to decreasing height.

GUIDED EXERCISE 2 | Pareto charts

This exercise is adapted from *The Deming Management Method* by Mary Walton. Suppose you want to arrive at college 15 minutes before your first class so that you can feel relaxed when you walk into class. An early arrival time also allows room for unexpected delays. However, you always find yourself arriving "just in time" or slightly late. What causes you to be late? Charlotte made a list of possible causes and then kept a checklist for 2 months (Table 2-12). On some days more than one item was checked because several events occurred that caused her to be late.

TABLE 2-12 **Causes for Lateness (September–October)**

Cause	Frequency
Snoozing after alarm goes off	15
Car trouble	5
Too long over breakfast	13
Last-minute studying	20
Finding something to wear	8
Talking too long with roommate	9
Other	3

(a) Make a Pareto chart showing the causes for lateness. Be sure to label the causes and draw the bars using the same vertical scale.

Figure 2-12 Pareto Chart: Conditions That Might Cause Lateness

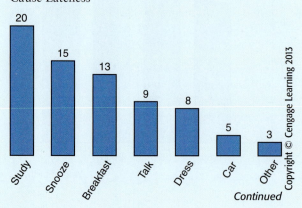

Continued

GUIDED EXERCISE 2   *continued*

(b) Looking at the Pareto chart, what recommendations do you have for Charlotte?

➡ According to the chart, rearranging study time or getting up earlier to allow for studying would cure her most frequent cause for lateness. Repairing the car might be important, but for getting to campus early, it would not be as effective as adjusting study time.

Circle graph or pie chart

Another popular pictorial representation of data is the *circle graph* or *pie chart*. It is relatively safe from misinterpretation and is especially useful for showing the division of a total quantity into its component parts. The total quantity, or 100%, is represented by the entire circle. Each wedge of the circle represents a component part of the total. These proportional segments are usually labeled with corresponding percentages of the total. Guided Exercise 3 shows how to make a circle graph.

Donut pie charts are a popular variation of pie charts. Problem 15 of this section discusses donut pie charts.

In a **circle graph** or **pie chart**, wedges of a circle visually display proportional parts of the total population that share a common characteristic.

GUIDED EXERCISE 3 | **Circle graph**

How long do we spend talking on the telephone after hours (at home after 5 P.M.)? The results from a recent survey of 500 people (as reported in *USA Today*) are shown in Table 2-13. We'll make a circle graph to display these data.

TABLE 2-13 Time Spent on Home Telephone after 5 P.M.

Time	Number	Fractional Part	Percentage	Number of Degrees
Less than ½ hour	296	296/500	59.2	59.2% × 360° ≈ 213°
½ hour to 1 hour	83	83/500	16.6	16.6% × 360° ≈ 60°
More than 1 hour	121	____	____	____
Total	____		____	____

Copyright © Cengage Learning 2013

(a) Fill in the missing parts in Table 2-13 for "More than 1 hour." Remember that the central angle of a circle is 360°. Round to the nearest degree.

➡ For "More than 1 hour," Fractional Part = 121/500; Percentage = 24.2%; Number of Degrees = 24.2% × 360° ≈ 87°. The symbol ≈ means "approximately equal."

(b) Fill in the totals. What is the total number of responses? Do the percentages total 100% (within rounding error)? Do the numbers of degrees total 360° (within rounding error)?

➡ The total number of responses is 500. The percentages total 100%. You must have such a total in order to create a circle graph. The numbers of degrees total 360°.

(c) Draw a circle graph. Divide the circle into pieces with the designated numbers of degrees. Label each piece, and show the percentage corresponding to each piece. The numbers of degrees are usually omitted from pie charts shown in newspapers, magazines, journals, and reports.

➡ **Figure 2-13** Hours on Home Telephone after 5 P.M.

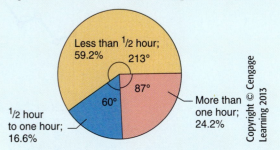

Suppose you begin an exercise program that involves walking or jogging for 30 minutes. You exercise several times a week but monitor yourself by logging the distance you cover in 30 minutes each Saturday. How do you display these data in a meaningful way? Making a bar chart showing the frequency of distances you cover might be interesting, but it does not really show how the distance you cover in 30 minutes has changed over time. A graph showing the distance covered on each date will let you track your performance over time.

Time-series graph

We will use a *time-series graph*. A time-series graph is a graph showing data measurements in chronological order. To make a time-series graph, we put time on the horizontal scale and the variable being measured on the vertical scale. In a basic time-series graph, we connect the data points by line segments.

> In a **time-series graph,** data are plotted in order of occurrence at regular intervals over a period of time.

EXAMPLE 5   TIME-SERIES GRAPH

Suppose you have been in the walking/jogging exercise program for 20 weeks, and for each week you have recorded the distance you covered in 30 minutes. Your data log is shown in Table 2-14.

TABLE 2-14   Distance (in Miles) Walked/Jogged in 30 Minutes

Week	1	2	3	4	5	6	7	8	9	10
Distance	1.5	1.4	1.7	1.6	1.9	2.0	1.8	2.0	1.9	2.0
Week	11	12	13	14	15	16	17	18	19	20
Distance	2.1	2.1	2.3	2.3	2.2	2.4	2.5	2.6	2.4	2.7

Copyright © Cengage Learning 2013

(a) Make a time-series graph.

SOLUTION: The data are appropriate for a time-series graph because they represent the same measurement (distance covered in a 30-minute period) taken at different times. The measurements are also recorded at equal time intervals (every week). To make our time-series graph, we list the weeks in order on the horizontal scale. Above each week, plot the distance covered that week on the vertical scale. Then connect the dots. Figure 2-14 shows the time-series graph. Be sure the scales are labeled.

FIGURE 2-14

Time-Series Graph of Distance (in Miles) Walked/Jogged in 30 Minutes

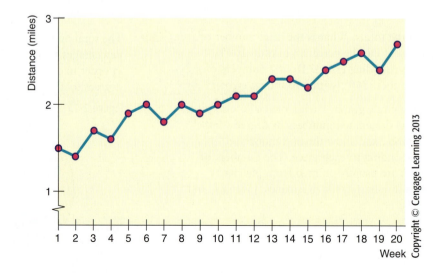

Copyright © Cengage Learning 2013

(b) From looking at Figure 2-14, can you detect any patterns?

> **SOLUTION:** There seems to be an upward trend in distance covered. The distances covered in the last few weeks are about a mile farther than those for the first few weeks. However, we cannot conclude that this trend will continue. Perhaps you have reached your goal for this training activity and now wish to maintain a distance of about 2.5 miles in 30 minutes.

Time series

Data sets composed of similar measurements taken at regular intervals over time are called *time series*. Time series are often used in economics, finance, sociology, medicine, and any other situation in which we want to study or monitor a similar measure over a period of time. A time-series graph can reveal some of the main features of a time series.

> **Time-series data** consist of measurements of the same variable for the same subject taken at regular intervals over a period of time.

CRITICAL THINKING

We've seen several styles of graphs. Which kinds are suitable for a specific data collection?

PROCEDURE

How to decide which type of graph to use

Bar graphs are useful for quantitative or qualitative data. With qualitative data, the frequency or percentage of occurrence can be displayed. With quantitative data, the measurement itself can be displayed, as was done in Figure 2-11. Watch that the measurement scale is consistent or that a jump scale squiggle is used.

Pareto charts identify the frequency of events or categories in decreasing order of frequency of occurrence.

Circle graphs display how a *total* is dispersed into several categories. The circle graph is very appropriate for qualitative data, or any data for which percentage of occurrence makes sense. Circle graphs are most effective when the number of categories or wedges is 10 or fewer.

Time-series graphs display how data change over time. It is best if the units of time are consistent in a given graph. For instance, measurements taken every day should not be mixed on the same graph with data taken every week.

For any graph: Provide a title, label the axes, and identify units of measure. As Edward Tufte suggests in his book *The Visual Display of Quantitative Information*, don't let artwork or skewed perspective cloud the clarity of the information displayed.

TECH NOTES   *Bar graphs, circle graphs, and time-series graphs*

TI-84Plus/TI-83Plus/TI-*n*spire (with TI-84Plus keypad) Only time-series. Place consecutive values 1 through the number of time segments in list L1 and corresponding data in L2. Press **Stat Plot** and highlight an *xy* line plot.

Excel 2010 First enter the data into the spreadsheet. Then click the **Insert** tab and select the type of chart you want to create. A variety of bar graphs, pie charts, and line graphs that can be used as time-series graphs are available. Use the **Design** tab and the **Layout** tab on the **Chart Tools** menu to access options such as title, axis labels, and so on, for your chart.

Minitab Use the menu selection **Graph**. Select the desired option and follow the instructions in the dialogue boxes.

VIEWPOINT   Do Ethical Standards Vary by the Situation?

The Lutheran Brotherhood conducted a national survey and found that nearly 60% of all U.S. adults claim that ethics vary by the situation, 33% claim that there is only one ethical standard, and 7% are not sure. How could you draw a circle graph to make a visual expression of Americans' views on ethical standards?

SECTION 2.2 PROBLEMS

aplia   Access end-of-section exercises at www.aplia.com

SECTION 2.3

Stem-and-Leaf Displays

FOCUS POINTS

- Construct a stem-and-leaf display from raw data.
- Use a stem-and-leaf display to visualize data distribution.
- Compare a stem-and-leaf display to a histogram.

Exploratory Data Analysis

EDA

Together with histograms and other graphics techniques, the stem-and-leaf display is one of many useful ways of studying data in a field called *exploratory data analysis* (often abbreviated as *EDA*). John W. Tukey wrote one of the definitive books on the subject, *Exploratory Data Analysis*. Another very useful reference for EDA techniques is the book *Applications, Basics, and Computing of Exploratory Data Analysis* by Paul F. Velleman and David C. Hoaglin. Exploratory data analysis techniques are particularly useful for detecting patterns and extreme data values. They are designed to help us explore a data set, to ask questions we had not thought of before, or to pursue leads in many directions.

EDA techniques are similar to those of an explorer. An explorer has a general idea of destination but is always alert for the unexpected. An explorer needs to assess situations quickly and often simplify and clarify them. An explorer makes pictures—that is, maps showing the relationships of landscape features. The aspects of rapid implementation, visual displays such as graphs and charts, data simplification, and robustness (that is, analysis that is not influenced much by extreme data values) are key ingredients of EDA techniques. In addition, these techniques are good for exploration because they require very few prior assumptions about the data.

EDA methods are especially useful when our data have been gathered for general interest and observation of subjects. For example, we may have data regarding the ages of applicants to graduate programs. We don't have a specific question in mind. We want to see what the data reveal. Are the ages fairly uniform or spread out? Are there exceptionally young or old applicants? If there are, we might look at other characteristics of these applicants, such as field of study. EDA methods help us quickly absorb some aspects of the data and then may lead us to ask specific questions to which we might apply methods of traditional statistics.

In contrast, when we design an experiment to produce data to answer a specific question, we focus on particular aspects of the data that are useful to us. If we want to determine the average highway gas mileage of a specific sports car, we use that model of car in well-designed tests. We don't need to worry about unexpected road conditions, poorly trained drivers, different fuel grades, sudden stops and starts, etc. Our experiment is designed to control outside factors. Consequently, we do not need to "explore" our data as much. We can often make valid assumptions about the data. Methods of traditional statistics will be very useful to analyze such data and answer our specific questions.

Stem-and-Leaf Display

In this text, we will introduce two EDA techniques: stem-and-leaf displays and, in Section 3.3, box-and-whisker plots. Let's first look at a stem-and-leaf display.

Stem-and-leaf display

> A **stem-and-leaf display** is a method of exploratory data analysis that is used to rank-order and arrange data into groups.

We know that frequency distributions and histograms provide a useful organization and summary of data. However, in a histogram, we lose most of the specific data values. A stem-and-leaf display is a device that organizes and groups data but allows us to recover the original data if desired. In the next example, we will make a stem-and-leaf display.

EXAMPLE 6   STEM-AND-LEAF DISPLAY

Many airline passengers seem weighted down by their carry-on luggage. Just how much weight are they carrying? The carry-on luggage weights in pounds were recorded for a random sample of 40 passengers returning from a vacation to Hawaii (see Table 2-15).

TABLE 2-15   **Weights of Carry-On Luggage in Pounds**

30	27	12	42	35	47	38	36	27	35
22	17	29	3	21	0	38	32	41	33
26	45	18	43	18	32	31	32	19	21
33	31	28	29	51	12	32	18	21	26

Stem and Leaf

To make a stem-and-leaf display, we break the digits of each data value into *two* parts. The left group of digits is called a *stem*, and the remaining group of digits on the right is called a *leaf*. We are free to choose the number of digits to be included in the stem.

The weights in our example consist of two-digit numbers. For a two-digit number, the stem selection is obviously the left digit. In our case, the tens digits will form the stems, and the units digits will form the leaves. For example, for the weight 12, the stem is 1 and the leaf is 2. For the weight 18, the stem is again 1, but the leaf is 8. In the stem-and-leaf display, we list each possible stem once on the left and all its leaves in the same row on the right, as in Figure 2-15(a). Finally, we order the leaves as shown in Figure 2-15(b).

Figure 2-15 shows a stem-and-leaf display for the weights of carry-on luggage. From the stem-and-leaf display in Figure 2-15(b), we see that two bags weighed 27 lb, one weighed 3 lb, one weighed 51 lb, and so on. We see that most of the weights were in the 30-lb range, only two were less than 10 lb, and six were over 40 lb. Note that the lengths of the lines containing the leaves give the visual impression that a sideways histogram would present.

As a final step, we need to indicate the scale. This is usually done by indicating the value represented by the stem and one leaf.

FIGURE 2-15

Stem-and-Leaf Displays of Airline Carry-On Luggage Weights

(a) Leaves Not Ordered

```
   3 | 2    represents 32 lb
Stem | Leaves
   0 | 3 0
   1 | 2 7 8 8 9 2 8
   2 | 7 7 2 9 1 6 1 8 9 1 6
   3 | 0 5 8 6 5 8 2 3 2 1 2 3 1 2
   4 | 2 7 1 5 3
   5 | 1
```

(b) Final Display with Leaves Ordered

```
   3 | 2    represents 32 lb
Stem | Leaves
   0 | 0 3
   1 | 2 2 7 8 8 8 9
   2 | 1 1 1 2 6 6 7 7 8 9 9
   3 | 0 1 1 2 2 2 2 3 3 5 5 6 8 8
   4 | 1 2 3 5 7
   5 | 1
```

There are no firm rules for selecting the group of digits for the stem. But whichever group you select, you must list all the possible stems from smallest to largest in the data collection.

PROCEDURE

HOW TO MAKE A STEM-AND-LEAF DISPLAY

1. Divide the digits of each data value into two parts. The leftmost part is called the *stem* and the rightmost part is called the *leaf*.

2. Align all the stems in a vertical column from smallest to largest. Draw a vertical line to the right of all the stems.

3. Place all the leaves with the same stem in the same row as the stem, and arrange the leaves in increasing order.

4. Use a label to indicate the magnitude of the numbers in the display. We include the decimal position in the label rather than with the stems or leaves.

LOOKING FORWARD

Ordering or sorting data is an essential first step to finding the median or center value of the data distribution. We will discuss the median in Section 3.1. The data also need to be ordered before we can construct a box-and-whisker plot (Section 3.3). After a little practice, you will find that stem-and-leaf diagrams provide a very efficient way of ordering data by hand.

GUIDED EXERCISE 4 | **Stem-and-leaf display**

What does it take to win at sports? If you're talking about basketball, one sportswriter gave the answer. He listed the winning scores of the conference championship games over the last 35 years. The scores for those games are:

132	118	124	109	104	101	125	83	99
131	98	125	97	106	112	92	120	103
111	117	135	143	112	112	116	106	117
119	110	105	128	112	126	105	102	

Michael J. Doolittle/The Image Works

To make a stem-and-leaf display, we'll use the first *two* digits as the stems (see Figure 2-16). Notice that the distribution of scores is fairly symmetrical.

(a) Use the first *two* digits as the stem. Then order the leaves. Provide a label that shows the meaning and units of the first stem and first leaf.

Figure 2-16 Winning Scores

```
08 | 3  represents 083 or 83 points
08 | 3
09 | 2 7 8 9
10 | 1 2 3 4 5 5 6 6 9
11 | 0 1 2 2 2 2 6 7 7 8 9
12 | 0 4 5 5 6 8
13 | 1 2 5
14 | 3
```

Copyright © Cengage Learning 2013

(b) Looking at the distribution, would you say that it is fairly symmetrical? Yes. Notice that stem 11 has the most data.

COMMENT Stem-and-leaf displays organize the data, let the data analyst spot extreme values, and are easy to create. In fact, they can be used to organize data so that frequency tables are easier to make. However, at this time, histograms are used more often in formal data presentations, whereas stem-and-leaf displays are used by data analysts to gain initial insights about the data.

TECH NOTES   *Stem-and-leaf display*

TI-84Plus/TI-83Plus/TI-*n*spire Does not support stem-and-leaf displays. You can sort the data by using keys **Stat ➤ Edit ➤ 2:SortA.**

Excel 2010 Enter your data and select the data you want to sort. On the **Home** ribbon, click the **Sort and Filter** button in the **Editing** group of the ribbon and select the desired sorting option.

Minitab Use the menu selections **Graph ➤ Stem-and-Leaf** and fill in the dialogue box.

Minitab Stem-and-Leaf Display (for Data in Guided Exercise 4)

```
Stem-and-Leaf of Scores      N=35
Leaf Unit=1.0

     1                    8    3
     5                    9    2789
    14                   10    123455669
   (11)                  11    01222267789
    10                   12    045568
     4                   13    125
     1                   14    3
```

Copyright © Cengage Learning 2013

The values shown in the left column represent depth. Numbers above the value in parentheses show the cumulative number of values from the top to the stem of the middle value. Numbers below the value in parentheses show the cumulative number of values from the bottom to the stem of the middle value. The number in parentheses shows how many values are on the same line as the middle value.

CRITICAL THINKING

Problems 5 and 6 show how to split a stem.

Problem 10 discusses back-to-back stem-and-leaf displays.

Stem-and-leaf displays show each of the original or truncated data values. By looking at the display "sideways," you can see the distribution shape of the data. If there are large gaps between stems containing leaves, especially at the top or bottom of the display, the data values in the first or last lines may be outliers. Outliers should be examined carefully to see if they are data errors or simply unusual data values. Someone very familiar with the field of study as well as the purpose of the study should decide how to treat outliers.

VIEWPOINT   What Does It Take to Win?

Scores for NFL Super Bowl games can be found at the NFL website. Visit the Brase/Brase statistics site at **http://www.cengagebrain.com** *and find the link to the NFL. Once at the NFL website, follow the links to the Super Bowl. Of special interest in football statistics is the spread, or difference, between scores of the winning and losing teams. If the spread is too large, the game can appear to be lopsided, and TV viewers become less interested in the game (and accompanying commercial ads). Make a stem-and-leaf display of the spread for the NFL Super Bowl games and analyze the results.*

Access end-of-section exercises at www.aplia.com

Chapter Review

SUMMARY

Organizing and presenting data are the main purposes of the branch of statistics called descriptive statistics. Graphs provide an important way to show how the data are distributed.

- Frequency tables show how the data are distributed within set classes. The classes are chosen so that they cover all data values and so that each data value falls within only one class. The number of classes and the class width determine the class limits and class boundaries. The number of data values falling within a class is the class frequency.

- A histogram is a graphical display of the information in a frequency table. Classes are shown on the horizontal axis, with corresponding frequencies on the vertical axis. Relative-frequency histograms show relative

frequencies on the vertical axis. Ogives show cumulative frequencies on the vertical axis. Dotplots are like histograms, except that the classes are individual data values.

- Bar graphs, Pareto charts, and pie charts are useful to show how quantitative or qualitative data are distributed over chosen categories.

- Time-series graphs show how data change over set intervals of time.

- Stem-and-leaf displays are effective means of ordering data and showing important features of the distribution.

Graphs aren't just pretty pictures. They help reveal important properties of the data distribution, including the shape and whether or not there are any outliers.

IMPORTANT WORDS & SYMBOLS

Section 2.1
Frequency distribution   34
Frequency table   34
Class width   35
Lower class limit   35
Upper class limit   35
Class frequency   36
Class midpoint   36
Class mark   36
Class boundaries   36
Relative frequency   36
Relative-frequency table   38
Histogram   38
Relative-frequency histogram   38
Mound-shaped symmetrical distribution   41
Uniform distribution   41
Skewed left   41
Skewed right   41

Bimodal distribution   41
Outlier   42
Cumulative frequency   42
Ogive   42

Section 2.2
Bar graph   44
Changing scale   45
Pareto chart   46
Circle graph or pie chart   47
Time-series graph   48
Time series   49

Section 2.3
EDA   50
Stem-and-leaf display   51
Stem   52
Leaf   52

VIEWPOINT | Personality Clash!

Karl Pearson and Sir Ronald Fisher are two very famous mathematicians who contributed a great deal to the understanding and practice of modern statistics. Each was an outstanding person in his own right; however, the men had a terrific personality clash in the world of statistics. In Chapter 4 we will study the Pearson product-moment correlation coefficient. In Chapter 9 we will examine two interpretations of statistical testing: the classical method of critical regions, favored by Pearson, and the P-value method, favored by Sir Ronald Fisher. For a fixed level of significance, the methods can be shown to be equivalent. However, the P-value method is favored strongly by research and technology and is therefore emphasized in this text. The personality battles of Pearson and Fisher are chronicled in the popular book The Lady Tasting Tea: How Statistics Revolutionized Science in the Twentieth Century *by David Salsburg.*

CHAPTER REVIEW PROBLEMS

1. *Critical Thinking* Consider these types of graphs: histogram, bar graph, Pareto chart, pie chart, and stem-and-leaf display.
   (a) Which are suitable for qualitative data?
   (b) Which are suitable for quantitative data?

2. *Critical Thinking* A consumer interest group is tracking the percentage of household income spent on gasoline over the past 30 years. Which graphical display would be more useful, a histogram or a time-series graph? Why?

3. *Critical Thinking* Describe how data outliers might be revealed in histograms and stem-and-leaf plots.

 4. *Expand Your Knowledge* How are dotplots and stem-and-leaf displays similar? How are they different?

5. *Focus Problem: Fuel Economy* Solve the focus problem at the beginning of this chapter.

6. *Criminal Justice: Prisoners* The time plot in Figure 2-18 gives the number of state and federal prisoners per 100,000 population (Source: *Statistical Abstract of the United States*, 120th edition).
   (a) Estimate the number of prisoners per 100,000 people for 1980 and for 1997.
   (b) *Interpretation* During the time period shown, there was increased prosecution of drug offenses, longer sentences for common crimes, and reduced access to parole. What does the time-series graph say about the prison population change per 100,000 people?
   (c) In 1997, the U.S. population was approximately 266,574,000 people. At the rate of 444 prisoners per 100,000 population, about how many prisoners were in the system? The projected U.S. population for the year 2020 is 323,724,000. If the rate of prisoners per 100,000 stays the same as in 1997, about how many prisoners do you expect will be in the system in 2020? To obtain the most recent information, visit the Brase/Brase statistics site at **http://www.cengagebrain.com** and find the link to the Census Bureau.

FIGURE 2-18

Number of State and Federal Prisoners per 100,000 Population

Statistical Abstract of the United States, 120th Edition

7. *IRS: Tax Returns* Almost everyone files (or will someday file) a federal income tax return. A research poll for TurboTax (a computer software package to aid in tax-return preparation) asked what aspect of filing a return people thought to be the most difficult. The results showed that 43% of the respondents said understanding the IRS jargon, 28% said knowing deductions, 10% said getting the right form, 8% said calculating the numbers, and 10% didn't know. Make a circle graph to display this information. *Note:* Percentages will not total 100% because of rounding.

8. *Law Enforcement: DUI* Driving under the influence of alcohol (DUI) is a serious offense. The following data give the ages of a random sample of 50 drivers arrested while driving under the influence of alcohol. This distribution is based on the age distribution of DUI arrests given in the *Statistical Abstract of the United States* (112th edition).

46	16	41	26	22	33	30	22	36	34
63	21	26	18	27	24	31	38	26	55
31	47	27	43	35	22	64	40	58	20
49	37	53	25	29	32	23	49	39	40
24	56	30	51	21	45	27	34	47	35

(a) Make a stem-and-leaf display of the age distribution.
(b) Make a frequency table using seven classes.
(c) Make a histogram showing class boundaries.
(d) Identify the shape of the distribution.

9. *Agriculture: Apple Trees* The following data represent trunk circumferences (in mm) for a random sample of 60 four-year-old apple trees at East Malling Agriculture Research Station in England (Reference: S. C. Pearce, University of Kent at Canterbury). *Note:* These data are also available for download at **http://www.cengagebrain.com**.

108	99	106	102	115	120	120	117	122	142
106	111	119	109	125	108	116	105	117	123
103	114	101	99	112	120	108	91	115	109
114	105	99	122	106	113	114	75	96	124
91	102	108	110	83	90	69	117	84	142
122	113	105	112	117	122	129	100	138	117

(a) Make a frequency table with seven classes showing class limits, class boundaries, midpoints, frequencies, and relative frequencies.
(b) Draw a histogram.

(c) Draw a relative-frequency histogram.

(d) Identify the shape of the distribution.

(e) Draw an ogive.

10. *Law: Corporation Lawsuits* Many people say the civil justice system is overburdened. Many cases center on suits involving businesses. The following data are based on a *Wall Street Journal* report. Researchers conducted a study of lawsuits involving 1908 businesses ranked in the *Fortune* 1000 over a 20-year period. They found the following distribution of civil justice cases brought before the federal courts involving the businesses:

Case Type	Number of Filings (in thousands)
Contracts	107
General torts (personal injury)	191
Asbestos liability	49
Other product liability	38
All other	21

Copyright © Cengage Learning 2013

Note: Contracts cases involve disputes over contracts between businesses.

(a) Make a Pareto chart of the cases. Which type of case occurs most frequently?

(b) Make a pie chart showing the percentage of cases of each type.

11. *Archaeology: Tree-Ring Data* The *Sand Canyon Archaeological Project* edited by W. D. Lipe and published by Crow Canyon Archaeological Center contains the stem-and-leaf diagram shown in Figure 2-19. The study uses tree rings to accurately determine the year in which a tree was cut. The figure gives the tree-ring-cutting dates for samples of timbers found in the architectural units at Sand Canyon Pueblo. The text referring to the figure says, "The three-digit numbers in the left column represent centuries and decades A.D. The numbers to the right represent individual years, with each number derived from an individual sample. Thus, **124|2 2 2** represents three samples dated to A.D. 1242." Use Figure 2-19 and the verbal description to answer the following questions.

(a) Which decade contained the most samples?

(b) How many samples had a tree-ring-cutting date between A.D. 1200 and A.D. 1239, inclusive?

(c) What are the dates of the longest interval during which no tree-cutting samples occurred? What might this indicate about new construction or renovation of the pueblo structures during this period?

FIGURE 2-19

Tree-Ring-Cutting Dates from Architectural Units of Sand Canyon Pueblo: *The Sand Canyon Archaeological Project*

```
119 | 5 6
120 | 0 0 1 2 3 3 3 3 3 3 3 3 3 3 3 3 3 3 3 3 3 3 3 3 3 3 3 3 3 3
120 |
121 | 2
121 | 5 5
122 | 0 0 1 1 1 1 2 2 3 4 4 4 4 4 4 4
122 | 5 8 9
123 | 0 1 2 3 3 4
123 | 5 5 5 5 5 5 5 5 5 5 5 5 5 5 5 6 8 8 9
124 | 1 2 2 2 2 2 2 2 2 2 2 2 2 2 2 2 2 2 2 2 2 2 2 3 4 4
124 | 5 6 8 9 9 9 9 9 9 9 9 9
125 | 0 0 0 0 0 0 0 0 0 0 0 0 0 0 0 1 1 1 1 1 1 1 2 2 2
125 |
126 | 0 0 0 1 2 2 2 2 2 2 2 2 2 2 2 2 2 2 4 4 4 4 4 4 4
126 | 5 5 5 6 6 7
127 | 0 1 1 1 1 4 4
```

The Sand Canyon Archaeological Project, edited by W.D. Lipe, published by Crow Canyon Archaeological Center.

12. *Interpretation* A Harris Poll surveyed 2085 U.S. adults regarding use of cell phones while driving. All the adults were asked their opinion regarding how dangerous it is for a driver to use a cell phone while driving. Graph (a) shows the percentage responding "very dangerous" by age group. Only the adults who drive and who have a cell phone were asked how often they talk on the cell phone while driving. Graph (b) shows the percentage responding "never" by age group.

Cell Phone Use and Driving

(a)

(b)

Copyright © Cengage Learning 2013

(a) What trend does this survey portray regarding age group and the opinion that using a cell phone while driving is very dangerous?

(b) How does the behavior of never using a cell phone while driving compare to the opinion that using a cell phone while driving is dangerous? Do you think that some of the differences in the behavior (never use a cell phone while driving) and the opinion (using a cell phone while driving is very dangerous) can be attributed to the differences in the survey population? Explain.

DATA HIGHLIGHTS: GROUP PROJECTS

Break into small groups and discuss the following topics. Organize a brief outline in which you summarize the main points of your group discussion.

1. Examine Figure 2-20, "Everyone Agrees: Slobs Make Worst Roommates." This is a clustered bar graph because two percentages are given for each response category: responses from men and responses from women. Comment about how the artistic rendition has slightly changed the format of a bar graph. Do the bars seem to have lengths that accurately reflect the relative percentages of the responses? In your own opinion, does the artistic rendition enhance or confuse the information? Explain. Which characteristic of "worst roommates" does the graphic seem to illustrate? Can this graph be considered a Pareto chart for men? for women? Why or why not? From the information given in the figure, do you think the survey just listed the four given annoying characteristics? Do you think a respondent could choose more than one characteristic? Explain your answer in terms of the percentages given and in terms of the explanation given in the graphic. Could this information also be displayed in one circle graph for men and another for women? Explain.

FIGURE 2-20

Everyone Agrees: Slobs Make Worst Roommates When asked what bothers them most about living with another person, men and women responded:

Sloppiness — 35% / 41%
Uneven sharing of chores — 32% / 15%
Irritating personal habits — 24% / 22%
Invasions of privacy — 9% / 22%

■ Men ■ Women

Source: Advantage Business Research for Mattel Compatibility

2. Examine Figure 2-21, "Global Teen Worries." How many countries were contained in the sample? The graph contains bars and a circle. Which bar is the longest? Which bar represents the greatest percentage? Is this a bar graph or not? If not, what changes would need to be made to put the information in a bar graph? Could the graph be made into a Pareto chart? Could it be made into a circle graph? Explain.

FIGURE 2-21

Global Teen Worries
Top sources of concern for teens, ages 15–18. Survey includes 41 countries.

Getting a good job — 70%
Parents' health — 63%
Losing someone they love — 57%
Personal health — 54%
Finishing education — 54%
Getting university degree — 52%

Source: BrainWaves Group's New World Teen Study

LINKING CONCEPTS: WRITING PROJECTS

Discuss each of the following topics in class or review the topics on your own. Then write a brief but complete essay in which you summarize the main points. Please include formulas and graphs as appropriate.

1. In your own words, explain the differences among histograms, relative-frequency histograms, bar graphs, circle graphs, time-series graphs, Pareto charts, and stem-and-leaf displays. If you have nominal data, which graphic displays might be useful? What if you have ordinal, interval, or ratio data?

2. What do we mean when we say a histogram is skewed to the left? to the right? What is a bimodal histogram? Discuss the following statement: "A bimodal histogram usually results if we draw a sample from two populations at once." Suppose you took a sample of weights of college football players and with this sample you included weights of cheerleaders. Do you think a histogram made from the combined weights would be bimodal? Explain.

3. Discuss the statement that stem-and-leaf displays are quick and easy to construct. How can we use a stem-and-leaf display to make the construction of a frequency table easier? How does a stem-and-leaf display help you spot extreme values quickly?

4. Go to the library and pick up a current issue of *The Wall Street Journal*, *Newsweek*, *Time*, *USA Today*, or other news medium. Examine each newspaper or magazine for graphs of the types discussed in this chapter. List the variables used, methods of data collection, and general types of conclusions drawn from the graphs. Another source for information is the Internet. Explore several websites and categorize the graphs you find as you did for the print medium. For interesting websites, visit the Brase/Brase statistics site at **http://www.cengagebrain.com** and find links to the Social Statistics Briefing Room, to law enforcement, and to golf.

USING TECHNOLOGY

Applications

Tables 2-17 and 2-18 show the first-round winning scores of the NCAA men's and women's basketball teams.

TABLE 2-17   Men's Winning First-Round NCAA Tournament Scores

95	70	79	99	83	72	79	101
69	82	86	70	79	69	69	70
95	70	77	61	69	68	69	72
89	66	84	77	50	83	63	58

Copyright © Cengage Learning 2013

TABLE 2-18   Women's Winning First-Round NCAA Tournament Scores

80	68	51	80	83	75	77	100
96	68	89	80	67	84	76	70
98	81	79	89	98	83	72	100
101	83	66	76	77	84	71	77

Copyright © Cengage Learning 2013

1. Use the software or method of your choice to construct separate histograms for the men's and women's winning scores. Try 5, 7, and 10 classes for each. Which number of classes seems to be the best choice? Why?

2. Use the same class boundaries for histograms of men's and of women's scores. How do the scores for the two groups compare? What general shape do the histograms follow?

3. Use the software or method of your choice to make stem-and-leaf displays for each set of scores. If your software does not make stem-and-leaf displays, sort the data first and then make a back-to-back display by hand. Do there seem to be any extreme values in either set? How do the data sets compare?

Technology Hints: Creating Histograms

The default histograms produced by the TI-84Plus/TI-83Plus/TI-*n*spire calculators, Minitab, and Excel all determine automatically the number of classes to use. To control the number of classes the technology uses, follow the key steps as indicated. The display screens are generated for data found in Table 2-1, "One-Way Commuting Distances (in Miles) for 60 Workers in Downtown Dallas."

TI-84Plus/TI-83Plus/TI-*n*spire (with TI-84Plus keypad)

Determine the class width for the number of classes you want and the lower class boundary for the first class. Enter the data in list L1.

Press **STATPLOT** and highlight On and the histogram plot.

Press **WINDOW** and set X min = lowest class boundary, Xscl = class width. Use appropriate values for the other settings.

Press **GRAPH. TRACE** gives boundaries and frequency.

Copyright © Cengage Learning 2013

Excel 2010

Determine the upper class boundaries for the five classes. Enter the data in one column. In a separate column, enter the upper class boundaries. Click the **Data** tab, and then click the **Data Analysis** button on the ribbon in the **Analysis** group. Select **Histogram** and click **OK**.

In the Histogram dialogue box, put the data range in the Input Range. Put the upper class boundaries range in the Bin Range. Then check **New Workbook** and **Chart Output** and click **OK**.

To make the bars touch, right click on a bar and select **Format Data Series...** Then under **Series Options** move the **Gap Width** slider to 0% (No Gap) and click **Close**.

Further adjustments to the histogram can be made by clicking the **Layout** tab and selecting options in the Labels or Axes section of the ribbon.

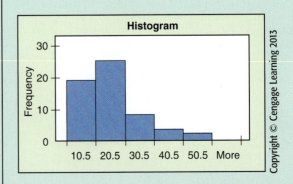

Copyright © Cengage Learning 2013

Minitab

Determine the class boundaries. Enter the data. Use the menu selection **Graph ➤ Histogram.**

Choose **Simple** and click **OK.** Select the graph variable and click **OK** to obtain a histogram with automatically selected classes. To set your own class boundaries, double click the displayed histogram. In the dialogue box, select **Binning.** Then choose **Cutpoint** and enter the class boundaries as cutpoint positions.

One-Way Commuting Distance

SPSS

The SPSS screen shot shows the default histogram created by the menu choices **Analyze ➤ Descriptive Statistics ➤ Frequencies.** In the dialogue box, move the variable containing the data into the variables window. Click **Charts** and select **Histograms.** Click the **Continue** button and then the **OK** button.

In SPSS Student Version 17 there are a number of additional ways to create histograms and other graphs. These options utilize the **Chart Builder** or **Graphboard Template Chooser** items in the **Graphs** menu. Under these options you can set the number of classes and the boundaries of the histogram. Specific instructions for using these options are provided in the *Technology Guide* for SPSS that accompanies this text.

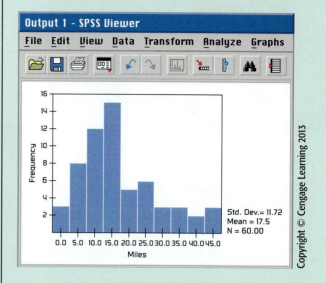

Note that SPSS requires that the measurement level of the data be set before using the **Chart Builder.** In SPSS, the **scale level** includes both interval and ratio levels of measurement. The **nominal** and **ordinal** levels are as described in this text.

Specific instructions for setting class boundaries (cutpoints) of a histogram are provided in the *Technology Guide* that accompanies this text.

3

PARAMOUNT TELEVISION/THE KOBAL COLLECTION/Picture Desk

Imagno/Getty

While the individual man is an insolvable puzzle, in the aggregate he becomes a mathematical certainty.

You can, for example, never foretell what any one man will do, but you can say with precision what an average number will be up to.

—ARTHUR CONAN DOYLE,
The Sign of Four

Sherlock Holmes spoke the words quoted here to his colleague, Dr. Watson, as the two were unraveling a mystery. The detective was implying that if a single member is drawn at random from a population, we cannot predict *exactly* what that member will look like. However, there are some "average" features of the entire population that an individual is likely to possess. The degree of certainty with which we would expect to observe such average features in any individual depends on our knowledge of the variations among individuals in the population. Sherlock Holmes has led us to two of the most important statistical concepts: average and variation.

AVERAGES AND VARIATION

Nancy Sheehan/PhotoEdit

PREVIEW QUESTIONS

What are commonly used measures of central tendency? What do they tell you? (SECTION 3.1)

How do variance and standard deviation measure data spread? Why is this important? (SECTION 3.2)

How do you make a box-and-whisker plot, and what does it tell about the spread of the data? (SECTION 3.3)

FOCUS PROBLEM

The Educational Advantage

Is it really worth all the effort to get a college degree? From a philosophical point of view, the love of learning is sufficient reason to get a college degree. However, the U.S. Census Bureau also makes another relevant point. Annually, college graduates (bachelor's degree) earn on average $23,291 more than high school graduates. This means college graduates earn about 83.4% more than high school graduates, and according to "Education Pays" on the next page, the gap in earnings is increasing. Furthermore, as the College Board indicates, for most Americans college remains relatively affordable.

After completing this chapter, you will be able to answer the following questions:

(a) Does a college degree *guarantee* someone an 83.4% increase in earnings over a high school degree? Remember, we are using only *averages* from census data.

(b) Using census data (not shown in "Education Pays"), it is estimated that the standard deviation of college-graduate earnings is about $8500. Compute a 75% Chebyshev confidence interval centered on the mean ($51,206) for bachelor's degree earnings.

(c) How much does college tuition cost? That depends, of course, on where you go to college. Construct a weighted average. Using the data from "College Affordable for Most," estimate midpoints for the cost intervals. Say 46% of tuitions cost about $4500;

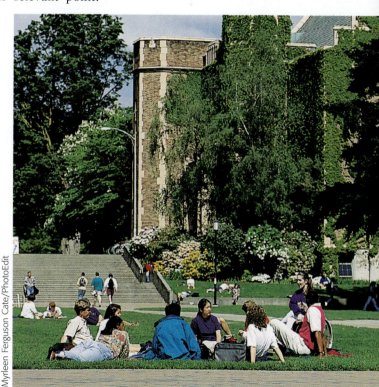

Myrleen Ferguson Cate/PhotoEdit

21% cost about $7500; 7% cost about $12,000; 8% cost about $18,000; 9% cost about $24,000; and 9% cost about $31,000. Compute the weighted average of college tuition charged at all colleges. (See Problem 9 in the Chapter Review Problems.)

Source: Census Bureau

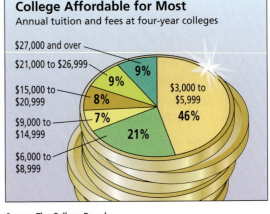

Source: The College Board

Measures of Central Tendency: Mode, Median, and Mean

FOCUS POINTS

- Compute mean, median, and mode from raw data.
- Interpret what mean, median, and mode tell you.
- Explain how mean, median, and mode can be affected by extreme data values.
- Define and compute a trimmed mean.
- Compute a weighted average.

The average price of an ounce of gold is $1500. The Zippy car averages 39 miles per gallon on the highway. A survey showed the average shoe size for women is size 9.

In each of the preceding statements, *one* number is used to describe the entire sample or population. Such a number is called an *average*. There are many ways to compute averages, but we will study only three of the major ones.

The easiest average to compute is the *mode*.

Average

Mode

> The **mode** of a data set is the value that occurs most frequently.

EXAMPLE 1   MODE

Count the letters in each word of this sentence and give the mode. The numbers of letters in the words of the sentence are

5   3   7   2   4   4   2   4   8   3   4   3   4

Scanning the data, we see that 4 is the mode because more words have 4 letters than any other number. For larger data sets, it is useful to order—or sort—the data before scanning them for the mode.

Not every data set has a mode. For example, if Professor Fair gives equal numbers of A's, B's, C's, D's, and F's, then there is no modal grade. In addition, the mode is not very stable. Changing just one number in a data set can change the mode dramatically. However, the mode is a useful average when we want to know the most frequently occurring data value, such as the most frequently requested shoe size.

Median

Another average that is useful is the *median*, or central value, of an ordered distribution. When you are given the median, you know there are an equal number of data values in the ordered distribution that are above it and below it.

PROCEDURE

HOW TO FIND THE MEDIAN

The **median** is the central value of an ordered distribution. To find it,

1. Order the data from smallest to largest.

2. For a distribution with an *odd* number of data values,

    Median = Middle data value

3. For a distribution with an *even* number of data values,

$$\text{Median} = \frac{\text{Sum of middle two values}}{2}$$

EXAMPLE 2   MEDIAN

What do barbecue-flavored potato chips cost? According to *Consumer Reports*, Vol. 66, No. 5, the prices per ounce in cents of the rated chips are

    19   19   27   28   18   35

(a) To find the median, we first order the data, and then note that there are an even number of entries. So the median is constructed using the two middle values.

    18   19   19   27   28   35
                 ↑   ↑
              middle values

$$\text{Median} = \frac{19 + 27}{2} = 23 \text{ cents}$$

(b) According to *Consumer Reports*, the brand with the lowest overall taste rating costs 35 cents per ounce. Eliminate that brand, and find the median price per ounce for the remaining barbecue-flavored chips. Again order the data. Note that there are an odd number of entries, so the median is simply the middle value.

    18   19   19   27   28
                 ↑
             middle value

    Median = middle value = 19 cents

lev dolgachov, 2009/Used under license from Shutterstock.com

(c) One ounce of potato chips is considered a small serving. Is it reasonable to budget about $10.45 to serve the barbecue-flavored chips to 55 people?

Yes, since the median price of the chips is 19 cents per small serving. This budget for chips assumes that there is plenty of other food!

The median uses the *position* rather than the specific value of each data entry. If the extreme values of a data set change, the median usually does not change. This is why the median is often used as the average for house prices. If one mansion costing several million dollars sells in a community of much lower-priced homes, the median selling price for houses in the community would be affected very little, if at all.

GUIDED EXERCISE 1 | **Median and mode**

Belleview College must make a report to the budget committee about the average credit hour load a full-time student carries. (A 12-credit-hour load is the minimum requirement for full-time status. For the same tuition, students may take up to 20 credit hours.) A random sample of 40 students yielded the following information (in credit hours):

17	12	14	17	13	16	18	20	13	12
12	17	16	15	14	12	12	13	17	14
15	12	15	16	12	18	20	19	12	15
18	14	16	17	15	19	12	13	12	15

(a) Organize the data from smallest to largest number of credit hours.

⇒
12 12 12 12 12 12 12 12 12 12
13 13 13 13 14 14 14 14 15 ⑮
⑮ 15 15 15 16 16 16 16 17 17
17 17 17 18 18 18 19 19 20 20

(b) Since there are an _____ (odd, even) number of values, we add the two middle values and divide by 2 to get the median. What is the median credit hour load?

⇒ There are an even number of entries. The two middle values are circled in part (a).

$$\text{Median} = \frac{15 + 15}{2} = 15$$

(c) What is the mode of this distribution? Is it different from the median? *Interpretation* If the budget committee is going to fund the college according to the average student credit hour load (more money for higher loads), which of these two averages do you think the college will report?

⇒ The mode is 12. It is different from the median. Since the median is higher, the college will probably report it and indicate that the average being used is the median.

Note: For small ordered data sets, we can easily scan the set to find the *location* of the median. However, for large ordered data sets of size n, it is convenient to have a formula to find the middle of the data set.

> For an ordered data set of size n,
>
> $$\text{Position of the middle value} = \frac{n + 1}{2}$$

For instance, if $n = 99$ then the middle value is the $(99 + 1)/2$ or 50th data value in the ordered data. If $n = 100$, then $(100 + 1)/2 = 50.5$ tells us that the two middle values are in the 50th and 51st positions.

Mean

What happens to the mean, median, and mode if you add a constant to each data value or multiply each data value by a constant? See Problems 12 and 13.

An average that uses the exact value of each entry is the *mean* (sometimes called the *arithmetic mean*). To compute the mean, we add the values of all the entries and then divide by the number of entries.

$$\text{Mean} = \frac{\text{Sum of all entries}}{\text{Number of entries}}$$

The mean is the average usually used to compute a test average.

EXAMPLE 3   MEAN

To graduate, Linda needs at least a B in biology. She did not do very well on her first three tests; however, she did well on the last four. Here are her scores:

| 58 | 67 | 60 | 84 | 93 | 98 | 100 |

Compute the mean and determine if Linda's grade will be a B (80 to 89 average) or a C (70 to 79 average).

SOLUTION:

$$\text{Mean} = \frac{\text{Sum of scores}}{\text{Number of scores}} = \frac{58 + 67 + 60 + 84 + 93 + 98 + 100}{7}$$

$$= \frac{560}{7} = 80$$

Since the average is 80, Linda will get the needed B.

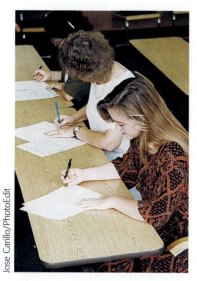

Summation symbol, Σ

COMMENT   When we compute the mean, we sum the given data. There is a convenient notation to indicate the sum. Let x represent any value in the data set. Then the notation

Σx (read "the sum of all given x values")

means that we are to sum all the data values. In other words, we are to sum all the entries in the distribution. The *summation symbol* Σ means *sum the following* and is capital sigma, the *S* of the Greek alphabet.

Formulas for the mean

The symbol for the mean of a *sample* distribution of x values is $\bar{x}$ (read "x bar"). If the data comprise the entire *population*, we use the symbol μ (lowercase Greek letter mu, pronounced "mew") to represent the mean.

PROCEDURE

When data consist of rates of change, the *harmonic mean* is the appropriate average to use. Problem 25 discusses the harmonic mean.

Problem 26 discusses the *geometric mean*, which is a useful average for percentages, ratios, or growth rates.

HOW TO FIND THE MEAN

1. Compute Σx; that is, find the sum of all the data values.
2. Divide the sum total by the number of data values.

Sample statistic $\bar{x}$ Population parameter μ

$$\bar{x} = \frac{\Sigma x}{n} \qquad\qquad \mu = \frac{\Sigma x}{N}$$

where n = number of data values in the sample

N = number of data values in the population

Jose Carillo/PhotoEdit

LOOKING FORWARD

In our future work with inferential statistics, we will use the mean $\bar{x}$ from a random sample to estimate the population parameter μ (Chapter 8) and to make decisions regarding the value of μ (Chapter 9).

CALCULATOR NOTE   It is very easy to compute the mean on *any* calculator: Simply add the data values and divide the total by the number of data. However, on calculators with a statistics mode, you place the calculator in that mode, *enter* the data, and then press the key for the mean. The key is usually designated $\bar{x}$. Because the formula for the population mean is the same as that for the sample mean, the same key gives the value for μ.

We have seen three averages: the mode, the median, and the mean. For later work, the mean is the most important. A disadvantage of the mean, however, is that it can be affected by exceptional values.

Resistant measure

A *resistant measure* is one that is not influenced by extremely high or low data values. The mean is not a resistant measure of center because we can make the mean as large as we want by changing the size of only one data value. The median, on the other hand, is more resistant. However, a disadvantage of the median is that it is not sensitive to the specific size of a data value.

Trimmed mean

A measure of center that is more resistant than the mean but still sensitive to specific data values is the *trimmed mean*. A trimmed mean is the mean of the data values left after "trimming" a specified percentage of the smallest and largest data values from the data set. Usually a 5% trimmed mean is used. This implies that we trim the lowest 5% of the data as well as the highest 5% of the data. A similar procedure is used for a 10% trimmed mean.

PROCEDURE

HOW TO COMPUTE A 5% TRIMMED MEAN

1. Order the data from smallest to largest.

2. Delete the bottom 5% of the data and the top 5% of the data. *Note: If the calculation of 5% of the number of data values does not produce a whole number, *round* to the nearest integer.

3. Compute the mean of the remaining 90% of the data.

GUIDED EXERCISE 2 | ## Mean and trimmed mean

Barron's Profiles of American Colleges, 19th edition, lists average class size for introductory lecture courses at each of the profiled institutions. A sample of 20 colleges and universities in California showed class sizes for introductory lecture courses to be

⑭	20	20	20	20	23	25	30	30	30
35	35	35	40	40	42	50	50	80	⑧⓪

(a) Compute the mean for the entire sample.   ➡   Add all the values and divide by 20:

$$\bar{x} = \frac{\Sigma x}{n} = \frac{719}{20} \approx 36.0$$

Continued

GUIDED EXERCISE 2   *continued*

(b) Compute a 5% trimmed mean for the sample.   ⟹   The data are already ordered. Since 5% of 20 is 1, we eliminate one data value from the bottom of the list and one from the top. These values are circled in the data set. Then we take the mean of the remaining 18 entries.

$$5\% \text{ trimmed mean} = \frac{\Sigma x}{n} = \frac{625}{18} \approx 34.7$$

(c) Find the median of the original data set.   ⟹   Note that the data are already ordered.

$$\text{Median} = \frac{30 + 35}{2} = 32.5$$

(d) Find the median of the 5% trimmed data set. Does the median change when you trim the data?   ⟹   The median is still 32.5. Notice that trimming the same number of entries from both ends leaves the middle position of the data set unchanged.

(e) Is the trimmed mean or the original mean closer to the median?   ⟹   The trimmed mean is closer to the median.

TECH NOTES   Minitab, Excel 2010, and TI-84Plus/TI-83Plus/TI-*n*spire calculators all provide the mean and median of a data set. Minitab and Excel 2010 also provide the mode. The TI-84Plus/TI-83Plus/TI-*n*spire calculators sort data, so you can easily scan the sorted data for the mode. Minitab provides the 5% trimmed mean, as does Excel 2010.

All this technology is a wonderful aid for analyzing data. However, *a measurement is worthless if you do not know what it represents or how a change in data values might affect it*. The defining formulas and procedures for computing the measures tell you a great deal about the measures. Even if you use a calculator to evaluate all the statistical measures, pay attention to the information the formulas and procedures give you about the components or features of the measurement.

CRITICAL THINKING

Data types and averages

In Chapter 1, we examined four levels of data: nominal, ordinal, interval, and ratio. The mode (if it exists) can be used with all four levels, including nominal. For instance, the modal color of all passenger cars sold last year might be blue. The median may be used with data at the ordinal level or above. If we ranked the passenger cars in order of customer satisfaction level, we could identify the median satisfaction level. For the mean, our data need to be at the interval or ratio level (although there are exceptions in which the mean of ordinal-level data is computed). We can certainly find the mean model year of used passenger cars sold or the mean price of new passenger cars.

Another issue of concern is that of taking the average of averages. For instance, if the values $520, $640, $730, $890, and $920 represent the mean monthly rents for five different apartment complexes, we can't say that $740 (the mean of the five numbers) is the mean monthly rent of all the apartments. We need to know the number of apartments in each complex before we can determine an average based on the number of apartments renting at each designated amount.

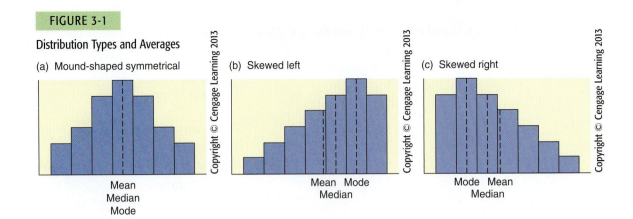

FIGURE 3-1

Distribution Types and Averages

(a) Mound-shaped symmetrical

(b) Skewed left

(c) Skewed right

Distribution shapes and averages

In general, when a data distribution is mound-shaped symmetrical, the values for the mean, median, and mode are the same or almost the same. For skewed-left distributions, the mean is less than the median and the median is less than the mode. For skewed-right distributions, the mode is the smallest value, the median is the next largest, and the mean is the largest. Figure 3-1 shows the general relationships among the mean, median, and mode for different types of distributions.

Weighted Average

Weighted average

Sometimes we wish to average numbers but we want to assign more importance, or weight, to some of the numbers. For instance, suppose your professor tells you that your grade will be based on a midterm and a final exam, each of which is based on 100 possible points. However, the final exam will be worth 60% of the grade and the midterm only 40%. How could you determine an average score that would reflect these different weights? The average you need is the *weighted average*.

$$\text{Weighted average} = \frac{\Sigma x w}{\Sigma w}$$

where x is a data value and w is the weight assigned to that data value. The sum is taken over all data values.

EXAMPLE 4   WEIGHTED AVERAGE

Suppose your midterm test score is 83 and your final exam score is 95. Using weights of 40% for the midterm and 60% for the final exam, compute the weighted average of your scores. If the minimum average for an A is 90, will you earn an A?

SOLUTION: By the formula, we multiply each score by its weight and add the results together. Then we divide by the sum of all the weights. Converting the percentages to decimal notation, we get

$$\text{Weighted average} = \frac{83(0.40) + 95(0.60)}{0.40 + 0.60}$$

$$= \frac{33.2 + 57}{1} = 90.2$$

Your average is high enough to earn an A.

TECH NOTES   The TI-84Plus/TI-83Plus/TI-*n*spire calculators directly support weighted averages. Both Excel 2010 and Minitab can be programmed to provide the averages.

TI-84Plus/TI-83Plus/TI-nspire (with TI-84 Plus keypad) Enter the data into one list, such as L1, and the corresponding weights into another list, such as L2. Then press **Stat ➤ Calc ➤ 1: 1-Var Stats.** Enter the list containing the data, followed by a comma and the list containing the weights.

VIEWPOINT   What's Wrong with Pitching Today?

One way to answer this question is to look at averages. Batting averages and average hits per game are shown for selected years from 1901 to 2000 (Source: The Wall Street Journal*).*

Year	1901	1920	1930	1941	1951	1961	1968	1976	1986	2000
B.A.	0.277	0.284	0.288	0.267	0.263	0.256	0.231	0.256	0.262	0.276
Hits	19.2	19.2	20.0	18.4	17.9	17.3	15.2	17.3	17.8	19.1

A quick scan of the averages shows that batting averages and average hits per game are virtually the same as they were almost 100 years ago. It seems there is nothing wrong with today's pitching! So what's changed? For one thing, the rules have changed! The strike zone is considerably smaller than it once was, and the pitching mound is lower. Both give the hitter an advantage over the pitcher. Even so, pitchers don't give up hits with any greater frequency than they did a century ago (look at the averages). However, modern hits go much farther, which is something a pitcher can't control.

SECTION 3.1 PROBLEMS

Access end-of-section exercises at www.aplia.com

SECTION 3.2

Measures of Variation

FOCUS POINTS

- Find the range, variance, and standard deviation.
- Compute the coefficient of variation from raw data. Why is the coefficient of variation important?
- Apply Chebyshev's theorem to raw data. What does a Chebyshev interval tell us?

An average is an attempt to summarize a set of data using just one number. As some of our examples have shown, an average taken by itself may not always be very meaningful. We need a statistical cross-reference that measures the spread of the data.

The *range* is one such measure of variation.

Range

The **range** is the difference between the largest and smallest values of a data distribution.

EXAMPLE 5   RANGE

A large bakery regularly orders cartons of Maine blueberries. The average weight of the cartons is supposed to be 22 ounces. Random samples of cartons from two suppliers were weighed. The weights in ounces of the cartons were

Supplier I:	17	22	22	22	27
Supplier II:	17	19	20	27	27

(a) Compute the range of carton weights from each supplier.

$$\text{Range} = \text{Largest value} - \text{Smallest value}$$
$$\text{Supplier I range} = 27 - 17 = 10 \text{ ounces}$$
$$\text{Supplier II range} = 27 - 17 = 10 \text{ ounces}$$

(b) Compute the mean weight of cartons from each supplier. In both cases the mean is 22 ounces.

(c) Look at the two samples again. The samples have the same range and mean. How do they differ? The bakery uses one carton of blueberries in each blueberry muffin recipe. It is important that the cartons be of consistent weight so that the muffins turn out right.

   Supplier I provides more cartons that have weights closer to the mean. Or, put another way, the weights of cartons from Supplier I are more clustered around the mean. The bakery might find Supplier I more satisfactory.

Blueberry patch

As we see in Example 5, although the range tells the difference between the largest and smallest values in a distribution, it does not tell us how much the other values vary from one another or from the mean.

Variance and Standard Deviation

Variance
Standard deviation

We need a measure of the distribution or spread of data around an expected value (either $\bar{x}$ or μ). *Variance* and *standard deviation* provide such measures. Formulas and rationale for these measures are described in the next Procedure display. Then, Examples and Guided Exercises show how to compute and interpret these measures.

   As we will see later, the formulas for variance and standard deviation differ slightly depending on whether we are using a sample or the entire population.

PROCEDURE

HOW TO COMPUTE THE SAMPLE VARIANCE AND SAMPLE STANDARD DEVIATION

Quantity	Description
x	The variable x represents a **data value** or outcome.
Mean $\bar{x} = \dfrac{\Sigma x}{n}$	This is the **average of the data values,** or what you "expect" to happen the next time you conduct the statistical experiment. Note that n is the sample size.
$x - \bar{x}$	This is the **difference** between what happened and what you expected to happen. This represents a "deviation" away from what you "expect" and is a measure of risk.

Continued

Sum of squares	$\Sigma(x - \bar{x})^2$	The expression $\Sigma(x - \bar{x})^2$ is called the **sum of squares.** The quantity $(x - \bar{x})$ is squared to make it non-negative. The sum is over all the data. If you don't square $(x - \bar{x})$, then the sum $\Sigma(x - \bar{x})$ is equal to 0 because the negative values cancel the positive values. This occurs even if some $(x - \bar{x})$ values are large, indicating a large deviation or risk.
	Sum of squares $\Sigma(x - \bar{x})^2$ or $\Sigma x^2 - \dfrac{(\Sigma x)^2}{n}$	This is an **algebraic simplification of the sum of squares** that is easier to compute. The **defining formula** for the sum of squares is the upper one. The **computation formula** for the sum of squares is the lower one. Both formulas give the same result.
Sample variance, s^2	**Sample variance** $s^2 = \dfrac{\Sigma(x - \bar{x})^2}{n - 1}$ or $s^2 = \dfrac{\Sigma x^2 - (\Sigma x)^2/n}{n - 1}$	This is the **sample variance, s^2.** The variance can be thought of as a kind of average of the $(x - \bar{x})^2$ values. However, for technical reasons, we divide the sum by the quantity $n - 1$ rather than n. This gives us the best mathematical estimate for the population variance. The **defining formula** for the variance is the upper one. The **computation formula** for the variance is the lower one. Both formulas give the same result.
Sample standard deviation, s	**Sample standard deviation** $s = \sqrt{\dfrac{\Sigma(x - \bar{x})^2}{n - 1}}$ or $s = \sqrt{\dfrac{\Sigma x^2 - (\Sigma x)^2/n}{n - 1}}$	This is the **sample standard deviation, s.** Why do we take the square root? Well, if the original x units were, say, days or dollars, then the s^2 units would be days squared or dollars squared (wow, what's that?). We take the square root to return to the original units of the data measurements. The standard deviation can be thought of as a measure of variability or risk. Larger values of s imply greater variability in the data. The **defining formula** for the standard deviation is the upper one. The **computation formula** for the standard deviation is the lower one. Both formulas give the same result.

COMMENT   Why is s called a *sample standard* deviation? First, it is computed from sample data. Then why do we use the word *standard* in the name? We know s is a measure of deviation or risk. You should be aware that there are other statistical measures of risk that we have not yet mentioned. However, s is the one that everyone uses, so it is called the "standard" (like standard time).

In statistics, the sample standard deviation and sample variance are used to describe the spread of data about the mean $\bar{x}$. Example 6 shows how to find these quantities by using the defining formulas. Guided Exercise 3 shows how to use the computation formulas.

As you will discover, for "hand" calculations, the computation formulas for s^2 and s are much easier to use. However, the defining formulas for s^2 and s emphasize the fact that the variance and standard deviation are based on the differences between each data value and the mean.

What happens to the value of the standard deviation if you add a constant to each data value or multiply each data value by a constant? See Problems 10 and 11.

Defining Formulas (Sample Statistic)

$$\text{Sample variance} = s^2 = \frac{\Sigma(x - \bar{x})^2}{n - 1} \tag{1}$$

$$\text{Sample standard deviation} = s = \sqrt{\frac{\Sigma(x - \bar{x})^2}{n - 1}} \tag{2}$$

where x is a member of the data set, $\bar{x}$ is the mean, and n is the number of data values. The sum is taken over all data values.

Computation Formulas (Sample Statistic)

$$\text{Sample variance} = s^2 = \frac{\Sigma x^2 - (\Sigma x)^2/n}{n - 1} \tag{3}$$

$$\text{Sample standard deviation} = s = \sqrt{\frac{\Sigma x^2 - (\Sigma x)^2/n}{n - 1}} \tag{4}$$

where x is a member of the data set, $\bar{x}$ is the mean, and n is the number of data values. The sum is taken over all data values.

EXAMPLE 6   SAMPLE STANDARD DEVIATION (DEFINING FORMULA)

Big Blossom Greenhouse was commissioned to develop an extra large rose for the Rose Parade. A random sample of blossoms from Hybrid A bushes yielded the following diameters (in inches) for mature peak blooms:

    2   3   3   8   10   10

Use the defining formula to find the sample variance and standard deviation.

SOLUTION: Several steps are involved in computing the variance and standard deviation. A table will be helpful (see Table 3-1). Since $n = 6$, we take the sum of the entries in Column I of Table 3-1 and divide by 6 to find the mean $\bar{x}$.

$$\bar{x} = \frac{\Sigma x}{n} = \frac{36}{6} = 6.0 \text{ inches}$$

TABLE 3-1   Diameters of Rose Blossoms (in inches)

Column I x	Column II $x - \bar{x}$	Column III $(x - \bar{x})^2$
2	$2 - 6 = -4$	$(-4)^2 = 16$
3	$3 - 6 = -3$	$(-3)^2 = 9$
3	$3 - 6 = -3$	$(-3)^2 = 9$
8	$8 - 6 = 2$	$(2)^2 = 4$
10	$10 - 6 = 4$	$(4)^2 = 16$
10	$10 - 6 = 4$	$(4)^2 = 16$
$\Sigma x = 36$		$\Sigma(x - \bar{x})^2 = 70$

Copyright © Cengage Learning 2013

Using this value for $\bar{x}$, we obtain Column II. Square each value in Column II to obtain Column III, and then add the values in Column III. To get the sample variance, divide the sum of Column III by $n - 1$. Since $n = 6$, $n - 1 = 5$.

$$s^2 = \frac{\Sigma(x - \bar{x})^2}{n - 1} = \frac{70}{5} = 14$$

Now obtain the sample standard deviation by taking the square root of the variance.

$$s = \sqrt{s^2} = \sqrt{14} \approx 3.74$$

(Use a calculator to compute the square root. Because of rounding, we use the approximately equals symbol, $\approx$.)

GUIDED EXERCISE 3 | ## Sample standard deviation (computation formula)

Big Blossom Greenhouse gathered another random sample of mature peak blooms from Hybrid B. The six blossoms had the following widths (in inches):

    5   5   5   6   7   8

(a) Again, we will construct a table so that we can find the mean, variance, and standard deviation more easily. In this case, what is the value of n? Find the sum of Column I in Table 3-2 and compute the mean.

 $n = 6$. The sum of Column I is $\Sigma x = 36$, so the mean is

$$\bar{x} = \frac{36}{6} = 6$$

TABLE 3-2 Complete Columns I and II

I x	II x^2
5	___
5	___
5	___
6	___
7	___
8	___
$\Sigma x =$ ___	$\Sigma x^2 =$ ___

TABLE 3-3 Completion of Table 3-2

I x	II x^2
5	25
5	25
5	25
6	36
7	49
8	64
$\Sigma x = 36$	$\Sigma x^2 = 224$

(b) What is the value of n? of $n - 1$? Use the computation formula to find the sample variance s^2. *Note:* Be sure to distinguish between Σx^2 and $(\Sigma x)^2$. For Σx^2, you square the x values first and then sum them. For $(\Sigma x)^2$, you sum the x values first and then square the result.

$n = 6; n - 1 = 5.$

$$s^2 = \frac{\Sigma x^2 - (\Sigma x)^2/n}{n - 1}$$

$$= \frac{224 - (36)^2/6}{5} = \frac{8}{5} = 1.6$$

$$s = \sqrt{s^2} = \sqrt{1.6} \approx 1.26$$

(c) Use a calculator to find the square root of the variance. Is this the standard deviation?

Yes.

Let's summarize and compare the results of Guided Exercise 3 and Example 6. The greenhouse found the following blossom diameters for Hybrid A and Hybrid B:

Hybrid A: Mean, 6.0 inches; standard deviation, 3.74 inches

Hybrid B: Mean, 6.0 inches; standard deviation, 1.26 inches

Interpretation In both cases, the means are the same: 6 inches. But the first hybrid has a larger standard deviation. This means that the blossoms of Hybrid A are less consistent than those of Hybrid B. If you want a rosebush that occasionally has 10-inch blooms and occasionally has 2-inch blooms, use Hybrid A. But if you want a bush that consistently produces roses close to 6 inches across, use Hybrid B.

ROUNDING NOTE Rounding errors cannot be completely eliminated, even if a computer or calculator does all the computations. However, software and calculator routines are designed to minimize the errors. If the mean is rounded, the value of the standard deviation will change slightly, depending on how much the mean is rounded. If you do your calculations "by hand" or re-enter intermediate values into a calculator, try to carry one or two more digits than occur in the original data. If your resulting answers vary slightly from those in this text, do not be overly concerned. The text answers are computer- or calculator-generated.

In most applications of statistics, we work with a random sample of data rather than the entire population of *all* possible data values. However, if we have data for the entire population, we can compute the *population mean μ*, *population variance σ²*, and *population standard deviation σ* (lowercase Greek letter sigma) using the following formulas:

Population mean, μ
Population variance, σ^2
Population standard deviation, σ

Population Parameters

$$\text{Population mean} = \mu = \frac{\Sigma x}{N}$$

$$\text{Population variance} = \sigma^2 = \frac{\Sigma(x - \mu)^2}{N}$$

$$\text{Population standard deviation} = \sigma = \sqrt{\frac{\Sigma(x - \mu)^2}{N}}$$

where N is the number of data values in the population and x represents the individual data values of the population.

What if you do not have access to raw data, but you do have a histogram or frequency table of the data and you want to estimate the mean and standard deviation? Problems 22–25 address grouped data, with the formulas shown before Problem 22.

We note that the formula for μ is the same as the formula for $\bar{x}$ (the sample mean) and that the formulas for σ^2 and σ are the same as those for s^2 and s (sample variance and sample standard deviation), except that the population size N is used instead of $n - 1$. Also, μ is used instead of $\bar{x}$ in the formulas for σ^2 and σ.

In the formulas for s and σ, we use $n - 1$ to compute s and N to compute σ. Why? The reason is that N (capital letter) represents the *population size*, whereas n (lowercase letter) represents the sample size. Since a random sample usually will not contain extreme data values (large or small), we divide by $n - 1$ in the formula for s to make s a little larger than it would have been had we divided by n. Courses in advanced theoretical statistics show that this procedure will give us the best possible estimate for the standard deviation σ. In fact, s is called the *unbiased estimate* for σ. If we have the population of all data values, then extreme data values are, of course, present, so we divide by N instead of $N - 1$.

Problem 26 discusses moving averages, and explores the difference in the standard deviations of raw data compared to the standard deviation of the moving average of the raw data.

Population size, N

COMMENT The computation formula for the population standard deviation is

$$\sigma = \sqrt{\frac{\Sigma x^2 - (\Sigma x)^2/N}{N}}$$

We've seen that the standard deviation (sample or population) is a measure of data spread. We will use the standard deviation extensively in later chapters.

TECH NOTES   Most scientific or business calculators have a statistics mode and provide the mean and sample standard deviation directly. The TI-84Plus/TI-83Plus/TI-nspire calculators, Excel 2010, and Minitab provide the median and several other measures as well.

Many technologies display only the sample standard deviation *s*. You can quickly compute σ if you know *s* by using the formula

$$\sigma = s\sqrt{\frac{n-1}{n}}$$

The mean given in displays can be interpreted as the sample mean $\bar{x}$ or the population mean μ as appropriate.

These three displays show output for the rose data from Guided Exercise 3.

TI-84Plus/TI-83Plus/TI-nspire (with TI-84Plus Keypad) Display Press STAT ➤ CALC ➤ 1:1-Var Stats. S_x is the sample standard deviation. σ_x is the population standard deviation.

```
1-Var Stats
 x̄=6
 Σx=36
 Σx²=224
 Sx=1.264911064
 σx=1.154700538
↓n=6
■
```

Copyright © Cengage Learning 2013

Excel 2010 Display Click the **Data** tab. Then in the **Analysis Group** click **Data Analysis**. Select **Descriptive Statistics**. In the dialogue box, fill in the input range with the column and row numbers containing your data. Then check the Summary Statistics box.

Column 1	
Mean	6
Standard Error	0.516398
Median	5.5
Mode	5
Standard Deviation	1.264911
Sample Variance	1.6
Kurtosis	−0.78125
Skewness	0.889391
Range	3
Minimum	5
Maximum	8
Sum	36
Count	6

Copyright © Cengage Learning 2013

Minitab Display Menu choices: Stat ➤ Basic Statistics ➤ Display Descriptive Statistics. StDev is the sample standard deviation. TrMean is the 5% trimmed mean.

N	Mean	Median	TrMean	StDev	SEMean
6	6.000	5.500	6.000	1.265	0.516

Minimum	Maximum	Q1	Q3
5.000	8.000	5.000	7.250

Copyright © Cengage Learning 2013

Now let's look at two immediate applications of the standard deviation. The first is the coefficient of variation, and the second is Chebyshev's theorem.

Coefficient of Variation

A disadvantage of the standard deviation as a comparative measure of variation is that it depends on the units of measurement. This means that it is difficult to use the standard deviation to compare measurements from different populations. For this reason, statisticians have defined the *coefficient of variation*, which expresses the standard deviation as a percentage of the sample or population mean.

Coefficient of variation, CV

> If $\bar{x}$ and s represent the sample mean and sample standard deviation, respectively, then the sample **coefficient of variation** CV is defined to be
>
> $$CV = \frac{s}{\bar{x}} \cdot 100\%$$
>
> If μ and σ represent the population mean and population standard deviation, respectively, then the population coefficient of variation CV is defined to be
>
> $$CV = \frac{\sigma}{\mu} \cdot 100\%$$

Notice that the numerator and denominator in the definition of CV have the same units, so CV itself has no units of measurement. This gives us the advantage of being able to directly compare the variability of two different populations using the coefficient of variation.

In the next Example and Guided Exercise, we will compute the CV of a population and of a sample and then compare the results.

EXAMPLE 7   COEFFICIENT OF VARIATION

The Trading Post on Grand Mesa is a small, family-run store in a remote part of Colorado. The Grand Mesa region contains many good fishing lakes, so the Trading Post sells spinners (a type of fishing lure). The store has a very limited selection of spinners. In fact, the Trading Post has only eight different types of spinners for sale. The prices (in dollars) are

| 2.10 | 1.95 | 2.60 | 2.00 | 1.85 | 2.25 | 2.15 | 2.25 |

Since the Trading Post has only eight different kinds of spinners for sale, we consider the eight data values to be the *population*.

(a) Use a calculator with appropriate statistics keys to verify that for the Trading Post data, $\mu \approx \$2.14$ and $\sigma \approx \$0.22$.

SOLUTION: Since the computation formulas for $\bar{x}$ and μ are identical, most calculators provide the value of $\bar{x}$ only. Use the output of this key for μ. The computation formulas for the sample standard deviation s and the population standard deviation σ are slightly different. Be sure that you use the key for σ (sometimes designated as σ_n or σ_x).

Ron Watts/Corbis

(b) Compute the CV of prices for the Trading Post and comment on the meaning of the result.

SOLUTION:

$$CV = \frac{\sigma}{\mu} \times 100\% = \frac{0.22}{2.14} \times 100\% = 10.28\%$$

Interpretation The coefficient of variation can be thought of as a measure of the spread of the data relative to the average of the data. Since the Trading Post is very small, it carries a small selection of spinners that are all priced similarly. The CV tells us that the standard deviation of the spinner prices is only 10.28% of the mean.

GUIDED EXERCISE 4 | **Coefficient of variation**

Cabela's in Sidney, Nebraska, is a very large outfitter that carries a broad selection of fishing tackle. It markets its products nationwide through a catalog service. A random sample of 10 spinners from Cabela's extensive spring catalog gave the following prices (in dollars):

   1.69   1.49   3.09   1.79   1.39   2.89   1.49   1.39   1.49   1.99

(a) Use a calculator with sample mean and sample standard deviation keys to compute $\bar{x}$ and s.

⟹ $\bar{x} = \$1.87$ and $s \approx \$0.62$.

(b) Compute the CV for the spinner prices at Cabela's.

⟹ $CV = \dfrac{s}{\bar{x}} \times 100\% = \dfrac{0.62}{1.87} \times 100\% = 33.16\%$

(c) *Interpretation* Compare the mean, standard deviation, and CV for the spinner prices at the Trading Post on Grand Mesa (Example 7) and Cabela's. Comment on the differences.

⟹ The CV for Cabela's is more than three times the CV for the Trading Post. Why? First, because of the remote location, the Trading Post tends to have somewhat higher prices (larger μ). Second, the Trading Post is very small, so it has a rather limited selection of spinners with a smaller variation in price.

Chebyshev's Theorem

From our earlier discussion about standard deviation, we recall that the spread or dispersion of a set of data about the mean will be small if the standard deviation is small, and it will be large if the standard deviation is large. If we are dealing with a symmetrical, bell-shaped distribution, then we can make very definite statements about the proportion of the data that must lie within a certain number of standard deviations on either side of the mean. This will be discussed in detail in Chapter 7 when we talk about normal distributions.

However, the concept of data spread about the mean can be expressed quite generally for *all data distributions* (skewed, symmetrical, or other shapes) by using the remarkable theorem of Chebyshev.

Chebyshev's theorem

Chebyshev's Theorem

For *any* set of data (either population or sample) and for any constant k greater than 1, the proportion of the data that must lie within k standard deviations on either side of the mean is *at least*

$$1 - \frac{1}{k^2}$$

> **Results of Chebyshev's Theorem**
>
> For *any* set of data:
> - *at least* 75% of the data fall in the interval from $\mu - 2\sigma$ to $\mu + 2\sigma$.
> - *at least* 88.9% of the data fall in the interval from $\mu - 3\sigma$ to $\mu + 3\sigma$.
> - *at least* 93.8% of the data fall in the interval from $\mu - 4\sigma$ to $\mu + 4\sigma$.

The results of Chebyshev's theorem can be derived by using the theorem and a little arithmetic. For instance, if we create an interval $k = 2$ standard deviations on either side of the mean, Chebyshev's theorem tells us that

$$1 - \frac{1}{2^2} = 1 - \frac{1}{4} = \frac{3}{4} \text{ or } 75\%$$

is the minimum percentage of data in the interval $\mu - 2\sigma$ to $\mu + 2\sigma$.

Notice that Chebyshev's theorem refers to the *minimum* percentage of data that must fall within the specified number of standard deviations of the mean. If the distribution is mound-shaped, an even *greater* percentage of data will fall into the specified intervals (see the empirical rule in Section 7.1).

Bob Daemmrich/The Image Works

EXAMPLE 8   CHEBYSHEV'S THEOREM

Students Who Care is a student volunteer program in which college students donate work time to various community projects, such as planting trees. Professor Gill is the faculty sponsor for this student volunteer program. For several years, Dr. Gill has kept a careful record of $x =$ total number of work hours volunteered by a student in the program each semester. For a random sample of students in the program, the mean number of hours was $\bar{x} = 29.1$ hours each semester, with a standard deviation of $s = 1.7$ hours each semester. Find an interval A to B for the number of hours volunteered into which at least 75% of the students in this program would fit.

SOLUTION: According to results of Chebyshev's theorem, at least 75% of the data must fall within 2 standard deviations of the mean. Because the mean is $\bar{x} = 29.1$ and the standard deviation is $s = 1.7$, the interval is

$$\bar{x} - 2s \text{ to } \bar{x} + 2s$$
$$29.1 - 2(1.7) \text{ to } 29.1 + 2(1.7)$$
$$25.7 \text{ to } 32.5$$

At least 75% of the students would fit into the group that volunteered from 25.7 to 32.5 hours each semester.

GUIDED EXERCISE 5 | *Chebyshev interval*

The *East Coast Independent News* periodically runs ads in its classified section offering a month's free subscription to those who respond. In this way, management can get a sense about the number of subscribers who read the classified section each day. Over a period of 2 years, careful records have been kept. The mean number of responses per ad is $\bar{x} = 525$, with a standard deviation $s = 30$.

Continued

GUIDED EXERCISE 5   *continued*

Determine a Chebyshev interval about the mean in which at least 88.9% of the data fall.

By Chebyshev's theorem, at least 88.9% of the data fall into the interval

$$\bar{x} - 3s \text{ to } \bar{x} + 3s$$

Because $\bar{x} = 525$ and $s = 30$, the interval is

$$525 - 3(30) \text{ to } 525 + 3(30)$$

or from 435 to 615 responses per ad.

LOOKING FORWARD

In Chapter 7 we will study normal distributions. For normal distributions, we can make much more precise statements about the percentage of data falling within 1, 2, or 3 standard deviations of the mean. In fact, the empirical rule (see Section 7.1) tells us that for normal distributions, about 68% of the data fall within 1 standard deviation of the mean, about 95% fall within 2 standard deviations of the mean, and almost all (99.7%) fall within 3 standard deviations of the mean.

CRITICAL THINKING

Averages such as the mean are often referred to in the media. However, an average by itself does not tell much about the way data are distributed about the mean. Knowledge about the standard deviation or variance, along with the mean, gives a much better picture of the data distribution.

Chebyshev's theorem tells us that no matter what the data distribution looks like, at least 75% of the data will fall within 2 standard deviations of the mean. As we will see in Chapter 7, when the distribution is mound-shaped and symmetrical, about 95% of the data are within 2 standard deviations of the mean. Data values beyond 2 standard deviations from the mean are less common than those closer to the mean.

Outlier

In fact, one indicator that a data value might be an *outlier* is that it is more than 2.5 standard deviations from the mean (Source: *Oxford Dictionary of Statistics*, Oxford University Press).

VIEWPOINT | Socially Responsible Investing

Make a difference and make money! Socially responsible mutual funds tend to screen out corporations that sell tobacco, weapons, and alcohol, as well as companies that are environmentally unfriendly. In addition, these funds screen out companies that use child labor in sweatshops. There are 68 socially responsible funds tracked by The Forum for Sustainable and Responsible Investment. For more information, visit the Brase/Brase statistics site at **http://www .cengagebrain.com** *and find the link to social investing.*

How do these funds rate compared to other funds? One way to answer this question is to study the annual percent returns of the funds using both the mean and standard deviation. (See Problem 20 of this section.)

SECTION 3.2
PROBLEMS

aplia™   Access end-of-section exercises at www.aplia.com

SECTION 3.3

Percentiles and Box-and-Whisker Plots

FOCUS POINTS

- Interpret the meaning of percentile scores.
- Compute the median, quartiles, and five-number summary from raw data.
- Make a box-and-whisker plot. Interpret the results.
- Describe how a box-and-whisker plot indicates spread of data about the median.

We've seen measures of central tendency and spread for a set of data. The arithmetic mean $\bar{x}$ and the standard deviation s will be very useful in later work. However, because they each utilize every data value, they can be heavily influenced by one or two extreme data values. In cases where our data distributions are heavily skewed or even bimodal, we often get a better summary of the distribution by utilizing relative positions of data rather than exact values.

Recall that the median is an average computed by using the relative position of the data. If we are told that 81 is the median score on a biology test, we know that after the data have been ordered, 50% of the data fall at or below the median value of 81. The median is an example of a *percentile*; in fact, it is the 50th percentile. The general definition of the *P*th percentile follows.

Percentile

> For whole numbers P (where $1 \le P \le 99$), the *P*th **percentile** of a distribution is a value such that $P\%$ of the data fall at or below it and $(100 - P)\%$ of the data fall at or above it.

In Figure 3-3, we see the 60th percentile marked on a histogram. We see that 60% of the data lie below the mark and 40% lie above it.

FIGURE 3-3

A Histogram with the 60th Percentile Shown

GUIDED EXERCISE 6 | **Percentiles**

You took the English achievement test to obtain college credit in freshman English by examination.

(a) If your score is at the 89th percentile, what percentage of scores are at or below yours?

⇨ The percentile means that 89% of the scores are at or below yours.

(b) If the scores ranged from 1 to 100 and your raw score is 95, does this necessarily mean that your score is at the 95th percentile?

⇨ No, the percentile gives an indication of relative position of the scores. The determination of your percentile has to do with the number of scores at or below yours. If everyone did very well and only 80% of the scores fell at or below yours, you would be at the 80th percentile even though you got 95 out of 100 points on the exam.

There are 99 percentiles, and in an ideal situation, the 99 percentiles divide the data set into 100 equal parts. (See Figure 3-4.) However, if the number of data elements is not exactly divisible by 100, the percentiles will not divide the data into equal parts.

FIGURE 3-4

Percentiles

There are several widely used conventions for finding percentiles. They lead to slightly different values for different situations, but these values are close together. For all conventions, the data are first *ranked* or ordered from smallest to largest. A natural way to find the *P*th percentile is to then find a value such that *P*% of the data fall at or below it. This will not always be possible, so we take the nearest value satisfying the criterion. It is at this point that there are a variety of processes to determine the exact value of the percentile.

We will not be very concerned about exact procedures for evaluating percentiles in general. However, *quartiles* are special percentiles used so frequently that we want to adopt a specific procedure for their computation.

Quartiles

Quartiles are those percentiles that divide the data into fourths. The *first quartile* Q_1 is the 25th percentile, the *second quartile* Q_2 is the median, and the *third quartile* Q_3 is the 75th percentile. (See Figure 3-5.)

FIGURE 3-5

Quartiles

Again, several conventions are used for computing quartiles, but the following convention utilizes the median and is widely adopted.

PROCEDURE

HOW TO COMPUTE QUARTILES

1. Order the data from smallest to largest.

2. Find the median. This is the second quartile.

3. The first quartile Q_1 is then the median of the lower half of the data; that is, it is the median of the data falling *below* the Q_2 position (and not including Q_2).

4. The third quartile Q_3 is the median of the upper half of the data; that is, it is the median of the data falling *above* the Q_2 position (and not including Q_2).

In short, all we do to find the quartiles is find three medians.

The median, or second quartile, is a popular measure of the center utilizing relative position. A useful measure of data spread utilizing relative position is the *interquartile range (IQR)*. It is simply the difference between the third and first quartiles.

Interquartile range, IQR

$$\text{Interquartile range} = Q_3 - Q_1$$

The interquartile range tells us the spread of the middle half of the data. Now let's look at an example to see how to compute all of these quantities.

EXAMPLE 9

QUARTILES

In a hurry? On the run? Hungry as well? How about an ice cream bar as a snack? Ice cream bars are popular among all age groups. *Consumer Reports* did a study of ice cream bars. Twenty-seven bars with taste ratings of at least "fair" were listed, and cost per bar was included in the report. Just how much does an ice cream bar cost? The data, expressed in dollars, appear in Table 3-4. As you can see, the cost varies quite a bit, partly because the bars are not of uniform size.

TABLE 3-4	Cost of Ice Cream Bars (in Dollars)						
0.99	1.07	1.00	0.50	0.37	1.03	1.07	1.07
0.97	0.63	0.33	0.50	0.97	1.08	0.47	0.84
1.23	0.25	0.50	0.40	0.33	0.35	0.17	0.38
0.20	0.18	0.16					

TABLE 3-5	**Ordered Cost of Ice Cream Bars (in Dollars)**						
0.16	0.17	0.18	0.20	0.25	0.33	0.33	0.35
0.37	0.38	0.40	0.47	0.50	0.50	0.50	0.63
0.84	0.97	0.97	0.99	1.00	1.03	1.07	1.07
1.07	1.08	1.23					

Copyright © Cengage Learning 2013

(a) Find the quartiles.

SOLUTION: We first order the data from smallest to largest. Table 3-5 shows the data in order. Next, we find the median. Since the number of data values is 27, there are an odd number of data, and the median is simply the center, or 14th, value. The value is shown boxed in Table 3-5.

$$\text{Median} = Q_2 = 0.50$$

There are 13 values below the median position, and Q_1 is the median of these values. It is the middle, or seventh, value and is shaded in Table 3-5.

$$\text{First quartile} = Q_1 = 0.33$$

There are also 13 values above the median position. The median of these is the seventh value from the right end. This value is also shaded in Table 3-5.

$$\text{Third quartile} = Q_3 = 1.00$$

(b) Find the interquartile range.

SOLUTION:
$$IQR = Q_3 - Q_1$$
$$= 1.00 - 0.33$$
$$= 0.67$$

This means that the middle half of the data has a cost spread of 67¢.

GUIDED EXERCISE 7 | **Quartiles**

Many people consider the number of calories in an ice cream bar as important as, if not more important than, the cost. The *Consumer Reports* article also included the calorie count of the rated ice cream bars (Table 3-6). There were 22 vanilla-flavored bars rated. Again, the bars varied in size, and some of the smaller bars had fewer calories.

Table 3-6 **Calories in Vanilla-Flavored Ice Cream Bars**

342	377	319	353	295
234	294	286	377	182
310	439	111	201	182
197	209	147	190	151
131	151			

Copyright © Cengage Learning 2013

(a) Our first step is to order the data.

Table 3-7 **Ordered Data**

111	131	147	151	151	182
182	190	197	201	209	234
286	294	295	310	319	342
353	377	377	439		

Copyright © Cengage Learning 2013

Continued

(b) There are 22 data values. Find the median.   ➡   Average the 11th and 12th data values, boxed together in Table 3-7.

$$\text{Median} = \frac{209 + 234}{2}$$

$$= 221.5$$

(c) How many values are below the median position? Find Q_1.   ➡   Since the median lies halfway between the 11th and 12th values, there are 11 values below the median position. Q_1 is the median of these values.

$Q_1 = 182$

(d) There are the same number of data above as below the median. Use this fact to find Q_3.   ➡ Q_3 is the median of the upper half of the data. There are 11 values in the upper portion.

$Q_3 = 319$

(e) Find the interquartile range and comment on its meaning.   ➡ $IQR = Q_3 - Q_1$

$$= 319 - 182$$

$$= 137$$

The middle portion of the data has a spread of 137 calories.

Box-and-Whisker Plots

Five-number summary

The quartiles together with the low and high data values give us a very useful *five-number summary* of the data and their spread.

> **Five-Number Summary**
>
> Lowest value, Q_1, median, Q_3, highest value

Box-and-whisker plot

We will use these five numbers to create a graphic sketch of the data called a *box-and-whisker plot*. Box-and-whisker plots provide another useful technique from exploratory data analysis (EDA) for describing data.

PROCEDURE

HOW TO MAKE A BOX-AND-WHISKER PLOT

1. Draw a vertical scale to include the lowest and highest data values.
2. To the right of the scale, draw a box from Q_1 to Q_3.
3. Include a solid line through the box at the median level.
4. Draw vertical lines, called *whiskers*, from Q_1 to the lowest value and from Q_3 to the highest value.

Whiskers

FIGURE 3-6

Box-and-Whisker Plot

Highest value
Q_3
Median
Q_1
Lowest value

Example 10 demonstrates the process of making a box-and-whisker plot.

EXAMPLE 10

BOX-AND-WHISKER PLOT

Using the data from Guided Exercise 7, make a box-and-whisker plot showing the calories in vanilla-flavored ice cream bars. Use the plot to make observations about the distribution of calories.

SOLUTION:

(a) In Guided Exercise 7, we ordered the data (see Table 3-7) and found the values of the median, Q_1, and Q_3. From this previous work we have the following five-number summary:

low value = 111; Q_1 = 182; median = 221.5; Q_3 = 319; high value = 439

(b) We select an appropriate vertical scale and make the plot (Figure 3-7).

(c) *Interpretation* A quick glance at the box-and-whisker plot reveals the following:

(i) The box tells us where the middle half of the data lies, so we see that half of the ice cream bars have between 182 and 319 calories, with an interquartile range of 137 calories.

(ii) The median is slightly closer to the lower part of the box. This means that the lower calorie counts are more concentrated. The calorie counts above the median are more spread out, indicating that the distribution is slightly skewed toward the higher values.

(iii) The upper whisker is longer than the lower, which again emphasizes skewness toward the higher values.

FIGURE 3-7

Box-and-Whisker Plot for Calories in Vanilla-Flavored Ice Cream Bars

COMMENT   In exploratory data analysis, *hinges* rather than quartiles are used to create the box. Hinges are computed in a manner similar to the method used to compute quartiles. However, in the case of an odd number of data values, the median itself is included in both the lower and upper halves of the data (see *Applications, Basics, and Computing of Exploratory Data Analysis* by Paul Velleman and David Hoaglin, Duxbury Press). This has the effect of shrinking the box and moving the ends of the box slightly toward the median. For an even number of data, the quartiles as we computed them equal the hinges.

GUIDED EXERCISE 8 | Box-and-whisker plot

The Renata College Development Office sent salary surveys to alumni who graduated 2 and 5 years ago. The voluntary responses received are summarized in the box-and-whisker plots shown in Figure 3-8.

(a) From Figure 3-8, estimate the median and extreme values of salaries of alumni graduating 2 years ago. In what range are the middle half of the salaries? The median seems to be about $44,000. The extremes are about $33,000 and $54,000. The middle half of the salaries fall between $40,000 and $47,000.

Continued

GUIDED EXERCISE 8   *continued*

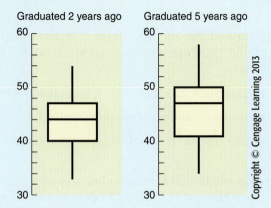

FIGURE 3-8

Box-and-Whisker Plots for Alumni Salaries
(in Thousands of Dollars)

(b) From Figure 3-8, estimate the median and extreme values of salaries of alumni graduating 5 years ago. What is the location of the middle half of the salaries?

⟹ The median seems to be $47,000. The extremes are $34,000 and $58,000. The middle half of the data is enclosed by the box with the low side at $41,000 and the high side at $50,000.

(c) *Interpretation* Compare the two box-and whisker plots and make comments about the salaries of alumni graduating 2 and 5 years ago.

⟹ The salaries of the alumni graduating 5 years ago have a larger range. They begin slightly higher than, and extend to about $4000 above, the salaries of those graduating 2 years ago. The middle half of the data is also more spread out, with higher boundaries and a higher median.

CRITICAL THINKING

Problem 12 discusses how to identify possible outliers on a box-and-whisker plot.

Box-and-whisker plots provide a graphic display of the spread of data about the median. The box shows the location of the middle half of the data. One-quarter of the data is located along each whisker.

To the extent that the median is centered in the box and the whiskers are about the same length, the data distribution is symmetrical around the median. If the median line is near one end of the box, the data are skewed toward the other end of the box.

We have developed the skeletal box-and-whisker plot. Other variations include *fences*, which are marks placed on either side of the box to represent various portions of data. Values that lie beyond the fences are *outliers*. Problem 12 of this section discusses some criteria for locating fences and identifying outliers.

TECH NOTES   *Box-and-whisker plot*

Both Minitab and the TI-84Plus/TI-83Plus/TI-*n*spire calculators support box-and-whisker plots. On the TI-84Plus/TI-83Plus/TI-*n*spire, the quartiles Q_1 and Q_3 are calculated as we calculate them in this text. In Minitab and Excel 2010, they are calculated using a slightly different process.

TI-84Plus/TI-83Plus/TI-*n*spire (with TI-84Plus Keypad) Press STATPLOT ➤ On. Highlight box plot. Use **Trace** and the arrow keys to display the values of the five-number summary. The display shows the plot for calories in ice cream bars.

Med = 221.5

Excel 2010 Does not produce box-and-whisker plots. However, each value of the five-number summary can be found. On the **Home** ribbon, click **Insert Function** f_x. In the dialogue box, select **Statistical** as the category and scroll to **Quartile.inc**. In the dialogue box, enter the data location and then enter the number of the value you want. For instance, enter 1 in the quartile box for the first quartile.

Minitab Press **Graph** ➤ **Boxplot**. In the dialogue box, set Display to IQRange Box.

VIEWPOINT | Is Shorter Higher?

Can you estimate a person's height from the pitch of his or her voice? Is a soprano shorter than an alto? Is a bass taller than a tenor? A statistical study of singers in the New York Choral Society provided information. For more information, visit the Brase/Brase statistics site at **http:// www.cengagebrain.com** *and find the link to DASL, the Carnegie Mellon University Data and Story Library. From Data Subjects, select Music and then Singers. Methods of this chapter can be used with new methods we will learn in Chapters 4 and 8–11 to examine such questions from a statistical point of view.*

SECTION 3.3 PROBLEMS   aplia   Access end-of-section exercises at www.aplia.com

Chapter Review

To characterize numerical data, we use measures both of center and of spread.

- Commonly used measures of center are the arithmetic mean, the median, and the mode. The weighted average and trimmed mean are also used as appropriate.

- Commonly used measures of spread are the variance, the standard deviation, and the range. The variance and standard deviation are measures of spread about the mean.

- Chebyshev's theorem enables us to estimate the data spread about the mean.

- The coefficient of variation lets us compare the relative spreads of different data sets.

- Other measures of data spread include percentiles, which indicate the percentage of data falling at or below the specified percentile value.

- Box-and-whisker plots show how the data are distributed about the median and the location of the middle half of the data distribution.

In later work, the average we will use most often is the mean; the measure of variation we will use most often is the standard deviation.

Section 3.1
Average 66
Mode 66
Median 67
Mean 69
Summation symbol, Σ 69
Sample mean, $\bar{x}$ 69
Population mean, μ 69
Resistant measure 70
Trimmed mean 70
Weighted average 72

Section 3.2
Range 73
Variance 74
Standard deviation 74
Sum of squares, $\Sigma(x - \bar{x})^2$ 75

Sample variance, s^2 75
Sample standard deviation, s 75
Population mean, μ 78
Population variance, σ^2 78
Population standard deviation, σ 78
Population size, N 78
Coefficient of variation, CV 80
Chebyshev's theorem 81
Outlier 83

Section 3.3
Percentile 84
Quartiles 85
Interquartile range, IQR 86
Five-number summary 88
Box-and-whisker plot 88
Whisker 88

VIEWPOINT | The Fujita Scale

How do you measure a tornado? Professor Fujita and Allen Pearson (director of the National Severe Storms Forecast Center) developed a measure based on wind speed and type of damage done by a tornado. The result is an excellent example of both descriptive and inferential statistical methods. For more information, visit the Brase/Brase statistics site at **http://www.cengagebrain .com** *and find the link to the tornado project. Then look up Fujita scale. If we group the data a little, the scale becomes*

FS	WS	%
F0 & F1	40–112	67
F2 & F3	113–206	29
F4 & F5	207–318	4

where FS represents Fujita scale; WS, wind speed in miles per hour; and %, percentage of all tornados. Out of 100 tornados, what would you estimate for the mean and standard deviation of wind speed?

CHAPTER REVIEW PROBLEMS

1. *Statistical Literacy*
   (a) What measures of variation indicate spread about the mean?
   (b) Which graphic display shows the median and data spread about the median?

2. *Critical Thinking* Look at the two histograms shown. Each involves the same number of data. The data are all whole numbers, so the height of each bar represents the number of values equal to the corresponding midpoint shown on the horizontal axis. Notice that both distributions are symmetrical.

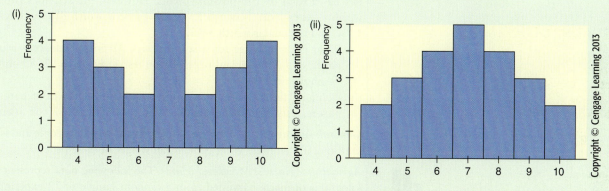

(a) Estimate the mode, median, and mean for each histogram.
(b) Which distribution has the larger standard deviation? Why?

3. *Critical Thinking* Consider the following Minitab display of two data sets:

Variable	N	Mean	SE Mean	StDev	Minimum	Q1	Median	Q3	Maximum
C1	20	20.00	1.62	7.26	7.00	15.00	20.00	25.00	31.00
C2	20	20.00	1.30	5.79	7.00	20.00	22.00	22.00	31.00

(a) What are the respective means? the respective ranges?
(b) Which data set seems more symmetrical? Why?
(c) Compare the interquartile ranges of the two sets. How do the middle halves of the data sets compare?

4. *Consumer: Radon Gas* "Radon: The Problem No One Wants to Face" is the title of an article appearing in *Consumer Reports*. Radon is a gas emitted from the ground that can collect in houses and buildings. At certain levels it can cause lung cancer. Radon concentrations are measured in picocuries per liter (pCi/L). A radon level of 4 pCi/L is considered "acceptable." Radon levels in a house vary from week to week. In one house, a sample of 8 weeks had the following readings for radon level (in pCi/L):

1.9	2.8	5.7	4.2	1.9	8.6	3.9	7.2

(a) Find the mean, median, and mode.
(b) Find the sample standard deviation, coefficient of variation, and range.

5. *Political Science: Georgia Democrats* How Democratic is Georgia? County-by-county results are shown for a recent election. For your convenience, the data

have been sorted in increasing order (Source: *County and City Data Book*, 12th edition, U.S. Census Bureau).

Percentage of Democratic Vote by County in Georgia

31	33	34	34	35	35	35	36	38	38	38	39	40	40	40	40
41	41	41	41	41	41	41	42	42	43	44	44	44	45	45	46
46	46	46	47	48	49	49	49	49	50	51	52	52	53	53	53
53	53	55	56	56	57	57	59	62	66	66	68				

(a) Make a box-and-whisker plot of the data. Find the interquartile range.

(b) *Grouped Data* Make a frequency table using five classes. Then estimate the mean and sample standard deviation using the frequency table. Compute a 75% Chebyshev interval centered about the mean.

(c) If you have a statistical calculator or computer, use it to find the actual sample mean and sample standard deviation. Otherwise, use the values $\Sigma x = 2769$ and $\Sigma x^2 = 132{,}179$ to compute the sample mean and sample standard deviation.

6. *Grades: Weighted Average* Professor Cramer determines a final grade based on attendance, two papers, three major tests, and a final exam. Each of these activities has a total of 100 possible points. However, the activities carry different weights. Attendance is worth 5%, each paper is worth 8%, each test is worth 15%, and the final is worth 34%.

(a) What is the average for a student with 92 on attendance, 73 on the first paper, 81 on the second paper, 85 on test 1, 87 on test 2, 83 on test 3, and 90 on the final exam?

(b) Compute the average for a student with the scores from part (a) on the papers, tests, and final exam, but with a score of only 20 on attendance.

7. *General: Average Weight* An elevator is loaded with 16 people and is at its load limit of 2500 pounds. What is the mean weight of these people?

8. *Agriculture: Harvest Weight of Maize* The following data represent weights in kilograms of maize harvested from a random sample of 72 experimental plots on St. Vincent, an island in the Caribbean (Reference: B. G. F. Springer, *Proceedings of the Caribbean Food Crops Society*, Vol. 10, pp. 147–152). *Note:* These data are also available for download at **http://www.cengagebrain.com**. For convenience, the data are presented in increasing order.

7.8	9.1	9.5	10.0	10.2	10.5	11.1	11.5	11.7	11.8
12.2	12.2	12.5	13.1	13.5	13.7	13.7	14.0	14.4	14.5
14.6	15.2	15.5	16.0	16.0	16.1	16.5	17.2	17.8	18.2
19.0	19.1	19.3	19.8	20.0	20.2	20.3	20.5	20.9	21.1
21.4	21.8	22.0	22.0	22.4	22.5	22.5	22.8	22.8	23.1
23.1	23.2	23.7	23.8	23.8	23.8	23.8	24.0	24.1	24.1
24.5	24.5	24.9	25.1	25.2	25.5	26.1	26.4	26.5	26.7
27.1	29.5								

(a) Compute the five-number summary.

(b) Compute the interquartile range.

(c) Make a box-and-whisker plot.

(d) *Interpretation* Discuss the distribution. Does the lower half of the distribution show more data spread than the upper half?

9. *Focus Problem: The Educational Advantage* Solve the focus problem at the beginning of this chapter.

10. *Agriculture: Bell Peppers* The pathogen *Phytophthora capsici* causes bell-pepper plants to wilt and die. A research project was designed to study the effect of soil water content and the spread of the disease in fields of bell pep-

pers (Source: *Journal of Agricultural, Biological, and Environmental Statistics*, Vol. 2, No. 2). It is thought that too much water helps spread the disease. The fields were divided into rows and quadrants. The soil water content (percentage of water by volume of soil) was determined for each plot. An important first step in such a research project is to give a statistical description of the data.

Soil Water Content for Bell-Pepper Study

15	14	14	14	13	12	11	11	11	11	10	11	13	16	10
9	15	12	9	10	7	14	13	14	8	9	8	11	13	13
15	12	9	10	9	9	16	16	12	10	11	11	12	15	6
10	10	10	11	9										

(a) Make a box-and-whisker plot of the data. Find the interquartile range.

(b) *Grouped Data* Make a frequency table using four classes. Then estimate the mean and sample standard deviation using the frequency table. Compute a 75% Chebyshev interval centered about the mean.

(c) If you have a statistical calculator or computer, use it to find the actual sample mean and sample standard deviation.

11. *Performance Rating: Weighted Average* A performance evaluation for new sales representatives at Office Automation Incorporated involves several ratings done on a scale of 1 to 10, with 10 the highest rating. The activities rated include new contacts, successful contacts, total contacts, dollar volume of sales, and reports. Then an overall rating is determined by using a weighted average. The weights are 2 for new contacts, 3 for successful contacts, 3 for total contacts, 5 for dollar value of sales, and 3 for reports. What would the overall rating be for a sales representative with ratings of 5 for new contacts, 8 for successful contacts, 7 for total contacts, 9 for dollar volume of sales, and 7 for reports?

DATA HIGHLIGHTS: GROUP PROJECTS

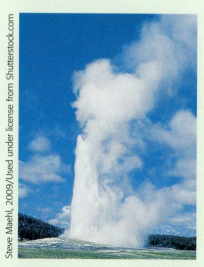

Old Faithful Geyser, Yellowstone National Park

Break into small groups and discuss the following topics. Organize a brief outline in which you summarize the main points of your group discussion.

1. *The Story of Old Faithful* is a short book written by George Marler and published by the Yellowstone Association. Chapter 7 of this interesting book talks about the effect of the 1959 earthquake on eruption intervals for Old Faithful Geyser. Dr. John Rinehart (a senior research scientist with the National Oceanic and Atmospheric Administration) has done extensive studies of the eruption intervals before and after the 1959 earthquake. Examine Figure 3-11. Notice the general shape. Is the graph more or less symmetrical? Does it have a single mode frequency? The mean interval between eruptions has remained steady at about 65 minutes for the past 100 years. Therefore, the 1959 earthquake did not significantly change the mean, but it did change the distribution of eruption intervals. Examine Figure 3-12. Would you say there are really two frequency modes, one shorter and the other longer? Explain. The overall mean is about the same for both graphs, but one graph has a much larger standard deviation (for eruption intervals) than the other. Do no calculations, just look at both graphs, and then explain which graph has the smaller and which has the larger standard deviation. Which distribution will have the larger coefficient of variation? In everyday terms, what would this mean if you were actually at Yellowstone waiting to see the next eruption of Old Faithful? Explain your answer.

2. Most academic advisors tell students to major in a field they really love. After all, it is true that money cannot buy happiness! Nevertheless, it is interesting to at least look at some of the higher-paying fields of study. After all, a field like

FIGURE 3-11

Typical Behavior of Old Faithful Geyser before 1959 Quake

FIGURE 3-12

Typical Behavior of Old Faithful Geyser after 1959 Quake

mathematics can be a lot of fun, once you get into it. We see that women's salaries tend to be less than men's salaries. However, women's salaries are rapidly catching up, and this benefits the entire workforce in different ways. Figure 3-13 shows the median incomes for college graduates with different majors. The employees in the sample are all at least 30 years old. Does it seem reasonable to assume that many of the employees are in jobs beyond the entry level? Explain. Compare the median incomes shown for all women aged 30 or older holding bachelor's degrees with the median incomes for men of similar age holding bachelor's degrees. Look at the particular majors listed. What percentage of men holding bachelor's degrees in mathematics make $52,316 or more? What percentage of women holding computer/information sciences degrees make $41,559 or more? How do median incomes for men and women holding engineering degrees compare? What about pharmacy degrees?

FIGURE 3-13

Lucrative Majors for Bachelor's Degrees
The U.S. median income for women and men (age 30 or older) with bachelor's degrees

Women
Median salary: $31,848

Major	Median income
Pharmacy	$47,567
Engineering	$46,389
Computer/ Information Sciences	$41,559
Physical Therapy/ related services	$40,491
Nursing	$40,096

Men
Median salary: $43,856

Major	Median income
Engineering	$52,998
Mathematics	$52,316
Physics	$51,819
Pharmacy	$50,805
Economics	$50,360

Source: Bureau of Labor Statistics

LINKING CONCEPTS: WRITING PROJECTS

Discuss each of the following topics in class or review the topics on your own. Then write a brief but complete essay in which you summarize the main points. Please include formulas and graphs as appropriate.

1. An average is an attempt to summarize a collection of data into just *one* number. Discuss how the mean, median, and mode all represent averages in this context. Also discuss the differences among these averages. Why is the mean a balance point? Why is the median a midway point? Why is the mode the most common data point? List three areas of daily life in which you think the mean, median, or mode would be the best choice to describe an "average."

2. Why do we need to study the variation of a collection of data? Why isn't the average by itself adequate? We have studied three ways to measure variation. The range, the standard deviation, and, to a large extent, a box-and-whisker plot all indicate the variation within a data collection. Discuss similarities and differences among these ways to measure data variation. Why would it seem reasonable to pair the median with a box-and-whisker plot and to pair the mean with the standard deviation? What are the advantages and disadvantages of each method of describing data spread? Comment on statements such as the following: (a) The range is easy to compute, but it doesn't give much information; (b) although the standard deviation is more complicated to compute, it has some significant applications; (c) the box-and-whisker plot is fairly easy to construct, and it gives a lot of information at a glance.

3. Why is the coefficient of variation important? What do we mean when we say that the coefficient of variation has no units? What advantage can there be in having no units? Why is *relative size* important?

   Consider robin eggs; the mean weight of a collection of robin eggs is 0.72 ounce and the standard deviation is 0.12 ounce. Now consider elephants; the mean weight of elephants in the zoo is 6.42 tons, with a standard deviation of 1.07 tons. The units of measurement are different and there is a great deal of difference between the weight of an elephant and that of a robin's egg. Yet the coefficient of variation is about the same for both. Comment on this from the viewpoint of the size of the standard deviation relative to that of the mean.

4. What is Chebyshev's theorem? Suppose you have a friend who knows very little about statistics. Write a paragraph or two in which you describe Chebyshev's theorem for your friend. Keep the discussion as simple as possible, but be sure to get the main ideas across to your friend. Suppose he or she asks, "What is this stuff good for?" and suppose you respond (a little sarcastically) that Chebyshev's theorem applies to everything from butterflies to the orbits of the planets! Would you be correct? Explain.

USING TECHNOLOGY

Raw Data

Application

Trade winds are one of the beautiful features of island life in Hawaii. The following data represent total air movement in miles per day over a weather station in Hawaii as determined by a continuous anemometer recorder. The period of observation is January 1 to February 15, 1971.

26	14	18	14
113	50	13	22
27	57	28	50
72	52	105	138
16	33	18	16
32	26	11	16
17	14	57	100
35	20	21	34
18	13	18	28
21	13	25	19
11	19	22	19
15	20		

Source: United States Department of Commerce, National Oceanic and Atmospheric Administration, Environmental Data Service. *Climatological Data, Annual Summary, Hawaii,* Vol. 67, No. 13. Asheville: National Climatic Center, 1971, pp. 11, 24.

(a) Use the computer to find the sample mean, median, and (if it exists) mode. Also, find the range, sample variance, and sample standard deviation.

(b) Use the five-number summary provided by the computer to make a box-and-whisker plot of total air movement over the weather station.

(c) Four data values are exceptionally high: 113, 105, 138, and 100. The strong winds of January 5 (113 reading) brought in a cold front that dropped snow on Haleakala National Park (at the 8000-foot elevation). The residents were so excited that they drove up to see the snow and caused such a massive traffic jam that the Park Service had to close the road. The winds of January 15, 16, and 28 (readings 105, 138, and 100) accompanied a storm with funnel clouds that did much damage. Eliminate these values (100, 105, 113, and 138) from the data bank and redo parts

(a) and (b). Compare your results with those previously obtained. Which average is most affected? What happens to the standard deviation? How do the two box-and-whisker plots compare?

Technology Hints: Raw Data

TI-84Plus/TI-83Plus/TI-nspire (with TI-84Plus keypad), Excel 2010, Minitab

The Tech Note in Section 3.2 gives brief instructions for finding summary statistics for raw data using the TI-84Plus/TI-83Plus/TI-nspire calculators, Excel 2010, and Minitab. The Tech Note in Section 3.3 gives brief instructions for constructing box-and-whisker plots using the TI-84Plus/TI-83Plus/TI-nspire calculators and Minitab.

SPSS

Many commands in SPSS provide an option to display various summary statistics. A direct way to display summary statistics is to use the menu choices **Analyze ➤ Descriptive Statistics ➤ Descriptives.** In the dialogue box, move the variable containing your data into the variables box. Click **Options...** and then check the summary statistics you wish to display. Click **Continue** and then **OK.** Notice that the median is not available. A more complete list of summary statistics is available with the menu choices **Analyze ➤ Descriptive Statistics ➤ Frequencies.** Click the **Statistics** button and check the summary statistics you wish to display.

For box-and-whisker plots, use the menu options **Graphs ➤ Legacy Dialogues ➤ Interactive ➤ Boxplot.** In the dialogue box, place the variable containing your data in the box along the vertical axis. After selecting the options you want, click **OK.**

SPSS Student Version 17 has other options for creating box-and-whisker plots. The *Technology Guide* for SPSS that accompanies this text has instructions for using the **Chart Builder** and **Graphboard Template Chooser** options found in the **Graphs** menu. Note that SPSS requires the appropriate level of measurement for processing data. In SPSS, the **scale level** corresponds to both interval and ratio level. The **nominal** and **ordinal** levels are as described in this text.

Cumulative Review Problems

CHAPTERS 1–3

Critical Thinking and Literacy

1. Consider the following measures: mean, median, variance, standard deviation, percentile.
   (a) Which measures utilize relative position of the data values?
   (b) Which measures utilize actual data values regardless of relative position?

2. Describe how the presence of possible outliers might be identified on
   (a) histograms.
   (b) dotplots.
   (c) stem-and-leaf displays.
   (d) box-and-whisker plots.

3. Consider two data sets, A and B. The sets are identical except that the high value of data set B is three times greater than the high value of data set A.
   (a) How do the medians of the two data sets compare?
   (b) How do the means of the two data sets compare?
   (c) How do the standard deviations of the two data sets compare?
   (d) How do the box-and-whisker plots of the two data sets compare?

4. You are examining two data sets involving test scores, set A and set B. The score 86 appears in both data sets. In which data set in each of the following situations does 86 represent a higher score? Explain.
   (a) The percentile rank of 86 is higher in set A than in set B.
   (b) The mean is 80 in both data sets, but set A has a higher standard deviation.

Applications

In west Texas, water is extremely important. The following data represent pH levels in groundwater for a random sample of 102 west Texas wells. A pH less than 7 is acidic and a pH above 7 is alkaline. Scanning the data, you can see that water in this region tends to be hard (alkaline). Too high a pH means the water is unusable or needs expensive treatment to make it usable (Reference: C. E. Nichols and V. E. Kane, Union Carbide technical report K/UR-1).

Dave G. Houser/Documentary Value/Corbis

These data are also available for download at **http://www.cengagebrain.com.** For convenience, the data are presented in increasing order.

x: pH of Groundwater in 102 West Texas Wells

7.0	7.0	7.0	7.0	7.0	7.0	7.0	7.0	7.1	7.1	7.1	7.1
7.1	7.1	7.1	7.1	7.1	7.1	7.2	7.2	7.2	7.2	7.2	7.2
7.2	7.2	7.2	7.2	7.3	7.3	7.3	7.3	7.3	7.3	7.3	7.3
7.3	7.3	7.3	7.4	7.4	7.4	7.4	7.4	7.4	7.4	7.4	7.4
7.5	7.5	7.5	7.5	7.5	7.5	7.5	7.5	7.6	7.6	7.6	7.6
7.6	7.6	7.6	7.6	7.6	7.7	7.7	7.7	7.7	7.7	7.7	7.8
7.8	7.8	7.8	7.8	7.9	7.9	7.9	7.9	7.9	8.0	8.1	8.1
8.1	8.1	8.1	8.1	8.1	8.2	8.2	8.2	8.2	8.2	8.2	8.2
8.4	8.5	8.6	8.7	8.8	8.8						

5. Write a brief description in which you outline how you would obtain a random sample of 102 west Texas water wells. Explain how random numbers would be used in the selection process.

6. Is the given data nominal, ordinal, interval, or ratio? Explain.

7. Make a stem-and-leaf display. Use five lines per stem so that leaf values 0 and 1 are on one line, 2 and 3 are on the next line, 4 and 5 are on the next, 6 and 7 are on the next, and 8 and 9 are on the last line of the stem.

8. Make a frequency table, histogram, and relative-frequency histogram using five classes. Recall that for decimal data, we "clear the decimal" to determine classes for whole-number data and then reinsert the decimal to obtain the classes for the frequency table of the original data.

9. Make an ogive using five classes.

10. Compute the range, mean, median, and mode for the given data.

11. (a) Verify that $\Sigma x = 772.9$ and $\Sigma x^2 = 5876.6$.
    (b) Compute the sample variance, sample standard deviation, and coefficient of variation for the given data. Is the sample standard deviation small relative to the mean pH?

12. Compute a 75% Chebyshev interval centered on the mean.

13. Make a box-and-whisker plot. Find the interquartile range.

Interpretation

Wow! In Problems 5–13 you constructed a lot of information regarding the pH of west Texas groundwater based on sample data. Let's continue the investigation.

14. Look at the histogram. Is the pH distribution for these wells symmetrical or skewed? Are lower or higher values more common?

15. Look at the ogive. What percentage of the wells have a pH less than 8.15? Suppose a certain crop can tolerate irrigation water with a pH between 7.35 and 8.55. What percentage of the wells could be used for such a crop?

16. Look at the stem-and-leaf plot. Are there any unusually high or low pH levels in this sample of wells? How many wells are neutral (pH of 7)?

17. Use the box-and-whisker plot to describe how the data are spread about the median. Are the pH values above the median more spread out than those below? Is this observation consistent with the skew of the histogram?

18. Suppose you are working for the regional water commissioner. You have been asked to submit a brief report about the pH level of groundwater in the west Texas region. Write such a report and include appropriate graphs.

4

Asia Images Group/Getty Images

4.1 Scatter Diagrams and Linear Correlation

4.2 Linear Regression and the Coefficient of Determination

Portrait of René Descartes c. 1649 (after) Frans Hals/Louvre/ Giraudon/ The Bridgeman Art Library

When it is not in our power to determine what is true, we ought to follow what is most probable.

—RENÉ DESCARTES

It is important to realize that statistics and probability do not deal in the realm of certainty. If there is any realm of human knowledge where genuine certainty exists, you may be sure that our statistical methods are not needed there. In most human endeavors, and in almost all of the natural world around us, the element of chance happenings cannot be avoided. When we cannot expect something with true certainty, we must rely on probability to be our guide. In this chapter, we will study regression, correlation, and forecasting. One of the tools we use is a scatter diagram or plot. René Decartes (1596–1650) was the first mathematician to systematically use rectangular coordinate plots. For this reason, such a coordinate axis is called a Cartesian axis.

For online student resources, visit the Brase/Brase, *Understanding Basic Statistics,* 6th edition website at
http://www.cengagebrain.com

CORRELATION AND REGRESSION

David Young-Wolff/PhotoEdit

PREVIEW QUESTIONS

How can you use a scatter diagram to visually estimate the degree of linear correlation of two random variables? (SECTION 4.1)

How do you compute the correlation coefficient and what does it tell you about the strength of the linear relationship between two random variables? (SECTION 4.1)

What is the least-squares criterion? How do you find the equation of the least-squares line? (SECTION 4.2)

What is the coefficient of determination and what does it tell you about explained variation of y in a random sample of data pairs (x, y)? (SECTION 4.2)

iofoto, 2010/Used under license from Shutterstock.com

FOCUS PROBLEM

Changing Populations and Crime Rate

Is the crime rate higher in neighborhoods where people might not know each other very well? Is there a relationship between crime rate and population change? If so, can we make predictions based on such a relationship? Is the relationship statistically significant? Is it possible to predict crime rates from population changes?

Denver is a city that has had a lot of growth and consequently a lot of population change in recent years. Sociologists studying population changes and crime rates could find a wealth of information in Denver statistics. Let x be a random variable representing percentage change in neighborhood population in the past few years, and let y be a random variable representing crime rate (crimes per 1000 population). A random sample of six Denver neighborhoods gave the following information (Source: *Neighborhood Facts*, Piton Foundation). To find out more about the Piton Foundation, visit the Brase/Brase statistics site at **http://www.cengagebrain.com/** and find the link to the Piton Foundation.

x	29	2	11	17	7	6
y	173	35	132	127	69	53

Using information presented in this chapter, you will be able to analyze the relationship between the variables x and y using the following tools:

- Scatter diagram
- Sample correlation coefficient and coefficient of determination
- Least-squares line equation
- Predictions for y using the least-squares line

(See Problem 10 in the Chapter Review Problems.)

SECTION 4.1   Scatter Diagrams and Linear Correlation

FOCUS POINTS
- Make a scatter diagram.
- Visually estimate the location of the "best-fitting" line for a scatter diagram.
- Use sample data to compute the sample correlation coefficient r.
- Investigate the meaning of the correlation coefficient r.

Paired data values

Studies of correlation and regression of two variables usually begin with a graph of *paired data values* (x, y). We call such a graph a *scatter diagram*.

Scatter diagram

Explanatory variable
Response variable

> A **scatter diagram** is a graph in which data pairs (x, y) are plotted as individual points on a grid with horizontal axis x and vertical axis y. We call x the **explanatory variable** and y the **response variable**.

By looking at a scatter diagram of data pairs, you can observe whether there seems to be a linear relationship between the x and y values.

EXAMPLE 1   SCATTER DIAGRAM

Phosphorus is a chemical used in many household and industrial cleaning compounds. Unfortunately, phosphorus tends to find its way into surface water, where it can kill fish, plants, and other wetland creatures. Phosphorus-reduction programs are required by law and are monitored by the Environmental Protection Agency (EPA) (Reference: *EPA report 832-R-93-005*).

A random sample of eight sites in a California wetlands study gave the following information about phosphorus reduction in drainage water. In this study, x is a random variable that represents phosphorus concentration (in 100 mg/L) at the inlet of a passive biotreatment facility and y is a random variable that represents total phosphorus concentration (in 100 mg/L) at the outlet of the passive biotreatment facility.

x	5.2	7.3	6.7	5.9	6.1	8.3	5.5	7.0
y	3.3	5.9	4.8	4.5	4.0	7.1	3.6	6.1

(a) Make a scatter diagram for these data.

SOLUTION: Figure 4-1 shows points corresponding to the given data pairs. These plotted points constitute the scatter diagram. To make the diagram, first

FIGURE 4-1

Phosphorus Concentration (100 mg/L)

scan the data and decide on an appropriate scale for each axis. Figure 4-1 shows the scatter diagram (points) along with a line segment showing the basic trend. Notice a "jump scale" on both axes.

(b) *Interpretation* Comment on the relationship between x and y shown in Figure 4-1.

SOLUTION: By inspecting the figure, we see that smaller values of x are associated with smaller values of y and larger values of x tend to be associated with larger values of y. Roughly speaking, the general trend seems to be reasonably well represented by an upward-sloping line segment, as shown in the diagram.

Of course, it is possible to draw many curves close to the points in Figure 4-1, but a straight line is the simplest and most widely used in elementary studies of paired data. We can draw many lines in Figure 4-1, but in some sense, the "best" line should be the one that comes closest to each of the points of the scatter diagram. To single out one line as the "best-fitting line," we must find a mathematical criterion for this line and a formula representing the line. This will be done in Section 4.2 using the *method of least squares*.

Introduction to linear correlation

Another problem precedes that of finding the "best-fitting line." That is the problem of determining how well the points of the scatter diagram are suited for fitting *any* line. Certainly, if the points are a very poor fit to any line, there is little use in trying to find the "best" line.

No linear correlation

If the points of a scatter diagram are located so that *no* line is realistically a "good" fit, we then say that the points possess *no linear correlation*. We see some examples of scatter diagrams for which there is no linear correlation in Figure 4-2.

FIGURE 4-2

Scatter Diagrams with No Linear Correlation

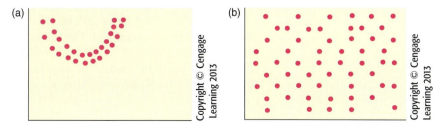

(a)

(b)

GUIDED EXERCISE 1 | *Scatter diagram*

A large industrial plant has seven divisions that do the same type of work. A safety inspector visits each division of 20 workers quarterly. The number x of work hours devoted to safety training and the number y of work hours lost due to industry-related accidents are recorded for each separate division in Table 4-1.

TABLE 4-1   **Safety Report**

Division	x	y
1	10.0	80
2	19.5	65
3	30.0	68
4	45.0	55
5	50.0	35
6	65.0	10
7	80.0	12

Copyright © Cengage Learning 2013

(a) Make a scatter diagram for these pairs. Place the x values on the horizontal axis and the y values on the vertical axis.

FIGURE 4-3   Scatter Diagram for Safety Report

(b) As the number of hours spent on safety training increases, what happens to the number of hours lost due to industry-related accidents?

In general, as the number of hours in safety training goes up, the number of hours lost due to accidents goes down.

(c) *Interpretation* Does a line fit the data reasonably well?

A line fits reasonably well.

(d) Draw a line that you think "fits best."

Use a downward-sloping line that lies close to the points. Later, you will find the equation of the line that is a "best fit."

If the points seem close to a straight line, we say the linear correlation is moderate to high, depending on how close the points lie to the line. If all the points do, in fact, lie on a line, then we have *perfect linear correlation*. In Figure 4-4, we see some diagrams with perfect linear correlation. In statistical applications, perfect linear correlation almost never occurs.

Perfect linear correlation

The variables x and y are said to have *positive correlation* if low values of x are associated with low values of y and high values of x are associated with high values of y. Figure 4-4 parts (a) and (c) show scatter diagrams in which the variables are positively correlated. On the other hand, if low values of x are associated with high values of y and high values of x are associated with low values of y, the variables are said to be *negatively correlated*. Figure 4-4 parts (b) and (d) show variables that are negatively correlated.

Positive correlation

Negative correlation

FIGURE 4-4

Scatter Diagrams with Moderate and Perfect Linear Correlation

Scatter diagrams for the same data look different from one another when they are graphed using different scales. Problem 19 at the end of this section explores how changing scales affects the look of a scatter diagram.

GUIDED EXERCISE 2 | *Scatter diagram and linear correlation*

Examine the scatter diagrams in Figure 4-5 and then answer the following questions.

FIGURE 4-5 Scatter Diagrams

(a) Which diagram has no linear correlation?

⇨ Figure 4-5(c) has no linear correlation. No straight-line fit should be attempted.

(b) Which has perfect linear correlation?

⇨ Figure 4-5(a) has perfect linear correlation and can be fitted exactly by a straight line.

(c) Which can be reasonably but not exactly fitted by a straight line?

⇨ Figure 4-5(b) can be reasonably fitted by a straight line.

TECH NOTES   The TI-84Plus/TI-83Plus/TI-*n*spire calculators, Excel 2010, and Minitab all produce scatter plots. For each technology, enter the *x* values in one column and the corresponding *y* values in another column. The displays following show the data from Guided Exercise 1 regarding safety training and hours lost because of accidents. Notice that the scatter plots do not necessarily show the origin.

TI-84Plus/TI-83Plus/TI-*nspire* (with TI-84Plus keypad) Enter the data into two columns. Use **Stat Plot** and choose the first type. Use option **9: ZoomStat** under **Zoom**. To check the scale, look at the settings displayed under **Window**.

Excel 2010 Enter the data into two columns. On the home screen, click the **Insert** tab. In the Chart Group, select **Scatter** and choose the first type. In the next ribbon, the Chart Layout Group offers options for including titles and axis labels. Right-clicking on data points provides other options, such as data labels. Changing the size of the diagram box changes the scale on the axes.

Minitab Enter the data into two columns. Use the menu selections **Stat ➤ Regression ➤ Fitted Line Plot**. The best-fit line is automatically plotted on the scatter diagram.

TI-84Plus/TI-83Plus/TI-*nspire* Display

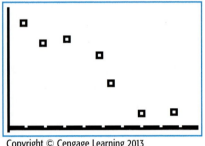

Copyright © Cengage Learning 2013

Excel Display

Copyright © Cengage Learning 2013

Minitab Display

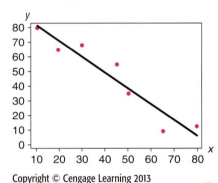

Copyright © Cengage Learning 2013

Sample Correlation Coefficient *r*

Looking at a scatter diagram to see whether a line best describes the relationship between the values of data pairs is useful. In fact, whenever you are looking for a relationship between two variables, making a scatter diagram is a good first step.

There is a mathematical measurement that describes the strength of the linear association between two variables. This measure is the *sample correlation coefficient r*. The full name for *r* is the *Pearson product-moment correlation coefficient*, named in honor of the English statistician Karl Pearson (1857–1936), who is credited with formulating *r*.

Sample correlation coefficient *r*

The **sample correlation coefficient *r*** is a numerical measurement that assesses the strength of a *linear* relationship between two variables *x* and *y*.

1. *r* is a unitless measurement between -1 and 1. In symbols, $-1 \leq r \leq 1$. If $r = 1$, there is perfect positive linear correlation. If $r = -1$, there is perfect negative linear correlation. If $r = 0$, there is no linear correlation. The closer *r* is to 1 or -1, the better a line describes the relationship between the two variables *x* and *y*.

2. Positive values of *r* imply that as *x* increases, *y* tends to increase. Negative values of *r* imply that as *x* increases, *y* tends to decrease.

3. The value of *r* is the same regardless of which variable is the explanatory variable and which is the response variable. In other words, the value of *r* is the same for the pairs (x, y) and the corresponding pairs (y, x).

4. The value of *r* does not change when either variable is converted to different units.

We'll develop the defining formula for r and then give a more convenient computation formula.

Development of Formula for *r*

If there is a *positive* linear correlation between variables x and y, then high values of x are paired with high values of y, and low values of x are paired with low values of y, Figure 4-6(a). In the case of *negative* linear correlation, high values of x are paired with low values of y, and low values of x are paired with high values of y. This relation is pictured in Figure 4-6(b). If there is *little or no linear correlation* between x and y, however, then we will find both high and low x values sometimes paired with high y values and sometimes paired with low y values. This relation is shown in Figure 4-6(c).

These observations lead us to the development of the formula for the sample correlation coefficient r. Taking *high* to mean "above the mean," we can express the relationships pictured in Figure 4-6 by considering the product

$$(x - \bar{x})(y - \bar{y})$$

If both x and y are high, both factors will be positive, and the product will be positive as well. The sign of this product will depend on the relative values of x and y compared with their respective means.

$$(x - \bar{x})(y - \bar{y}) \begin{cases} \text{is positive if } x \text{ and } y \text{ are both "high"} \\ \text{is positive if } x \text{ and } y \text{ are both "low"} \\ \\ \text{is negative if } x \text{ is "low," but } y \text{ is "high"} \\ \text{is negative if } x \text{ is "high," but } y \text{ is "low"} \end{cases}$$

In the case of positive linear correlation, most of the products $(x - \bar{x})(y - \bar{y})$ will be positive, and so will the sum over all the data pairs

$$\Sigma(x - \bar{x})(y - \bar{y})$$

For negative linear correlation, the products will tend to be negative, so the sum also will be negative. On the other hand, in the case of little or no linear correlation, the sum will tend to 0.

One trouble with the preceding sum is that it increases or decreases depending on the units of x and y. Because we want r to be unitless, we standardize both x and y of a data pair by dividing each factor $(x - \bar{x})$ by the sample standard deviation s_x and each factor $(y - \bar{y})$ by s_y. Finally, we take an average of all the

FIGURE 4-6

Patterns for Linear Correlation

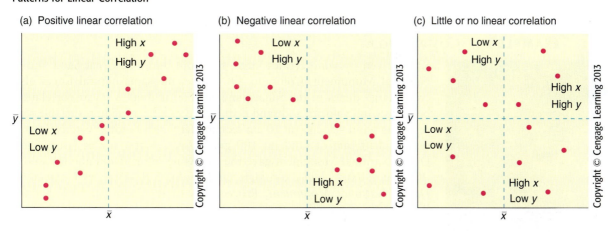

(a) Positive linear correlation   (b) Negative linear correlation   (c) Little or no linear correlation products. For technical reasons, we take the average by dividing by $n - 1$ instead of by n. This process leads us to the desired measurement, r.

$$r = \frac{1}{n-1} \Sigma \left[\frac{(x - \bar{x})}{s_x} \cdot \frac{(y - \bar{y})}{s_y} \right] \tag{1}$$

Computation Formula for *r*

The defining formula for r shows how the mean and standard deviation of each variable in the data pair enter into the formulation of r. However, the defining formula is technically difficult to work with because of all the subtractions and products. A computation formula for r uses the raw data values of x and y directly.

PROCEDURE

HOW TO COMPUTE THE SAMPLE CORRELATION COEFFICIENT *r*

Requirements
Obtain a random sample of n data pairs (x, y).

Procedure

1. Using the data pairs, compute Σx, Σy, Σx^2, Σy^2, and Σxy.

2. With n = sample size, Σx, Σy, Σx^2, Σy^2, and Σxy, you are ready to compute the sample correlation coefficient r using the computation formula

$$r = \frac{n\Sigma xy - (\Sigma x)(\Sigma y)}{\sqrt{n\Sigma x^2 - (\Sigma x)^2}\sqrt{n\Sigma y^2 - (\Sigma y)^2}} \tag{2}$$

Be careful! The notation Σx^2 means first square x and then calculate the sum, whereas $(\Sigma x)^2$ means first sum the x values and then square the result.

Note: For inferences (Section 11.4), the data pairs should have a *bivariate normal variation*. That is, for a fixed value of x, the y values should have a normal distribution (or at least a mound-shaped and symmetrical distribution) and vice versa for a fixed y value. Chapter 7 discusses normal distributions.

When *x* and *y* values of the data pairs are exchanged, the sample correlation coefficient *r* remains the same. Problem 20 explores this result.

Interpretation It can be shown mathematically that r is always a number between $+1$ and -1 ($-1 \leq r \leq +1$). Table 4-2 gives a quick summary of some basic facts about r.

For most applications, you will use a calculator or computer software to compute r directly. However, to build some familiarity with the structure of the sample correlation coefficient, it is useful to do some calculations for yourself. Example 2 and Guided Exercise 3 show how to use the computation formula to compute r.

EXAMPLE 2   COMPUTING *r*

David Muench/Terra/Corbis

Sand driven by wind creates large, beautiful dunes at Great Sand Dunes National Monument, Colorado. Of course, the same natural forces also create large dunes in the Sahara and Arabia. Is there a linear correlation between wind velocity and sand drift rate? Let x be a random variable representing wind velocity (in 10 cm/sec) and let y be a random variable representing drift rate of sand (in 100 gm/cm/sec). A test site at Great Sand Dunes National Monument gave the following information about x and y (Reference: *Hydrologic, Geologic, and Biologic Research at Great Sand Dunes National Monument*, Proceedings of the National Park Service Research Symposium):

| TABLE 4-2 | Some Facts about the Correlation Coefficient |

If *r* Is	Then	The Scatter Diagram Might Look Something Like
0	There is no linear correlation among the points of the scatter diagram.	Copyright © Cengage Learning 2013
1 or −1	There is a perfect linear correlation between *x* and *y* values; all points lie on the best-fit line.	*r* = −1 Copyright © Cengage Learning 2013 *r* = 1 Copyright © Cengage Learning 2013
Between 0 and 1 (0 < *r* < 1)	The *x* and *y* values have a *positive correlation*. By this we mean that *large x* values are associated with *large y* values, and *small x* values are associated with *small y* values.	Copyright © Cengage Learning 2013 As we go from left to right, the best-fit line goes *up*.
Between −1 and 0 (−1 < *r* < 0)	The *x* and *y* values have a *negative correlation*. By this we mean that *large x* values are associated with *small y* values, and *small x* values are associated with *large y* values.	Copyright © Cengage Learning 2013 As we go from left to right, the best-fit line goes *down*.

Copyright © Cengage Learning 2013

x	70	115	105	82	93	125	88
y	3	45	21	7	16	62	12

Copyright © Cengage Learning 2013

(a) Construct a scatter diagram. Do you expect *r* to be positive?

SOLUTION: Figure 4-7 displays the scatter diagram. From the scatter diagram, it appears that as *x* values increase, *y* values also tend to increase. Therefore, *r* should be positive.

FIGURE 4-7

Wind Velocity (10 cm/sec) and Drift Rate of Sand (100 gm/cm/sec)

Copyright © Cengage Learning 2013

TABLE 4-3	Computation Table			
x	y	x^2	y^2	xy
70	3	4900	9	210
115	45	13,225	2025	5175
105	21	11,025	441	2205
82	7	6724	49	574
93	16	8649	256	1488
125	62	15,625	3844	7750
88	12	7744	144	1056
$\Sigma x = 678$	$\Sigma y = 166$	$\Sigma x^2 = 67,892$	$\Sigma y^2 = 6768$	$\Sigma xy = 18,458$

Copyright © Cengage Learning 2013

(b) Compute r using the computation formula [formula (2)].

SOLUTION: To find r, we need to compute Σx, Σx^2, Σy, Σy^2, and Σxy. It is convenient to organize the data in a table of five columns (Table 4-3) and then sum the entries in each column. Of course, many calculators give these sums directly. Using the computation formula for r, the sums from Table 4-3, and $n = 7$, we have

$$r = \frac{n\Sigma xy - (\Sigma x)(\Sigma y)}{\sqrt{n\Sigma x^2 - (\Sigma x)^2} \sqrt{n\Sigma y^2 - (\Sigma y)^2}} \qquad (2)$$

$$= \frac{7(18,458) - (678)(166)}{\sqrt{7(67,892) - (678)^2} \sqrt{7(6768) - (166)^2}} \approx \frac{16,658}{(124.74)(140.78)} \approx 0.949$$

Note: Using a calculator to compute r directly gives 0.949 to three places after the decimal.

(c) *Interpretation* What does the value of r tell you?

SOLUTION: Since r is very close to 1, we have an indication of a strong positive linear correlation between wind velocity and drift rate of sand. In other words, we expect that higher wind speeds tend to mean greater drift rates. Because r is so close to 1, the association between the variables appears to be linear.

Because it is quite a task to compute r for even seven data pairs, the use of columns as in Table 4-3 is extremely helpful. Your value for r should always be between -1 and 1, inclusive. Use a scatter diagram to get a rough idea of the value of r. If your computed value of r is outside the allowable range, or if it disagrees quite a bit with the scatter diagram, recheck your calculations. Be sure you distinguish between expressions such as (Σx^2) and $(\Sigma x)^2$. Negligible rounding errors may occur, depending on how you (or your calculator) round.

GUIDED EXERCISE 3 | *Computing r*

In one of the Boston city parks, there has been a problem with muggings in the summer months. A police cadet took a random sample of 10 days (out of the 90-day summer) and compiled the following data. For each day, x represents the number of police officers on duty in the park and y represents the number of reported muggings on that day.

x	10	15	16	1	4	6	18	12	14	7
y	5	2	1	9	7	8	1	5	3	6

Copyright © Cengage Learning 2013

Continued

GUIDED EXERCISE 3   *continued*

(a) Construct a scatter diagram of x and y values.

 Figure 4-8 shows the scatter diagram.

FIGURE 4-8 Scatter Diagram for Number of Police Officers versus Number of Muggings

(b) From the scatter diagram, do you think the computed value of r will be positive, negative, or 0? Explain.

r will be negative. The general trend is that large x values are associated with small y values, and vice versa. From left to right, the best-fit line goes down.

(c) Complete **TABLE 4-4**.

x	y	x^2	y^2	xy
10	5	100	25	50
15	2	225	4	30
16	1	256	1	16
1	9	1	81	9
4	7	16	49	28
6	8	___	___	___
18	1	___	___	___
12	5	___	___	___
14	3	___	___	___
7	6	49	36	42
$\Sigma x = 103$	$\Sigma y = 47$	$\Sigma x^2 =$ ___	$\Sigma y^2 =$ ___	$\Sigma xy =$ ___
$(\Sigma x)^2 =$ ___		$(\Sigma y)^2 =$ ___		

Copyright © Cengage Learning 2013

TABLE 4-5   Completion of Table 4-4

x	y	x^2	y^2	xy
6	8	36	64	48
18	1	324	1	18
12	5	144	25	60
14	3	196	9	42
		$\Sigma x^2 = 1347$	$\Sigma y^2 = 295$	$\Sigma xy = 343$
$(\Sigma x)^2 = 10{,}609$		$(\Sigma y)^2 = 2209$		

Copyright © Cengage Learning 2013

(d) Compute r. Alternatively, find the value of r directly by using a calculator or computer software.

$$r = \frac{n\Sigma xy - (\Sigma x)(\Sigma y)}{\sqrt{n\Sigma x^2 - (\Sigma x)^2}\sqrt{n\Sigma y^2 - (\Sigma y)^2}}$$

$$= \frac{10(343) - (103)(47)}{\sqrt{10(1347) - (103)^2}\sqrt{10(295) - (47)^2}}$$

$$\approx \frac{-1411}{(53.49)(27.22)} \approx -0.969$$

(e) *Interpretation* What does the value of r tell you about the relationship between the number of police officers and the number of muggings in the park?

There is a strong negative linear relationship between the number of police officers and the number of muggings. It seems that the more officers there are in the park, the lower the number of muggings.

LOOKING FORWARD

If the scatter diagram and the value of the sample correlation coefficient r indicate a linear relationship between the data pairs, how do we find a suitable linear equation for the data? This process, called linear regression, is presented in Section 4.2.

TECH NOTES Most calculators that support two-variable statistics provide the value of the sample correlation coefficient r directly. Statistical software provides r, r^2, or both.

TI-84Plus/TI-83Plus/TI-*n*spire (with TI-84Plus keypad) First use **CATALOG**, find **DiagnosticOn**, and press **Enter** twice. Then, when you use **STAT**, **CALC**, option **8:LinReg(a + bx)**, the value of r will be given (data from Example 2). In Section 4.2 we will discuss the line $y = a + bx$ and the meaning of r^2.

Excel 2010 Excel gives the value of the sample correlation coefficient r in several outputs. One way to find the value of r is to click **Insert function** (f_x). Then in the dialogue box, select **Statistical** for the category and **Correl** for the function.

Minitab Use the menu selection **Stat ➤ Basic Statistics ➤ Correlation**.

```
LinReg
  y=a+bx
  a=-79.97763496
  b=1.070565553
  r²=.8997719968
  r=.9485631222
■
```

Copyright © Cengage Learning 2013

LOOKING FORWARD

When the data are ranks (without ties) instead of measurements, the Pearson product-moment correlation coefficient can be reduced to a simpler equation called the Spearman rank correlation coefficient. This coefficient is used with nonparametric methods and is discussed in *Understandable Statistics*, 10th edition.

CRITICAL THINKING

Sample correlation compared to population correlation

Cautions about Correlation

The correlation coefficient can be thought of as a measure of how well a linear model fits the data points on a scatter diagram. The closer r is to 1 or -1, the better a line "fits" the data. Values of r close to 0 indicate a poor fit to any line.

Usually a scatter diagram does not contain *all* possible data points that could be gathered. Most scatter diagrams represent only a *random sample* of data pairs taken from a very large population of all possible pairs. Because r is computed on the basis of a random sample of (x, y) pairs, we expect the values of r to vary from one sample to the next (much as the sample mean $\bar{x}$ varies from sample to sample). This brings up the question of the *significance* of r. Or, put another way, what are the chances that our random sample of data pairs indicates a high correlation when, in fact, the population's x and y values are not so strongly correlated? Right now, let's just say that the significance of r is a separate issue that will be treated in Section 11.4, where we test the *population correlation coefficient* ρ (Greek letter *rho*, pronounced "row").

r = **sample correlation coefficient** computed from a random sample of (x, y) data pairs.

ρ = **population correlation coefficient** computed from all population data pairs (x, y).

Population correlation coefficient, ρ

Problem 21 demonstrates an informal process for determining whether or not r is significant. Problem 22 explores the effect of sample size on the significance of r.

There is a less formal way to address the significance of r using a table of "critical values" or "cutoff values" based on the r distribution and the number of data pairs. Problem 21 at the end of this section discusses this method.

The value of the sample correlation coefficient r and the strength of the linear relationship between variables is computed based on the sample data. The situation may change for measurements larger than or smaller than the data values included in the sample. For instance, for infants there may be a high positive correlation between age in months and weight. However, that correlation might not apply for people ages 20 to 30 years.

Causation

The correlation coefficient is a mathematical tool for measuring the strength of a linear relationship between two variables. As such, it makes no implication about cause or effect. The fact that two variables tend to increase or decrease together does not mean that a change in one is *causing* a change in the other. A strong correlation between x and y is sometimes due to other (either known or unknown) variables. Such variables are called *lurking*

Lurking variables

variables.

In ordered pairs (x, y), x is called the **explanatory variable** and y is called the **response variable**. When r indicates a linear correlation between x and y, changes in values of y tend to respond to changes in values of x according to a linear model. A **lurking variable** is a variable that is neither an explanatory nor a response variable but may be responsible for changes in both x and y.

EXAMPLE 3   CAUSATION AND LURKING VARIABLES

Over a period of years, the population of a certain town increased. It was observed that during this period the correlation between x, the number of people attending church, and y, the number of people in the city jail, was $r = 0.90$. Does going to church *cause* people to go to jail? Is there a *lurking variable* that might cause both variables x and y to increase?

SOLUTION: We hope church attendance does not cause people to go to jail! During this period, there was an increase in population. Therefore, it is not too surprising that both the number of people attending church and the number of people in jail increased. The high correlation between x and y is likely due to the lurking variable of population increase.

Correlation between averages

The correlation between two variables consisting of averages is usually higher than the correlation between two variables representing corresponding raw data. One reason is that the use of averages reduces the variation that exists between individual measurements (see Section 7.5 and the central limit theorem). A high correlation based on two variables consisting of averages does not necessarily imply a high correlation between two variables consisting of individual measurements. See Problem 23 at the end of this section.

Problem 23 at the end of this section explores the correlation of averages.

VIEWPOINT Low on Credit, High on Cost!!!

How do you measure automobile insurance risk? One way is to use a little statistics and customer credit ratings. Insurers say statistics show that drivers who have a history of bad credit are more likely to be in serious car accidents. According to a high-level executive at Allstate Insurance Company, financial instability is an extremely powerful predictor of future insurance losses. In short, there seems to be a strong correlation between bad credit ratings and auto insurance claims. Consequently, insurance companies want to charge higher premiums to customers with bad credit ratings. Consumer advocates object strongly because they say bad credit does not cause automobile accidents, and more than 20 states prohibit or restrict the use of credit ratings to determine auto insurance premiums. Insurance companies respond by saying that your best defense is to pay your bills on time!

SECTION 4.1 PROBLEMS

aplia   Access end-of-section exercises at www.aplia.com

PROCEDURE

HOW TO USE TABLE 4-6 TO TEST ρ

1. First compute r from a random sample of n data pairs (x, y).
2. Find the table entry in the row headed by n and the column headed by your choice of α. Your choice of α is the risk you are willing to take of mistakenly concluding that $\rho \neq 0$ when, in fact, $\rho = 0$.
3. Compare $|r|$ to the table entry.

   (a) If $|r| \geq$ table entry, then there is sufficient evidence to conclude that $\rho \neq 0$, and we say that r is **significant**. In other words, we conclude that there is some population correlation between the two variables x and y.

   (b) If $|r| <$ table entry, then the evidence is insufficient to conclude that $\rho \neq 0$, and we say that r is **not significant**. We do not have enough evidence to conclude that there is any correlation between the two variables x and y.

TABLE 4-6   **Critical Values for Correlation Coefficient r**

n	$\alpha = 0.05$	$\alpha = 0.01$	n	$\alpha = 0.05$	$\alpha = 0.01$	n	$\alpha = 0.05$	$\alpha = 0.01$
3	1.00	1.00	13	0.53	0.68	23	0.41	0.53
4	0.95	0.99	14	0.53	0.66	24	0.40	0.52
5	0.88	0.96	15	0.51	0.64	25	0.40	0.51
6	0.81	0.92	16	0.50	0.61	26	0.39	0.50
7	0.75	0.87	17	0.48	0.61	27	0.38	0.49
8	0.71	0.83	18	0.47	0.59	28	0.37	0.48
9	0.67	0.80	19	0.46	0.58	29	0.37	0.47
10	0.63	0.76	20	0.44	0.56	30	0.36	0.46
11	0.60	0.73	21	0.43	0.55			
12	0.58	0.71	22	0.42	0.54			

Linear Regression and the Coefficient of Determination

FOCUS POINTS

- State the least-squares criterion.
- Use sample data to find the equation of the least-squares line. Graph the least-squares line.
- Use the least-squares line to predict a value of the response variable *y* for a specified value of the explanatory variable *x*.
- Explain the difference between interpolation and extrapolation.
- Explain why extrapolation beyond the sample data range might give results that are misleading or meaningless.
- Use r^2 to determine *explained* and *unexplained* variation of the response variable *y*.

In Denali National Park, Alaska, the wolf population is dependent on a large, strong caribou population. In this wild setting, caribou are found in very large herds. The well-being of an entire caribou herd is not threatened by wolves. In fact, it is thought that wolves keep caribou herds strong by helping prevent over-population. Can the caribou population be used to predict the size of the wolf population?

Let *x* be a random variable that represents the fall caribou population (in hundreds) in Denali National Park, and let *y* be a random variable that represents the late-winter wolf population in the park. A random sample of recent years gave the following information (Reference: U.S. Department of the Interior, National Biological Service):

x	30	34	27	25	17	23	20
y	66	79	70	60	48	55	60

Looking at the scatter diagram in Figure 4-9, we can ask some questions:

1. Do the data indicate a linear relationship between *x* and *y*?

2. Can you find an equation for the best-fitting line relating *x* and *y*? Can you use this relationship to predict the size of the wolf population when you know the size of the caribou population?

3. What fractional part of the variability in *y* can be associated with the variability in *x*? What fractional part of the variability in *y* is not associated with a corresponding variability in *x*?

FIGURE 4-9

Caribou and Wolf Populations

FIGURE 4-10

Least-Squares Criterion

Σd^2 is as small as possible   Caribou (100's) *x*

The first step in answering these questions is to try to express the relationship as a mathematical equation. There are many possible equations, but the simplest and most widely used is the linear equation, or the equation of a straight line. Because we will be using this line to predict the y values from the x values, we call x the *explanatory variable* and y the *response variable*.

Our job is to find the linear equation that "best" represents the points of the scatter diagram. For our criterion of best-fitting line, we use the *least-squares criterion*, which states that the line we fit to the data points must be such that *the sum of the squares of the vertical distances from the points to the line be made as small as possible*. The least-squares criterion is illustrated in Figure 4-10.

Least-squares criterion

The sum of the squares of the vertical distances from the data points (x, y) to the line is made as small as possible.

In Figure 4-10, d represents the difference between the y coordinate of the data point and the corresponding y coordinate on the line. Thus, if the data point lies above the line, d is positive, but if the data point is below the line, d is negative. As a result, the sum of the d values can be small even if the points are widely spread in the scatter diagram. However, the squares d^2 cannot be negative. By minimizing the sum of the squares, we are, in effect, not allowing positive and negative d values to "cancel out" one another in the sum. It is in this way that we can meet the least-squares criterion of minimizing the sum of the squares of the vertical distances between the points and the line over *all* points in the scatter diagram.

We use the notation $\hat{y} = a + bx$ for the *least-squares line*. A little algebra tells us that b is the slope and a is the y intercept of the line. In this context, $\hat{y}$ (read "y hat") represents the value of the response variable y estimated using the least-squares line and a given value of the explanatory variable x.

Techniques of calculus can be applied to show that a and b may be computed using the following procedure.

(margin notes)
Explanatory variable
Response variable
Least-squares criterion

Least-squares line

PROCEDURE

Problem 21 demonstrates that for the same data set, the least-squares line for predicting y is essentially different from the least-squares line for predicting x.

Slope, b

y intercept, a

HOW TO FIND THE EQUATION OF THE LEAST-SQUARES LINE
$\hat{y} = a + bx$

Requirements

Obtain a random sample of n data pairs (x, y), where x is the *explanatory variable* and y is the *response variable*.

Procedure

1. Using the data pairs, compute Σx, Σy, Σx^2, Σy^2, and Σxy. Then compute the sample means $\bar{x}$ and $\bar{y}$.

2. With n = sample size, Σx, Σy, Σx^2, Σy^2, Σxy, $\bar{x}$, and $\bar{y}$, you are ready to compute the slope b and y intercept a using the computation formulas

$$\text{Slope: } b = \frac{n\Sigma xy - (\Sigma x)(\Sigma y)}{n\Sigma x^2 - (\Sigma x)^2} \qquad (3)$$

$$\text{y intercept: } a = \bar{y} - b\bar{x} \qquad (4)$$

Be careful! The notation Σx^2 means first square x and then calculate the sum, whereas $(\Sigma x)^2$ means first sum the x values and then square the result.

3. The equation of the least-squares line computed from your sample data is

$$\hat{y} = a + bx \tag{5}$$

Note: For inferences (Section 11.4), the data pairs should have a *bivarate normal variation*. That is, for a fixed value of x, the y values should have a normal distribution (or at least a mound-shaped and symmetrical distribution) and vice versa for a fixed y value. Chapter 7 discusses normal distributions.

COMMENT The computation formulas for the slope of the least-squares line, the sample correlation coefficient r, and the standard deviations s_x and s_y use many of the same sums. There is, in fact, a relationship between the sample correlation coefficient r and the slope of the least-squares line b. In instances where we know r, s_x, and s_y, we can use the following formula to compute b:

$$b = r\left(\frac{s_y}{s_x}\right) \tag{6}$$

COMMENT In other mathematics courses, the slope-intercept form of the equation of a line is usually given as $y = mx + b$, where m refers to the slope of the line and b to the y coordinate of the y intercept. In statistics, when there is only one explanatory variable, it is common practice to use the letter b to designate the slope of the least-squares line and the letter a to designate the y coordinate of the intercept. For example, these are the symbols used on the TI-84Plus/TI-83Plus/TI-*n*spire calculators as well as on many other calculators.

Using the formulas to find the values of *a* and *b*

For most applications, you can use a calculator or computer software to compute a and b directly. However, to build some familiarity with the structure of the computation formulas, it is useful to do some calculations yourself. Example 4 shows how to use the computation formulas to find the values of a and b and the equation of the least-squares line $\hat{y} = a + bx$.

Note: If you are using your calculator to find the values of a and b directly, you may omit the discussion regarding use of the formulas. Go to the margin header "Using the values of a and b to construct the equation of the least-squares line."

EXAMPLE 4   LEAST-SQUARES LINE

Let's find the least-squares equation relating the variables x = size of caribou population (in hundreds) and y = size of wolf population in Denali National Park. Use x as the explanatory variable and y as the response variable.

(a) Use the computation formulas to find the slope b of the least-squares line and the y intercept a.

SOLUTION: Table 4-7 gives the data values x and y along with the values x^2, y^2, and xy. First compute the sample means.

$$\bar{x} = \frac{\Sigma x}{n} = \frac{176}{7} \approx 25.14 \text{ and } \bar{y} = \frac{\Sigma y}{n} = \frac{438}{7} \approx 62.57$$

Next compute the slope b.

$$b = \frac{n\Sigma xy - (\Sigma x)(\Sigma y)}{n\Sigma x^2 - (\Sigma x)^2} = \frac{7(11{,}337) - (176)(438)}{7(4628) - (176)^2} = \frac{2271}{1420} \approx 1.60$$

Joe McDonald/Encyclopedia/Corbis

TABLE 4-7	Sums for Computing b, $\bar{x}$, and $\bar{y}$			
x	y	x^2	y^2	xy
30	66	900	4356	1980
34	79	1156	6241	2686
27	70	729	4900	1890
25	60	625	3600	1500
17	48	289	2304	816
23	55	529	3025	1265
20	60	400	3600	1200
$\Sigma x = 176$	$\Sigma y = 438$	$\Sigma x^2 = 4628$	$\Sigma y^2 = 28{,}026$	$\Sigma xy = 11{,}337$

Copyright © Cengage Learning 2013

Use the values of b, $\bar{x}$, and $\bar{y}$ to compute the y intercept a.

$$a = \bar{y} - b\bar{x} \approx 62.57 - 1.60(25.14) \approx 22.35$$

Note that calculators give the values $b \approx 1.599$ and $a \approx 22.36$. These values differ slightly from those you computed using the formulas because of rounding.

Using the values of *a* and *b* to construct the equation of the least-squares line

(b) Use the values of a and b (either computed or obtained from a calculator) to find the equation of the least-squares line.

SOLUTION:

$$\hat{y} = a + bx$$
$$\hat{y} \approx 22.35 + 1.60x$$

Graphing the least-squares line

(c) Graph the equation of the least-squares line on a scatter diagram.

SOLUTION: To graph the least-squares line, we have several options available. The slope-intercept method of algebra is probably the quickest, but may not always be convenient if the intercept is not within the range of the sample data values. It is just as easy to select two x values in the range of the x data values and then use the least-squares line to compute two corresponding $\hat{y}$ values.

In fact, we already have the coordinates of one point on the least-squares line. By the formula for the intercept [formula (4)], the point $(\bar{x}, \bar{y})$ is always on the least-squares line. For our example, $(\bar{x}, \bar{y}) = (25.14, 62.57)$.

> The point $(\bar{x}, \bar{y})$ is always on the least-squares line.

Another x value within the data range is $x = 34$. Using the least-squares line to compute the corresponding $\hat{y}$ value gives

$$\hat{y} \approx 22.35 + 1.60(34) \approx 76.75$$

We place the two points $(25.14, 62.57)$ and $(34, 76.75)$ on the scatter diagram (using a different symbol than that used for the sample data points) and connect the points with a line segment (Figure 4-11).

Meaning of slope

In the equation $\hat{y} = a + bx$, the slope b tells us how many units $\hat{y}$ changes for each unit change in x. In Example 4 regarding size of wolf and caribou populations,

$$\hat{y} \approx 22.35 + 1.60x$$

The slope 1.60 tells us that if the number of caribou (in hundreds) changes by 1 (hundred), then we expect the sustainable wolf population to change by 1.60.

FIGURE 4-11

Caribou and Wolf Populations

In other words, our model says that an increase of 100 caribou will increase the predicted wolf population by 1.60. If the caribou population decreases by 400, we predict that the sustainable wolf population will decrease by 6.4.

Marginal change

> The slope of the least-squares line tells us how many units the response variable is expected to change for each unit change in the explanatory variable. The number of units change in the response variable for each unit change in the explanatory variable is called the **marginal change** of the response variable.

Some points in the data set have a strong influence on the equation of the least-squares line.

Influential points

> A point is **influential** if removing it would substantially change the equation of the least-squares line or other calculations associated with linear regression. An influential point often has an *x* value near the extreme high or low value of the data set.

Figure 4-12 shows two scatter diagrams produced in Excel. Figure 4-12(a) has an influential point. Figure 4-12(b) shows the scatter diagram with the influential point removed. Notice that the equations for the least-squares lines are quite different.

FIGURE 4-12

(a) Influential Point Present   (b) Influential Point Removed

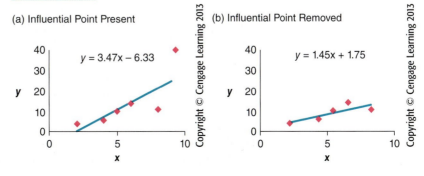

If a data set has an influential point, look at the influential point carefully to make sure it is not the result of a data collection error or a data-recording error. A valid influential point affects the equation of the least-squares line. The group project in Data Highlights at the end of this chapter further explores influential points.

CRITICAL THINKING

Predicting $\hat{y}$ for a specified x

Using the Least-Squares Line for Prediction

Making predictions is one of the main applications of linear regression. In other words, you use the equation of the least-squares line to predict the $\hat{y}$ value for a specified x value. The accuracy of the prediction depends on several components.

How well does the least-squares line fit the original data points?
The accuracy of the prediction depends on how well the least-squares line fits the original raw data points. Here are some tools to assess the fit of the line:

- Look at the scatter diagram, taking into account the scale of each axis.
- See if there are any influential points.
- Consider the value of the sample correlation coefficient r. The closer r is to 1 or -1, the better the least-squares line fits the data.
- Consider the value of the coefficient of determination r^2 (as discussed later in this section).
- Look at the residuals and a residual plot (see Problem 19 for a discussion of residual plots).

Residual

Problems 19 and 20 at the end of this section show how to make a residual plot.

> The **residual** is the difference between the y value in a specified data pair (x, y) and the value $\hat{y} = a + bx$ predicted by the least-squares line for the same x.
>
> $y - \hat{y}$ is the **residual**.

If the residuals seem random about 0, the least-squares line provides a reasonable model for the data. Later in this section you will see the residual used in the development of the *coefficient of determination*, another important measurement associated with linear regression.

Does the prediction involve interpolation or extrapolation?

Interpolation
Extrapolation

Another issue that affects the validity of predictions is whether you are *interpolating* or *extrapolating*.

> Predicting $\hat{y}$ values for x values that are **between** observed x values in the data set is called **interpolation**.

> Predicting $\hat{y}$ values for x values that are **beyond** observed x values in the data set is called **extrapolation**. Extrapolation may produce unrealistic forecasts.

The least-squares line may not reflect the relationship between x and y for values of x outside the data range. For example, there is a fairly high correlation between height and age for boys ages 1–10 years. In general, the older the boy, the taller the boy. A least-squares line based on such data would give good predictions of height for boys of ages between 1 and 10. However, it would be fairly meaningless to use the same linear regression line to predict the height of a 20-year-old or 50-year-old man.

The data are sample data.

Another consideration when working with predictions is the fact that the least-squares line is based on sample data. Each different sample will produce a slightly different equation for the least-squares line.

The least-squares line uses x as the explanatory variable and y as the response variable.

One more important fact about predictions: The least-squares line is developed with x as the explanatory variable and y as the response variable. This model can be used only to predict y values from specified x values. If you wish to begin with y values and predict corresponding x values, you must start all over and compute a new equation. Such an equation would be developed using a model with x as the response variable and y as the explanatory variable. (See Problem 21 at the end of this section.) Note that the equation for predicting x values *cannot* be derived from the least-squares line predicting y simply by solving the equation for x.

> The least-squares line developed with x as the explanatory variable and y as the response variable can be used only to predict y values from specified x values.

Example 5 shows how to use the least-squares line for predictions.

EXAMPLE 5 PREDICTIONS

We continue with Example 4 regarding size of the wolf population as it relates to size of the caribou population. Suppose you want to predict the size of the wolf population when the size of the caribou population is 21 (hundred).

Joe McDonald/Corbis

(a) In the least-squares model developed in Example 4, which is the explanatory variable and which is the response variable? Can you use the equation to predict the size of the wolf population for a specified size of caribou population?

SOLUTION: The least-squares line $\hat{y} \approx 22.35 + 1.60x$ was developed using x = size of caribou population (in hundreds) as the explanatory variable and y = size of wolf population as the response variable. We can use the equation to predict the y value for a specified x value.

(b) The sample data pairs have x values ranging from 17 (hundred) to 34 (hundred) for the size of the caribou population. To predict the size of the wolf population when the size of the caribou population is 21 (hundred), will you be interpolating or extrapolating?

SOLUTION: Interpolating, since 21 (hundred) falls within the range of sample x values.

(c) Predict the size of the wolf population when the caribou population is 21 (hundred).

SOLUTION: Using the least-squares line from Example 4 and the value 21 in place of x gives

$$\hat{y} \approx 22.35 + 1.60x \approx 22.35 + 1.60(21) \approx 55.95$$

Rounding up to a whole number gives a prediction of 56 for the size of the wolf population.

GUIDED EXERCISE 4 | *Least-squares line*

The Quick Sell car dealership has been using 1-minute spot ads on a local TV station. The ads always occur during the evening hours and advertise the different models and price ranges of cars on the lot that week. During a 10-week period, a Quick Sell dealer kept a weekly record of the number x of TV ads versus the number y of cars sold. The results are given in Table 4-8.

The manager decided that Quick Sell can afford only 12 ads per week. At that level of advertisement, how many cars can Quick Sell expect to sell each week? We'll answer this question in several steps.

TABLE 4-8

x	y
6	15
20	31
0	10
14	16
25	28
16	20
28	40
18	25
10	12
8	15

(a) Draw a scatter diagram for the data.

➡ The scatter diagram is shown in Figure 4-13. The plain red dots in Figure 4-13 are the points of the scatter diagram. Notice that the least-squares line is also shown, with two extra points used to position that line.

FIGURE 4-13   Scatter Diagram and Least-Squares Line for Table 4-8

(b) Look at formulas (3)–(5) pertaining to the least-squares line. Two of the quantities that we need to find b are Σx and Σxy. List the others.

➡ We also need n, Σy, Σx^2, and $(\Sigma x)^2$.

(c) Complete Table 4-9(a).

➡ The missing table entries are shown in Table 4-9(b).

TABLE 4-9(a)

x	y	x^2	xy
6	15	36	90
20	31	400	620
0	10	0	0
14	16	196	224
25	28	625	700
16	20	256	320
28	40	___	___
18	25	___	___
10	12	___	___
8	15	64	120
$\Sigma x = 145$	$\Sigma y = 212$	$\Sigma x^2 = $ ___	$\Sigma xy = $ ___

TABLE 4-9(b)

x^2	xy
$(28)^2 = 784$	$28(40) = 1120$
$(18)^2 = 324$	$18(25) = 450$
$(10)^2 = 100$	$10(12) = 120$
$\Sigma x^2 = 2785$	$\Sigma xy = 3764$

(d) Compute the sample means $\bar{x}$ and $\bar{y}$

➡️

$$\bar{x} = \frac{\Sigma x}{n} = \frac{145}{10} = 14.5$$

$$\bar{y} = \frac{\Sigma y}{n} = \frac{212}{10} = 21.2$$

(e) Compute a and b for the equation $\hat{y} = a + bx$ of the least-squares line. Alternatively, find the values of a and b directly by using a calculator or computer software.

➡️

$$b = \frac{n\Sigma xy - (\Sigma x)(\Sigma y)}{n\Sigma x^2 - (\Sigma x)^2}$$

$$= \frac{10(3764) - (145)(212)}{10(2785) - (145)^2} = \frac{6900}{6825} \approx 1.01$$

$$a = \bar{y} - b\bar{x}$$

$$\approx 21.2 - 1.01(14.5) \approx 6.56$$

(f) What is the equation of the least-squares line $\hat{y} = a + bx$?

➡️

Using the values of a and b computed in part (e) or obtained directly from a calculator,

$$\hat{y} \approx 6.56 + 1.01x$$

(g) Plot the least-squares line on your scatter diagram.

➡️

The least-squares line goes through the point $(\bar{x}, \bar{y}) = (14.5, 21.2)$. To get another point on the line, select a value for x and compute the corresponding $\hat{y}$ value using the equation $\hat{y} = 6.56 + 1.01x$. For $x = 20$, we get $\hat{y} = 6.56 + 1.01(20) = 26.8$, so the point $(20, 26.8)$ is also on the line. The least-squares line is shown in Figure 4-13.

(h) Read the $\hat{y}$ value for $x = 12$ from your graph. Then use the equation of the least-squares line to calculate $\hat{y}$ when $x = 12$. How many cars can the manager expect to sell if 12 ads per week are aired on TV?

➡️

The graph gives $\hat{y} \approx 19$. From the equation, we get

$$\hat{y} \approx 6.56 + 1.01x$$

$$\approx 6.56 + 1.01(12)$$

$$\approx 18.68$$

To the nearest whole number, the manager can expect to sell 19 cars when 12 ads are aired on TV each week.

(i) *Interpretation* How reliable do you think the prediction is? Explain. (Guided Exercise 5 will show that $r \approx 0.919$.)

➡️

The prediction should be fairly reliable. The prediction involves interpolation, and the scatter diagram shows that the data points are clustered around the least-squares line. From Guided Exercise 5 we have the information that r is high. Of course, other variables might affect the value of $\hat{y}$ for $x = 12$.

TECH NOTES   When we have more data pairs, it is convenient to use a technology tool such as the TI-84Plus/TI-83Plus/TI-*n*spire calculators, Excel 2010, or Minitab to find the equation of the least-squares line. The displays show results for the data of Guided Exercise 4 regarding car sales and ads.

TI-84Plus/TI-83Plus/TI-*n*spire (with TI-84Plus keypad) Press STAT, choose Calculate, and use option 8:LinReg (a + bx). For a graph showing the scatter plot and the least-squares line, press the STAT PLOT key, turn on a plot, and highlight the first type. Then press the Y = key. To enter the equation of the least-squares line, press VARS, select 5:Statistics, highlight EQ, and then highlight 1:RegEQ. Press ENTER. Finally, press ZOOM and choose 9:ZoomStat.

Excel 2010 There are several ways to find the equation of the least-squares line in Excel. One way is to make a scatter plot. On the home screen, click the **Insert** tab. In the Chart Group, select **Scatter** and choose the first type. In the next ribbon, the Chart Layout Group offers options for including titles and axis labels. Right-click on any data point and select **Add Trendline**. In the dialogue box, select **linear** and check **Display Equation on Chart**. To display the value of the coefficient of determination, check **Display R-squared Value on Chart**.

TI-84Plus/TI-83/TI-*n*spire Display

Excel Display

Minitab There are a number of ways to generate the least-squares line. One way is to use the menu selection **Stat ➤ Regression ➤ Fitted Line Plot**. The least-squares equation is shown with the diagram.

Coefficient of Determination

There is another way to answer the question "How good is the least-squares line as an instrument of regression?" The *coefficient of determination* r^2 is the square of the sample correlation coefficient r.

Coefficient of determination, r^2

> **Coefficient of determination r^2**
>
> 1. Compute the sample correlation coefficient r using the procedure from Section 4.1. Then simply compute r^2, the sample coefficient of determination.
>
> 2. *Interpretation* The value r^2 is the ratio of **explained variation** over total variation. That is, r^2 is the fractional amount of total variation in y that can be explained by using the linear model $\hat{y} = a + bx$.
>
> 3. *Interpretation* Furthermore, $1 - r^2$ is the fractional amount of total variation in y that is **unexplained variation** due to random chance or to the possibility of lurking variables that influence y.

Explained variation

Unexplained variation

In other words, the coefficient of determination r^2 is a measure of the proportion of variation in y that is explained by the regression line, using x as the explanatory variable. If $r = 0.90$, then $r^2 = 0.81$ is the coefficient of determination. We can say that about 81% of the (variation) behavior of the y variable can be explained by the corresponding (variation) behavior of the x variable if we use the equation of the least-squares line. The remaining 19% of the (variation) behavior of the y variable is due to random chance or to the possibility of lurking variables that influence y.

GUIDED EXERCISE 5 | *Coefficient of determination r^2*

In Guided Exercise 4, we looked at the relationship between x = number of 1-minute spot ads on TV advertising different models of cars and y = number of cars sold each week by the sponsoring car dealership.

(a) Using the sums found in Guided Exercise 4, compute the sample correlation coefficient r.
$n = 10$, $\Sigma x = 145$, $\Sigma y = 212$, $\Sigma x^2 = 2785$, and $\Sigma xy = 3764$. You also need $\Sigma y^2 = 5320$.

$$r = \frac{n\Sigma xy - (\Sigma x)(\Sigma y)}{\sqrt{n\Sigma x^2 - (\Sigma x)^2}\ \sqrt{n\Sigma y^2 - (\Sigma y)^2}}$$

$$r = \frac{10(3764) - (145)(212)}{\sqrt{10(2785) - (145)^2}\ \sqrt{10(5320) - (212)^2}}$$

$$\approx \frac{6900}{(82.61)(90.86)}$$

$$\approx 0.919$$

(b) Compute the coefficient of determination r^2.

$r^2 \approx 0.845$.

(c) *Interpretation* What percentage of the variation in the number of car sales can be explained by the variation in the number of ads and the least-squares line?

84.5%.

(d) *Interpretation* What percentage of the variation in the number of car sales is not explained by the ads and the least-squares line?

100% − 84.5%, or 15.5%.

VIEWPOINT | It's Freezing!

Can you use average temperatures in January to predict how bad the rest of the winter will be? Can you predict the number of days with freezing temperatures for the entire calendar year using conditions in January? How good would such a forecast be for predicting growing season or number of frost-free days? Methods of this section can help you answer such questions. For more information, visit the Brase/Brase statistics site at **http://www.cengagebrain.com** *and find the link to Temperatures.*

SECTION 4.2 PROBLEMS | aplia™   Access end-of-section exercises at www.aplia.com

Chapter Review

SUMMARY

This chapter discusses linear-regression models and inferences related to these models.

- A scatter diagram of data pairs (x, y) gives a graphical display of the relationship (if any) between x and y data. We are looking for a linear relationship.

- For data pairs (x, y), x is called the *explanatory variable* and is plotted along the horizontal axis. The *response variable* y is plotted along the vertical axis.

- The Pearson product-moment *correlation coefficient r* gives a numerical measurement assessing the strength of a linear relationship between x and y. It is based on a random sample of (x, y) data pairs.

- The value of r ranges from -1 to 1, with 1 indicating perfect positive linear correlation, -1 indicating perfect negative linear correlation, and 0 indicating no linear correlation.

- If the scatter diagram and sample correlation coefficient r indicate a linear relationship between x and y values of the data pairs, we use the least-squares criterion to develop the equation of the least-squares line

$$\hat{y} = a + bx$$

where $\hat{y}$ is the value of y predicted by the least-squares line for a given x value, a is the y intercept, and b is the slope.

- The *coefficient of determination r^2* is a value that measures the proportion of variation in y explained by the least-squares line, the linear regression model, and the variation in the explanatory variable x.

- The difference $y - \hat{y}$ between the y value in the data pair (x, y) and the corresponding predicted value $\hat{y}$ for the same x is called the *residual*.

IMPORTANT WORDS & SYMBOLS

VIEWPOINT   Living Arrangements

Male, female, married, single, living alone, living with friends or relatives—all these categories are of interest to the U.S. Census Bureau. In addition to these categories, there are others, such as age, income, and health needs. How strongly correlated are these variables? Can we use one or more of these variables to predict the others? How good is such a prediction? Methods of this chapter can help you answer such questions. For more information regarding such data, visit the Brase/Brase statistics site at **http://www.cengagebrain.com** *and find the link to the Census Bureau.*

CHAPTER REVIEW PROBLEMS

1. *Statistical Literacy* Suppose the scatter diagram of a random sample of data pairs (x, y) shows no linear relationship between x and y. Do you expect the value of the sample correlation coefficient r to be close to 1, -1, or 0?

2. *Critical Thinking* Suppose you and a friend each take different random samples of data pairs (x, y) from the same population. Assume the samples are the same size. Based on your samples, you compute $r = 0.83$. Based on her sample, your friend computes $r = 0.79$. Is your friend's value for r wrong? Explain.

3. *Statistical Literacy* When using the least-squares line for prediction, are results usually more reliable for extrapolation or interpolation?

4. *Statistical Literacy* Suppose that for $x = 3$, the predicted value is $\hat{y} = 6$. The data pair $(3, 8)$ is part of the sample data. What is the value of the residual for $x = 3$?

In Problems 5–10,
   (a) Draw a scatter diagram for the data.
   (b) Find $\bar{x}$, $\bar{y}$, b, and the equations of the least squares line. Plot the line on the scatter diagram of part (a).
   (c) Find the sample correlation coefficient r and the coefficient of determination r^2. What percentage of variation in y is explained by the least-squares model?

5. *Desert Ecology: Wildlife* Bighorn sheep are beautiful wild animals found throughout the western United States. Data for this problem are based on information taken from *The Desert Bighorn*, edited by Monson and Sumner (University of Arizona Press). Let x be the age of a bighorn sheep (in years), and let y be the mortality rate (percentage that die) for this age group. For example, $x = 1$, $y = 14$ means that 14% of the bighorn sheep between 1 and 2 years old die. A random sample of Arizona bighorn sheep gave the following information:

x	1	2	3	4	5
y	14	18.9	14.4	19.6	20.0

Complete parts (a) through (c), given $\Sigma x = 15$, $\Sigma y = 86.9$, $\Sigma x^2 = 55$, $\Sigma y^2 = 1544.73$, and $\Sigma xy = 273.4$.

6. *Sociology: Job Changes* A sociologist is interested in the relation between $x =$ number of job changes and $y =$ annual salary (in thousands of dollars) for people living in the Nashville area. A random sample of 10 people employed in Nashville provided the following information:

x (Number of job changes)	4	7	5	6	1	5	9	10	10	3
y (Salary in $1,000)	33	37	34	32	32	38	43	37	40	33

Complete parts (a) through (c), given $\Sigma x = 60$, $\Sigma y = 359$, $\Sigma x^2 = 442$, $\Sigma y^2 = 13{,}013$, and $\Sigma xy = 2231$.
   (d) If someone had $x = 2$ job changes, what does the least-squares line predict for y, the annual salary?

7. *Medicine: Fat Babies* Modern medical practice tells us not to encourage babies to become too fat. Is there a positive correlation between the weight x of a 1-year-old baby and the weight y of the mature adult (30 years old)? A random sample of medical files produced the following information for 14 females:

x (lb)	21	25	23	24	20	15	25	21	17	24	26	22	18	19
y (lb)	125	125	120	125	130	120	145	130	130	130	130	140	110	115

Complete parts (a) through (c), given $\Sigma x = 300$, $\Sigma y = 1775$, $\Sigma x^2 = 6572$, $\Sigma y^2 = 226{,}125$, and $\Sigma xy = 38{,}220$.

(d) If a female baby weighs 20 pounds at 1 year, what do you predict she will weigh at 30 years of age?

8. *Sales: Insurance* Dorothy Kelly sells life insurance for the Prudence Insurance Company. She sells insurance by making visits to her clients' homes. Dorothy believes that the number of sales should depend, to some degree, on the number of visits made. For the past several years, she has kept careful records of the number of visits x she makes each week and the number of people y who buy insurance that week. For a random sample of 15 such weeks, the x and y values are:

x	11	19	16	13	28	5	20	14	22	7	15	29	8	25	16
y	3	11	8	5	8	2	5	6	8	3	5	10	6	10	7

Complete parts (a) through (c), given $\Sigma x = 248$, $\Sigma y = 97$, $\Sigma x^2 = 4856$, $\Sigma y^2 = 731$, and $\Sigma xy = 1825$.

(d) In a week during which Dorothy makes 18 visits, how many people do you predict will buy insurance from her?

9. *Marketing: Coupons* Each box of Healthy Crunch breakfast cereal contains a coupon entitling you to a free package of garden seeds. At the Healthy Crunch home office, they use the weight of incoming mail to determine how many of their employees are to be assigned to collecting coupons and mailing out seed packages on a given day. (Healthy Crunch has a policy of answering all its mail on the day it is received.)

   Let x = weight of incoming mail (in pounds) and y = number of employees required to process the mail in one working day. A random sample of 8 days gave the following data:

x (lb)	11	20	16	6	12	18	23	25
y (Number of employees)	6	10	9	5	8	14	13	16

Complete parts (a) through (c), given $\Sigma x = 131$, $\Sigma y = 81$, $\Sigma x^2 = 2435$, $\Sigma y^2 = 927$, and $\Sigma xy = 1487$.

(d) If Healthy Crunch receives 15 pounds of mail, how many employees should be assigned mail duty that day?

10. *Focus Problem: Changing Population and Crime Rate* Let x be a random variable representing percentage change in neighborhood population in the past few years, and let y be a random variable representing crime rate (crimes per 1000 population). A random sample of six Denver neighborhoods gave the following information (Source: *Neighborhood Facts*, Piton Foundation):

x	29	2	11	17	7	6
y	173	35	132	127	69	53

Complete parts (a) through (c), given $\Sigma x = 72$, $\Sigma y = 589$, $\Sigma x^2 = 1340$, $\Sigma y^2 = 72{,}277$, and $\Sigma xy = 9499$.

(d) For a neighborhood with $x = 12\%$ change in population in the past few years, predict the change in the crime rate (per 1,000 residents).

DATA HIGHLIGHTS: GROUP PROJECTS

Break into small groups and discuss the following topics. Organize a brief outline in which you summarize the main points of your group discussion.

Scatter diagrams! Are they really useful? Scatter diagrams give a first impression of a data relationship and help us assess whether a linear relation provides a reasonable model for the data. In addition, we can spot *influential points*. A data point with an extreme x value can heavily influence the position of the least-squares line. In this project, we look at data sets with an influential point.

x	1	4	5	9	10	15
y	3	7	6	10	12	4

(a) Compute r and b, the slope of the least-squares line. Find the equation of the least-squares line, and sketch the line on the scatter diagram.

(b) Notice the point boxed in blue in Figure 4-14. Does it seem to lie away from the linear pattern determined by the other points? The coordinates of that point are (15, 4). Is it an influential point? Remove that point from the model and recompute r, b, and the equation of the least-squares line. Sketch this least-squares line on the diagram. How does the removal of the influential point affect the values of r and b and the position of the least-squares line?

FIGURE 4-14

Scatter Diagram

FIGURE 4-15

Scatter Diagram

Copyright © Cengage Learning 2013

(c) Consider the scatter diagram of Figure 4-15. Is there an influential point? If you remove the influential point, will the slope of the new least-squares line be larger or smaller than the slope of the line from the original data? Will the correlation coefficient r be larger or smaller?

**LINKING CONCEPTS:
WRITING PROJECTS**

Discuss each of the following topics in class or review the topics on your own. Then write a brief but complete essay in which you summarize the main points. Please include formulas and graphs as appropriate.

1. What do we mean when we say that two variables have a strong positive (or negative) linear correlation? What would a scatter diagram for these variables look like? Is it possible that two variables could be strongly related somehow but have a low *linear* correlation? Explain and draw a scatter diagram to demonstrate your point.

2. What do we mean by the least-squares criterion? Give a very general description of how the least-squares criterion is involved in the construction of the least-squares line. Why do we say the least-squares line is the "best-fitting" line for the data set?

3. Use the Internet or go to the library and find a magazine or journal article in your field of major interest to which the content of this chapter could be applied. List the variables used, method of data collection, and general type of information and conclusions drawn.

USING TECHNOLOGY

Simple Linear Regression (One Explanatory Variable)

The data in this section are taken from this source:

Based on King, Cuchlaine A. M. *Physical Geography*. Oxford: Basil Blackwell, 1980, pp. 77–86, 196–206.

Throughout the world, natural ocean beaches are beautiful sights to see. If you have visited natural beaches, you may have noticed that when the gradient or dropoff is steep, the grains of sand tend to be larger. In fact, a man-made beach with the "wrong" size granules of sand tends to be washed away and eventually replaced when the proper size grain is selected by the action of the ocean and the gradient of the bottom. Since man-made beaches are expensive, grain size is an important consideration.

In the data that follow, x = median diameter (in millimeters) of granules of sand and y = gradient of beach slope (in degrees) on natural ocean beaches.

x	y
0.17	0.63
0.19	0.70
0.22	0.82
0.235	0.88
0.235	1.15
0.30	1.50
0.35	4.40
0.42	7.30
0.85	11.30

1. Find the sample mean and standard deviation for x and y.

2. Make a scatter plot. Would you expect a moderately high correlation and a good fit for the least-squares line?

3. Find the equation of the least-squares line, and graph the line on the scatter plot.

4. Find the sample correlation coefficient r and the coefficient of determination r^2. Is r significant at the 1% level of significance (two-tailed test)?

5. Suppose you have a truckload of sifted sand in which the median size of granules is 0.38 mm. If you want to put this sand on a beach and you don't want the sand to wash away, then what does the least-squares line predict for the angle of the beach? *Note:* Heavy storms that produce abnormal waves may also wash out the sand. However, in the long run, the size of sand granules that remain on the beach or that are brought back to the beach by wave action is determined to a large extent by the angle at which the beach drops off.

6. Suppose we now have a truckload of sifted sand in which the median size of the granules is 0.45 mm. Repeat Problem 5.

Technology Hints (Simple Regression)

TI-84Plus/TI-83Plus/TI-*n*spire (with TI-84Plus keypad)

Be sure to set **DiagnosticOn** (under **Catalog**).

(a) Scatter diagram: Use **STAT PLOT,** select the first type, use **ZOOM** option **9:ZoomStat.**

(b) Least-squares line and r: Use **STAT, CALC,** option **8:LinReg(a + bx).**

(c) Graph least-squares line and predict: Press **Y=.** Then, under **VARS,** select **5:Statistics,** then select **EQ,** and finally select item **1:RegEQ.** Press enter. This sequence of steps will automatically set $Y_1 =$ your regression equation. Press **GRAPH.** To find a predicted value, when the graph is showing press the **CALC** key and select item **1:Value.** Enter the x value, and the corresponding y value will appear.

Excel 2010

(a) Scatter plot, least-squares line, r^2: On home screen, click **Insert tab.** In the Charts group, select **Scatter** and choose the first type. Once plot is displayed, right-click on any data point. Select **trend line.** Under options, check display line and display r^2.

(b) Prediction: Use **Insert function** $\boxed{f_x}$ ➤ **Statistical** ➤ **Forecast.**

(c) Coefficient r: Use $\boxed{f_x}$ ➤ **Statistical** ➤ **Correl.**

Minitab

(a) Scatter plot, least-squares line, r^2, S_e: (See Section 11.4.) Use menu selection **Stat** ➤ **Regression** ➤ **Fitted line plot.** The value of S_e is displayed as the value of s.

(b) Coefficient r: Use menu selection **Stat** ➤ **Basic Statistics** ➤ **Correlation.**

SPSS

SPSS offers several options for finding the correlation coefficient r and the equation of the least-squares line. First enter the data in the data editor and label the variables appropriately in the variable view window. Use the menu choices **Analyze** ➤ **Regression** ➤ **Linear** and select dependent and independent variables. The output includes the correlation coefficient, the standard error of estimate, the constant, and the coefficient of the dependent variable with corresponding t values and P values for two-tailed tests (as discussed in Section 11.4). The display shows the results for the data in this chapter's Focus Problem regarding crime rate and percentage change in population.

SPSS Display

Model Summary

Model	R	R Square	Adjusted R Square	Std. Error of the Estimate
1	.927[a]	.859	.823	22.59076

a. Predictors: (Constant), % change in population

Coefficients[a]

Model		Unstandardized Coefficients B	Unstandardized Coefficients Std. Error	Standardized Coefficients Beta	t	Sig.
1	(Constant)	36.881	15.474		2.383	.076
	% change in population	5.107	1.035	.927	4.932	.008

a. Dependent Variable: Crime rate per 1,000

With the menu choices **Graph ➤ Legacy Dialogues ➤ Interactive ➤ Scatterplot**, SPSS produces a scatter diagram with the least-squares line, least-squares equation, coefficient of determination r^2, and optional prediction bands (as discussed in Section 11.4). In the dialogue box, move the dependent variable to the box along the vertical axis and the independent variable to the box along the horizontal axis. Click the "fit" tab, highlight Regression as method, and check the box to include the constant in the equation. For optional prediction band, check individual, enter the confidence level, and check total. The following display shows a scatter diagram for the data in this chapter's Focus Problem regarding crime rate and percentage change in population.

SPSS Display for Focus Problem

5

Nathan Griffith/Comet/Corbis

North Wind/North Wind Picture Archive

We see that the theory of probabilities is at bottom only common sense reduced to calculation; it makes us appreciate with exactitude what reasonable minds feel by a sort of instinct, often without being able to account for it.

—PIERRE-SIMON LAPLACE

This quote explains how the great mathematician Pierré-Simon Laplace (1749–1827) described the theory of mathematical probability. The discovery of the mathematical theory of probability was shared by two Frenchmen: Blaise Pascal and Pierre de Fermat. These 17th-century scholars were attracted to the subject by the inquiries of the Chevalier de Méré, a gentleman gambler.

Although the first applications of probability were to games of chance and gambling, today the subject seems to pervade almost every aspect of modern life. Everything from the orbits of spacecraft to the social behavior of woodchucks is described in terms of probabilities.

ELEMENTARY PROBABILITY THEORY

David Young-Wolff/PhotoEdit

PREVIEW QUESTIONS

Why would anyone study probability? Hint: *Most big issues in life involve uncertainty.* (SECTION 5.1)

What are the basic definitions and rules of probability? (SECTION 5.2)

What are counting techniques, trees, permutations, and combinations? (SECTION 5.3)

FOCUS PROBLEM

How Often Do Lie Detectors Lie?

James Burke is an educator who is known for his interesting science-related radio and television shows aired by the British Broadcasting Corporation. His book *Chances: Risk and Odds in Everyday Life* (Virginia Books, London) contains a great wealth of fascinating information about probabilities. The following quote is from Professor Burke's book:

> *If I take a polygraph test and lie, what is the risk I will be detected?* According to some studies, there's about a 72 percent chance you will be caught by the machine.

> *What is the risk that if I take a polygraph test it will incorrectly say that I lied?* At least 1 in 15 will be thus falsely accused.

Both of these statements contain conditional probabilities, which we will study in Section 5.2. Information from that section will enable us to answer the following:

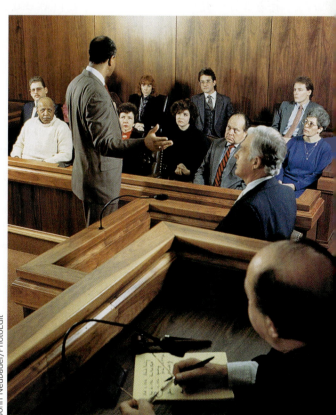

John Neubauer/PhotoEdit

Suppose a person answers 10% of a long battery of questions with lies. Assume that the remaining 90% of the questions are answered truthfully.

1. Estimate the percentage of answers the polygraph will *wrongly* indicate as lies.

2. Estimate the percentage of answers the polygraph will *correctly* indicate as lies.

(See Problem 25 in Section 5.2.)

SECTION 5.1

What Is Probability?

FOCUS POINTS
- Assign probabilities to events.
- Explain how the law of large numbers relates to relative frequencies.
- Apply basic rules of probability in everyday life.
- Explain the relationship between statistics and probability.

We encounter statements given in terms of probability all the time. An excited sports announcer claims that Sheila has a 90% chance of breaking the world record in the upcoming 100-yard dash. Henry figures that if he guesses on a true–false question, the probability of getting it right is 1/2. The Right to Health lobby claims that the probability is 0.40 of getting an erroneous report from a medical laboratory at a low-cost health center. It is consequently lobbying for a federal agency to license and monitor all medical laboratories.

When we use probability in a statement, we're using a *number between 0 and 1* to indicate the likelihood of an event.

Probability is a numerical measure between 0 and 1 that describes the likelihood that an event will occur. Probabilities closer to 1 indicate that the event is more likely to occur. Probabilities closer to 0 indicate that the event is less likely to occur.

Probability of an event A, P(A)

$P(A)$, read "P of A," denotes the **probability of event A.**
If $P(A) = 1$, event A is certain to occur.
If $P(A) = 0$, event A is certain not to occur.

It is important to know what probability statements mean and how to determine probabilities of events, because probability is the language of inferential statistics.

Probability assignments

Intuition

1. A probability assignment based on **intuition** incorporates past experience, judgment, or opinion to estimate the likelihood of an event.

Relative frequency

2. A probability assignment based on **relative frequency** uses the formula

$$\text{Probability of event} = \text{relative frequency} = \frac{f}{n} \qquad (1)$$

where f is the frequency of the event occurrence in a sample of n observations.

Equally likely outcomes

3. A probability assignment based on **equally likely outcomes** uses the formula

$$\text{Probability of event} = \frac{\text{Number of outcomes favorable to event}}{\text{Total number of outcomes}} \qquad (2)$$

EXAMPLE 1   PROBABILITY ASSIGNMENT

Consider each of the following events, and determine how the probability is assigned.

(a) A sports announcer claims that Sheila has a 90% chance of breaking the world record in the 100-yard dash.

SOLUTION: It is likely the sports announcer used intuition based on Sheila's past performance.

(b) Henry figures that if he guesses on a true–false question, the probability of getting it right is 0.50.

SOLUTION: In this case there are two possible outcomes: Henry's answer is either correct or incorrect. Since Henry is guessing, we assume the outcomes are equally likely. There are $n = 2$ equally likely outcomes, and only one is correct. By formula (2),

$$P(\text{correct answer}) = \frac{\text{Number of favorable outcomes}}{\text{Total number of outcomes}} = \frac{1}{2} = 0.50$$

(c) The Right to Health lobby claims that the probability of getting an erroneous medical laboratory report is 0.40, based on a random sample of 200 laboratory reports, of which 80 were erroneous.

SOLUTION: Formula (1) for relative frequency gives the probability, with sample size $n = 200$ and number of errors $f = 80$:

$$P(\text{error}) = \text{relative frequency} = \frac{f}{n} = \frac{80}{200} = 0.40$$

We've seen three ways to assign probabilities: intuition, relative frequency, and—when outcomes are equally likely—a formula. Which do we use? Most of the time it depends on the information that is at hand or that can be feasibly obtained. Our choice of methods also depends on the particular problem. In Guided Exercise 1, you will see three different situations, and you will decide the best way to assign the probabilities. *Remember, probabilities are numbers between 0 and 1, so don't assign probabilities outside this range.*

GUIDED EXERCISE 1 | Determine a probability

Assign a probability to the indicated event on the basis of the information provided. Indicate the technique you used: intuition, relative frequency, or the formula for equally likely outcomes.

(a) A random sample of 500 students at Hudson College were surveyed and it was determined that 375 wear glasses or contact lenses. Estimate the probability that a Hudson College student selected at random wears corrective lenses.

 In this case we are given a sample size of 500, and we are told that 375 of these students wear corrective lenses. It is appropriate to use a relative frequency for the desired probability:

$$P(\text{student wears corrective lenses}) = \frac{f}{n} = \frac{375}{500} = 0.75$$

Continued

(b) The Friends of the Library hosts a fundraising barbecue. George is on the cleanup committee. There are four members on this committee, and they draw lots to see who will clean the grills. Assuming that each member is equally likely to be drawn, what is the probability that George will be assigned the grill-cleaning job?

There are four people on the committee, and each is equally likely to be drawn. It is appropriate to use the formula for equally likely events. George can be drawn in only one way, so there is only one outcome favorable to the event.

$$P(\text{George}) = \frac{\text{No. of favorable outcomes}}{\text{Total no. of outcomes}} = \frac{1}{4} = 0.25$$

(c) Joanna photographs whales for Sea Life Adventure Films. On her next expedition, she is to film blue whales feeding. Based on her knowledge of the habits of blue whales, she is almost certain she will be successful. What specific number do you suppose she estimates for the probability of success?

Since Joanna is almost certain of success, she should make the probability close to 1. We could say $P(\text{success})$ is above 0.90 but less than 1. This probability assignment is based on intuition.

The technique of using the relative frequency of an event as the probability of that event is a common way of assigning probabilities and will be used a great deal in later chapters. The underlying assumption we make is that if events occurred a certain percentage of times in the past, they will occur about the same percentage of times in the future. In fact, this assumption can be strengthened to a very general statement called the *law of large numbers*.

Law of large numbers

Law of large numbers

In the long run, as the sample size increases and increases, the relative frequencies of outcomes get closer and closer to the theoretical (or actual) probability value.

The law of large numbers is the reason businesses such as health insurance, automobile insurance, and gambling casinos can exist and make a profit.

No matter how we compute probabilities, it is useful to know what outcomes are possible in a given setting. For instance, if you are going to decide the probability that Hardscrabble will win the Kentucky Derby, you need to know which other horses will be running.

To determine the possible outcomes for a given setting, we need to define a *statistical experiment*.

Statistical experiment

Event

Simple event

Sample space

A **statistical experiment** or **statistical observation** can be thought of as any random activity that results in a definite outcome.

An **event** is a collection of one or more outcomes of a statistical experiment or observation.

A **simple event** is one particular outcome of a statistical experiment.

The set of all simple events constitutes the **sample space** of an experiment.

EXAMPLE 2   USING A SAMPLE SPACE

Human eye color is controlled by a single pair of genes (one from the father and one from the mother) called a *genotype*. Brown eye color, B, is dominant over blue eye color, ℓ. Therefore, in the genotype Bℓ, consisting of one brown gene B and one blue gene ℓ, the brown gene dominates. A person with a Bℓ genotype has brown eyes.

If both parents have brown eyes and have genotype Bℓ, what is the probability that their child will have blue eyes? What is the probability the child will have brown eyes?

SOLUTION: To answer these questions, we need to look at the sample space of all possible eye-color genotypes for the child. They are given in Table 5-1.

TABLE 5-1   Eye Color Genotypes for Child

Father	Mother	
	B	ℓ
B	BB	Bℓ
ℓ	ℓB	ℓℓ

Copyright © Cengage Learning 2013

According to genetics theory, the four possible genotypes for the child are equally likely. Therefore, we can use formula (2) to compute probabilities. Blue eyes can occur only with the ℓℓ genotype, so there is only one outcome favorable to blue eyes. By formula (2),

$$P(\text{blue eyes}) = \frac{\text{Number of favorable outcomes}}{\text{Total number of outcomes}} = \frac{1}{4}$$

Brown eyes occur with the three remaining genotypes: BB, Bℓ, and ℓB. By formula (2),

$$P(\text{brown eyes}) = \frac{\text{Number of favorable outcomes}}{\text{Total number of outcomes}} = \frac{3}{4}$$

GUIDED EXERCISE 2   *Using a sample space*

Professor Gutierrez is making up a final exam for a course in literature of the Southwest. He wants the last three questions to be of the true–false type. To guarantee that the answers do not follow his favorite pattern, he lists all possible true–false combinations for three questions on slips of paper and then picks one at random from a hat.

(a) Finish listing the outcomes in the given sample space.

TTT	FTT	TFT	____
TTF	FTF	TFF	____

➡ The missing outcomes are FFT and FFF.

(b) What is the probability that all three items will be false? Use the formula

$$P(\text{all F}) = \frac{\text{No. of favorable outcomes}}{\text{Total no. of outcomes}}$$

➡ There is only one outcome, FFF, favorable to all false, so

$$P(\text{all F}) = \frac{1}{8}$$

(c) What is the probability that exactly two items will be true?

➡ There are three outcomes that have exactly two true items: TTF, TFT, and FTT. Thus,

$$P(\text{two T}) = \frac{\text{No. of favorable outcomes}}{\text{Total no. of outcomes}} = \frac{3}{8}$$

There is another important point about probability assignments of simple events:

> The **sum** of the probabilities of all simple events in a sample space must equal 1.

We can use this fact to determine the probability that an event will not occur. For instance, if you think the probability is 0.65 that you will win a tennis match, you assume the probability is 0.35 that your opponent will win.

The *complement* of an event A is the event that A *does not occur*. We use the notation A^c to designate the complement of event A. Figure 5-1 shows the event A and its complement A^c.

Notice that the two distinct events A and A^c make up the entire sample space. Therefore, the sum of their probabilities is 1.

Complement of an event A, A^c

> The **complement of event** A is the event that A *does not occur*. A^c designates the complement of event A. Furthermore,
>
> 1. $P(A) + P(A^c) = 1$
> 2. $P(\text{event } A \text{ does } not \text{ occur}) = P(A^c) = 1 - P(A)$ (3)

EXAMPLE 3   COMPLEMENT OF AN EVENT

The probability that a college student who has not received a flu shot will get the flu is 0.45. What is the probability that a college student will *not* get the flu if the student has not had the flu shot?

SOLUTION: In this case, we have

$$P(\text{will get flu}) = 0.45$$

$$P(\text{will } not \text{ get flu}) = 1 - P(\text{will get flu}) = 1 - 0.45 = 0.55$$

FIGURE 5-1

Event A and Its Complement A^c

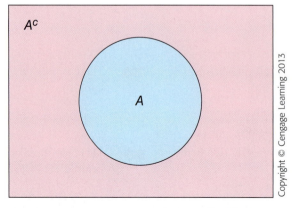

Sample space

Copyright © Cengage Learning 2013

GUIDED EXERCISE 3 | ## Complement of an event

A veterinarian tells you that if you breed two cream-colored guinea pigs, the probability that an offspring will be pure white is 0.25. What is the probability that an offspring will not be pure white?

(a) P(pure white) + P(*not* pure white) = _____   ⟹   1.

(b) P(*not* pure white) = _____   ⟹   1 − 0.25, or 0.75.

LOOKING FORWARD

The complement rule has special application in Chapter 6. There we study probabilities associated with binomial experiments. These are experiments that have only two outcomes, "success" and "failure." One feature of these experiments is that the outcomes can be categorized according to the number of "successes" out of n independent trials.

> Simple Events of Sample Space of a Binomial Experiment with n trials
> 0 successes, 1 success, 2 successes, 3 successes, ..., n successes

A convenient way to compute the probability of "at least one" success is to observe that the complement of "at least one success" is the event "0 successes." Then, by the complement rule

$$P(\text{at least one success}) = 1 - P(0 \text{ successes})$$

Problems 8, 9, and 10 of this section show a similar use of the complement rule.

Sometimes, instead of being told the probability of an event, we are given the odds for the event. Problems 19 and 20 discuss odds in favor of and odds against an event.

Summary: Some important facts about probability

1. A **statistical experiment** or **statistical observation** is any random activity that results in a definite outcome. A **simple event** consists of one and only one outcome of the experiment. The **sample space** is the set of all simple events. An **event** A is any subset of the sample space.

2. The probability of an event A is denoted by $P(A)$.

3. The probability of an event is a number between 0 and 1. The closer to 1 the probability is, the more likely it is the event will occur. The closer to 0 the probability is, the less likely it is the event will occur.

4. The sum of the probabilities of all simple events in a sample space is 1.

5. Probabilities can be assigned by using intuition, relative frequencies, or the formula for equally likely outcomes. Additional ways to assign probabilities will be introduced in later chapters.

6. The **complement** of an event A is denoted by A^c. So, A^c is the event that A does not occur.

7. $P(A) + P(A^c) = 1$.

Interpreting Probabilities

As stressed in item 3 in the summary about probabilities, the closer the probability is to 1, the more likely the event is to occur. The closer the probability is to 0, the less likely the event is to occur. However, just because the probability of an event is very high, it is not a certainty that the event will occur. Likewise, even though the probability of an event is very low, the event might still occur.

Events with low probability but big consequences are of special concern. Such events are sometimes referred to as "black swan events," as discussed in the very popular book *The Black Swan* by Nassim Nicholas Taleb. In his varied careers Taleb has been a stock trader, author, and professor at the University of Massachusetts. His special interests are mathematical problems of luck, uncertainty, probability, and general knowledge. In *The Black Swan*, Taleb discusses "the impact of the highly improbable." If an event has (a) little impact, bearing, or meaning on a person's life and (b) very low probability, then (c) it seems reasonable to ignore the event.

However, some of the really big mistakes in a person's life can result from misjudging either (a) the size of an event's impact or (b) the likelihood the event will occur. An event of great importance cannot be ignored even if it has a low probability of occurrence.

So what should a person do?

Taleb describes himself as a hyperskeptic when others are gullible, and gullible when others are skeptical. In particular, he recommends being skeptical about (mentally) confirming an event when the errors of confirmation are costly. Lots and lots of data should not be thought of as confirming an event when a simple instance can disconfirm an important and costly event. Taleb recommends that one be skeptical in the face of wild randomness (and big consequences) and gullible when the randomness (and consequences) are mild. One way to estimate the "randomness" of a situation is to estimate its probability. Estimating the "consequences" of an event in your own life must be a personal decision.

Probability Related to Statistics

We conclude this section with a few comments on the nature of statistics versus probability. Although statistics and probability are closely related fields of mathematics, they are nevertheless separate fields. It can be said that probability is the medium through which statistical work is done. In fact, if it were not for probability theory, inferential statistics would not be possible.

Put very briefly, probability is the field of study that makes statements about what will occur when samples are drawn from a *known population*. Statistics is the field of study that describes how samples are to be obtained and how inferences are to be made about *unknown populations*.

A simple but effective illustration of the difference between these two subjects can be made by considering how we treat the following examples:

Example of a probability application

Condition: We *know* the exact makeup of the *entire* population.

Example: Given 3 green marbles, 5 red marbles, and 4 white marbles in a bag, draw 6 marbles at random from the bag. What is the probability that none of the marbles is red?

Example of a statistical application

Condition: We have only *samples* from an otherwise *unknown* population.

Example: Draw a random sample of 6 marbles from the (unknown) population of all marbles in a bag and observe the colors. Based on the sample results, make a conjecture about the colors and numbers of marbles in the entire population of all marbles in the bag.

In another sense, probability and statistics are like flip sides of the same coin. On the probability side, you know the overall description of the population. The central problem is to compute the likelihood that a specific outcome will happen. On the statistics side, you know only the results of a sample drawn from the population. The central problem is to describe the sample (descriptive statistics) and to draw conclusions about the population based on the sample results (inferential statistics).

In statistical work, the inferences we draw about an unknown population are not claimed to be absolutely correct. Since the population remains unknown (in a theoretical sense), we must accept a "best guess" for our conclusions and act using the most probable answer rather than absolute certainty.

Probability is the topic of this chapter. However, we will not study probability just for its own sake. Probability is a wonderful field of mathematics, but we will study mainly the ideas from probability that are needed for a proper understanding of statistics.

VIEWPOINT | What Makes a Good Teacher?

A survey of 735 students at nine colleges in the United States was taken to determine instructor behaviors that help students succeed. Data from this survey can be found by visiting the Brase/Brase statistics site at **http://www.cengagebrain.com** *and finding the link to DASL, the Carnegie Mellon University Data and Story Library. Once at the DASL site, select Data Subjects, then Psychology, and then Instructor Behavior. You can estimate the probability of how a student would respond (very positive, neutral, very negative) to different instructor behaviors. For example, more than 90% of the students responded "very positive" to the instructor's use of real-world examples in the classroom.*

SECTION 5.1 PROBLEMS   aplia   Access end-of-section exercises at www.aplia.com

---

SECTION 5.2

Some Probability Rules—Compound Events

FOCUS POINTS

- Compute probabilities of general compound events.
- Compute probabilities involving independent events or mutually exclusive events.
- Use survey results to compute conditional probabilities.

Conditional Probability and Multiplication Rules

You roll two dice. What is the probability that you will get a 5 on each die? Consider a collection of 6 balls identical in every way except in color. There are 3 green balls, 2 blue balls, and 1 red ball. You draw two balls at random from the collection *without replacing* the first ball before drawing the second. What is the probability that both balls will be green?

It seems that these two problems are nearly alike. They are alike in the sense that in each case, you are to find the probability of two events occurring *together*. In the first problem, you are to find

P(5 on 1st die *and* 5 on 2nd die)

In the second, you want

P(green ball on 1st draw *and* green ball on 2nd draw)

The two problems differ in one important aspect, however. In the dice problem, the outcome of a 5 on the first die does not have any effect on the probability of getting a 5 on the second die. Because of this, the events are *independent*.

Independent events

> Two events are **independent** if the occurrence or nonoccurrence of one event does *not* change the probability that the other event will occur.

In the problem concerning a collection of colored balls, the probability that the first ball is green is 3/6, since there are 6 balls in the collection and 3 of them are green. If you get a green ball on the first draw, the probability of getting a green ball on the second draw is changed to 2/5, because one green ball has already been drawn and only 5 balls remain. Therefore, the two events in the ball-drawing problem are *not* independent. They are, in fact, *dependent*, since the outcome of the first draw changes the probability of getting a green ball on the second draw.

Dependent events

Probability of *A and B*

Why does the *independence* or *dependence* of two events matter? The type of events determines the way we compute the probability of the two events happening together. If two events A and B are *independent*, then we use formula (4) to compute the probability of the event A *and* B:

> **Multiplication rule for independent events**
>
> $$P(A \text{ and } B) = P(A) \cdot P(B) \qquad (4)$$

If the events are *dependent*, then we must take into account the changes in the probability of one event caused by the occurrence of the other event. The notation P(A, *given* B) denotes the probability that event A will occur *given* that event B has occurred. This is called a *conditional probability*. We read P(A, *given* B) as "probability of A given B." If A and B are dependent events, then $P(A) \neq P(A, \text{given } B)$, because the occurrence of event B has changed the probability that event A will occur. A standard notation for P(A, *given* B) is $P(A \mid B)$.

Conditional probability
$P(A \mid B)$

> **General multiplication rule for any events**
>
> $$P(A \text{ and } B) = P(A) \cdot P(B \mid A) \qquad (5)$$
> $$P(A \text{ and } B) = P(B) \cdot P(A \mid B) \qquad (6)$$

We will use either formula (5) or formula (6) according to the information available.

Multiplication rules of probability

Formulas (4), (5), and (6) constitute the *multiplication rules of probability*. They help us compute the probability of events happening together when the sample space is too large for convenient reference or when it is not completely known.

Note: For conditional probability, observe that the multiplication rule

$$P(A \text{ and } B) = P(B) \cdot P(A \mid B)$$

can be solved for $P(A \mid B)$, leading to

Conditional probability (when *P(B)* ≠ 0)

$$P(A \mid B) = \frac{P(A \text{ and } B)}{P(B)}$$

We will see some applications of this formula in later chapters.

Let's use the multiplication rules to complete the dice and ball-drawing problems just discussed. We'll compare the results with those obtained by using the sample space directly.

EXAMPLE 4   MULTIPLICATION RULE, INDEPENDENT EVENTS

Suppose you are going to throw two fair dice. What is the probability of getting a 5 on each die?

SOLUTION USING THE MULTIPLICATION RULE: The two events are independent, so we should use formula (4). *P*(5 on 1st die *and* 5 on 2nd die) = *P*(5 on 1st) · *P*(5 on 2nd). To finish the problem, we need to compute the probability of getting a 5 when we throw one die.

There are six faces on a die, and on a fair die each is equally likely to come up when you throw the die. Only one face has five dots, so by formula (2) for equally likely outcomes,

$$P(5 \text{ on die}) = \frac{1}{6}$$

Now we can complete the calculation.

$$P(5 \text{ on 1st die } and \text{ 5 on 2nd die}) = P(5 \text{ on 1st}) \cdot P(5 \text{ on 2nd})$$
$$= \frac{1}{6} \cdot \frac{1}{6} = \frac{1}{36}$$

SOLUTION USING THE SAMPLE SPACE: The first task is to write down the sample space. Each die has six equally likely outcomes, and each outcome of the second die can be paired with each of the first. The sample space is shown in Figure 5-2. The total number of outcomes is 36, and only one is favorable to a 5 on the first die *and* a 5 on the second. The 36 outcomes are equally likely, so by formula (2) for equally likely outcomes,

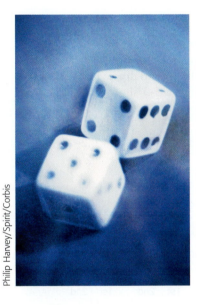

Philip Harvey/Spirit/Corbis

FIGURE 5-2

Sample Space for Two Dice

Copyright © Cengage Learning 2013

$$P(\text{5 on 1st } and \text{ 5 on 2nd}) = \frac{1}{36}$$

The two methods yield the same result. The multiplication rule was easier to use because we did not need to look at all 36 outcomes in the sample space for tossing two dice.

EXAMPLE 5   MULTIPLICATION RULE, DEPENDENT EVENTS

Consider a collection of 6 balls that are identical except in color. There are 3 green balls, 2 blue balls, and 1 red ball. Compute the probability of drawing 2 green balls from the collection if the first ball is *not replaced* before the second ball is drawn.

MULTIPLICATION-RULE METHOD: These events are *dependent*. The probability of a green ball on the first draw is 3/6, but on the second draw the probability of a green ball is only 2/5 if a green ball was removed on the first draw. By the multiplication rule for dependent events,

$$P(\text{green ball on 1st draw } and \text{ green ball on 2nd draw)} = P(\text{green on 1st}) \cdot P(\text{green on 2nd | green on 1st})$$

$$= \frac{3}{2} \cdot \frac{2}{5} = \frac{1}{5} = 0.2$$

SAMPLE-SPACE METHOD: Each of the 6 possible outcomes for the 1st draw must be paired with each of the 5 possible outcomes for the second draw. This means that there are a total of 30 possible pairs of balls. Figure 5-3 shows all the possible pairs of balls. In 6 of the pairs, both balls are green.

$$P(\text{green ball on 1st draw } and \text{ green ball on 2nd draw}) = \frac{6}{30} = 0.2$$

Again, the two methods agree.

FIGURE 5-3

Sample Space for Drawing Two Balls without Replacement from a Collection of 3 Green Balls, 2 Blue Balls, and 1 Red Ball

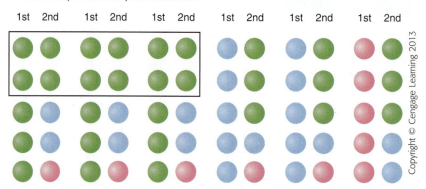

Copyright © Cengage Learning 2013

The multiplication rules apply whenever we wish to determine the probability of two events happening *together*. To indicate "together," we use *and* between the events. But before you use a multiplication rule to compute the probability of *A and B*, you must determine if *A* and *B* are independent or dependent events.

PROCEDURE	HOW TO USE THE MULTIPLICATION RULES

1. First determine whether A and B are independent events. If $P(A) = P(A \mid B)$, then the events are independent.

2. If A and B are independent events,

$$P(A \text{ and } B) = P(A) \cdot P(B) \tag{4}$$

3. If A and B are any events,

$$P(A \text{ and } B) = P(A) \cdot P(B \mid A) \tag{5}$$

$$P(A \text{ and } B) = P(B) \cdot P(A \mid B) \tag{6}$$

Let's practice using the multiplication rule.

GUIDED EXERCISE 4 | **Multiplication rule**

Andrew is 55, and the probability that he will be alive in 10 years is 0.72. Ellen is 35, and the probability that she will be alive in 10 years is 0.92. Assuming that the life span of one will have no effect on the life span of the other, what is the probability they will both be alive in 10 years?

(a) Are these events dependent or independent?

⟹ Since the life span of one does not affect the life span of the other, the events are independent.

(b) Use the appropriate multiplication rule to find P(Andrew alive in 10 years *and* Ellen alive in 10 years).

⟹ We use the rule for independent events:

$P(A \text{ and } B) = P(A) \cdot P(B)$

P(Andrew alive *and* Ellen alive)

$= P$(Andrew alive) $\cdot P$(Ellen alive)

$= (0.72)(0.92) \approx 0.66$

GUIDED EXERCISE 5 | **Dependent events**

A quality-control procedure for testing Ready-Flash disposable cameras consists of drawing two cameras at random from each lot of 100 without replacing the first camera before drawing the second. If both are defective, the entire lot is rejected. Find the probability that both cameras are defective if the lot contains 10 defective cameras. Since we are drawing the cameras at random, assume that each camera in the lot has an equal chance of being drawn.

(a) What is the probability of getting a defective camera on the first draw?

⟹ The sample space consists of all 100 cameras. Since each is equally likely to be drawn and there are 10 defective ones,

$$P(\text{defective camera}) = \frac{10}{100} = \frac{1}{10}$$

Continued

(b) The first camera drawn is not replaced, so there are only 99 cameras for the second draw. What is the probability of getting a defective camera on the second draw if the first camera was defective?

⟹ If the first camera was defective, then there are only 9 defective cameras left among the 99 remaining cameras in the lot.

$P(\text{def. cam. on 2nd draw} \mid \text{def. cam. on 1st})$

$$= \frac{9}{99} = \frac{1}{11}$$

(c) Are the probabilities computed in parts (a) and (b) different? Does drawing a defective camera on the first draw change the probability of getting a defective camera on the second draw? Are the events dependent?

⟹ The answer to all these questions is yes.

(d) Use the formula for dependent events

$$P(A \text{ and } B) = P(A) \cdot P(B \mid A)$$

to compute $P(\text{1st camera defective and 2nd camera defective})$.

⟹ $P(\text{1st defective and 2nd defective}) = \dfrac{1}{10} \cdot \dfrac{1}{11}$

$$= \frac{1}{110}$$

$$\approx 0.009$$

More than two independent events

The multiplication rule for independent events extends to *more than two independent events*. For instance, if the probability that a single seed will germinate is 0.85, and you plant 3 seeds, the probability that they will all germinate (assuming seed germinations are independent) is

$$P(\text{1st germinates and 2nd germinates and 3rd germinates}) = (0.85)(0.85)(0.85)$$

$$\approx 0.614$$

Addition Rules

One of the multiplication rules can be used any time we are trying to find the probability of two events happening *together*. Pictorially, we are looking for the probability of the shaded region in Figure 5-4(a).

Probability of *A or B*

Another way to combine events is to consider the possibility of one event *or* another occurring. For instance, if a sports car saleswoman gets an extra bonus if she sells a convertible or a car with leather upholstery, she is interested in the

FIGURE 5-4

(a) Event *A* and *B*

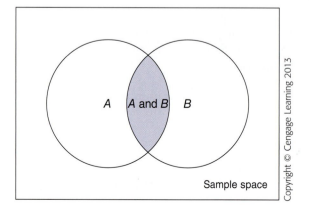

Sample space

(b) Event *A* or *B*

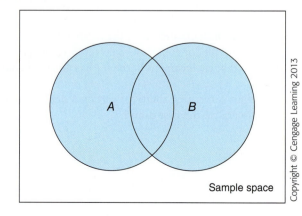

Sample space

Copyright © Cengage Learning 2013

probability that you will buy a car that is a convertible *or* that has leather upholstery. Of course, if you bought a convertible with leather upholstery, that would be fine, too. Pictorially, the shaded portion of Figure 5-4(b) represents the outcomes satisfying the *or* condition. Notice that the condition *A or B* is satisfied by any one of the following conditions:

1. Any outcome in *A* occurs.

2. Any outcome in *B* occurs.

3. Any outcome in both *A* and *B* occurs.

It is important to distinguish between the *or* combinations and the *and* combinations because we apply different rules to compute their probabilities.

GUIDED EXERCISE 6 | **Combining events**

Indicate how each of the following pairs of events are combined. Use either the *and* combination or the *or* combination.

(a) Satisfying the humanities requirement by taking a course in the history of Japan or by taking a course in classical literature

➡️ Use the *or* combination.

(b) Buying new tires and aligning the tires

➡️ Use the *and* combination.

(c) Getting an A not only in psychology but also in biology

➡️ Use the *and* combination.

(d) Having at least one of these pets: cat, dog, bird, rabbit

➡️ Use the *or* combination.

Once you decide that you are to find the probability of an *or* combination rather than an *and* combination, what formula do you use? It depends on the situation. In particular, it depends on whether or not the events being combined share any outcomes. Example 6 illustrates two situations.

EXAMPLE 6   PROBABILITY OF EVENTS COMBINED WITH *OR*

Consider an introductory statistics class with 31 students. The students range from freshmen through seniors. Some students are male and some are female. Figure 5-5 shows the sample space of the class.

FIGURE 5-5

Sample Space for Statistics Class

F Designates Female   M Designates Male

Freshman	Sophomore	Junior	Senior
FFFFF FFFF MMM MMM	FFF MMM MM	FFFF MM	F M
15 students	8 students	6 students	2 students

Copyright © Cengage Learning 2013

(a) Suppose we select one student at random from the class. Find the probability that the student is either a freshman or a sophomore.

Since there are 15 freshmen out of 31 students, $P(\text{freshman}) = \dfrac{15}{31}$.

Since there are 8 sophomores out of 31 students, $P(\text{sophomore}) = \dfrac{8}{31}$.

$$P(\text{freshman } or \text{ sophomore}) = \frac{15}{31} + \frac{8}{31} = \frac{23}{31} \approx 0.742$$

Notice that we can simply add the probability of freshman to the probability of sophomore to find the probability that a student selected at random will be either a freshman or a sophomore. No student can be both a freshman and a sophomore at the same time.

(b) Select one student at random from the class. What is the probability that the student is either a male or a sophomore? Here we note that

$$P(\text{sophomore}) = \frac{8}{31} \quad P(\text{male}) = \frac{14}{31} \quad P(\text{sophomore } and \text{ male}) = \frac{5}{31}$$

If we simply add $P(\text{sophomore})$ and $P(\text{male})$, we're including $P(\text{sophomore } and \text{ male})$ twice in the sum. To compensate for this double summing, we simply subtract $P(\text{sophomore } and \text{ male})$ from the sum. Therefore,

$$P(\text{sophomore } or \text{ male}) = P(\text{sophomore}) + P(\text{male}) - P(\text{sophomore } and \text{ male})$$

$$= \frac{8}{31} + \frac{14}{31} - \frac{5}{31} = \frac{17}{31} \approx 0.548$$

Mutually exclusive events

We say the events A and B are *mutually exclusive* or *disjoint* if they cannot occur together. This means that A and B have no outcomes in common or, put another way, that $P(A \text{ and } B) = 0$.

> Two events are **mutually exclusive** or **disjoint** if they cannot occur together. In particular, events A and B are mutually exclusive if $P(A \text{ and } B) = 0$.

Formula (7) is the *addition rule for mutually exclusive events A and B*.

Addition rules

> **Addition rule for *mutually exclusive* events *A* and *B***
>
> $P(A \text{ or } B) = P(A) + P(B)$ (7)

If the events are not mutually exclusive, we must use the more general formula (8), which is the *general addition rule for any events A and B*.

> **General addition rule for any events *A* and *B***
>
> $P(A \text{ or } B) = P(A) + P(B) - P(A \text{ and } B)$ (8)

You may ask: Which formula should I use? The answer is: Use formula (7) only if you know that A and B are mutually exclusive (i.e., cannot occur together); if you do not know whether A and B are mutually exclusive, then use formula (8).

Formula (8) is valid either way. Notice that when A and B are mutually exclusive, then $P(A \text{ and } B) = 0$, so formula (8) reduces to formula (7).

PROCEDURE

How to use the addition rules

1. First determine whether A and B are mutually exclusive events.
   If $P(A \text{ and } B) = 0$, then the events are mutually exclusive.
2. If A and B are mutually exclusive events,
$$P(A \text{ or } B) = P(A) + P(B) \tag{7}$$
3. If A and B are any events,
$$P(A \text{ or } B) = P(A) + P(B) - P(A \text{ and } B) \tag{8}$$

GUIDED EXERCISE 7 | **Mutually exclusive events**

The Cost Less Clothing Store carries remaindered pairs of slacks. If you buy a pair of slacks in your regular waist size without trying them on, the probability that the waist will be too tight is 0.30 and the probability that it will be too loose is 0.10.

(a) Are the events "too tight" and "too loose" mutually exclusive?

➡ The waist cannot be both too tight and too loose at the same time, so the events are mutually exclusive.

(b) If you choose a pair of slacks at random in your regular waist size, what is the probability that the waist will be too tight or too loose?

➡ Since the events are mutually exclusive,

$P(\text{too tight } or \text{ too loose})$

$= P(\text{too tight}) + P(\text{too loose})$

$= 0.30 + 0.10$

$= 0.40$

GUIDED EXERCISE 8 | **General addition rule**

Professor Jackson is in charge of a program to prepare people for a high-school equivalency exam. Records show that 80% of the students need work in math, 70% need work in English, and 55% need work in both areas.

(a) Are the events "needs math" and "needs English" mutually exclusive?

➡ These events are not mutually exclusive, since some students need both. In fact,

$P(\text{needs math } and \text{ needs English}) = 0.55$

(b) Use the appropriate formula to compute the probability that a student selected at random needs math or needs English.

➡ Since the events are not mutually exclusive, we use formula (8):

$P(\text{needs math } or \text{ needs English})$

$= P(\text{needs math}) + P(\text{needs English})$

$\quad - P(\text{needs math } and \text{ English})$

$= 0.80 + 0.70 - 0.55$

$= 0.95$

More than two mutually exclusive events

The addition rule for mutually exclusive events can be extended to apply to the situation in which we have more than two events, all of which are mutually exclusive of all the other events.

EXAMPLE 7   MUTUALLY EXCLUSIVE EVENTS

Laura is playing Monopoly. On her next move she needs to throw a sum bigger than 8 on the two dice in order to land on her own property and pass Go. What is the probability that Laura will throw a sum bigger than 8?

SOLUTION: When two dice are thrown, the largest sum that can come up is 12. Consequently, the only sums larger than 8 are 9, 10, 11, and 12. These outcomes are mutually exclusive, since only one of these sums can possibly occur on one throw of the dice. The probability of throwing more than 8 is the same as

$$P(9 \text{ or } 10 \text{ or } 11 \text{ or } 12)$$

Since the events are mutually exclusive,

$$P(9 \text{ or } 10 \text{ or } 11 \text{ or } 12) = P(9) + P(10) + P(11) + P(12)$$
$$= \frac{4}{36} + \frac{3}{36} + \frac{2}{36} + \frac{1}{36}$$
$$= \frac{10}{36} = \frac{5}{18}$$

To get the specific values of $P(9)$, $P(10)$, $P(11)$, and $P(12)$, we used the sample space for throwing two dice (see Figure 5-2 on page 147). There are 36 equally likely outcomes—for example, those favorable to 9 are 6, 3; 3, 6; 5, 4; and 4, 5. So $P(9) = 4/36$. The other values can be computed in a similar way.

LOOKING FORWARD

Chapters 6 and 7 involve several probability distributions. In these chapters, most of the events of interest are mutually exclusive as well as independent. This means we can compute probabilities by using the extended addition rule for mutually exclusive events and the extended multiplication rule for independent events.

Further Examples

Most of us have been asked to participate in a survey. Schools, retail stores, news media, and government offices all conduct surveys. There are many types of surveys, and it is not our intention to give a general discussion of this topic. Let us study a very popular method called the *simple tally survey*. Such a survey consists of questions to which the responses can be recorded in the rows and columns of a table called a *contingency table*. These questions are appropriate to the information you want and are designed to cover the *entire* population of interest. In addition, the questions should be designed so that you can partition the sample space of responses into distinct (that is, mutually exclusive) sectors.

If the survey includes responses from a reasonably large random sample, then the results should be representative of your population. In this case, you can estimate simple probabilities, conditional probabilities, and the probabilities of some combinations of events directly from the results of the survey.

EXAMPLE 8   SURVEY

At Hopewell Electronics, all 140 employees were asked about their political affiliations. The employees were grouped by type of work, as executives or production workers. The results with row and column totals are shown in Table 5-2.

TABLE 5-2   **Employee Type and Political Affiliation**

Employee Type	Political Affiliation			Row Total
	Democrat (*D*)	Republican (*R*)	Independent (*I*)	
Executive (*E*)	5	34	9	48
Production worker (*PW*)	63	21	8	92
Column Total	68	55	17	140 Grand Total

Copyright © Cengage Learning 2013

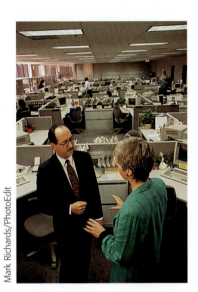

Mark Richards/PhotoEdit

Suppose an employee is selected at random from the 140 Hopewell employees. Let us use the following notation to represent different events of choosing: *E* = executive; *PW* = production worker; *D* = Democrat; *R* = Republican; *I* = Independent.

(a) Compute $P(D)$ and $P(E)$.

SOLUTION: To find these probabilities, we look at the *entire* sample space.

$$P(D) = \frac{\text{Number of Democrats}}{\text{Number of employees}} = \frac{68}{140} \approx 0.486$$

$$P(E) = \frac{\text{Number of executives}}{\text{Number of employees}} = \frac{48}{140} \approx 0.343$$

(b) Compute $P(D \mid E)$.

SOLUTION: For the conditional probability, we restrict our attention to the portion of the sample space satisfying the condition of being an executive.

$$P(D \mid E) = \frac{\text{Number of executives who are Democrats}}{\text{Number of executives}} = \frac{5}{48} \approx 0.104$$

(c) Are the events *D* and *E* independent?

SOLUTION: One way to determine if the events *D* and *E* are independent is to see if $P(D) = P(D \mid E)$ [or equivalently, if $P(E) = P(E \mid D)$]. Since $P(D) \approx 0.486$ and $P(D \mid E) \approx 0.104$, we see that $P(D) \neq P(D \mid E)$. This means that the events *D* and *E* are *not* independent. The probability of event *D* depends on whether or not event *E* has occurred.

(d) Compute $P(D \ and \ E)$.

SOLUTION: This probability is not conditional, so we must look at the entire sample space.

$$P(D \ and \ E) = \frac{\text{Number of executives who are Democrats}}{\text{Total number of employees}} = \frac{5}{140} \approx 0.036$$

Let's recompute this probability using the rules of probability for dependent events.

$$P(D \ and \ E) = P(E) \cdot P(D \mid E) = \frac{48}{140} \cdot \frac{5}{48} = \frac{5}{140} \approx 0.036$$

The results using the rules are consistent with those using the sample space.

(e) Compute $P(D \text{ or } E)$.

SOLUTION: From part (d), we know that events D and E are not mutually exclusive, because $P(D \text{ and } E) \neq 0$. Therefore,

$$P(D \text{ or } E) = P(D) + P(E) - P(D \text{ and } E)$$

$$= \frac{68}{140} + \frac{48}{140} - \frac{5}{140} = \frac{111}{140} \approx 0.793$$

GUIDED EXERCISE 9   Survey

Using Table 5-2 on page 155, let's consider other probabilities regarding the types of employees at Hopewell and their political affiliations. This time let's consider the production worker and the affiliation of Independent. Suppose an employee is selected at random from the group of 140.

(a) Compute $P(I)$ and $P(PW)$.

$$P(I) = \frac{\text{No. of Independents}}{\text{Total no. of employees}}$$

$$= \frac{17}{140} \approx 0.121$$

$$P(PW) = \frac{\text{No. of production workers}}{\text{Total no. of employees}}$$

$$= \frac{92}{140} \approx 0.657$$

(b) Compute $P(I \mid PW)$. This is a conditional probability. Be sure to restrict your attention to production workers, since that is the condition given.

$$P(I \mid PW) = \frac{\text{No. of Independent production workers}}{\text{No. of production workers}}$$

$$= \frac{8}{92} \approx 0.087$$

(c) Compute $P(I \text{ and } PW)$. In this case, look at the entire sample space and the number of employees who are both Independent and in production.

$$P(I \text{ and } PW) = \frac{\text{No. of Independent production workers}}{\text{Total no. of employees}}$$

$$= \frac{8}{140} \approx 0.057$$

(d) Use the multiplication rule for dependent events to calculate $P(I \text{ and } PW)$. Is the result the same as that of part (c)?

By the multiplication rule,

$$P(I \text{ and } PW) = P(PW) \cdot P(I \mid PW)$$

$$= \frac{92}{140} \cdot \frac{8}{92} = \frac{8}{140} \approx 0.057$$

The results are the same.

(e) Compute $P(I \text{ or } PW)$. Are the events mutually exclusive?

Since the events are not mutully exclusive,

$$P(I \text{ or } PW) = P(I) + P(PW) - P(I \text{ and } PW)$$

$$= \frac{17}{140} + \frac{92}{140} - \frac{8}{140}$$

$$= \frac{101}{140} \approx 0.721$$

As you apply probability to various settings, keep the following rules in mind:

Summary of basic probability rules

A statistical experiment or statistical observation is any random activity that results in a recordable outcome. The sample space is the set of all simple events that are the outcomes of the statistical experiment and cannot be broken into other "simpler" events. A general event is any subset of the sample space. The notation $P(A)$ designates the probability of event A.

1. $P(\text{entire sample space}) = 1$

2. For any event A: $0 \leq P(A) \leq 1$

3. A^c designates the **complement** of A: $P(A^c) = 1 - P(A)$

4. Events A and B are **independent events** if $P(A) = P(A \mid B)$.

5. Multiplication rules:

   General: $P(A \text{ and } B) = P(A) \cdot P(B \mid A)$

   $\qquad P(A \text{ and } B) = P(B) \cdot P(A \mid B)$

   Independent events: $P(A \text{ and } B) = P(A) \cdot P(B)$

6. Conditional probability: $P(A \mid B) = \dfrac{P(A \text{ and } B)}{P(B)}$

7. Events A and B are **mutually exclusive** if $P(A \text{ and } B) = 0$.

8. Addition rules:

   General: $P(A \text{ or } B) = P(A) + P(B) - P(A \text{ and } B)$

   Mutually exclusive events: $P(A \text{ or } B) = P(A) + P(B)$

CRITICAL THINKING

Translating events described by common English phrases into events described using *and, or, complement,* or *given* takes a bit of care. Table 5-3 shows some common phrases and their corresponding translations into symbols.

TABLE 5-3   **English Phrases and Corresponding Symbolic Translations**

Consider the following events for a person selected at random from the general population:

A = person is taking college classes

B = person is under 30 years old

Phrase	Symbolic Expression
1. The probability that a person is under 30 years old and is taking college classes is 40%.	$P(B \text{ and } A) = 0.40$ or $P(A \text{ and } B) = 0.40$
2. The probability that a person under 30 years old is taking college classes is 45%.	$P(A \mid B) = 0.45$
3. The probability is 45% that a person is taking college classes if the person is under 30.	$P(A \mid B) = 0.45$
4. The probability that a person taking college classes is under 30 is 0.60.	$P(B \mid A) = 0.60$
5. The probability that a person is not taking college classes or is under 30 years old is 0.75.	$P(A^c \text{ or } B) = 0.75$

In this section, we have studied some important rules that are valid in all probability spaces. The rules and definitions of probability are not only interesting, but also have extensive *applications* in our everyday lives.

VIEWPOINT   The Psychology of Odors

The Smell and Taste Treatment Research Foundation of Chicago collected data on the time required to complete a maze while subjects were smelling different scents. Data for this survey can be found by visiting the Brase/Brase statistics site at **http://www.cengagebrain.com** *and finding the link to DASL, the Carnegie Mellon University Data and Story Library. Once at the DASL site, select Data Subjects, then Psychology, and then Scents. You can estimate conditional probabilities regarding response times for smokers, nonsmokers, and types of scents.*

SECTION 5.2 PROBLEMS

 aplia   Access end-of-section exercises at www.aplia.com

SECTION 5.3

Trees and Counting Techniques

FOCUS POINTS

- Organize outcomes in a sample space using tree diagrams.
- Compute number of ordered arrangements of outcomes using permutations.
- Compute number of (nonordered) groupings of outcomes using combinations.
- Explain how counting techniques relate to probability in everyday life.

When outcomes are equally likely, we compute the probability of an event by using the formula

$$P(A) = \frac{\text{Number of outcomes favorable to the event } A}{\text{Number of outcomes in the sample space}}$$

The probability formula requires that we be able to determine the number of outcomes in the sample space. In the problems we have done in previous sections, this task has not been difficult because the number of outcomes was small or the sample space consisted of fairly straightforward events. The tools we present in this section will help you count the number of possible outcomes in larger sample spaces or those formed by more complicated events.

When an outcome of an experiment is composed of a series of events, the multiplication rule gives us the *total number* of outcomes.

Multiplication rule of counting

> **Multiplication rule of counting**
>
> Consider the series of events E_1 through E_m, where n_1 is the number of possible outcomes for event E_1, n_2 is the number of possible outcomes for event E_2, and n_m designates the number of possible outcomes for event E_m. Then the product
>
> $$n_1 \times n_2 \times \cdots \times n_m$$
>
> gives the total number of possible outcomes for the series of events E_1, followed by E_2, up through event E_m.

EXAMPLE 9   MULTIPLICATION RULE

Jacqueline is in a nursing program and is required to take a course in psychology and one in physiology (A and P) next semester. She also wants to take Spanish II. If there are two sections of psychology, two of A and P, and three of Spanish II, how many different class schedules can Jacqueline choose from? (Assume that the times of the sections do not conflict.)

SOLUTION: Creating a class schedule can be considered an experiment with a series of three events. There are two possible outcomes for the psychology section, two for the A and P section, and three for the Spanish II section. By the multiplication rule, the total number of class schedules possible is

$$2 \times 2 \times 3 = 12$$

Tree diagram

A *tree diagram* gives a visual display of the total number of outcomes of an experiment consisting of a series of events. From a tree diagram, we can determine not only the total number of outcomes, but also the individual outcomes.

EXAMPLE 10   TREE DIAGRAM

Using the information from Example 9, let's make a tree diagram that shows all the possible course schedules for Jacqueline.

SOLUTION: Figure 5-6 shows the tree diagram. Let's study the diagram. There are two branches from Start. These branches indicate the two possible choices for psychology sections. No matter which section of psychology Jacqueline chooses, she can choose from the two available A and P sections. Therefore, we have two

FIGURE 5-6

Tree Diagram for Selecting Class Schedules

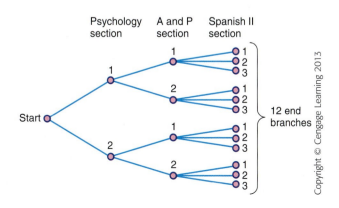

branches leading from *each* psychology branch. Finally, after the psychology and A and P sections are selected, there are three choices for Spanish II. That is why there are three branches from *each* A and P section.

The tree ends with a total of 12 branches. The number of end branches tells us the number of possible schedules. The outcomes themselves can be listed from the tree by following each series of branches from start to end. For instance, the top branch from Start generates the schedules shown in Table 5-4. The other six schedules can be listed in a similar manner, except they begin with the second section of psychology.

TABLE 5-4	Schedules Utilizing Section 1 of Psychology	
Psychology Section	A and P Section	Spanish II Section
1	1	1
1	1	2
1	1	3
1	2	1
1	2	2
1	2	3

Copyright © Cengage Learning 2013

GUIDED EXERCISE 10 | **Tree diagram and multiplication rule**

Louis plays three tennis matches. Use a tree diagram to list the possible win and loss sequences Louis can experience for the set of three matches.

(a) On the first match Louis can win or lose. From Start, indicate these two branches.

FIGURE 5-7 *W* = Win, *L* = Lose

(b) Regardless of whether Louis wins or loses the first match, he plays the second and can again win or lose. Attach branches representing these two outcomes to *each* of the first-match results.

FIGURE 5-8

(c) Louis may win or lose the third match. Attach branches representing these two outcomes to *each* of the second-match results.

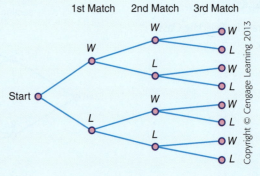

FIGURE 5-9

Continued

GUIDED EXERCISE 10   *continued*

(d) How many possible win–loss sequences are there for the three matches?

➡ Since there are eight branches at the end, there are eight sequences.

(e) Complete this list of win–loss sequences.

1st	2nd	3rd
W	W	W
W	W	L
W	L	W
W	L	L
___	___	___
___	___	___
___	___	___
___	___	___

➡ The last four sequences all involve a loss on Match 1.

1st	2nd	3rd
L	W	W
L	W	L
L	L	W
L	L	L

(f) Use the multiplication rule to compute the total number of outcomes for the three matches.

➡ The number of outcomes for a series of three events, each with two outcomes, is

$$2 \times 2 \times 2 = 8$$

Tree diagrams help us display the outcomes of an experiment involving several stages. If we label each branch of the tree with an appropriate probability, we can use the tree diagram to help us compute the probability of an outcome displayed on the tree. One of the easiest ways to illustrate this feature of tree diagrams is to use an experiment of drawing balls out of an urn. We do this in the next example.

EXAMPLE 11   TREE DIAGRAM AND PROBABILITY

Suppose there are five balls in an urn. They are identical except for color. Three of the balls are red and two are blue. You are instructed to draw out one ball, note its color, and set it aside. Then you are to draw out another ball and note its color. What are the outcomes of the experiment? What is the probability of each outcome?

SOLUTION: The tree diagram in Figure 5-10 will help us answer these questions. Notice that since you did not replace the first ball before drawing the second one, the two stages of the experiment are dependent. The probability associated with

FIGURE 5-10

Tree Diagram for Urn Experiment

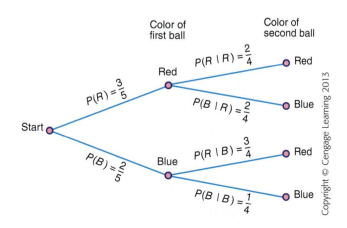

the color of the second ball depends on the color of the first ball. For instance, on the top branches, the color of the first ball drawn is red, so we compute the probabilities of the color of the second ball accordingly. The tree diagram helps us organize the probabilities.

From the diagram, we see that there are four possible outcomes to the experiment. They are

RR = red on 1st *and* red on 2nd

RB = red on 1st *and* blue on 2nd

BR = blue on 1st *and* red on 2nd

BB = blue on 1st *and* blue on 2nd

To compute the probability of each outcome, we will use the multiplication rule for dependent events. As we follow the branches for each outcome, we will find the necessary probabilities.

$$P(R \text{ on 1st } and \text{ } R \text{ on 2nd}) = P(R) \cdot P(R \mid R) = \frac{3}{5} \cdot \frac{2}{4} = \frac{3}{10}$$

$$P(R \text{ on 1st } and \text{ } B \text{ and 2nd}) = P(R) \cdot P(B \mid R) = \frac{3}{5} \cdot \frac{2}{4} = \frac{3}{10}$$

$$P(B \text{ on 1st } and \text{ } R \text{ on 2nd}) = P(B) \cdot P(R \mid B) = \frac{2}{5} \cdot \frac{3}{4} = \frac{3}{10}$$

$$P(B \text{ on 1st } and \text{ } B \text{ and 2nd}) = P(B) \cdot P(B \mid B) = \frac{2}{5} \cdot \frac{1}{4} = \frac{1}{10}$$

Notice that the probabilities of the outcomes in the sample space add to 1, as they should.

Sometimes when we consider *n* items, we need to know the number of different *ordered arrangements* of the *n* items that are possible. The multiplication rules can help us find the number of possible ordered arrangements. Let's consider the classic example of determining the number of different ways in which eight people can be seated at a dinner table. For the first chair at the head of the table, there are eight choices. For the second chair, there are seven choices, since one person is already seated. For the third chair, there are six choices, since two people are already seated. By the time we get to the last chair, there is only one person left for that seat. We can view each arrangement as an outcome of a series of eight events. Event 1 is *fill the first chair*, event 2 is *fill the second chair*, and so forth. The multiplication rule will tell us the number of different outcomes.

Choices for	1st	2nd	3rd	4th	5th	6th	7th	8th	chair position
	↓	↓	↓	↓	↓	↓	↓	↓	
	(8)	(7)	(6)	(5)	(4)	(3)	(2)	(1)	= 40,320

In all, there are 40,320 different seating arrangements for eight people. It is no wonder that it takes a little time to seat guests at a dinner table!

The multiplication pattern shown above is not unusual. In fact, it is an example of the multiplication indicated by the *factorial notation* 8!.

Factorial notation

! is read "factorial"

8! is read "8 factorial"

8! = 8 · 7 · 6 · 5 · 4 · 3 · 2 · 1

In general, *n*! indicates the product of *n* with each of the positive counting numbers less than *n*. *By special definition*, 0!=1.

Factorial notation

For a counting number n,
$$0! = 1$$
$$1! = 1$$
$$n! = n(n-1)(n-2)\ \cdots\ 1$$

GUIDED EXERCISE 11 | **Factorial**

(a) Evaluate $3!$.

    $3! = 3 \cdot 2 \cdot 1 = 6$

(b) In how many different ways can three objects be arranged in order? How many choices do you have for the first position? for the second position? for the third position?

You have three choices for the first position, two for the second position, and one for the third position. By the multiplication rule, you have

$$(3)(2)(1) = 3! = 6 \text{ arrangements.}$$

We have considered the number of ordered arrangements of n objects taken as an entire group. But what if we don't arrange the entire group? Specifically, we considered a dinner party for eight and found the number of ordered seating arrangements for all eight people. However, suppose you have an open house and have only five chairs. How many ways can five of the eight people seat themselves in the chairs? The formula we use to compute this number is called the *permutation formula*. As we see in the next example, the *permutations rule* is really another version of the multiplication rule.

Permutations rule

Counting rule for permutations

The number of ways to *arrange in order* n distinct objects, taking them r at a time, is

$$P_{n,r} = \frac{n!}{(n-r)!} \qquad\qquad (9)$$

where n and r are whole numbers and $n \geq r$. Another commonly used notation for permutations is nPr.

EXAMPLE 12   PERMUTATIONS RULE

Let's compute the number of possible ordered seating arrangements for eight people in five chairs.

SOLUTION: In this case, we are considering a total of $n = 8$ different people, and we wish to arrange $r = 5$ of these people. Substituting into formula (9), we have

$$P_{n,r} = \frac{n!}{(n-r)!}$$

$$P_{8,5} = \frac{8!}{(8-5)!} = \frac{8!}{3!} = \frac{40,320}{6} = 6720$$

Using the multiplication rule, we get the same results:

Chair	1		2		3		4		5		
Choices	8	×	7	×	6	×	5	×	4	=	6720

The permutations rule has the advantage of using factorials. Most scientific calculators have a factorial key (!) as well as a permutations key (nPr) (see Tech Notes).

TECH NOTES   Most scientific calculators have a factorial key, often designated $x!$ or $n!$. Many of these same calculators have the permutation function built in, often labeled nPr. They also have the combination function, which is discussed next. The combination function is often labeled nCr.

TI-84Plus/TI-83Plus/TI-*n*spire (with TI-84Plus keypad) The factorial, permutation, and combination functions are all under **MATH**, then **PRB**.

Excel 2010 Click on **Insert Function** $\boxed{f_x}$, then select **all** for the category. **Fact** gives factorials, **Permut** gives permutations, and **Combin** gives combinations.

In each of our previous counting formulas, we have taken the *order* of the objects or people into account. However, suppose that in your political science class you are given a list of 10 books. You are to select 4 to read during the semester. The order in which you read the books is not important. We are interested in the *different groupings* or *combinations* of 4 books from among the 10 on the list. The next formula tells us how to compute the number of different combinations.

> **Counting rule for combinations**
>
> The number of *combinations* of n objects taken r at a time is
>
> $$C_{n,r} = \frac{n!}{r!(n-r)!} \tag{10}$$
>
> where n and r are whole numbers and $n \geq r$. Other commonly used notations for combinations include nCr and $\binom{n}{r}$.

Notice the difference between the concepts of permutations and combinations. When we consider permutations, we are considering groupings *and order*. When we consider combinations, we are considering only the number of different groupings. For combinations, order within the groupings is not considered. As a result, the number of combinations of n objects taken r at a time is generally smaller than the number of permutations of the same n objects taken r at a time. In fact, the combinations formula is simply the permutations formula with the number of permutations of each distinct group divided out. In the formula for combinations, notice the factor of $r!$ in the denominator.

Combinations rule

Now let's look at an example in which we use the *combinations rule* to compute the number of *combinations* of 10 books taken 4 at a time.

EXAMPLE 13   COMBINATIONS

In your political-science class, you are assigned to read any 4 books from a list of 10 books. How many different groups of 4 are available from the list of 10?

SOLUTION: In this case, we are interested in *combinations*, rather than permutations, of 10 books taken 4 at a time. Using $n = 10$ and $r = 4$, we have

$$C_{n,r} = \frac{n!}{r!(n-r)!} = \frac{10!}{4!(10-4)!} = 210$$

There are 210 different groups of 4 books that can be selected from the list of 10. An alternate solution method is to use the combinations key (often nCr or $C_{n,r}$) on a calculator.

LOOKING FORWARD

We will see the combinations rule again in Section 6.2 when we discuss the formula for the binomial probability distribution.

PROCEDURE

HOW TO DETERMINE THE NUMBER OF OUTCOMES OF AN EXPERIMENT

1. If the experiment consists of a series of stages with various outcomes, use the multiplication rule or a tree diagram.

2. If the outcomes consist of ordered subgroups of r items taken from a group of n items, use the permutations rule, $P_{n,r}$.

$$P_{n,r} = \frac{n!}{(n-r)!} \qquad\qquad (9)$$

3. If the outcomes consist of nonordered subgroups of r items taken from a group of n items, use the combinations rule, $C_{n,r}$.

$$C_{n,r} = \frac{n!}{r!(n-r)!} \qquad\qquad (10)$$

GUIDED EXERCISE 12 | **Permutations and combinations**

The board of directors at Belford Community Hospital has 12 members.

(i) Three officers—president, vice president, and treasurer—must be elected from among the members. How many different slates of officers are possible? We will view a slate of officers as a list of three people, with the president listed first, the vice president listed second, and the treasurer listed third. For instance, if Mr. Acosta, Ms. Hill, and Mr. Smith wish to be on a slate together, there are several different slates possible, depending on the person listed for each office. Not only are we asking for the number of different groups of three names for a slate, we are also concerned about order.

(a) Do we use the permutations rule or the combinations rule? What is the value of n? What is the value of r?

➡ We use the permutations rule, since order is important. The size of the group from which the slates of officers are to be selected is n. The size of each slate is r.

$n = 12$ and $r = 3$

(b) Use the permutations rule with $n = 12$ and $r = 3$ to compute $P_{12,3}$.

➡ $P_{n,r} = \dfrac{n!}{(n-r)!} = \dfrac{12!}{(12-3!)} = 1320$

An alternative is to use the permutations key on a calculator.

Continued

GUIDED EXERCISE 12 *continued*

(ii) Three members from the group of 12 on the board of directors at Belford Community Hospital will be selected to go to a convention (all expenses paid) in Hawaii. How many different groups of 3 are there?

(c) Do we use the permutations rule or the combinations rule? What is the value of n? What is the value of r?

We use the combinations rule, because order is not important. The size of the board is $n = 12$ and the size of each group going to the convention is $r = 3$.

(d) Use the combinations rule with $n = 12$ and $r = 3$ to compute $C_{12,3}$.

$$C_{n,r} = \frac{n!}{r!(n-r)!} = \frac{12!}{3!(12-3)!} = 220$$

An alternative is to use the combinations key on a calculator.

VIEWPOINT | Powerball

Powerball is a multistate lottery game that consists of drawing five distinct whole numbers from the numbers 1 through 59 in any order. Then one more number from the numbers 1 through 39 is selected as the Powerball number (this number could be one of the original five). Powerball numbers are drawn every Wednesday and Saturday. If you match all six numbers, you win the jackpot, which is worth at least $10 million. Use methods of this section to show that there are 195,249,054 possible Powerball plays. For more information about the game of Powerball and the probability of winning different prizes, visit the Brase/Brase statistics site at **http:// www.cengagebrain.com** *and find the link to the Multi-State Lottery Association. Then select Powerball.*

SECTION 5.3 PROBLEMS

Access end-of-section exercises at www.aplia.com

Chapter Review

SUMMARY

In this chapter we explored basic features of probability.

- The probability of an event A is a number between 0 and 1, inclusive. The more likely the event, the closer the probability of the event is to 1.

- Three main ways to determine the probability of an event are the method of relative frequency, the method of equally likely outcomes, and intuition. Other important ways will be discussed later.

- The law of large numbers indicates that as the number of trials of a statistical experiment or observation increases, the relative frequency of a designated event becomes closer to the theoretical probability of that event.

- Events are mutually exclusive if they cannot occur together. Events are independent if the occurrence of one event does not change the probability of the occurrence of the other.

- Conditional probability is the probability that one event will occur, given that another event has occurred.

- The complement rule gives the probability that an event will not occur. The addition rule gives the probability that at least one of two specified events will occur. The multiplication rule gives the probability that two events will occur together.

- To determine the probability of equally likely events, we need to know how many outcomes are possible. Devices such as tree diagrams and counting rules—such as the multiplication rule of counting, the permutations rule, and the combinations rule—help us determine the total number of outcomes of a statistical experiment or observation.

In most of the statistical applications of later chapters, we will use the addition rule for mutually exclusive events and the multiplication rule for independent events.

IMPORTANT WORDS & SYMBOLS

VIEWPOINT | Deathday and Birthday

Can people really postpone death? If so, how much can the timing of death be influenced by psychological, social, or other influential factors? One special event is a birthday. Do famous people try to postpone their deaths until an important birthday? Both Thomas Jefferson and John Adams died on July 4, 1826, when the United States was celebrating its 50th birthday. Is this only a strange coincidence, or is there an unexpected connection between birthdays and deathdays? The probability associated with a death rate decline of famous people just before important birthdays has been studied by Professor D. P. Phillips of the State University of New York and is presented in the book Statistics: A Guide to the Unknown, *edited by J. M. Tanur.*

CHAPTER REVIEW PROBLEMS

1. *Statistical Literacy* Consider the following two events for an individual:

   A = owns a cell phone B = owns a laptop computer

   Translate each event into words.
   (a) A^c
   (b) A *and* B
   (c) A *or* B
   (d) $A \mid B$
   (e) $B \mid A$

2. *Statistical Literacy* If two events A and B are mutually exclusive, what is the value of $P(A \text{ and } B)$?

3. *Statistical Literacy* If two events A and B are independent, how do the probabilities $P(A)$ and $P(A \mid B)$ compare?

4. *Interpretation* You are considering 2 facial cosmetic surgeries. These are elective surgeries and their outcomes are independent. The probability of success for each surgery is 0.90. What is the probability of success for both surgeries? If the probability of success for both surgeries is less than 0.85, you will decide not to have the surgeries. Will you have the surgeries or not?

5. *Interpretation* You are applying for two jobs, and you estimate the probability of getting an offer for the first job is 0.70 while the probability of getting an offer for the second job is 0.80. Assume the job offers are independent.
   (a) Compute the probability of getting offers for both jobs. How does this probability compare to the probability of getting each individual job offer?
   (b) Compute the probability of getting an offer for either the first or the second job. How does this probability compare to the probability of getting each individual job offer? Does it seem worthwhile to apply for both jobs? Explain.

6. *Critical Thinking* You are given the information that $P(A) = 0.30$ and $P(B) = 0.40$.
   (a) Do you have enough information to compute $P(A \text{ or } B)$? Explain.
   (b) If you know that events A and B are mutually exclusive, do you have enough information to compute $P(A \text{ or } B)$? Explain.

7. *Critical Thinking* You are given the information that $P(A) = 0.30$ and $P(B) = 0.40$.
   (a) Do you have enough information to compute $P(A \text{ and } B)$? Explain.
   (b) If you know that events A and B are independent, do you have enough information to compute $P(A \text{ and } B)$? Explain.

8. *Critical Thinking* For a class activity, your group has been assigned the task of generating a quiz question that requires use of the formula for conditional probability to compute $P(B \mid A)$. Your group comes up with the following question: "If $P(A \text{ and } B) = 0.40$ and $P(A) = 0.20$, what is the value of $P(B \mid A)$?" What is wrong with this question? *Hint:* Consider the answer you get when using the correct formula, $P(B \mid A) = P(A \text{ and } B)/P(A)$.

9. *Salary Raise: Women* Does it pay to ask for a raise? A national survey of heads of households showed the percentage of those who asked for a raise and the percentage who got one (*USA Today*). According to the survey, of the women interviewed, 24% had asked for a raise, and of those women who had asked for a raise, 45% received the raise. If a woman is selected at random from the survey population of women, find the following probabilities: P(woman asked for a raise); P(woman received raise, *given* she asked for one); P(woman asked for raise *and* received raise).

10. *Salary Raise: Men* According to the same survey quoted in Problem 9, of the men interviewed, 20% had asked for a raise and 59% of the men who had asked for a

raise received the raise. If a man is selected at random from the survey population of men, find the following probabilities: P(man asked for a raise); P(man received raise, *given* he asked for one); P(man asked for raise *and* received raise).

11. *General: Thumbtack* Drop a thumbtack and observe how it lands.
    (a) Describe how you could use a relative frequency to estimate the probability that a thumbtack will land with its flat side down.
    (b) What is the sample space of outcomes for the thumbtack?
    (c) How would you make a probability assignment to this sample space if, when you drop 500 tacks, 340 land flat side down?

12. *Survey: Reaction to Poison Ivy* Allergic reactions to poison ivy can be miserable. Plant oils cause the reaction. Researchers at the Allergy Institute did a study to determine the effects of washing the oil off within 5 minutes of exposure. A random sample of 1000 people with known allergies to poison ivy participated in the study. Oil from the poison ivy plant was rubbed on a patch of skin. For 500 of the subjects, it was washed off *within* 5 minutes. For the other 500 subjects, the oil was washed off *after* 5 minutes. The results are summarized in Table 5-5.

    TABLE 5-5   **Time Within Which Oil Was Washed Off**

Reaction	Within 5 Minutes	After 5 Minutes	Row Total
None	420	50	470
Mild	60	330	390
Strong	20	120	140
Column Total	500	500	1000

    Copyright © Cengage Learning 2013

    Let's use the following notation for the various events: W = washing oil off within 5 minutes, A = washing oil off after 5 minutes, N = no reaction, M = mild reaction, S = strong reaction. Find the following probabilities for a person selected at random from this sample of 1000 subjects.
    (a) $P(N)$, $P(M)$, $P(S)$
    (b) $P(N \mid W)$, $P(S \mid W)$
    (c) $P(N \mid A)$, $P(S \mid A)$
    (d) $P(N \text{ and } W)$, $P(M \text{ and } W)$
    (e) $P(N \text{ or } M)$. Are the events N = no reaction and M = mild reaction mutually exclusive? Explain.
    (f) Are the events N = no reaction and W = washing oil off within 5 minutes independent? Explain.

13. *General: Two Dice* In a game of craps, you roll two fair dice. Whether you win or lose depends on the sum of the numbers appearing on the tops of the dice. Let x be the random variable that represents the sum of the numbers on the tops of the dice.
    (a) What values can x take on?
    (b) What is the probability distribution of these x values (that is, what is the probability that $x = 2$, 3, etc.)?

14. *Academic: Passing French* Class records at Rockwood College indicate that a student selected at random has probability 0.77 of passing French 101. For the student who passes French 101, the probability is 0.90 that he or she will pass French 102. What is the probability that a student selected at random will pass both French 101 and French 102?

15. *Combination: City Council* There is money to send two of eight city council members to a conference in Honolulu. All want to go, so they decide to choose the members to go to the conference by a random process. How many different combinations of two council members can be selected from the eight who want to go to the conference?

16. | *Basic Computation* Compute: (a) $P_{7,2}$ (b) $C_{7,2}$ (c) $P_{3,3}$ (d) $C_{4,4}$

17. | *Counting: Exam Answers* There are five multiple-choice questions on an exam, each with four possible answers. Determine the number of possible answer sequences for the five questions. Only one of the sets can contain all five correct answers. If you are guessing, so that you are as likely to choose one sequence of answers as another, what is the probability of getting all five answers correct?

18. | *Scheduling: College Courses* A student must satisfy the literature, social science, and philosophy requirements this semester. There are four literature courses to select from, three social science courses, and two philosophy courses. Make a tree diagram showing all the possible sequences of literature, social science, and philosophy courses.

19. | *General: Combination Lock* To open a combination lock, you turn the dial to the right and stop at a number; then you turn it to the left and stop at a second number. Finally, you turn the dial back to the right and stop at a third number. If you used the correct sequence of numbers, the lock opens. If the dial of the lock contains 10 numbers, 0 through 9, determine the number of different combinations possible for the lock. Note: The same number can be reused.

20. | *General: Combination Lock* You have a combination lock. Again, to open it you turn the dial to the right and stop at a first number; then you turn it to the left and stop at a second number. Finally, you turn the dial to the right and stop at a third number. Suppose you remember that the three numbers for your lock are 2, 9, and 5, but you don't remember the order in which the numbers occur. How many sequences of these three numbers are possible?

DATA HIGHLIGHTS: GROUP PROJECTS

Break into small groups and discuss the following topics. Organize a brief outline in which you summarize the main points of your group discussion.

1. Look at Figure 5-11, "Who's Cracking the Books?"

   (a) Does the figure show the probability distribution of grade records for male students? for female students? Describe all the grade-record probability distributions shown in Figure 5-11. Find the probability that a male student selected at random has a grade record showing mostly A's.

   (b) Is the probability distribution shown for all students making mostly A's? Explain your answer. *Hint:* Do the percentages shown for mostly A's add

FIGURE 5-11

Who's Cracking the Books?

Undergraduate Grade Record by Student Characteristic

Student characteristic		C's and D's or lower	B's and C's	Mostly B's	A's and B's	Mostly A's
Gender:	Men	38.8 %	16.6 %	22.6 %	9.6 %	12.4 %
	Women	29.4	16.2	26.2	12.0	16.2
Class level:	Graduating senior	15.8 %	21.9 %	34.5 %	14.7 %	13.0 %
	All other class levels	35.4	15.8	23.6	10.5	14.7
Age:	18 or younger	42.6 %	14.7 %	23.4 %	9.4 %	10.0 %
	19 to 23	38.1	19.0	25.1	9.4	8.3
	24 to 29	33.3	16.9	24.7	10.3	14.9
	30 to 39	23.1	13.2	25.9	14.8	23.0
	40 and older	20.1	10.1	22.0	14.8	33.0

Source: U.S. Department of Education

up to 1? Can Figure 5-11 be used to determine the probability that a student selected at random has mostly A's? Can it be used to determine the probability that a female student selected at random has mostly A's? What is the probability?

(c) Can we use the information shown in the figure to determine the probability that a graduating senior has grades consisting of mostly B's or higher? What is the probability?

(d) Does Figure 5-11 give sufficient information to determine the probability that a student selected at random is in the age range 19 to 23 *and* has grades that are mostly B's? What is the probability that a student selected at random has grades that are mostly B's, *given* that he or she is in the age range 19 to 23?

(e) Suppose that 65% of the students at State University are between 19 and 23 years of age. What is the probability that a student selected at random is in this age range *and* has grades that are mostly B's?

2. Consider the information given in Figure 5-12, "Vulnerable Knees." What is the probability that an orthopedic case selected at random involves knee problems? Of those cases, estimate the probability that the case requires full knee replacement. Compute the probability that an orthopedic case selected at random involves a knee problem *and* requires a full knee replacement. Next, look at the probability distribution for ages of patients requiring full knee replacements. Medicare insurance coverage begins when a person reaches age 65. What is the probability that the age of a person receiving a knee replacement is 65 or older?

FIGURE 5-12

Vulnerable Knees

Source: American Academy of Orthopedic Surgeons

LINKING CONCEPTS: WRITING PROJECTS

Discuss each of the following topics in class or review the topics on your own. Then write a brief but complete essay in which you summarize the main points. Please include formulas as appropriate.

1. Discuss the following concepts and give examples from everyday life in which you might encounter each concept. *Hint:* For instance, consider the "experiment" of arriving for class. Some possible outcomes are not arriving (that is, missing class), arriving on time, and arriving late.

   (a) Sample space.

   (b) Probability assignment to a sample space. In your discussion, be sure to include answers to the following questions:

      (i) Is there more than one valid way to assign probabilities to a sample space? Explain and give an example.

      (ii) How can probabilities be estimated by relative frequencies? How can probabilities be computed if events are equally likely?

2. Discuss the concepts of mutually exclusive events and independent events. List several examples of each type of event from everyday life.

   (a) If A and B are mutually exclusive events, does it follow that A and B *cannot* be independent events? Give an example to demonstrate your answer. *Hint:* Discuss an election where only one person can win the election. Let A be the event that party A's candidate wins, and let B be the event that party B's candidate wins. Does the outcome of one event determine the outcome of the other event? Are A and B mutually exclusive events?

   (b) Discuss the conditions under which $P(A \text{ and } B) = P(A) \cdot P(B)$ is true. Under what conditions is this not true?

   (c) Discuss the conditions under which $P(A \text{ or } B) = P(A) + P(B)$ is true. Under what conditions is this not true?

3. Although we learn a good deal about probability in this course, the main emphasis is on statistics. Write a few paragraphs in which you talk about the distinction between probability and statistics. In what types of problems would probability be the main tool? In what types of problems would statistics be the main tool? Give some examples of both types of problems. What kinds of outcomes or conclusions do we expect from each type of problem?

USING TECHNOLOGY

Demonstration of the Law of Large Numbers

Computers can be used to simulate experiments. With packages such as Excel 2010, Minitab, and SPSS, programs using random-number generators can be designed (see the *Technology Guides*) to simulate activities such as tossing a die.

The following printouts show the results of the simulations for tossing a die 6, 500, 50,000, 500,000, and 1,000,000 times. Notice how the relative frequencies of the outcomes approach the theoretical probabilities of 1/6 or 0.16667 for each outcome. Do you expect the same results every time the simulation is done? Why or why not?

Results of tossing one die 6 times

Outcome	Number of Occurrences	Relative Frequency
⚀	0	0.00000
⚁	1	0.16667
⚂	2	0.33333
⚃	0	0.00000
⚄	1	0.16667
⚅	2	0.33333

Results of tossing one die 500 times

Outcome	Number of Occurrences	Relative Frequency
⚀	87	0.17400
⚁	83	0.16600
⚂	91	0.18200
⚃	69	0.13800
⚄	87	0.17400
⚅	83	0.16600

Results of tossing one die 50,000 times

Outcome	Number of Occurrences	Relative Frequency
⚀	8,528	0.17056
⚁	8,354	0.16708
⚂	8,246	0.16492
⚃	8,414	0.16828
⚄	8,178	0.16356
⚅	8,280	0.16560

Results of tossing one die 500,000 times

Outcome	Number of Occurrences	Relative Frequency
⚀	83,644	0.16729
⚁	83,368	0.16674
⚂	83,398	0.16680
⚃	83,095	0.16619
⚄	83,268	0.16654
⚅	83,227	0.16645

Results of tossing one die 1,000,000 times

Outcome	Number of Occurrences	Relative Frequency
⚀	166,643	0.16664
⚁	166,168	0.16617
⚂	167,391	0.16739
⚃	165,790	0.16579
⚄	167,243	0.16724
⚅	166,765	0.16677

6

AP Photo/Matt Rourke

6.1 Introduction to Random Variables and Probability Distributions

6.2 Binomial Probabilities

6.3 Additional Properties of the Binomial Distribution

National Portrait Gallery, Smithsonian Institution/Art Resource, NY

Education is the key to unlock the golden door of freedom.

—George Washington Carver

George Washington Carver (1859–1943) won international fame for agricultural research. After graduating from Iowa State Agricultural College, he was appointed a faculty member in the Iowa State Botany Department. Carver took charge of the greenhouse and started a fungus collection that later included more than 20,000 species. This collection brought him professional acclaim in the field of botany.

At the invitation of his friend Booker T. Washington, Carver joined the faculty of the Tuskegee Institute, where he spent the rest of his long and distinguished career. Carver's creative genius accounted for more than 300 inventions from peanuts, 118 inventions from sweet potatoes, and 75 inventions from pecans.

Gathering and analyzing data were important components of Carver's work. Methods you will learn in this course are widely used in research in every field, including agriculture.

For online student resources, visit the Brase/Brase, *Understanding Basic Statistics,* 6th edition website at http://www.cengagebrain.com

The Binomial Probability Distribution and Related Topics

Andresr/istockphoto.com

PREVIEW QUESTIONS

What is a random variable? How do you compute μ and σ for a discrete random variable? (SECTION 6.1)

Many of life's experiences consist of some successes together with some failures. Suppose you make n attempts to succeed at a certain project. How can you use the binomial probability distribution to compute the probability of r successes? (SECTION 6.2)

How do you compute μ and σ for the binomial distribution? (SECTION 6.3)

FOCUS PROBLEM

Personality Preference Types: Introvert or Extrovert?

Isabel Briggs Myers was a pioneer in the study of personality types. Her work has been used successfully in counseling, educational, and industrial settings. In the book *Manual: A Guide to the Development and Use of the Myers-Briggs Type Indicator*, by Myers and McCaulley, it was reported that based on a very large sample (2282 professors), approximately 45% of all university professors are extroverted.

After completing this chapter, you will be able to answer the following questions. Suppose you have classes with six different professors.

(a) What is the probability that all six are extroverts?

(b) What is the probability that none of your professors is an extrovert?

(c) What is the probability that at least two of your professors are extroverts?

(d) In a group of six professors selected at random, what is the *expected number* of extroverts? What is the *standard deviation* of the distribution? (See Problem 18 of Section 6.3.)

Mitch Wojnarowicz/The Image Works

COMMENT   Both extroverted and introverted professors can be excellent teachers.

SECTION 6.1

Introduction to Random Variables and Probability Distributions

FOCUS POINTS
- Distinguish between discrete and continuous random variables.
- Graph discrete probability distributions.
- Compute μ and σ for a discrete probability distribution.

Random Variables

For our purposes, we say that a *statistical experiment* or *observation* is any process by which measurements are obtained. For instance, you might count the number of eggs in a robin's nest or measure daily rainfall in inches. It is common practice to use the letter x to represent the quantitative result of an experiment or observation. As such, we call x a variable.

Random variable

Discrete random variable

Continuous random variable

A quantitative variable x is a **random variable** if the value that x takes on in a given experiment or observation is a chance or random outcome.

A **discrete random variable** can take on only a finite number of values or a countable number of values.

A **continuous random variable** can take on any of the countless number of values in a line interval.

The distinction between discrete and continuous random variables is important because of the different mathematical techniques associated with the two kinds of random variables.

In most of the cases we will consider, a *discrete random variable* will be the result of a count. The number of students in a statistics class is a discrete random variable. Values such as 15, 25, 50, and 250 are all possible. However, 25.5 students is not a possible value for the number of students.

Most of the *continuous random variables* we will see will occur as the result of a measurement on a continuous scale. For example, the air pressure in an automobile tire represents a continuous random variable. The air pressure could, in theory, take on any value from 0 lb/in^2 (psi) to the bursting pressure of the tire. Values such as 20.126 psi, 20.12678 psi, and so forth are possible.

GUIDED EXERCISE 1 | Discrete or continuous random variables

Which of the following random variables are discrete and which are continuous?

(a) *Measure* the time it takes a student selected at random to register for the fall term.

⟹   Time can take on any value, so this is a continuous random variable.

(b) *Count* the number of bad checks drawn on Upright Bank on a day selected at random.

⟹   The number of bad checks can be only a whole number such as 0, 1, 2, 3, etc. This is a discrete variable.

(c) *Measure* the amount of gasoline needed to drive your car 200 miles.

⟹   We are measuring volume, which can assume any value, so this is a continuous random variable.

(d) Pick a random sample of 50 registered voters in a district and find the number who voted in the last county election.

⟹   This is a count, so the variable is discrete.

Probability Distribution of a Discrete Random Variable

A random variable has a probability distribution whether it is discrete or continuous.

Probability distribution

A **probability distribution** is an assignment of probabilities to each distinct value of a discrete random variable or to each interval of values of a continuous random variable.

Features of the probability distribution of a discrete random variable

1. The probability distribution has a probability assigned to *each* distinct value of the random variable.

2. The sum of all the assigned probabilities must be 1.

EXAMPLE 1   DISCRETE PROBABILITY DISTRIBUTION

Dr. Mendoza developed a test to measure boredom tolerance. He administered it to a group of 20,000 adults between the ages of 25 and 35. The possible scores were 0, 1, 2, 3, 4, 5, and 6, with 6 indicating the highest tolerance for boredom. The test results for this group are shown in Table 6-1.

(a) If a subject is chosen at random from this group, the probability that he or she will have a score of 3 is 6000/20,000, or 0.30. In a similar way, we can use relative frequencies to compute the probabilities for the other scores (Table 6-2). These probability assignments make up the probability distribution. Notice that the scores are mutually exclusive: No one subject has two scores. The sum of the probabilities of all the scores is 1.

(b) The graph of this distribution is simply a relative-frequency histogram (see Figure 6-1) in which the height of the bar over a score represents the probability of that score. Since each bar is one unit wide, the area of the

TABLE 6-1	Boredom Tolerance Test Scores for 20,000 Subjects
Score	Number of Subjects
0	1400
1	2600
2	3600
3	6000
4	4400
5	1600
6	400

TABLE 6-2	Probability Distribution of Scores on Boredom Tolerance Test
Score x	Probability $P(x)$
0	0.07
1	0.13
2	0.18
3	0.30
4	0.22
5	0.08
6	0.02
	$\Sigma P(x) = 1$

FIGURE 6-1

Graph of the Probability Distribution of Test Scores

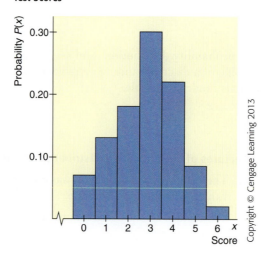

bar over a score equals the height and thus represents the probability of that score. Since the sum of the probabilities is 1, the area under the graph is also 1.

(c) The Topnotch Clothing Company needs to hire someone with a score on the boredom tolerance test of 5 or 6 to operate the fabric press machine. Since the scores 5 and 6 are mutually exclusive, the probability that someone in the group who took the boredom tolerance test made either a 5 or a 6 is the sum

$$P(5 \text{ or } 6) = P(5) + P(6)$$
$$= 0.08 + 0.02 = 0.10$$

Notice that to find $P(5 \text{ or } 6)$, we could have simply added the *areas* of the bars over 5 and over 6. One out of 10 of the group who took the boredom tolerance test would qualify for the position at Topnotch Clothing.

GUIDED EXERCISE 2 | **Discrete probability distribution**

One of the elementary tools of cryptanalysis (the science of code breaking) is the use of relative frequencies of occurrence of different letters in the alphabet to break standard English alphabet codes. Large samples of plain text, such as newspaper stories, generally yield about the same relative frequencies for letters. A sample 1000 letters long yielded the information in Table 6-3.

(a) Use the relative frequencies to compute the omitted probabilities in Table 6-3.

➡ Table 6-4 shows the completion of Table 6-3.

Continued

GUIDED EXERCISE 2   *continued*

TABLE 6-3 Frequencies of Letters in a 1000-Letter Sample

Letter	Freq.	Prob.	Letter	Freq.	Prob.
A	73	——	N	78	0.078
B	9	0.009	O	74	——
C	30	0.030	P	27	0.027
D	44	0.044	Q	3	0.003
E	130	——	R	77	0.077
F	28	0.028	S	63	0.063
G	16	0.016	T	93	0.093
H	35	0.035	U	27	——
I	74	——	V	13	0.013
J	2	0.002	W	16	0.016
K	3	0.003	X	5	0.005
L	35	0.035	Y	19	0.019
M	25	0.025	Z	1	0.001

Source: From *Elementary Cryptanalysis: A Mathematical Approach,* by Abraham Sinkov. Copyright © 1968 and renewed 1996 by Yale University. Used by permission of Random House, Inc. © Mathematical Association of America 2011. All rights reserved.

TABLE 6-4 Entries for Table 6-3

Letter	Relative Frequency	Probability
A	$\dfrac{73}{1,000}$	0.073
E	$\dfrac{130}{1,000}$	0.130
I	$\dfrac{74}{1,000}$	0.074
O	$\dfrac{74}{1,000}$	0.074
U	$\dfrac{27}{1,000}$	0.027

(b) Do the probabilities of all the individual letters add up to 1?

➡ Yes.

(c) If a letter is selected at random from a newspaper story, what is the probability that the letter will be a vowel?

➡ If a letter is selected at random,
$$P(a, e, i, o, \text{or } u) = P(a) + P(e) + P(i) + P(o) + P(u)$$
$$= 0.073 + 0.130 + 0.074 + 0.074 + 0.027$$
$$= 0.378$$

Mean and standard deviation of a discrete probability distribution

A probability distribution can be thought of as a relative-frequency distribution based on a very large *n*. As such, it has a mean and standard deviation. If we are referring to the probability distribution of a *population*, then we use the Greek letters μ for the mean and σ for the standard deviation. When we see the Greek letters used, we know the information given is from the *entire population* rather than just a sample. If we have a sample probability distribution, we use $\bar{x}$ (*x* bar) and *s*, respectively, for the mean and standard deviation.

The **mean** and the **standard deviation of a discrete population probability distribution** are found by using these formulas:

$$\mu = \Sigma x P(x); \ \mu \text{ is called the expected value of } x$$

$$\sigma = \sqrt{\Sigma (x - \mu)^2 P(x)}; \ \sigma \text{ is called the standard deviation of } x$$

where *x* is the value of a random variable,
    $P(x)$ is the probability of that variable, and
    the sum Σ is taken for all the values of the random variable.

Note: μ is the *population mean* and σ is the underlying *population standard deviation* because the sum Σ is taken over *all* values of the random variable (i.e., the entire sample space).

Expected value

The mean of a probability distribution is often called the *expected value* of the distribution. This terminology reflects the idea that the mean represents a "central point" or "cluster point" for the entire distribution. Of course, the mean or expected value is an average value, and as such, it *need not be a point of the sample space*.

The standard deviation is often represented as a measure of *risk*. A larger standard deviation implies a greater likelihood that the random variable x is different from the expected value μ.

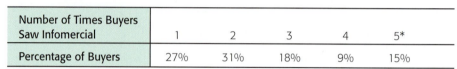

EXAMPLE 2

EXPECTED VALUE, STANDARD DEVIATION

Are we influenced to buy a product by an ad we saw on TV? The National Infomercial Marketing Association determined the number of times *buyers* of a product had watched a TV infomercial *before* purchasing the product. The results are shown here:

Number of Times Buyers Saw Infomercial	1	2	3	4	5*
Percentage of Buyers	27%	31%	18%	9%	15%

*This category was "5 or more," but will be treated as 5 in this example.

We can treat the information shown as an estimate of the probability distribution because the events are mutually exclusive and the sum of the percentages is 100%. Compute the mean and standard deviation of the distribution.

SOLUTION: We put the data in the first two columns of a computation table and then fill in the other entries (see Table 6-5). The average number of times a buyer views the infomercial before purchase is

$$\mu = \Sigma x P(x) = 2.54 \text{ (sum of column 3)}$$

To find the standard deviation, we take the square root of the sum of column 6:

$$\sigma = \sqrt{\Sigma(x - \mu)^2 P(x)} \approx \sqrt{1.869} \approx 1.37$$

TABLE 6-5   **Number of Times Buyers View Infomercial before Making Purchase**

x (Number of Viewings)	$P(x)$	$xP(x)$	$x - \mu$	$(x - \mu)^2$	$(x - \mu)^2 P(x)$
1	0.27	0.27	−1.54	2.372	0.640
2	0.31	0.62	−0.54	0.292	0.091
3	0.18	0.54	0.46	0.212	0.038
4	0.09	0.36	1.46	2.132	0.192
5	0.15	0.75	2.46	6.052	0.908
		$\mu = \Sigma x P(x) = 2.54$			$\Sigma(x - \mu)^2 P(x) = 1.869$

CALCULATOR NOTE Some calculators, including the TI-84Plus/TI-83Plus/TI-$n$spire (with TI-84Plus keypad) models, accept fractional frequencies. If yours does, you can get μ and σ directly by entering the outcomes in list L_1 with corresponding probabilities in list L_2. Then use **1-Var Stats L_1, L_2**.

GUIDED EXERCISE 3 | **Expected value**

At a carnival, you pay $2.00 to play a coin-flipping game with three fair coins. On each coin one side has the number 0 and the other side has the number 1. You flip the three coins at one time and you win $1.00 for every 1 that appears on top. Are your expected earnings equal to the cost to play? We'll answer this question in several steps.

(a) In this game, the random variable of interest counts the number of 1's that show. What is the sample space for the values of this random variable?

→ The sample space is {0, 1, 2, 3}, since any of these numbers of 1's can appear.

(b) There are eight equally likely outcomes for throwing three coins. They are 000, 001, 010, 011, 100, 101, _____, and _____.

→ 110 and 111.

(c) Complete Table 6-6.

TABLE 6-6

Number of 1's, x	Frequency	P(x)	xP(x)
0	1	0.125	0
1	3	0.375	_____
2	3	_____	_____
3	_____	_____	_____

Copyright © Cengage Learning 2013

→ **TABLE 6-7 Completion of Table 6-6**

Number of 1's, x	Frequency	P(x)	xP(x)
0	1	0.125	0
1	3	0.375	0.375
2	3	0.375	0.750
3	1	0.125	0.375

Copyright © Cengage Learning 2013

(d) The expected value is the sum

$$\mu = \Sigma x P(x)$$

Sum the appropriate column of Table 6-6 to find this value.

Interpretation Are your expected earnings less than, equal to, or more than the cost of the game?

→ The expected value can be found by summing the last column of Table 6-7. The expected value is $1.50. It costs $2.00 to play the game; the expected value is less than the cost. The carnival is making money. In the long run, the carnival can expect to make an average of about 50 cents per player.

We have seen probability distributions of discrete variables and the formulas to compute the mean and standard deviation of a discrete population probability distribution. Probability distributions of continuous random variables are similar except that the probability assignments are made to intervals of values rather than to specific values of the random variable. We will see an important example of a discrete probability distribution, the binomial distribution, in the next section, and one of a continuous probability distribution in Chapter 7 when we study the normal distribution.

We conclude this section with some useful information about combining random variables.

VIEWPOINT | The Rosetta Project

Around 196 B.C., Egyptian priests inscribed a decree on a granite slab affirming the rule of 13-year-old Ptolemy V. The proclamation was in Egyptian hieroglyphics with a translation in a form of ancient Greek. By 1799, the meaning of Egyptian hieroglyphics had been lost for many centuries. However, Napoleon's troops discovered the granite slab (the Rosetta Stone). Linguists then used the Rosetta Stone and their knowledge of ancient Greek to unlock the meaning of the Egyptian hieroglyphics.

Linguistic experts say that because of industrialization and globalization, by the year 2100 as many as 90% of the world's languages may be extinct. To help preserve some of these languages for future generations, 1000 translations of the first three chapters of Genesis have been inscribed in tiny text onto 3-inch nickel discs and encased in hardened glass balls that are expected to last at least 1000 years. Why Genesis? Because it is the most translated text in the world. The Rosetta Project is sending the discs to libraries and universities all over the world. It is very difficult to send information into the future. However, if in the year 2500 linguists are using the "Rosetta Discs" to unlock the meaning of a lost language, you may be sure they will use statistical methods of cryptanalysis (see Guided Exercise 2). To find out more about the Rosetta Project, visit the Brase/Brase statistics site at **http://www.cengagebrain.com** *and find the link to the Rosetta Project site.*

**SECTION 6.1
PROBLEMS**

 aplia™   Access end-of-section exercises at www.aplia.com

SECTION 6.2

Binomial Probabilities

FOCUS POINTS

- List the defining features of a binomial experiment.
- Compute binomial probabilities using the formula $P(r) = C_{n,r} p^r q^{n-r}$.
- Use the binomial table to find $P(r)$.
- Use the binomial probability distribution to solve real-world applications.

Binomial Experiment

On a TV game show, each contestant has a try at the wheel of fortune. The wheel of fortune is a roulette wheel with 36 slots, one of which is gold. If the ball lands in the gold slot, the contestant wins $50,000. No other slot pays. What is the probability that the game show will have to pay the fortune to three contestants out of 100?

In this problem, the contestant and the game show sponsors are concerned about only two outcomes from the wheel of fortune: The ball lands on the gold, or the ball does not land on the gold. This problem is typical of an entire class of problems that are characterized by the feature that there are exactly two possible outcomes (for each trial) of interest. These problems are called *binomial experiments*, or *Bernoulli experiments* after the Swiss mathematician Jacob Bernoulli, who studied them extensively in the late 1600s.

Binomial experiments

Features of a binomial experiment

Number of trials, *n*

Independent trials

Success, *S*
Failure, *F*

Probability of success, $P(S) = p$
Probability of failure,
$\quad P(F) = q = 1 - p$

Number of successes, *r*

Features of a binomial experiment

1. There is a *fixed number of trials*. We denote this number by the letter *n*.

2. The *n* trials are *independent* and repeated under identical conditions.

3. Each trial has only *two outcomes*: success, denoted by *S*, and failure, denoted by *F*.

4. For each individual trial, the *probability of success is the same*. We denote the probability of success by *p* and that of failure by *q*. Since each trial results in either success or failure, $p + q = 1$ and $q = 1 - p$.

5. The central problem of a binomial experiment is to find the *probability of r successes out of n trials*.

EXAMPLE 3   BINOMIAL EXPERIMENT

Let's see how the wheel of fortune problem meets the criteria of a binomial experiment. We'll take the criteria one at a time.

SOLUTION:

1. Each of the 100 contestants has a trial at the wheel, so there are $n = 100$ trials in this problem.

2. Assuming that the wheel is fair, the *trials are independent*, since the result of one spin of the wheel has no effect on the results of other spins.

3. We are interested in only two outcomes on each spin of the wheel: The ball either lands on the gold, or it does not. Let's call landing on the gold *success* (*S*) and not landing on the gold *failure* (*F*). In general, the assignment of the terms *success* and *failure* to outcomes does not imply good or bad results. These terms are assigned simply for the user's convenience.

4. On each trial the probability *p* of success (landing on the gold) is 1/36, since there are 36 slots and only one of them is gold. Consequently, the probability of failure is

$$q = 1 - p = 1 - \frac{1}{36} = \frac{35}{36}$$

on each trial.

5. We want to know the probability of 3 successes out of 100 trials, so $r = 3$ in this example. It turns out that the probability the quiz show will have to pay the fortune to 3 contestants out of 100 is about 0.23. Later in this section we'll see how this probability was computed.

Any time we make selections from a population *without replacement, we do not have independent trials*. However, replacement is often not practical. If the number of trials is quite small with respect to the population, we *almost* have independent trials, and we can say the situation is *closely approximated* by a binomial experiment. For instance, suppose we select 20 tuition bills at random from a collection of 10,000 bills issued at one college and observe if each bill is in error or not. If 600 of the 10,000 bills are in error, then the probability that the first one selected is in error is 600/10,000, or 0.0600. If the first is in error, then the probability that the second is in error is 599/9999, or 0.0599. Even if the first 19 bills selected are in error, the probability that the 20th is also in error is 581/9981, or 0.0582. All these probabilities round to 0.06, and we can say that the independence condition is approximately satisfied.

GUIDED EXERCISE 4 | **Binomial experiment**

Let's analyze the following binomial experiment to determine p, q, n, and r:

According to the *Textbook of Medical Physiology*, 5th edition, by Arthur Guyton, 9% of the population has blood type B. Suppose we choose 18 people at random from the population and test the blood type of each. What is the probability that three of these people have blood type B? *Note:* Independence is approximated because 18 people is an extremely small sample with respect to the entire population.

(a) In this experiment, we are observing whether or not a person has type B blood. We will say we have a success if the person has type B blood. What is failure?

➡ Failure occurs if a person does not have type B blood.

(b) The probability of success is 0.09, since 9% of the population has type B blood. What is the probability of failure, q?

➡ The probability of failure is
$$q = 1 - p$$
$$= 1 - 0.09 = 0.91.$$

(c) In this experiment, there are $n = \underline{\hspace{1cm}}$ trials.

➡ In this experiment, $n = 18$.

(d) We wish to compute the probability of 3 successes out of 18 trials. In this case, $r = \underline{\hspace{1cm}}$.

➡ In this case, $r = 3$.

Next, we will see how to compute the probability of r successes out of n trials when we have a binomial experiment.

Computing Probabilities for a Binomial Experiment Using the Binomial Distribution Formula

A model with three trials

The central problem of a binomial experiment is finding the probability of r successes out of n trials. Now we'll see how to find these probabilities.

Suppose you are taking a timed final exam. You have three multiple-choice questions left to do. Each question has four suggested answers, and only one of the answers is correct. You have only 5 seconds left to do these three questions, so you decide to mark answers on the answer sheet without even reading the questions. Assuming that your answers are randomly selected, what is the probability that you get zero, one, two, or all three questions correct?

This is a binomial experiment. Each question can be thought of as a trial, so there are $n = 3$ trials. The possible outcomes on each trial are success S, indicating a correct response, or failure F, meaning a wrong answer. The trials are independent—the outcome of any one trial does not affect the outcome of the others.

What is the *probability of success* on any one question? Since you are guessing and there are four answers from which to select, the probability of a correct answer is 0.25. The probability q of a wrong answer is then 0.75. In short, we have a binomial experiment with $n = 3$, $p = 0.25$, and $q = 0.75$.

Now, what are the possible outcomes in terms of success or failure for these three trials? Let's use the notation SSF to mean success on the first question, success on the second, and failure on the third. There are eight possible combinations of S's and F's. They are

SSS SSF SFS FSS SFF FSF FFS FFF

To compute the probability of each outcome, we can use the multiplication law because the trials are independent. For instance, the probability of success on the first two questions and failure on the last is

$$P(SSF) = P(S) \cdot P(S) \cdot P(F) = p \cdot p \cdot q = p^2 q = (0.25)^2(0.75) \approx 0.047$$

In a similar fashion, we can compute the probability of each of the eight outcomes. These are shown in Table 6-8, along with the number of successes r associated with each outcome.

TABLE 6-8   **Outcomes for a Binomial Experiment with $n = 3$ Trials**

Outcome	Probability of Outcome	r (Number of Successes)
SSS	$P(SSS) = P(S)P(S)P(S) = p^3 = (0.25)^3 \approx 0.016$	3
SSF	$P(SSF) = P(S)P(S)P(F) = p^2q = (0.25)^2(0.75) \approx 0.047$	2
SFS	$P(SFS) = P(S)P(F)P(S) = p^2q = (0.25)^2(0.75) \approx 0.047$	2
FSS	$P(FSS) = P(F)P(S)P(S) = p^2q = (0.25)^2(0.75) \approx 0.047$	2
SFF	$P(SFF) = P(S)P(F)P(F) = pq^2 = (0.25)(0.75)^2 \approx 0.141$	1
FSF	$P(FSF) = P(F)P(S)P(F) = pq^2 = (0.25)(0.75)^2 \approx 0.141$	1
FFS	$P(FFS) = P(F)P(F)P(S) = pq^2 = (0.25)(0.75)^2 \approx 0.141$	1
FFF	$P(FFF) = P(F)P(F)P(F) = q^3 = (0.75)^3 \approx 0.422$	0

Copyright © Cengage Learning 2013

Now we can compute the probability of r successes out of three trials for $r = 0, 1, 2,$ or 3. Let's compute $P(1)$. The notation $P(1)$ stands for the probability of one success. For three trials, there are three different outcomes that show exactly one success. They are the outcomes *SFF, FSF,* and *FFS*. Since the outcomes are mutually exclusive, we can add the probabilities. So,

$$P(1) = P(SFF \text{ or } FSF \text{ or } FFS) = P(SFF) + P(FSF) + P(FFS)$$
$$= pq^2 + pq^2 + pq^2$$
$$= 3pq^2$$
$$= 3(0.25)(0.75)^2$$
$$= 0.422$$

In the same way, we can find $P(0)$, $P(2)$, and $P(3)$. These values are shown in Table 6-9.

TABLE 6-9   **$P(r)$ for $n = 3$ Trials, $p = 0.25$**

r (Number of Successes)	$P(r)$ (Probability of r Successes in 3 Trials)	$P(r)$ for $p = 0.25$
0	$P(0) = P(FFF) = q^3$	0.422
1	$P(1) = P(SFF) + P(FSF) + P(FFS) = 3pq^2$	0.422
2	$P(2) = P(SSF) + P(SFS) + P(FSS) = 3p^2q$	0.141
3	$P(3) = P(SSS) = p^3$	0.016

Copyright © Cengage Learning 2013

We have done quite a bit of work to determine your chances of $r = 0, 1, 2,$ or 3 successes on three multiple-choice questions if you are just guessing. Now we see that there is only a small chance (about 0.016) that you will get them all correct.

Table 6-9 can be used as a model for computing the probability of r successes out of only *three* trials. How can we compute the probability of 7 successes out of 10 trials? We can develop a table for $n = 10$, but this would be a tremendous task because there are 1024 possible combinations of successes and failures on 10 trials. Fortunately, mathematicians have given us a direct formula to compute the probability of r successes for any number of trials.

General formula for binomial probability distribution

Formula for the binomial probability distribution

$$P(r) = \frac{n!}{r!(n-r)!} p^r q^{n-r} = C_{n,r} p^r q^{n-r}$$

where n = number of trials

p = probability of success on each trial

$q = 1 - p$ = probability of failure on each trial

r = random variable representing the number of successes out of n trials $(0 \leq r \leq n)$

! = factorial notation. Recall from Section 5.3 that the factorial symbol $n!$ designates the product of all the integers between 1 and n. For instance, $4! = 4 \cdot 3 \cdot 2 \cdot 1 = 24$. Special cases are $1! = 1$ and $0! = 1$.

$C_{n,r} = \dfrac{n!}{r!(n-r)!}$ is the binomial coefficient. Many calculators have a key designated nCr that gives the value of $C_{n,r}$ directly.

Binomial coefficient, $C_{n,r}$

Note: The binomial coefficient $C_{n,r}$ represents the number of combinations of n distinct objects (n = number of trials in this case) taken r at a time (r = number of successes). For more information about $C_{n,r}$, see Section 5.3.

Let's look more carefully at the formula for $P(r)$. There are two main parts. The expression $p^r q^{n-r}$ is the probability of getting one outcome with r successes and $n - r$ failures. The binomial coefficient $C_{n,r}$ counts the number of outcomes that have r successes and $n - r$ failures. For instance, in the case of $n = 3$ trials, we saw in Table 6-8 that the probability of getting an outcome with one success and two failures was pq^2. This is the value of $p^r q^{n-r}$ when $r = 1$ and $n = 3$. We also observed that there were three outcomes with one success and two failures, so $C_{3,1}$ is 3.

Now let's take a look at an application of the binomial distribution formula in Example 4.

EXAMPLE 4   COMPUTE $P(r)$ USING THE BINOMIAL DISTRIBUTION FORMULA

Privacy is a concern for many users of the Internet. One survey showed that 59% of Internet users are somewhat concerned about the confidentiality of their e-mail. Based on this information, what is the probability that for a random sample of 10 Internet users, 6 are concerned about the privacy of their e-mail?

SOLUTION:

(a) This is a binomial experiment with 10 trials. If we assign success to an Internet user being concerned about the privacy of e-mail, the probability of success is 59%. We are interested in the probability of 6 successes. We have

$$n = 10 \quad p = 0.59 \quad q = 0.41 \quad r = 6$$

FIGURE 6-2

TI-84Plus/TI-83Plus/TI-*n*spire (with TI-84Plus keypad) Display

```
10 nCr 6*.59^6*.
41^(10-6)
         .2503034245
```

By the formula,

$$P(6) = C_{10,6}(0.59)^6(0.41)^{10-6}$$
$$= 210(0.59)^6(0.41)^4 \quad \text{Use a calculator or the}$$
$$\text{formula for } c_{n,r}.$$
$$\approx 210(0.0422)(0.0283) \quad \text{Use a calculator.}$$
$$\approx 0.25$$

There is a 25% chance that *exactly* 6 of the 10 Internet users are concerned about the privacy of e-mail.

(b) Many calculators have a built-in combinations function. On the TI-84Plus/TI-83Plus/TI-*n*spire (with TI-84Plus keypad) calculators, press the **MATH** key and select **PRB**. The combinations function is designated nCr. Figure 6-2 displays the process for computing P(6) directly on these calculators.

Using a Binomial Distribution Table

In many cases we will be interested in the probability of a range of successes. In such cases, we need to use the addition rule for mutually exclusive events. For instance, for $n = 6$ and $p = 0.50$,

$$P(4 \text{ or fewer successes}) = P(r \le 4)$$
$$= P(r = 4 \text{ or } 3 \text{ or } 2 \text{ or } 1 \text{ or } 0)$$
$$= P(4) + P(3) + P(2) + P(1) + P(0)$$

It would be a bit of a chore to use the binomial distribution formula to compute all the required probabilities. Table 2 of the Appendix gives values of $P(r)$ for selected p values and values of n through 20. To use the table, find the appropriate section for n, and then use the entries in the columns headed by the p values and the rows headed by the r values.

Table 6-10 is an excerpt from Table 2 of the Appendix showing the section for $n = 6$. Notice that all possible r values between 0 and 6 are given as row headers. The value $p = 0.50$ is one of the column headers. For $n = 6$ and $p = 0.50$, you can find the value of $P(4)$ by looking at the entry in the row headed by 4 and the column headed by 0.50. Notice that $P(4) = 0.234$ (to three digits).

TABLE 6-10   **Excerpt from Table 2 of the Appendix for $n = 6$**

							P							
n	*r*	.01	.05	.10	...	.30	...	.50	...	.70	...	.85	.90	.95
6	0	.941	.735	.531	...	.118	...	.016	...	.001	...	.000	.000	.000
	1	.057	.232	.354	...	.303	...	.094	...	.010	...	.000	.000	.000
	2	.001	.031	.098	...	.324	...	.234	...	.060	...	.006	.001	.000
	3	.000	.002	.015	...	.185	...	.312	...	.185	...	.042	.015	.002
	4	.000	.000	.001	...	.060	...	.234	...	.324		.176	.098	.031
	5	.000	.000	.000	...	.010	...	.094	...	.303	...	.399	.354	.232
	6	.000	.000	.000	...	.001	...	.016	...	.118	...	.377	.531	.735

Likewise, you can find other values of $P(r)$ from the table. In fact, for $n = 6$ and $p = 0.50$,

$$P(r \le 4) = P(4) + P(3) + P(2) + P(1) + P(0)$$
$$= 0.234 + 0.312 + 0.234 + 0.094 + 0.016 = 0.890$$

Alternatively, to compute $P(r \le 4)$ for $n = 6$, you can use the fact that the total of all $P(r)$ values for r between 0 and 6 is 1 and the complement rule. Since the complement of the event $r \le 4$ is the event $r \ge 5$, we have

$$P(r \le 4) = 1 - P(5) - P(6)$$
$$= 1 - 0.094 - 0.016 = 0.890$$

Note: In Table 2 of the Appendix, probability entries of 0.000 do not mean the probability is exactly zero. Rather, to three digits after the decimal, the probability rounds to 0.000.

CRITICAL THINKING

In Chapter 5, we saw the complement rule of probability. As we saw in the previous discussion, this rule provides a useful strategy to simplify binomial probability computations for a range of successes. For example, in a binomial experiment with $n = 7$ trials, the sample space for the number of successes r is

  0 1 2 3 4 5 6 7

Notice that $r = 0$ successes is the complement of $r \ge 1$ successes. By the complement rule,

$$P(r \ge 1) = 1 - P(r = 0)$$

It is faster to compute or look up $P(r = 0)$ and subtract than it is to compute or look up all the probabilities $P(r = 1)$ through $P(r = 7)$.

Likewise, the outcome $r \le 2$ is the complement of the outcome $r \ge 3$, so

$$P(r \ge 3) = 1 - P(r \le 2) = 1 - P(r = 2) - P(r = 1) - P(r = 0)$$

Before you use the complement rule for computing probabilities, be sure the outcomes used comprise complementary events. Complements can be expressed in several ways. For instance, the complement of the event $r \le 2$ can be expressed as event $r > 2$ or event $r \ge 3$.

Because of rounding in the binomial probability table, probabilities computed by using the addition rule directly might differ slightly from corresponding probabilities computed by using the complement rule.

EXAMPLE 5   USING THE BINOMIAL DISTRIBUTION TABLE TO FIND $P(r)$

A biologist is studying a new hybrid tomato. It is known that the seeds of this hybrid tomato have probability 0.70 of germinating. The biologist plants six seeds.

(a) What is the probability that *exactly* four seeds will germinate?

SOLUTION: This is a binomial experiment with $n = 6$ trials. Each seed planted represents an independent trial. We'll say germination is success, so the probability for success on each trial is 0.70.

  $n = 6$ $p = 0.70$ $q = 0.30$ $r = 4$

We wish to find $P(4)$, the probability of exactly four successes.

In Table 2 of the Appendix, find the section with $n = 6$ (excerpt is given in Table 6-10). Then find the entry in the column headed by $p = 0.70$ and the row headed by $r = 4$. This entry is 0.324.

  $P(4) = 0.324$

(b) What is the probability that *at least* four seeds will germinate?

SOLUTION: In this case, we are interested in the probability of four or more seeds germinating. This means we are to compute $P(r \geq 4)$. Since the events are mutually exclusive, we can use the addition rule:

$$P(r \geq 4) = P(r = 4 \ or \ r = 5 \ or \ r = 6) = P(4) + P(5) + P(6)$$

We already know the value of $P(4)$. We need to find $P(5)$ and $P(6)$.

Use the same part of the table but find the entries in the row headed by the *r* value 5 and then the *r* value 6. Be sure to use the column headed by the value of *p*, 0.70.

$$P(5) = 0.303 \ and \ P(6) = 0.118$$

Now we have all the parts necessary to compute $P(r \geq 4)$.

$$P(r \geq 4) = P(4) + P(5) + P(6)$$
$$= 0.324 + 0.303 + 0.118$$
$$= 0.745$$

In Guided Exercise 5 you'll practice using the formula for $P(r)$ in one part, and then you'll use Table 2 of the Appendix for $P(r)$ values in the second part.

GUIDED EXERCISE 5 | ## Find P(r)

A rarely performed and somewhat risky eye operation is known to be successful in restoring the eyesight of 30% of the patients who undergo it. A team of surgeons has developed a new technique for this operation that has been successful in four of six operations. Does it seem likely that the new technique is much better than the old? We'll use the binomial probability distribution to answer this question. We'll compute the probability of at least four successes in six trials for the old technique.

(a) Each operation is a binomial trial. In this case,
$n =$ ____, $p =$ ____, and $q =$ ____, and $r =$ ____.

➡ $n = 6$, $p = 0.30$, $q = 1 - 0.30 = 0.70$, and $r = 4$.

(b) Use your values of *n*, *p*, and *q* to compute $P(4)$ from the formula

$$P(r) = C_{n, \, r}p^{r}q^{n-r}$$

➡ $P(4) = C_{6,4}(0.30)^{4}(0.70)^{2}$
$= 15(0.0081)(0.490)$
≈ 0.060

(c) Compute the probability of *at least* four successes out of the six trials.

$$P(r \geq 4) = P(r = 4 \ or \ r = 5 \ or \ r = 6)$$
$$= P(4) + P(5) + P(6)$$

Use Table 2 of the Appendix to find values of $P(4)$, $P(5)$, and $P(6)$. Then use these values to compute $P(r \geq 4)$.

➡ To find $P(4)$, $P(5)$, and $P(6)$ in Table 2, we look in the section labeled $n = 6$. Then we find the column headed by $p = 0.30$. To find $P(4)$, we use the row labeled $r = 4$. For the values of $P(5)$ and $P(6)$, we look in the same column but in the rows headed by $r = 5$ and $r = 6$, respectively.

$$P(r \geq 4) = P(4) + P(5) + P(6)$$
$$= 0.060 + 0.010 + 0.001 = 0.071$$

(d) *Interpetation* Under the older operation technique, the probability that at least four patients out of six regain their eyesight is _____. Does it seem that the new technique is better than the old? Would you encourage the surgical team to do more work on the new technique?

➡ It seems the new technique is better than the old since, by pure chance, the probability of four or more successes out of six trials is only 0.071 for the old technique. This means one of the following two things may be happening:

(i) The new method is no better than the old method, and our surgeons have encountered a rare event (probability 0.071), or

(ii) The new method is in fact better. We think it is worth encouraging the surgeons to do more work on the new technique.

Using Technology to Compute Binomial Probabilities

Some calculators and computer-software packages support the binomial distribution. In general, these technologies will provide both the probability $P(r)$ for an exact number of successes r and the cumulative probability $P(r \leq k)$, where k is a specified value less than or equal to the number of trials n. Note that most of the technologies use the letter x instead of r for the random variable denoting the number of successes out of n trials.

TECH NOTES   The software packages Minitab and Excel 2010, as well as the TI-84Plus/TI-83Plus/TI-$n$spire calculators, include built-in binomial probability distribution options. These options give the probability $P(r)$ of a specific number of successes r, as well as the cumulative total probability for r or fewer successes.

TI-84Plus/TI-83Plus/TI-$n$spire (with TI-84Plus keypad) Press the **DISTR** key and scroll to **binompdf** (n, p, r). Enter the number of trials n, the probability of success on a single trial p, and the number of successes r. This gives $P(r)$. For the cumulative probability that there are r or fewer successes, use **binomcdf** (n, p, r).

$P(r = 4)$

$P(r \leq 4)$

```
binompdf(6,.3,4)

              .059535
binomcdf(6,.3,4)

              .989065
```

Copyright © Cengage Learning 2013

Excel 2010 Click **Insert Function** (f_x). In the dialogue box, select **Statistical** for the category, and then select **BINOM.DIST**. In the next dialogue box, fill in the values r, n, and p. For $P(r)$, use false; for P(at least r successes), use true.

Minitab First, enter the r values $0, 1, 2, \ldots, n$ in a column. Then use menu choice **Calc** ➤ **Probability Distribution** ➤ **Binomial**. In the dialogue box, select Probability for $P(r)$ or Cumulative for P(at least r successes). Enter the number of trials n, the probability of success p, and the column containing the r values. A sample printout is shown in Problem 25 at the end of this section.

LOOKING FORWARD

There are several ways to find the probability of r successes out of n binomial trials. In particular, we have used the general formula for the binomial probability distribution, a table of binomial probabilities such as Table 2 of the Appendix, or results from calculator or computer software that is based on the formula. However, depending on the number of trials and the size of the probability of success, the calculations can become tedious, or rounding errors can become an issue (*see* Linking Concepts, Problem 2). In Section 7.6 we will see how to use the *normal* probability distribution to approximate the binomial distribution in the case of sufficiently large numbers of trials n. A *Poisson* probability distribution can be used to approximate the binomial distribution when the probability of success on a single trial p is very small and the number of trials is 100 or more (See Section 5.4 of *Understandable Statistics*).

Common expressions and corresponding inequalities

Many times we are asked to compute the probability of a range of successes. For instance, in a binomial experiment with n trials, we may be asked to compute the probability of four or more successes. Table 6-11 shows how common English expressions such as "four or more successes" translate to inequalities involving r.

TABLE 6-11	Common English Expressions and Corresponding Inequalities (consider a Binomial Experiment with n Trials and r Successes)	
Expression	**Inequality**
Four or more successes At least four successes No fewer than four successes Not less than four successes	$r \geq 4$ That is, $r = 4, 5, 6, \ldots, n$
Four or fewer successes At most four successes No more than four successes The number of successes does not exceed four	$r \leq 4$ That is, $r = 0, 1, 2, 3,$ or 4
More than four successes The number of successes exceeds four	$r > 4$ That is, $r = 5, 6, 7, \ldots, n$
Fewer than four successes The number of successes is not as large as four	$r < 4$ That is, $r = 0, 1, 2,$ or 3

Copyright © Cengage Learning 2013

Interpretation Often we are not interested in the probability of a *specific number* of successes out of n binomial trials. Rather, we are interested in a minimum number of successes, a maximum number of successes, or a range of a number of successes. In other words, we are interested in the probability of *at least* a certain number of successes or *no more than* a certain number of successes, or the probability that the number of successes is between two given values.

For instance, suppose engineers have determined that at least 3 of 5 rivets on a bridge connector need to hold. If the probability that a single rivet holds is 0.80, and the performances of the rivets are independent, then the engineers are interested in the probability that *at least* 3 of the rivets hold. Notice that $P(r \geq 3$ rivets hold$) = 0.943$, while $P(r = 3$ rivets hold$) = 0.205$. The probability of *at least* 3 successes is much higher than the probability of *exactly* 3 successes. Safety concerns require that 3 of the rivets hold. However, there is a greater margin of safety if more than 3 rivets hold.

As you consider binomial experiments, determine whether you are interested in a *specific number* of successes or a *range* of successes.

Sampling without Replacement: Use of the Hypergeometric Probability Distribution

If the population is relatively small and we draw samples without replacement, the assumption of independent trials is not valid and we should not use the binomial distribution.

Hypergeometric probability distribution

The *hypergeometric distribution* is a probability distribution of a random variable that has two outcomes when sampling is done without replacement. This is the distribution that is appropriate when the sample size is so small that sampling without replacement results in trials that are not even approximately independent. A discussion of the hypergeometric distribution can be found in Appendix I of *Understandable Statistics*, 10th edition.

SECTION 6.2 PROBLEMS	aplia™ Access end-of-section exercises at www.aplia.com

SECTION 6.3

Additional Properties of the Binomial Distribution

FOCUS POINTS
- Make histograms for binomial distributions.
- Compute μ and σ for a binomial distribution.
- Compute the minimum number of trials n needed to achieve a given probability of success $P(r)$.

Graphing a Binomial Distribution

Any probability distribution may be represented in graphic form. How should we graph the binomial distribution? Remember, the binomial distribution tells us the probability of r successes out of n trials. Therefore, we'll place values of r along the horizontal axis and values of $P(r)$ on the vertical axis. The binomial distribution is a *discrete* probability distribution because r can assume only whole-number values such as 0, 1, 2, 3, ... Therefore, a histogram is an appropriate graph of a binomial distribution.

PROCEDURE

HOW TO GRAPH A BINOMIAL DISTRIBUTION

1. Place r values on the horizontal axis.
2. Place $P(r)$ values on the vertical axis.
3. Construct a bar over each r value extending from $r - 0.5$ to $r + 0.5$. The height of the corresponding bar is $P(r)$.

Let's look at an example to see exactly how we'll make these histograms.

EXAMPLE 6   GRAPH OF A BINOMIAL DISTRIBUTION

A waiter at the Green Spot Restaurant has learned from long experience that the probability that a lone diner will leave a tip is only 0.7. During one lunch hour, the waiter serves six people who are dining by themselves. Make a graph of the binomial probability distribution that shows the probabilities that 0, 1, 2, 3, 4, 5, or all 6 lone diners leave tips.

SOLUTION: This is a binomial experiment with $n = 6$ trials. Success is achieved when the lone diner leaves a tip, so the probability of success is 0.7 and that of failure is 0.3:

$$n = 6 \quad p = 0.7 \quad q = 0.3$$

We want to make a histogram showing the probability of r successes when $r = 0, 1, 2, 3, 4, 5,$ or 6. It is easier to make the histogram if we first make a table of r values and the corresponding $P(r)$ values (Table 6-12). We'll use Table 2 of the Appendix to find the $P(r)$ values for $n = 6$ and $p = 0.70$.

FIGURE 6-3

Graph of the Binomial Distribution for $n = 6$ and $p = 0.7$

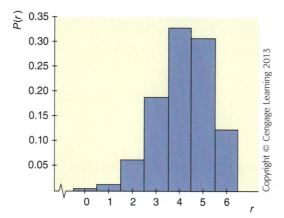

Copyright © Cengage Learning 2013

TABLE 6-12	Binomial Distribution for $n = 6$ and $p = 0.70$
r	$P(r)$
0	0.001
1	0.010
2	0.060
3	0.185
4	0.324
5	0.303
6	0.118

Copyright © Cengage Learning 2013

Mark Richards/PhotoEdit

To construct the histogram, we'll put r values on the horizontal axis and $P(r)$ values on the vertical axis. Our bars will be one unit wide and will be centered over the appropriate r value. The height of the bar over a particular r value tells the probability of that r (see Figure 6-3).

The probability of a particular value of r is given not only by the height of the bar over that r value but also by the *area* of the bar. Each bar is only one unit wide, so its area (area = height times width) equals its height. Since the area of each bar represents the probability of the r value under it, the sum of the areas of the bars must be 1. In this example, the sum turns out to be 1.001. It is not exactly equal to 1 because of rounding error.

Guided Exercise 6 illustrates another binomial distribution with $n = 6$ trials. The graph will be different from that of Figure 6-3 because the probability of success p is different.

GUIDED EXERCISE 6 | *Graph of a binomial distribution*

Jim enjoys playing basketball. He figures that he makes about 50% of the field goals he attempts during a game. Make a histogram showing the probability that Jim will make 0, 1, 2, 3, 4, 5, or 6 shots out of six attempted field goals.

Continued

(a) This is a binomial experiment with $n =$ _____ trials. In this situation, we'll say success occurs when Jim makes an attempted field goal. What is the value of p?

➡ In this example, $n = 6$ and $p = 0.5$.

(b) Use Table 2 of the Appendix to complete Table 6-13 of $P(r)$ values for $n = 6$ and $p = 0.5$.

TABLE 6-13

r	P(r)
0	0.016
1	0.094
2	0.234
3	_____
4	_____
5	_____
6	_____

Copyright © Cengage Learning 2013

➡ **TABLE 6-14 Completion of Table 6-13**

r	P(r)
.	.
.	.
.	.
3	0.312
4	0.234
5	0.094
6	0.016

Copyright © Cengage Learning 2013

(c) Use the values of $P(r)$ given in Table 6-14 to complete the histogram in Figure 6-4.

FIGURE 6-4 Beginning of Graph of Binomial Distribution for $n = 6$ and $p = 0.5$

➡ **FIGURE 6-5** Completion of Figure 6-4

(d) The area of the bar over $r = 2$ is 0.234. What is the area of the bar over $r = 4$? How does the probability that Jim makes exactly two field goals out of six compare with the probability that he makes exactly four field goals out of six?

➡ The area of the bar over $r = 4$ is also 0.234. Jim is as likely to make two out of six field goals attempted as he is to make four out of six.

In Example 6 and Guided Exercise 6, we see the graphs of two binomial distributions associated with $n = 6$ trials. The two graphs are different because the probability of success p is different in the two cases. In Example 6, $p = 0.7$ and the graph is skewed to the left—that is, the left tail is longer. In Guided Exercise 6, p is equal to 0.5 and the graph is symmetrical—that is, if we fold it in half, the two halves coincide exactly. Whenever *p equals 0.5, the graph of the binomial distribution will be symmetrical no matter how many trials we have.* In Chapter 7, we will see that if the number of trials n is quite large, the binomial distribution is almost symmetrical over the bars containing most of the area even when p is not close to 0.5.

Mean and Standard Deviation of a Binomial Distribution

Two other features that help describe the graph of any distribution are the balance point of the distribution and the spread of the distribution about that balance point. The *balance point* is the mean μ of the distribution, and the *measure of spread* that is most commonly used is the standard deviation σ. The mean μ is the *expected value* of the number of successes.

For the binomial distribution, we can use two special formulas to compute the mean μ and the standard deviation σ. These are easier to use than the general formulas in Section 6.1 for μ and σ of any discrete probability distribution.

PROCEDURE

Mean of a binomial distribution

Standard deviation of a binomial distribution

HOW TO COMPUTE μ AND σ FOR A BINOMIAL DISTRIBUTION

$\mu = np$ is the **expected number of successes** for the random variable r

$\sigma = \sqrt{npq}$ is the **standard deviation** for the random variable r

where
  r is a random variable representing the number of successes in a binomial distribution,
  n is the number of trials,
  p is the probability of success on a single trial, and
  $q = 1 - p$ is the probability of failure on a single trial.

EXAMPLE 7   COMPUTE μ AND σ

Let's compute the mean and standard deviation for the distribution of Example 6 that describes that probabilities of lone diners leaving tips at the Green Spot Restaurant.

SOLUTION: In Example 6,

$$n = 6 \quad p = 0.7 \quad q = 0.3$$

For the binomial distribution,

$$\mu = np = 6(0.7) = 4.2$$

The balance point of the distribution is at $\mu = 4.2$. The standard deviation is given by

$$\sigma = \sqrt{npq} = \sqrt{6(0.7)(0.3)} = \sqrt{1.26} \approx 1.12$$

The mean μ is not only the balance point of the distribution; it is also the *expected value* of r. Specifically, in Example 6, the waiter can expect 4.2 lone diners out of 6 to leave a tip. (The waiter would probably round the expected value to 4 tippers out of 6.)

GUIDED EXERCISE 7 | *Expected value and standard deviation*

When Jim (of Guided Exercise 6) shoots field goals in basketball games, the probability that he makes a shot is only 0.5.

Continued

GUIDED EXERCISE 7   *continued*

(a) The mean of the binomial distribution is the expected value of r successes out of n trials. Out of six throws, what is the expected number of goals Jim will make?

➡ The expected value is the mean μ:

$$\mu = np = 6(0.5) = 3$$

Jim can expect to make three goals out of six tries.

(b) For six trials, what is the standard deviation of the binomial distribution of the number of successful field goals Jim makes?

➡ $\sigma = \sqrt{npq} = \sqrt{6(0.5)(0.5)} = \sqrt{1.5} \approx 1.22$

CRITICAL THINKING

Unusual Values

Chebyshev's theorem tells us that no matter what the data distribution looks like, at least 75% of the data will fall within 2 standard deviations of the mean. As we will see in Chapter 7, when the distribution is mound-shaped and symmetrical, about 95% of the data are within 2 standard deviations of the mean. Data values beyond 2 standard deviations from the mean are less common than those closer to the mean.

In fact, one indicator that a data value might be an outlier is that it is more than 2.5 standard deviations from the mean (Source: *Statistics*, by G. Upton and I. Cook, Oxford University Press).

Unusual values

For a binomial distribution, it is unusual for the number of successes r to be higher than $\mu + 2.5\sigma$ or lower than $\mu - 2.5\sigma$.

We can use this indicator to determine whether a specified number of successes out of n trials in a binomial experiment is unusual.

For instance, consider a binomial experiment with 20 trials for which the probability of success on a single trial is $p = 0.70$. The expected number of successes is $\mu = 14$, with a standard deviation of $\sigma \approx 2$. A number of successes above 19 or below 9 would be considered unusual. However, such numbers of successes are possible.

VIEWPOINT   Kodiak Island, Alaska

Kodiak Island is famous for its giant brown bears. The sea surrounding the island is also famous for its king crabs. The state of Alaska's Department of Fish and Game has collected a huge amount of data regarding ocean latitude, ocean longitude, and size of king crabs. Of special interest to commercial fishing skippers is the size of crabs. Those too small must be returned to the sea. To find locations and sizes of king crab catches near Kodiak Island, visit the Brase/Brase statistics site at **http://www.cengagebrain.com** *and find the link to the StatLib site hosted by the Department of Statistics at Carnegie Mellon University. Once at StatLib, go to crab data. From this information, it is possible to use methods of this chapter and Chapter 8 to estimate the proportion of legal crab in a sea skipper's catch.*

SECTION 6.3
PROBLEMS

Access end-of-section exercises at www.aplia.com

Chapter Review

SUMMARY

This chapter discusses random variables and important probability distributions associated with discrete random variables.

- The value of a *random variable* is determined by chance.

- Random variables are either *discrete* or *continuous*.

- A probability distribution of a discrete random variable x consists of all distinct values of x and the corresponding probabilities $P(x)$. For each $x, 0 \leq P(x) \leq 1$ and $\Sigma P(x) = 1$.

- A discrete probability distribution can be displayed visually by a *probability histogram* in which the values of the random variable x are displayed on the horizontal axis, the height of each bar is $P(x)$, and each bar is 1 unit wide.

- For discrete probability distributions,

$$\mu = \Sigma x P(x) \text{ and } \sigma = \sqrt{\Sigma (x - \mu)^2 P(x)}$$

- The mean μ is called the *expected value* of the probability distribution.

- A *binomial experiment* consists of a fixed number n of independent trials repeated under identical conditions. There are two outcomes for each trial, called *success* and *failure*. The probability p of success on each trial is the same.

- The number of successes r in a binomial experiment is the random variable for the binomial probability distribution. Probabilities can be computed using a formula or probability distribution outputs from a computer or calculator. Some probabilities can be found in Table 2 of the Appendix.

- For a binomial distribution,

$$\mu = np \text{ and } \sigma = \sqrt{npq},$$

where $q = 1 - p$.

- For a binomial experiment, the number of successes is usually within the interval from $\mu - 2.5\sigma$ to $\mu + 2.5\sigma$. A number of successes outside this range of values is unusual but can occur.

It is important to check the conditions required for the use of the binomial probability distribution.

IMPORTANT WORDS & SYMBOLS

VIEWPOINT | What's Your Type?

Are students and professors really compatible? One way of answering this question is to look at the Myers-Briggs Type Indicator for personality preferences. What is the probability that your professor is introverted and judgmental? What is the probability that you are extroverted and perceptive? Are most of the leaders in student government extroverted and judgmental? Is it true that members of Phi Beta Kappa have personality types more like those of the professors? We will consider questions such as these in more detail in Chapters 8 (estimation) and 9 (hypothesis testing), where we will continue our work with binomial probabilities. In the meantime, you can find many answers regarding careers, probability, and personality types in Applications of the Myers-Briggs Type Indicator in Higher Education, *edited by J. Provost and S. Anchors.*

CHAPTER REVIEW PROBLEMS

1. | *Statistical Literacy* What are the requirements for a probability distribution?

2. | *Statistical Literacy* List the criteria for a binomial experiment. What does the random variable of a binomial experiment measure?

3. | *Critical Thinking* For a binomial probability distribution, it is unusual for the number of successes to be less than $\mu - 2.5\sigma$ or greater than $\mu + 2.5\sigma$.
   (a) For a binomial experiment with 10 trials for which the probability of success on a single trial is 0.2, is it unusual to have more than five successes? Explain.
   (b) If you were simply guessing on a multiple-choice exam consisting of 10 questions with five possible responses for each question, would you be likely to get more than half of the questions correct? Explain.

4. | *Critical Thinking* Consider a binomial experiment. If the number of trials is increased, what happens to the expected value? to the standard deviation? Explain.

5. | *Probability Distribution: Auto Leases* The Consumer Bankers Association released a report showing the lengths of automobile leases for new automobiles. The results are:

Lease Length in Months	Percentage of Leases
13–24	12.7%
25–36	37.1%
37–48	28.5%
49–60	21.5%
More than 60	0.2%

   (a) Use the midpoint of each class, and call the midpoint of the last class 66.5 months, for purposes of computing the expected lease term. Also find the standard deviation of the distribution.
   (b) Sketch a graph of the probability distribution for the duration of new auto leases.

6. | *Ecology: Predator and Prey* Isle Royale, an island in Lake Superior, has provided an important study site of wolves and their prey. In the National Park Service Scientific Monograph Series 11, *Wolf Ecology and Prey Relationships on Isle Royale*, Peterson gives results of many wolf–moose studies. Of special interest is the study of the number of moose killed by wolves. In the period from

georgesankar.com/Alamy

1958 to 1974, there were 296 moose deaths identified as wolf kills. The age distribution of the kills is as follows.

Age of Moose in Years	Number Killed by Wolves
Calf (0.5 yr)	112
1–5	53
6–10	73
11–15	56
16–20	2

(a) For each age group, compute the probability that a moose in that age group is killed by a wolf.
(b) Consider all ages in a class equal to the class midpoint. Find the expected age of a moose killed by a wolf and the standard deviation of the ages.

7. *Insurance: Auto* State Farm Insurance studies show that in Colorado, 55% of the auto insurance claims submitted for property damage are submitted by males under 25 years of age. Suppose 10 property damage claims involving automobiles are selected at random.
(a) Let r be the number of claims made by males under age 25. Make a histogram for the r-distribution probabilities.
(b) What is the probability that six or more claims are made by males under age 25?
(c) What is the expected number of claims made by males under age 25? What is the standard deviation of the r-probability distribution?

8. *Quality Control: Pens* A stationery store has decided to accept a large shipment of ball-point pens if an inspection of 20 randomly selected pens yields no more than two defective pens.
(a) Find the probability that this shipment is accepted if 5% of the total shipment is defective.
(b) Find the probability that this shipment is not accepted if 15% of the total shipment is defective.

9. *Criminal Justice: Inmates* According to *Harper's Index*, 50% of all federal inmates are serving time for drug dealing. A random sample of 16 federal inmates is selected.
(a) What is the probability that 12 or more are serving time for drug dealing?
(b) What is the probability that 7 or fewer are serving time for drug dealing?
(c) What is the expected number of inmates serving time for drug dealing?

10. *Airlines: On-Time Arrivals* *Consumer Reports* rated airlines and found that 80% of the flights involved in the study arrived on time (that is, within 15 minutes of scheduled arrival time). Assuming that the on-time arrival rate is representative of the entire commercial airline industry, consider a random sample of 200 flights. What is the expected number that will arrive on time? What is the standard deviation of this distribution?

11. *Agriculture: Grapefruit* It is estimated that 75% of a grapefruit crop is good; the other 25% have rotten centers that cannot be detected until the grapefruit are cut open. The grapefruit are sold in sacks of 10. Let r be the number of good grapefruit in a sack.
(a) Make a histogram of the probability distribution of r.
(b) What is the probability of getting no more than one bad grapefruit in a sack? What is the probability of getting at least one good grapefruit in a sack?
(c) What is the expected number of good grapefruit in a sack?
(d) What is the standard deviation of the r-probability distribution?

12. *Restaurants: Reservations* The Orchard Café has found that about 5% of the diners who make reservations don't show up. If 82 reservations have been made, how many diners can be expected to show up? Find the standard deviation of this distribution.

13. *College Life: Student Government* The student government claims that 85% of all students favor an increase in student fees to buy indoor potted plants for the classrooms. A random sample of 12 students produced 2 in favor of the project. What is the probability that 2 or fewer in the sample will favor the project, assuming the student government's claim is correct? *Interpretation* Do the data support the student government's claim, or does it seem that the percentage favoring the increase in fees is less than 85%?

DATA HIGHLIGHTS: GROUP PROJECTS

Break into small groups and discuss the following topics. Organize a brief outline in which you summarize the main points of your group discussion.

1. Powerball! Imagine, you could win a jackpot worth at least $10 million. Some jackpots have been worth more than $250 million! Powerball is a multistate lottery. To play Powerball, you purchase a $1 ticket. On the ticket you select five distinct white balls (numbered 1 through 59) and then one red Powerball (numbered 1 through 39). The red Powerball number may be any of the numbers 1 through 39, including any such numbers you selected for the white balls. Every Wednesday and Saturday there is a drawing. If your chosen numbers match those drawn, you win! Figure 6-7 shows all the prizes and the probability of winning each prize, and specifies how many numbers on your ticket must match those drawn to win the prize. The Multi-State Lottery Association maintains a website that displays the results of each drawing, as well as a history of the results of previous drawings. For updated Powerball data, visit the Brase/Brase statistics site at **http://www.cengagebrain.com** and find the link to the Multi-State Lottery Association.

FIGURE 6-7

Match	Approximate Probability	Prize
5 white balls + Powerball	0.00000000512	Jackpot*
5 white balls	0.000000195	$200,000
4 white balls + Powerball	0.00000138	$10,000
4 white balls	0.0000525	$100
3 white balls + Powerball	0.0000733	$100
3 white balls	0.00279	$7
2 white balls + Powerball	0.00127	$7
1 white ball + Powerball	0.00810	$4
0 white balls + Powerball	0.0162	$3
Overall chance of winning	0.0285	

*The jackpot will be divided equally (if necessary) among multiple winners and is paid in 30 annual installments or in a reduced lump sum.

(a) Assume the jackpot is $10 million and there will be only one jackpot winner. Figure 6-7 lists the prizes and the probability of winning each prize. What is the probability of *not winning* any prize? Consider all the prizes and their respective probabilities, and the prize of $0 (no win) and its probability. Use all these values to estimate your expected winnings μ if you play one ticket. How much do you effectively contribute to the state in which you purchased the ticket (ignoring the overhead cost of operating Powerball)?

(b) Suppose the jackpot increased to $25 million (and there was to be only one winner). Compute your expected winnings if you buy one ticket. Does the probability of winning the jackpot change because the jackpot is higher?

(c) Pretend that you are going to buy 10 Powerball tickets when the jackpot is $10 million. Use the random-number table to select your numbers. Check the Multi-State Lottery Association website (or any other Powerball site) for the most recent drawing results to see if you would have won a prize.

(d) The probability of winning *any* prize is about 0.0285. Suppose you decide to buy five tickets. Use the binomial distribution to compute the probability of winning (any prize) at least once. *Note:* You will need to use the binomial distribution formula. Carry at least three digits after the decimal.

2. Would you like to travel in space, if given a chance? According to Opinion Research for Space Day Partners, if your answer is yes, you are not alone. Forty-four percent of adults surveyed agreed that they would travel in space if given a chance. Look at Figure 6-8, and use the information presented to answer the following questions.

FIGURE 6-8

Source: Opinion Research for Space Day Partners

(a) According to Figure 6-8, the probability that an adult selected at random agrees with the statement that humanity should explore planets is 64%. Round this probability to 65%, and use this estimate with the binomial distribution table to determine the probability that of 10 adults selected at random, at least half would agree that humanity should explore planets.

(b) Does space exploration have an impact on daily life? Find the probability that of 10 adults selected at random, at least 9 would agree that space exploration does have an impact on daily life. *Hint:* Use the formula for the binomial distribution.

(c) In a room of 35 adults, what is the expected number who would travel in space, given a chance? What is the standard deviation?

LINKING CONCEPTS: WRITING PROJECTS

Discuss each of the following topics in class or review the topics on your own. Then write a brief but complete essay in which you summarize the main points. Please include formulas and graphs as appropriate.

1. Discuss what we mean by a binomial experiment. As you can see, a binomial process or binomial experiment involves a lot of assumptions! For example, all the trials are supposed to be independent and repeated under identical conditions. Is this always true? Can we always be completely certain that the probability of success does not change from one trial to the next? In the real world, there is almost nothing we can be absolutely sure about, so the *theoretical* assumptions of the binomial probability distribution often will not be completely satisfied. Does that mean we cannot use the binomial distribution to solve practical problems? Looking at this chapter, the answer seems to be that we can indeed use the binomial distribution even if not all the assumptions are *exactly* met. We find in practice that the conclusions are sufficiently accurate for our intended application. List three applications of the binomial distribution for which you think that although some of the assumptions are not exactly met, there is adequate reason to apply the binomial distribution anyhow.

2. Why do we need to learn the formula for the binomial probability distribution? Using the formula repeatedly can be very tedious. To cut down on tedious calculations, most people will use a binomial table such as the one found in the Appendix of this book.

   (a) However, there are many applications for which a table in the back of *any* book is not adequate. For instance, compute

   $$P(r = 3) \text{ where } n = 5 \text{ and } p = 0.735$$

   Can you find the result in the table? Do the calculation by using the formula. List some other situations in which a table might not be adequate to solve a particular binomial distribution problem.

   (b) The formula itself also has limitations. For instance, consider the difficulty of computing

   $$P(r \geq 285) \text{ where } n = 500 \text{ and } p = 0.6$$

   What are some of the difficulties you run into? Consider the calculation of $P(r = 285)$. You will be raising 0.6 and 0.4 to very high powers; this will give you very, very small numbers. Then you need to compute $C_{500,285}$, which is a very, very large number. When combining extremely large and extremely small numbers in the same calculation, most accuracy is lost unless you carry a huge number of significant digits. If this isn't tedious enough, consider the steps you need to compute

   $$P(r \geq 285) = P(r = 285) + P(r = 286) + \cdots + P(r = 500)$$

   Does it seem clear that we need a better way to estimate $P(r \geq 285)$? In Chapter 7, you will learn a much better way to estimate binomial probabilities when the number of trials is large.

3. In Chapter 3, we learned about means and standard deviations. In Section 6.1, we learned that probability distributions also can have a mean and standard deviation. Discuss what is meant by the expected value and standard deviation of a binomial distribution. How does this relate back to the material we learned in Chapter 3 and Section 6.1?

4. In Chapter 2, we looked at the shapes of distributions. Review the concepts of skewness and symmetry; then categorize the following distributions as to skewness or symmetry:

   (a) A binomial distribution with $n = 11$ trials and $p = 0.50$

   (b) A binomial distribution with $n = 11$ trials and $p = 0.10$

   (c) A binomial distribution with $n = 11$ trials and $p = 0.90$

   In general, does it seem true that binomial probability distributions in which the probability of success is close to 0 are skewed right, whereas those with probability of success close to 1 are skewed left?

USING TECHNOLOGY

Binomial Distributions

Although tables of binomial probabilities can be found in most libraries, such tables are often inadequate. Either the value of p (the probability of success on a trial) you are looking for is not in the table, or the value of n (the number of trials) you are looking for is too large for the table. In Chapter 7, we will study the normal approximation to the binomial. This approximation is a great help in many practical applications. Even so, we sometimes use the formula for the binomial probability distribution on a computer or graphing calculator to compute the probability we want.

Applications

The following percentages were obtained over many years of observation by the U.S. Weather Bureau. All data listed are for the month of December.

Location	Long-Term Mean % of Clear Days in Dec.
Juneau, Alaska	18%
Seattle, Washington	24%
Hilo, Hawaii	36%
Honolulu, Hawaii	60%
Las Vegas, Nevada	75%
Phoenix, Arizona	77%

Source: Adapted from *Local Climatological Data,* U.S. Weather Bureau publication, "Normals, Means, and Extremes" table.

In the locations listed, the month of December is a relatively stable month with respect to weather. Since weather patterns from one day to the next are more or less the same, it is reasonable to use a binomial probability model.

1. Let r be the number of clear days in December. Since December has 31 days, $0 \le r \le 31$. Using appropriate computer software or calculators available to you, find the probability $P(r)$ for each of the listed locations when $r = 0, 1, 2, \ldots, 31$.

2. For each location, what is the expected value of the probability distribution? What is the standard deviation?

You may find that using cumulative probabilities and appropriate subtraction of probabilities, rather than addition of probabilities, will make finding the solutions to Applications 3–7 easier.

3. Estimate the probability that Juneau will have at most 7 clear days in December.

4. Estimate the probability that Seattle will have from 5 to 10 (including 5 and 10) clear days in December.

5. Estimate the probability that Hilo will have at least 12 clear days in December.

6. Estimate the probability that Phoenix will have 20 or more clear days in December.

7. Estimate the probability that Las Vegas will have from 20 to 25 (including 20 and 25) clear days in December.

Technology Hints

TI-84Plus/TI-83Plus/TI-*n*spire (with TI-84 Plus keypad), Excel 2010, Minitab

The Tech Note in Section 6.2 gives specific instructions for binomial distribution functions on the TI-84Plus/TI-83Plus/TI-*n*spire (with TI-84Plus keypad) calculators, Excel 2010, and Minitab.

SPSS

In SPSS, the function **PDF.BINOM(q,n,p)** gives the probability of q successes out of n trials, where p is the probability of success on a single trial. In the data editor, name a variable r and enter values 0 through n. Name another variable Prob_r. Then use the menu choices **Transform ➤ Compute**. In the dialogue box, use Prob_r for the target variable. In the function group, select **PDF and Noncentral PDF**. In the function box, select **PDF.BINOM(q,n,p)**. Use the variable r for q and appropriate values for n and p. Note that the function **CDF.BINOM(q,n,p)**, from the **CDF and Noncentral CDF** group, gives the cumulative probability of 0 through q successes.

Cumulative Review Problems

CHAPTERS 4–6

The Hill of Tara is located in south-central Meath, not far from Dublin, Ireland. Tara is of great cultural and archaeological importance, since it is by legend the seat of the ancient high kings of Ireland. For more information, see *Tara: An Archaeological Survey*, by Conor Newman, Royal Irish Academy, Dublin.

Magnetic surveying is one technique used by archaeologists to determine anomalies arising from variations in magnetic susceptibility. Unusual changes in magnetic susceptibility might (or might not) indicate an important archaeological discovery. Let x be a random variable that represents a magnetic susceptibility (MS) reading for a randomly chosen site on the Hill of Tara. A random sample of 120 sites gave the readings shown in Table A below.

TABLE A **Magnetic Susceptibility Readings, centimeter-gram-second $\times 10^{-6}$**

Comment	Magnetic Susceptibility	Number of Readings	Estimated Probability
"cool"	$0 \le x < 10$	30	30/120 = 0.25
"neutral"	$10 \le x < 20$	54	54/120 = 0.45
"warm"	$20 \le x < 30$	18	18/20 = 0.15
"very interesting"	$30 \le x < 40$	12	12/120 = 0.10
"hot spot"	$40 \le x$	6	6/120 = 0.05

Answers may vary slightly due to rounding.

1. *Statistical Literacy: Sample Space* What is a statistical experiment? How could the magnetic susceptibility intervals $0 \le x < 10, 10 \le x < 20$, and so on, be considered events in the sample space of all possible readings?

2. *Statistical Literacy: Probability* What is probability? What do we mean by relative frequency as a probability estimate for events? What is the law of large numbers? How would the law of large numbers apply in this context?

3. *Statistical Literacy: Probability Distribution* Do the probabilities shown in Table A add up to 1? Why should they total to 1?

4. *Probability Rules* For a site chosen at random, estimate the following probabilities.
   (a) $P(0 \le x < 30)$ (b) $P(10 \le x < 40)$
   (c) $P(x < 20)$ (d) $P(x \ge 20)$
   (e) $P(30 \le x)$ (f) $P(x \text{ not less than } 10)$
   (g) $P(0 \le x < 10 \text{ or } 40 \le x)$
   (h) $P(40 \le x \text{ and } 20 \le x)$

5. *Discrete Probability Distribution* Consider the midpoint of each interval. Assign the value 45 as the midpoint for the interval $40 \le x$. The midpoints constitute the sample space for a discrete random variable. Using Table A, compute the expected value μ and the standard deviation σ.

Midpoint x	5	15	25	35	45
$P(x)$					

6. *Binomial Distribution* Suppose a reading between 30 and 40 is called "very interesting" from an archaeological point of view. Let us say you take readings at $n = 12$ sites chosen at random. Let r be a binomial random variable that represents the number of "very interesting" readings from these 12 sites.

(a) Let us call "very interesting" a binomial success. Use Table A to find p, the probability of success on a single trial, where $p = P(\text{success}) = P(30 \leq x < 40)$.

(b) What is the expected value μ and standard deviation σ for the random variable r?

(c) What is the probability that you will find *at least* one "very interesting" reading in the 12 sites?

(d) What is the probability that you will find *fewer than* three "very interesting" readings in the 12 sites?

7. *Linear Regression: Blood Glucose* Let x be a random variable that represents blood glucose level after a 12-hour fast. Let y be a random variable representing blood glucose level 1 hour after drinking sugar water (after the 12-hour fast). Units are in mg/10 mL. A random sample of eight adults gave the following information. (Reference: *American Journal of Clinical Nutrition*, Vol. 19, pp. 345–351.)

$$\Sigma x = 63.8; \Sigma x^2 = 521.56; \Sigma y = 90.7;$$
$$\Sigma y^2 = 1070.87; \Sigma xy = 739.65$$

x	6.2	8.4	7.0	7.5	8.1	6.9	10.0	9.7
y	9.8	10.7	10.3	11.9	14.2	7.0	14.6	12.2

(a) Draw a scatter diagram for the data.

(b) Find the equation of the least-squares line and graph it on the scatter diagram.

(c) Find the sample correlation coefficient r and the sample coefficient of determination r^2. Explain the meaning of r^2 in the context of the application.

(d) If $x = 9.0$, use the least-squares line to predict y.

Halaska/PhotoLibrary/Getty Images

7

Historical Pictures/Stock Montage

One cannot escape the feeling that these mathematical formulas have an independent existence and an intelligence of their own, that they are wiser than we are, wiser even than their discoverers, that we get more out of them than was originally put into them.

—HEINRICH HERTZ

How can it be that mathematics, a product of human thought independent of experience, is so admirably adapted to the objects of reality?

—ALBERT EINSTEIN

For online student resources, visit the Brase/Brase, *Understanding Basic Statistics,* 6th edition website at **http://www.cengagebrain.com**

Heinrich Hertz (1857–1894) was a pioneer in the study of radio waves. His work and the later work of Maxwell and Marconi led the way to modern radio, television, and radar. Albert Einstein is world renowned for his great discoveries in relativity and quantum mechanics. Everyone who has worked in both mathematics and real-world applications cannot help but marvel at how the "pure thought" of the mathematical sciences can predict and explain events in other realms. In this chapter, we will study the most important type of probability distribution in all of mathematical statistics: the normal distribution. Why is the normal distribution so important? Two of the reasons are that it applies to a wide variety of situations and that other distributions tend to become normal under certain conditions.

NORMAL CURVES AND SAMPLING DISTRIBUTIONS

PREVIEW QUESTIONS

What are some characteristics of a normal distribution? What does the empirical rule tell you about data spread around the mean? (SECTION 7.1)

Can you compare apples and oranges, or maybe elephants and butterflies? In most cases, the answer is no—unless you first standardize your measurements. What are a standard normal distribution and a standard z score? (SECTION 7.2)

How do you convert any normal distribution to a standard normal distribution? How do you find probabilities of "standardized events"? (SECTION 7.3)

As humans, our experiences are finite and limited. Consequently, most of the important decisions in our lives are based on sample (incomplete) information. What is a probability sampling distribution? How will sampling distributions help us make good decisions based on incomplete information? (SECTION 7.4)

There is an old saying: All roads lead to Rome. In statistics, we could recast this saying: All probability distributions average out to be normal distributions (as the sample size increases). How can we take advantage of this in our study of sampling distributions? (SECTION 7.5)

The binomial and normal distributions are two of the most important probability distributions in statistics. Under certain limiting conditions, the binomial can be thought to evolve into the normal distribution. How can you apply this concept in the real world? (SECTION 7.6)

Many issues in life come down to success or failure. In most cases, we will not be successful all the time, so proportions of successes are very important. What is the probability sampling distribution for proportions? (SECTION 7.6)

Dana White/PhotoEdit

FOCUS PROBLEM

Impulse Buying

The Food Marketing Institute, *Progressive Grocer*, New Products News, and the Point of Purchase Advertising Institute are organizations that analyze supermarket sales. One of the interesting discoveries was that the average amount of impulse buying in a grocery store is very time-dependent. As reported in *The Denver Post*, "When you dilly dally in a store for

Syracuse Newspapers/David Lassman/The Image Works Image

10 unplanned minutes, you can kiss nearly $20 good-bye." For this reason, it is in the best interest of the supermarket to keep you in the store longer. In the *Post* article, it was pointed out that long checkout lines (near end-aisle displays), "samplefest" events of free tasting, video kiosks, magazine and book sections, and so on, help keep customers in the store longer. On average, a single customer who strays from his or her grocery list can plan on impulse spending of $20 for every 10 minutes spent wandering about in the supermarket.

Let x represent the dollar amount spent on supermarket impulse buying in a 10-minute (unplanned) shopping interval. Based on the *Post* article, the mean of the x distribution is about $20 and the (estimated) standard deviation is about $7.

(a) Consider a random sample of $n = 100$ customers, each of whom has 10 minutes of unplanned shopping time in a supermarket. From the central limit theorem, what can you say about the probability distribution of $\bar{x}$, the *average* amount spent by these customers due to impulse buying? Is the $\bar{x}$ distribution approximately normal? What are the mean and standard deviation of the $\bar{x}$ distribution? Is it necessary to make any assumption about the x distribution? Explain.

(b) What is the probability that $\bar{x}$ is between $18 and $22?

(c) Let us assume that x has a distribution that is approximately normal. What is the probability that x is between $18 and $22?

(d) In part (b), we used $\bar{x}$, the *average* amount spent, computed for 100 customers. In part (c), we used x, the amount spent by only *one* individual customer. The answers to parts (b) and (c) are very different. Why would this happen? In this example, $\bar{x}$ is a much more predictable or reliable statistic than x. Consider that almost all marketing strategies and sales pitches are designed for the *average* customer and *not* the *individual* customer. How does the central limit theorem tell us that the average customer is much more predictable than the individual customer? (See Problem 18 of Section 7.5.)

SECTION 7.1

Graphs of Normal Probability Distributions

FOCUS POINTS

- Graph a normal curve and summarize its important properties.
- Apply the empirical rule to solve real-world problems.

Normal distribution

One of the most important examples of a continuous probability distribution is the *normal distribution*. This distribution was studied by the French mathematician Abraham de Moivre (1667–1754) and later by the German mathematician Carl Friedrich Gauss (1777–1855), whose work is so important that the normal distribution is sometimes called *Gaussian*. The work of these mathematicians provided a foundation on which much of the theory of statistical inference is based.

Applications of a normal probability distribution are so numerous that some mathematicians refer to it as "a veritable Boy Scout knife of statistics." However, before we can apply it, we must examine some of the properties of a normal distribution.

A rather complicated formula, presented later in this section, defines a normal distribution in terms of μ and σ, the mean and standard deviation of the population distribution. It is only through this formula that we can verify if a distribution is normal. However, we can look at the graph of a normal distri-

A Normal Curve

Another continuous probability distribution is discussed in the section problems. Problems 12 and 13 discuss the *uniform probability distribution*.

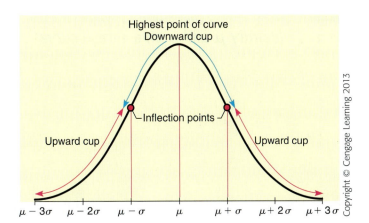

Normal curve

bution and get a good pictorial idea of some of the essential features of any normal distribution.

The graph of a normal distribution is called a *normal curve*. It possesses a shape very much like the cross section of a pile of dry sand. Because of its shape, blacksmiths would sometimes use a pile of dry sand in the construction of a mold for a bell. Thus the normal curve is also called a *bell-shaped curve* (see Figure 7-1).

We see that a general normal curve is smooth and symmetrical about the vertical line extending upward from the mean μ. Notice that the highest point of the curve occurs over μ. If the distribution were graphed on a piece of sheet metal, cut out, and placed on a knife edge, the balance point would be at μ. We also see that the curve tends to level out and approach the horizontal (x axis) like a glider making a landing. However, in mathematical theory, such a glider would never quite finish its landing because a normal curve never touches the horizontal axis.

The parameter σ controls the spread of the curve. The curve is quite close to the horizontal axis at $\mu + 3\sigma$ and $\mu - 3\sigma$. Thus, if the standard deviation σ is large, the curve will be more spread out; if it is small, the curve will be more peaked. Figure 7-1 shows the normal curve cupped downward for an interval on either side of the mean μ. Then it begins to cup upward as we go to the lower part of the bell. The exact places where the *transition* between the upward and downward cupping occur are above the points $\mu + \sigma$ and $\mu - \sigma$. In the terminology of calculus, transition points such as these are called *inflection points*.

Downward cup
Upward cup

Symmetry of normal curves

Important properties of a normal curve

1. The curve is bell-shaped, with the highest point over the mean μ.

2. The curve is symmetrical about a vertical line through μ.

3. The curve approaches the horizontal axis but never touches or crosses it.

4. The inflection (transition) points between cupping upward and downward occur above $\mu + \sigma$ and $\mu - \sigma$.

5. The area under the entire curve is 1.

The parameters that control the shape of a normal curve are the mean μ and the standard deviation σ. When both μ and σ are specified, a specific normal curve is determined. In brief, μ locates the balance point and σ determines the extent of the spread.

GUIDED EXERCISE 1 | *Identify μ and σ on a normal curve*

Look at the normal curves in Figure 7-2.

FIGURE 7-2

Copyright © Cengage Learning 2013

(a) Do these distributions have the same mean? If so, what is it?

➡ The means are the same, since both graphs have the high point over 6. $\mu = 6$.

(b) One of the curves corresponds to a normal distribution with $\sigma = 3$ and the other to one with $\sigma = 1$. Which curve has which σ?

➡ Curve A has $\sigma = 1$ and curve B has $\sigma = 3$. (Since curve B is more spread out, it has the larger σ value.)

COMMENT   The normal distribution curve is always above the horizontal axis. The area beneath the curve and above the axis is exactly 1. As such, the normal distribution curve is an example of a *density curve*. The formula used to generate the shape of the normal distribution curve is called the *Normal Density Function*. If *x* is a normal random variable with mean μ and standard deviation σ, the formula for the normal density function is

Normal Density Function

$$f(x) = \frac{e^{(-1/2)((x-\mu)/\sigma)^2}}{\sigma\sqrt{2\pi}}$$

In this text, we will not use this formula explicitly. However, we will use tables of areas based on the normal density function.

The total area under any normal curve studied in this book will *always* be 1. The graph of the normal distribution is important because the portion of the *area* under the curve above a given interval represents the *probability* that a measurement will lie in that interval.

In Section 3.2, we studied Chebyshev's theorem. This theorem gives us information about the *smallest* proportion of data that lies within 2, 3, or *k* standard deviations of the mean. This result applies to *any* distribution. However, for normal distributions, we can get a much more precise result, which is given by the *empirical rule*.

Empirical rule

Empirical rule

For a distribution that is symmetrical and bell-shaped (in particular, for a normal distribution):

Approximately 68% of the data values will lie within 1 standard deviation on each side of the mean.

Approximately 95% of the data values will lie within 2 standard deviations on each side of the mean.

Approximately 99.7% (or almost all) of the data values will lie within 3 standard deviations on each side of the mean.

The preceding statement is called the *empirical rule* because, for symmetrical, bell-shaped distributions, the given percentages are observed in practice. Furthermore, for the normal distribution, the empirical rule is a direct consequence of the very nature of the distribution (see Figure 7-3). Notice that the empirical rule is a stronger statement than Chebyshev's theorem in that it gives *definite percentages*, not just lower limits. Of course, the empirical rule applies only to normal or symmetrical, bell-shaped distributions, whereas Chebyshev's theorem applies to all distributions.

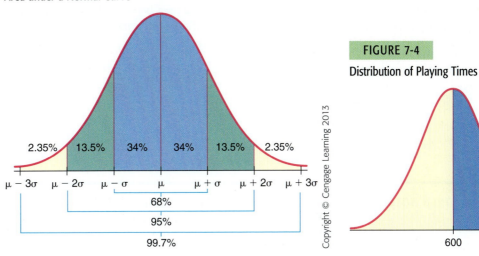

FIGURE 7-3

Area under a Normal Curve

FIGURE 7-4

Distribution of Playing Times

EXAMPLE 1   EMPIRICAL RULE

The playing life of a Sunshine radio is normally distributed with mean $\mu = 600$ hours and standard deviation $\sigma = 100$ hours. What is the probability that a radio selected at random will last from 600 to 700 hours?

SOLUTION: The probability that the playing life will be between 600 and 700 hours is equal to the percentage of the total area under the curve that is shaded in Figure 7-4. Since $\mu = 600$ and $\mu + \sigma = 600 + 100 = 700$, we see that the shaded area is simply the area between μ and $\mu + \sigma$. The area from μ to $\mu + \sigma$ is 34% of the total area. This tells us that the probability a Sunshine radio will last between 600 and 700 playing hours is about 0.34.

GUIDED EXERCISE 2 | *Empirical rule*

The yearly wheat yield per acre on a particular farm is normally distributed with mean $\mu = 35$ bushels and standard deviation $\sigma = 8$ bushels.

(a) Shade the area under the curve in Figure 7-5 (next page) that represents the probability that an acre will yield between 19 and 35 bushels.

➡️ See Figure 7-6.

(b) Is the area the same as the area between $\mu - 2\sigma$ and μ?

➡️ Yes, since $\mu = 35$ and $\mu - 2\sigma = 35 - 2(8) = 19$.

Continued

GUIDED EXERCISE 2 *continued*

FIGURE 7-5

Bushels

Copyright © Cengage Learning 2013

FIGURE 7-6 Completion of Figure 7-5

Copyright © Cengage Learning 2013

(c) Use Figure 7-3 to find the percentage of area over the interval between 19 and 35.

⇨ The area between the values $\mu - 2\sigma$ and μ is 47.5% of the total area.

(d) What is the probability that the yield will be between 19 and 35 bushels per acre?

⇨ It is 47.5% of the total area, which is 1. Therefore, the probability is 0.475 that the yield will be between 19 and 35 bushels.

TECH NOTES

We can graph normal distributions using the TI-84Plus/TI-83Plus/TI-*n*spire calculators, Excel 2010, and Minitab. In each technology, set the range of *x* values between $\mu - 3.5\sigma$ and $\mu + 3.5\sigma$. Then use the built-in normal density functions to generate the corresponding *y* values.

TI-84Plus/TI-83Plus/TI-*n*spire (with TI-84Plus keypad) Press the Y = key. Then, under **DISTR**, select **1:normalpdf** (x,μ,σ) and fill in desired μ and σ values. Press the **WINDOW** key. Set **Xmin** to $\mu - 3\sigma$ and **Xmax** to $\mu + 3\sigma$. Finally, press the **ZOOM** key and select option **0:ZoomFit**.

Excel 2010 In one column, enter *x* values from $\mu - 3.5\sigma$ to $\mu + 3.5\sigma$ in increments of 0.2σ. In the next column, generate *y* values by using the ribbon choice **Insert Function** $\boxed{f_x}$. In the dialogue box, select **Statistical** for the Category and then for the Function, select **NORMDIST** $(x,\mu,\sigma,\text{false})$. Next click the **Insert** tab and in the **Charts** group, click **Scatter**. Select the scatter diagram with smooth lines (third grouping).

Minitab In one column, enter *x* values from -3.5σ to 3.5σ in increments of 0.2σ. In the next column, enter *y* values by using the menu choices **Calc ➤ Probability Distribution ➤ Normal**. Fill in the dialogue box. Next, use menu choices **Graph ➤ Plot**. Fill in the dialogue box. Under Display, select connect.

P(x)

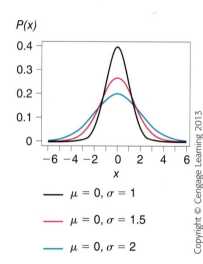

— $\mu = 0, \sigma = 1$

— $\mu = 0, \sigma = 1.5$

— $\mu = 0, \sigma = 2$

Copyright © Cengage Learning 2013

LOOKING FORWARD

Normal probability distributions will be used extensively in our later work. For instance, when we repeatedly take samples of the same size from a distribution and compute the same mean for each sample, we'll find that the sample means follow a distribution that is normal or approximately normal (Section 7.5). Also, when the number of trials is sufficiently large, the binomial distribution can be approximated by a normal distribution (Section 7.6). The distribution of the sample proportion of successes in a fixed number of binomial trials also can be approximated by a normal distribution (Section 7.6).

COMMENT   Uniform Distribution

Normal distributions are the central topic of this chapter. Normal distributions are very important in general probability and statistics. However, there are other, more specialized distributions that also have many applications. One such distribution is the *uniform distribution*. To learn more about this distribution, please see Problems 12 and 13 at the end of this section.

VIEWPOINT | In Control? Out of Control?

If you care about quality, you also must care about control! Dr. Walter Shewhart invented control charts when he was working for Bell Telephone Laboratories. The great contribution of control charts is that they separate variation into two sources: (1) random or chance causes (in control) and (2) special or assignable causes (out of control). A process is said to be in statistical control *when it is no longer afflicted with special or assignable causes. The performance of a process that is in statistical control is predictable. Predictability and quality control tend to be closely associated.*

(Source: Adapted from the classic text *Statistical Method from the Viewpoint of Quality Control,* by W. A. Shewhart, with foreword by W. E. Deming, Dover Publications.)

SECTION 7.1 PROBLEMS   Access end-of-section exercises at www.aplia.com

---

SECTION 7.2

Standard Units and Areas under the Standard Normal Distribution

FOCUS POINTS

- Given μ and σ, convert raw data to z scores.
- Given μ and σ, convert z scores to raw data.
- Graph the standard normal distribution and find areas under the standard normal curve.

z Scores and Raw Scores

Normal distributions vary from one another in two ways: The mean μ may be located anywhere on the x axis, and the bell shape may be more or less spread according to the size of the standard deviation σ. The differences among the normal distributions cause difficulties when we try to compute the area under the curve in a specified interval of x values and, hence, the probability that a measurement will fall into that interval.

It would be a futile task to try to set up a table of areas under the normal curve for each different μ and σ combination. We need a way to standardize the distributions so that we can use *one* table of areas for *all* normal distributions. We achieve this standardization by considering how many standard deviations a measurement lies from the mean. In this way, we can compare a value in one normal distribution with a value in another, different normal distribution. The next situation shows how this is done.

Suppose Tina and Jack are in two different sections of the same course. Each section is quite large, and the scores on the midterm exams of each section follow a normal distribution. In Tina's section, the average (mean) was 64 and her score

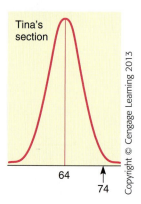

FIGURE 7-11

Distributions of Midterm Scores

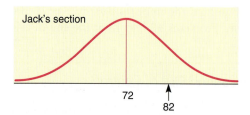

was 74. In Jack's section, the mean was 72 and his score was 82. Both Tina and Jack were pleased that their scores were each 10 points above the average of each respective section. However, the fact that each was 10 points above average does not really tell us how each did *with respect to the other students in the section.* In Figure 7-11, we see the normal distribution of grades for each section.

Tina's 74 was higher than most of the other scores in her section, while Jack's 82 is only an upper-middle score in his section. Tina's score is far better with respect to her class than Jack's score is with respect to his class.

The preceding situation demonstrates that it is not sufficient to know the difference between a measurement (*x* value) and the mean of a distribution. We need also to consider the spread of the curve, or the standard deviation. What we really want to know is the number of standard deviations between a measurement and the mean. This "distance" takes both μ and σ into account.

We can use a simple formula to compute the number *z* of standard deviations between a measurement *x* and the mean μ of a normal distribution with standard deviation σ:

$$\left(\begin{array}{c} \text{Number of standard deviations} \\ \text{between the measurement and} \\ \text{the mean} \end{array} \right) = \left(\frac{\text{Difference between the}}{\begin{array}{c}\text{measurement and the mean} \\ \hline \text{Standard deviation}\end{array}} \right)$$

z value, *z* score, or standard score

The *z* **value** or *z* **score** (also known as the **standard score**) gives the number of standard deviations between the original measurement *x* and the mean μ of the *x* distribution.

$$z = \frac{x - \mu}{\sigma}$$

TABLE 7-1

x Values and Corresponding z Values

x Value in Original Distribution	Corresponding z Value or Standard Unit
$x = \mu$	$z = 0$
$x > \mu$	$z > 0$
$x < \mu$	$z < 0$

Standard units

The mean is a special value of a distribution. Let's see what happens when we convert $x = \mu$ to a *z* value:

$$z = \frac{x - \mu}{\sigma} = \frac{\mu - \mu}{\sigma} = 0$$

The mean of the original distribution is always zero, in *standard units.* This makes sense because the mean is zero standard variations from itself.

An *x* value in the original distribution that is *above* the mean μ has a corresponding *z* value that is *positive.* Again, this makes sense because a measurement above the mean would be a positive number of standard deviations from the mean. Likewise, an *x* value *below* the mean has a *negative z* value. (See Table 7-1.)

Note

Unless otherwise stated, in the remainder of the book we will take the word *average* to be either the sample arithmetic mean $\bar{x}$ or the population mean μ.

EXAMPLE 2   STANDARD SCORE

A pizza parlor franchise specifies that the average (mean) amount of cheese on a large pizza should be 8 ounces and the standard deviation only 0.5 ounce. An inspector picks out a large pizza at random in one of the pizza parlors and finds that it is made with 6.9 ounces of cheese. Assume that the amount of cheese on a pizza follows a normal distribution. If the amount of cheese is below the mean by more than 3 standard deviations, the parlor will be in danger of losing its franchise.

How many standard deviations from the mean is 6.9? Is the pizza parlor in danger of losing its franchise?

SOLUTION: Since we want to know the number of standard deviations from the mean, we want to convert 6.9 to standard z units.

$$z = \frac{x - \mu}{\sigma} = \frac{6.9 - 8}{0.5} = -2.20$$

Interpretation The amount of cheese on the selected pizza is only 2.20 standard deviations below the mean. The fact that z is negative indicates that the amount of cheese is 2.20 standard deviations *below* the mean. The parlor will not lose its franchise based on this sample.

Raw score, *x*

We have seen how to convert from x measurements to standard units z. We can easily reverse the process to find the original *raw score x* if we know the mean and standard deviation of the original x distribution. Simply solve the z score formula for x.

> Given an x distribution with mean μ and standard deviation σ, the **raw score x** corresponding to a z score is
>
> $$x = z\sigma + \mu$$

GUIDED EXERCISE 3 | Standard score and raw score

Rod figures that it takes an average (mean) of 17 minutes with a standard deviation of 3 minutes to drive from home, park the car, and walk to an early morning class.

(a) One day it took Rod 21 minutes to get to class. How many standard deviations from the average is that? Is the z value positive or negative? Explain why it should be either positive or negative.

  The number of standard deviations from the mean is given by the z value:

$$z = \frac{x - \mu}{\sigma} = \frac{21 - 17}{3} \approx 1.33$$

The z value is positive. We should expect a positive z value, since 21 minutes is *more* than the mean of 17.

(b) What commuting time corresponds to a standard score of $z = -2.5$? *Interpretation* Could Rod count on making it to class in this amount of time or less?

$x = z\sigma + \mu$
$= (-2.5)(3) + 17$
$= 9.5$ minutes

No, commute times at or less than 2.5 standard deviations below the mean are rare.

LOOKING FORWARD

The basic structure of the formula for the standard score of a distribution is very general. When we verbalize the formula, we see it is

$$z = \frac{\text{measurement} - \text{mean of the distribution}}{\text{standard deviation of the distribution}}$$

We will see this general formula used again and again. In particular, when we look at sampling distributions for the mean (Section 7.4) and when we use the normal approximation of the binomial distribution (Section 7.6), we'll see this formula. We'll also see it when we discuss the sampling distribution for proportions (Section 7.6). Further uses occur in computations for confidence intervals (Chapter 8) and hypothesis testing (Chapter 9).

Standard Normal Distribution

If the original distribution of *x values is normal*, then the corresponding *z values have a normal distribution as well*. The *z* distribution has a mean of 0 and a standard deviation of 1. The normal curve with these properties has a special name:

Standard normal distribution

> The **standard normal distribution** is a normal distribution with mean $\mu = 0$ and standard deviation $\sigma = 1$ (Figure 7-12).

Any normal distribution of *x* values can be converted to the standard normal distribution by converting all *x* values to their corresponding *z* values. The resulting standard distribution will always have mean $\mu = 0$ and standard deviation $\sigma = 1$.

Areas under the Standard Normal Curve

Area under the standard normal curve

We have seen how to convert *any* normal distribution to the *standard* normal distribution. We can change any *x* value to a *z* value and back again. But what is the advantage of all this work? The advantage is that there are extensive tables that show the *area under the standard normal curve* for almost any interval along the *z* axis. The areas are important because each area is equal to the *probability* that the measurement of an item selected at random falls in this interval. Thus, the *standard* normal distribution can be a tremendously helpful tool.

FIGURE 7-12

The Standard Normal Distribution
($\mu = 0, \sigma = 1$)

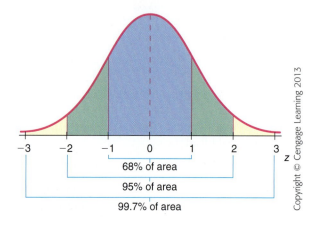

68% of area

95% of area

99.7% of area

Copyright © Cengage Learning 2013

Using a Standard Normal Distribution Table

Using a table to find areas and probabilities associated with the standard normal distribution is a fairly straightforward activity. However, it is important to first observe the range of z values for which areas are given. This range is usually depicted in a picture that accompanies the table.

Left-tail style table

In this text, *we will use the left-tail style table.* This style table gives cumulative areas to the left of a specified z. Determining other areas under the curve utilizes the fact that the area under the entire curve is 1. Taking advantage of the symmetry of the normal distribution is also useful. The procedures you learn for using the left-tail style normal distribution table apply directly to cumulative normal distribution areas found on calculators and in computer software packages such as Excel 2010 and Minitab.

EXAMPLE 3   STANDARD NORMAL DISTRIBUTION TABLE

Use Table 3 of the Appendix to find the described areas under the standard normal curve.

(a) Find the area under the standard normal curve to the left of $z = -1.00$.

SOLUTION: First, shade the area to be found on the standard normal distribution curve, as shown in Figure 7-13. Notice that the z value we are using is negative. This means that we will look at the portion of Table 3 of the Appendix for which the z values are negative. In the upper left corner of the table we see the letter z. The column under z gives us the units value and tenths value for z. The other column headings indicate the hundredths value of z. Table entries give areas under the standard normal curve to the left of the listed z values. To find the area to the left of $z = -1.00$, we use the row headed by -1.0 and then move to the column headed by the hundredths position, .00. This entry is shaded in Table 7-2. We see that the area is 0.1587.

Copyright © Cengage Learning 2013

TABLE 7-2	**Excerpt from Table 3 of the Appendix Showing Negative z Values**					
z	.00	.01	...	.07	.08	.09
−3.4	.0003	.0003	...	.0003	.0003	.0002
⋮						
−1.1	.1357	.1335	...	.1210	.1190	.1170
−1.0	.1587	.1562	...	.1423	.1401	.1379
−0.9	.1841	.1814	...	.1660	.1635	.1611
⋮						
−0.0	.5000	.4960	...	.4721	.4681	.4641

Copyright © Cengage Learning 2013

(b) Find the area to the left of $z = 1.18$, as illustrated in Figure 7-14.

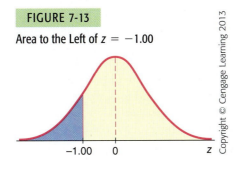

FIGURE 7-13

Area to the Left of $z = -1.00$

Copyright © Cengage Learning 2013

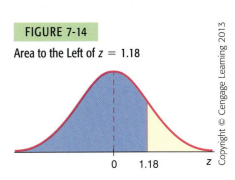

FIGURE 7-14

Area to the Left of $z = 1.18$

Copyright © Cengage Learning 2013

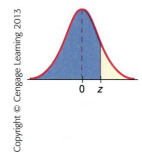

TABLE 7-3	Excerpt from Table 3 of the Appendix Showing Positive z Values					
z	.00	.01	.02	...	.08	.09
0.0	.5000	.5040	.5080	...	.5319	.5359
:						
0.9	.8159	.8186	.8212	...	.8365	.8389
1.0	.8413	.8438	.8461	...	.8599	.8621
1.1	.8643	.8665	.8686	...	.8810	.8830
:						
3.4	.9997	.9997	.9997	...	.9997	.9998

Copyright © Cengage Learning 2013

SOLUTION: In this case, we are looking for an area to the left of a positive z value, so we look in the portion of Table 3 that shows positive z values. Again, we first sketch the area to be found on a standard normal curve, as shown in Figure 7-14. Look in the row headed by 1.1 and move to the column headed by .08. The desired area is shaded (see Table 7-3). We see that the area to the left of 1.18 is 0.8810.

GUIDED EXERCISE 4 | ## Using the standard normal distribution table

Table 3, Areas of a Standard Normal Distribution, is located in the Appendix as well as in the endpapers of the text. Spend a little time studying the table, and then answer these questions.

(a) As z values increase, do the areas to the left of z increase?

> Yes. As z values increase, we move to the right on the normal curve, and the areas increase.

(b) If a z value is negative, is the area to the left of z less than 0.5000?

> Yes. Remember that a negative z value is on the left side of the standard normal distribution. The entire left half of the normal distribution has area 0.5, so any area to the left of $z = 0$ will be less than 0.5.

(c) If a z value is positive, is the area to the left of z greater than 0.5000?

> Yes. Positive z values are on the right side of the standard normal distribution, and any area to the left of a positive z value includes the entire left half of the normal distribution.

Using Table 3 to find other areas

Table 3 gives areas under the standard normal distribution that are to the *left of a z value*. How do we find other areas under the standard normal curve?

PROCEDURE

HOW TO USE A LEFT-TAIL STYLE STANDARD NORMAL DISTRIBUTION TABLE

1. For areas to the left of a specified z value, use the table entry directly.
2. For areas to the right of a specified z value, look up the table entry for z and subtract the area from 1.
   Note: Another way to find the same area is to use the symmetry of the normal curve and look up the table entry for $-z$.
3. For areas between two z values, z_1 and z_2 (where $z_2 > z_1$), *subtract* the table area for z_1 from the table area for z_2.

FIGURE 7-15

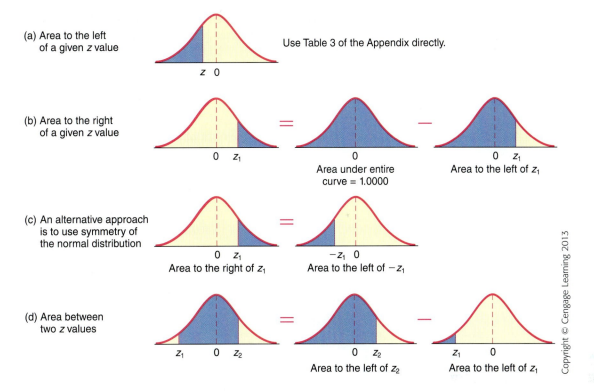

(a) Area to the left of a given *z* value — Use Table 3 of the Appendix directly.

(b) Area to the right of a given *z* value — Area under entire curve = 1.0000 — Area to the left of z_1

(c) An alternative approach is to use symmetry of the normal distribution — Area to the right of z_1 — Area to the left of $-z_1$

(d) Area between two *z* values — Area to the left of z_2 — Area to the left of z_1

Figure 7-15 illustrates the procedure for using Table 3, Areas of a Standard Normal Distribution, to find any specified area under the standard normal distribution. Again, it is useful to sketch the area in question before you use Table 3.

COMMENT   Notice that the *z* values shown in Table 3 of the Appendix are formatted to the hundredths position. It is convenient to *round or format z values to the hundredths position* before using the table. The areas are all given to four places after the decimal, so give your answers to four places after the decimal.

COMMENT   The smallest *z* value shown in Table 3 is -3.49, while the largest value is 3.49. These values are, respectively, far to the left and far to the right on the standard normal distribution, with very little area beyond either value. We will follow the common convention of treating any area to the left of a *z* value smaller than -3.49 as 0.000. Similarly, we will consider any area to the right of a *z* value greater than 3.49 as 0.000. We understand that there is some area in these extreme tails. However, these areas are each less than 0.0002. Now let's get real about this! Some very specialized applications, beyond the scope of this book, do need to measure areas and corresponding probabilities in these extreme tails. But in most practical applications, *we follow the convention of treating the areas in the extreme tails as zero.*

Convention for using Table 3 of the Appendix

1. Treat any area to the left of a *z* value smaller than -3.49 as 0.000.
2. Treat any area to the left of a *z* value greater than 3.49 as 1.000.

EXAMPLE 4   USING A TABLE TO FIND AREAS

Use Table 3 of the Appendix to find the specified areas.

(a) Find the area between $z = 1.00$ and $z = 2.70$.

SOLUTION: First, sketch a diagram showing the area (see Figure 7-16). Because we are finding the area between two z values, we subtract corresponding table entries.

$$(\text{Area between } 1.00 \text{ and } 2.70) = (\text{Area left of } 2.70) - (\text{Area left of } 1.00)$$
$$= 0.9965 - 0.8413$$
$$= 0.1552$$

(b) Find the area to the right of $z = 0.94$.

SOLUTION: First, sketch the area to be found (see Figure 7-17).

$$(\text{Area right of } 0.94) = (\text{Area under entire curve}) - (\text{Area left of } 0.94)$$
$$= 1.000 - 0.8264$$
$$= 0.1736$$

Alternatively,

$$(\text{Area right of } 0.94) = (\text{Area left of } -0.94)$$
$$= 0.1736$$

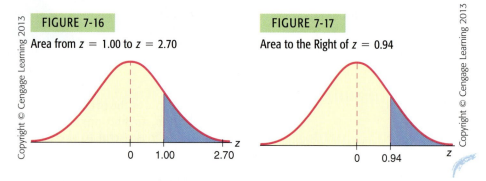

FIGURE 7-16
Area from $z = 1.00$ to $z = 2.70$

FIGURE 7-17
Area to the Right of $z = 0.94$

Copyright © Cengage Learning 2013

Probabilities associated with the standard normal distribution

We have practiced the skill of finding areas under the standard normal curve for various intervals along the z axis. This skill is important because *the probability that z lies in an interval is given by the area* under the standard normal curve above that interval.

Because the normal distribution is continuous, there is no area under the curve exactly over a specific z. Therefore, probabilities such as $P(z \geq z_1)$ are the same as $P(z > z_1)$. When dealing with probabilities or areas under a normal curve that are specified with inequalities, *strict inequality* symbols can be used *interchangeably* with *inequality* or *equal* symbols.

GUIDED EXERCISE 5 | **Probabilities associated with the standard normal distribution**

Let z be a random variable with a standard normal distribution.

(a) $P(z \geq 1.15)$ refers to the probability that z values lie to the right of 1.15. Shade the corresponding area under the standard normal curve (Figure 7-18) and find $P(z \geq 1.15)$.

➡ **FIGURE 7-18** Area to Be Found

Copyright © Cengage Learning 2013

$$P(z \geq 1.15) = 1.000 - P(z \leq 1.15) = 1.000 - 0.8749 = 0.1251$$

Alternatively,

$$P(z \geq 1.15) = P(z \leq -1.15) = 0.1251$$

GUIDED EXERCISE 5

(b) Find $P(-1.78 \leq z \leq 0.35)$.
First, sketch the area under the standard normal curve corresponding to the area (Figure 7-19).

➡ **FIGURE 7-19 Area to Be Found**

$$P(-1.78 \leq z \leq 0.35) = P(z \leq 0.35) - P(z \leq -1.78)$$
$$= 0.6368 - 0.0375 = 0.5993$$

TECH NOTES   The TI-84Plus/TI-83Plus/TI-*n*spire calculators, Excel 2010, and Minitab all provide cumulative areas under any normal distribution, including the standard normal. The Tech Note of Section 7.3 shows examples.

VIEWPOINT | Mighty Oaks from Little Acorns Grow!

Just how big is that acorn? What if we compare it with other acorns? Is that oak tree taller than an average oak tree? How does it compare with other oak trees? What do you mean, this oak tree has a larger geographic range? Compared with what? Answers to questions such as these can be given only if we resort to standardized statistical units. Can you compare a single oak tree with an entire forest of oak trees? The answer is yes, if you use standardized z scores. For more information about sizes of acorns, oak trees, and geographic locations, visit the Brase/Brase statistics site at **http://www.cengagebrain.com** *and find the link to DASL, the Carnegie Mellon University Data and Story Library. From the DASL site, find Biology under Data Subjects, and select Acorns.*

SECTION 7.2
PROBLEMS

aplia.   Access end-of-section exercises at www.aplia.com

SECTION 7.3 | ## Areas under Any Normal Curve

FOCUS POINTS
- Compute the probability of "standardized events."
- Find a *z* score from a given normal probability (inverse normal).
- Use the inverse normal to solve guarantee problems.

Normal Distribution Areas

In many applied situations, the original normal curve is not the standard normal curve. Generally, there will not be a table of areas available for the original normal curve. This does not mean that we cannot find the probability that a measurement x will fall into an interval from a to b. What we must do is *convert* the original measurements x, a, and b to z values.

PROCEDURE

To compute a z score, we need to know the standard deviation of the distribution. But what if we don't know σ? If we have a range of data values, we can estimate σ. Problems 31–34 show how this is done.

HOW TO WORK WITH NORMAL DISTRIBUTIONS

To find areas and probabilities for a random variable x that follows a normal distribution with mean μ and standard deviation σ, convert x values to z values using the formula

$$z = \frac{x - \mu}{\sigma}$$

Then use Table 3 of the Appendix to find corresponding areas and probabilities.

EXAMPLE 5

NORMAL DISTRIBUTION PROBABILITY

Let x have a normal distribution with $\mu = 10$ and $\sigma = 2$. Find the probability that an x value selected at random from this distribution is between 11 and 14. In symbols, find $P(11 \leq x \leq 14)$.

SOLUTION: Since probabilities correspond to areas under the distribution curve, we want to find the area under the x curve above the interval from $x = 11$ to $x = 14$. To do so, we will convert the x values to standard z values and then use Table 3 of the Appendix to find the corresponding area under the standard curve.

We use the formula

$$z = \frac{x - \mu}{\sigma}$$

to convert the given x interval to a z interval.

$$z_1 = \frac{11 - 10}{2} = 0.50 \quad \text{(Use } x = 11, \mu = 10, \sigma = 2.\text{)}$$

$$z_2 = \frac{14 - 10}{2} = 2.00 \quad \text{(Use } x = 14, \mu = 10, \sigma = 2.\text{)}$$

The corresponding areas under the x and z curves are shown in Figure 7-20. From Figure 7-20 we see that

$$P(11 \leq x \leq 14) = P(0.50 \leq z \leq 2.00)$$
$$= P(z \leq 2.00) - P(z \leq 0.50)$$
$$= 0.9772 - 0.6915 \quad \text{(From Table 3 of the Appendix)}$$
$$= 0.2857$$

Interpretation The probability is 0.2857 that an x value selected at random from a normal distribution with mean 10 and standard deviation 2 lies between 11 and 14.

FIGURE 7-20

Corresponding Areas under the x Curve and z Curve

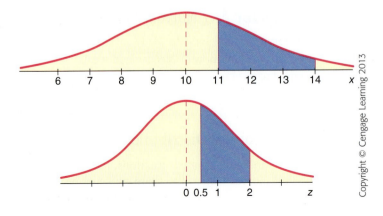

GUIDED EXERCISE 6 | **Normal distribution probability**

Sunshine Blu-ray players have a functional life that is normally distributed with a mean of 2.3 years and a standard deviation of 0.4 year. What is the probability that a Blu-ray player will break down during the guarantee period of 2 years?

(a) Let x represent the life of a Blu-ray player. The statement that the player breaks during the 2-year guarantee period means the life is less than 2 years, or $x \leq 2$. Convert this to a statement about z.

$$z = \frac{x - \mu}{\sigma} = \frac{2 - 2.3}{0.4} = -0.75$$

So $x \leq 2$ means $z \leq -0.75$.

(b) Indicate the area to be found in Figure 7-21. Does this area correspond to the probability that $z \leq -0.75$?

See Figure 7-22.
Yes, the shaded area does correspond to the probability that $z \leq -0.75$.

FIGURE 7-21

Copyright © Cengage Learning 2013

FIGURE 7-22 $z \leq -0.75$

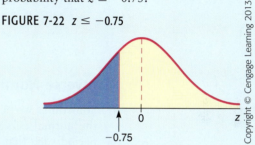

Copyright © Cengage Learning 2013

(c) Use Table 3 of the Appendix to find $P(z \leq -0.75)$.

$P(z \leq -0.75) = 0.2266$.

(d) *Interpretation* What is the probability that a Blu-ray player will break before the end of the guarantee period?
[*Hint:* $P(x \leq 2) = P(z \leq -0.75)$.]

The probability is
$$P(x \leq 2) = P(z \leq -0.75)$$
$$= 0.2266$$

This means that the company will repair or replace about 23% of the Blu-ray players.

TECH NOTES   The TI-84Plus/TI-83Plus/TI-*n*spire calculators, Excel 2010, and Minitab all provide areas under any normal distribution. Excel 2010 and Minitab give the left-tail area to the left of a specified x value. The TI-84Plus/TI-83Plus/TI-*n*spire has you specify an interval from a lower bound to an upper bound and provides the area under the normal curve for that interval. For example, to solve Guided Exercise 6 regarding the probability that a Blu-ray player will break during the guarantee period, we find $P(x \leq 2)$ for a normal distribution with $\mu = 2.3$ and $\sigma = 0.4$.

TI-84Plus/TI-83Plus/TI-*n*spire (with TI-84Plus keypad) Press the **DISTR** key, select **2:normalcdf** (lower bound, upper bound, μ, σ) and press Enter. Type in the specified values. For a left-tail area, use a lower bound setting at about 4 standard deviations below the mean. Likewise, for a right-tail area, use an upper bound setting about 4 standard deviations above the mean. For our example, use a lower bound of $\mu - 4\sigma = 2.3 - 4(0.4) = 0.7$.

```
normalcdf(.7,2,2.3,
.4)
            .2265955934
```

Excel 2010 Select **Insert Function** (f_x). In the dialogue box, select **Statistical** for the Category and then for the Function, select **NORM.DIST**. Fill in the dialogue box using True for cumulative.

f_x	=NORM.DIST(2,2.3,0.4, TRUE)		
	C	D	E
0.226627			

Minitab Use the menu selection **Calc ➤ Probability Distribution ➤ Normal**. Fill in the dialogue box, marking cumulative.

```
Cumulative Distribution Function
Normal with mean = 2.3 and
standard deviation  = 0.4
     x      P(X <= x)
    2.0      0.2266
```

Inverse Normal Distribution

Finding z or x, given a probability

Sometimes we need to find z or x values that correspond to a given area under the normal curve. This situation arises when we want to specify a guarantee period such that a given percentage of the total products produced by a company last at least as long as the duration of the guarantee period. In such cases, we use the standard normal distribution table "in reverse." When we look up an area and find the corresponding z value, we are using the *inverse normal probability distribution*.

Inverse normal probability distribution

EXAMPLE 6   FIND x, GIVEN PROBABILITY

Magic Video Games, Inc., sells an expensive video-games package. Because the package is so expensive, the company wants to advertise an impressive guarantee for the life expectancy of its computer control system. The guarantee policy will refund the full purchase price if the computer fails during the guarantee period. The research department has done tests that show that the mean life for the computer is 30 months, with standard deviation of 4 months. The computer life is normally distributed. How long can the guarantee period be if management does not want to refund the purchase price on more than 7% of the Magic Video Games packages?

SOLUTION: Let us look at the distribution of lifetimes for the computer control system, and shade the portion of the distribution in which the computer lasts fewer months than the guarantee period. (See Figure 7-23.)

FIGURE 7-23

7% of the Computers Have a Lifetime Less Than the Guarantee Period

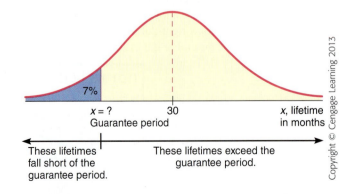

TABLE 7-4	Excerpt from Table 3 of the Appendix				
z	.00	...	.07	.08	.09
:					
−1.4	.0808		.0708	.0694	.0681

↑
0.0700

Copyright © Cengage Learning 2013

If a computer system lasts fewer months than the guarantee period, a full-price refund will have to be made. The lifetimes requiring a refund are in the shaded region in Figure 7-23. This region represents 7% of the total area under the curve.

We can use Table 3 of the Appendix to find the z value such that 7% of the total area under the *standard* normal curve lies to the left of the z value. Then we convert the z value to its corresponding x value to find the guarantee period.

We want to find the z value with 7% of the area under the standard normal curve to the left of z. Since we are given the area in a left tail, we can use Table 3 of the Appendix directly to find z. The area value is 0.0700. However, this area is not in our table, so we use the closest area, which is 0.0694, and the corresponding z value of $z = −1.48$ (see Table 7-4).

To translate this value back to an x value (in months), we use the formula

$$x = z\sigma + \mu$$

$$= −1.48(4) + 30 \quad \text{(Use } \sigma = 4 \text{ months and } \mu = 30 \text{ months.)}$$

$$= 24.08 \text{ months}$$

Interpretation The company can guarantee the Magic Video Games package for $x = 24$ months. For this guarantee period, they expect to refund the purchase price of no more than 7% of the video games packages.

Example 6 had us find a z value corresponding to a given area to the left of z. What if the specified area is to the right of z or between $−z$ and z? Figure 7-24 shows us how to proceed.

FIGURE 7-24

Inverse Normal: Use Table 3 of the Appendix to Find z Corresponding to a Given Area A $(0 < A < 1)$

(a) **Left-tail case:**
The given area A is to the left of z.

  or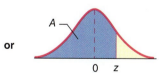

For the left-tail case, look up the number A in the body of the table and use the corresponding z value.

(b) **Right-tail case:**
The given area A is to the right of z.

  or

For the right-tail case, look up the number $1 − A$ in the body of the table and use the corresponding z value.

(c) **Center case:**
The given area A is symmetric and centered above $z = 0$. Half of A lies to the left and half lies to the right of $z = 0$.

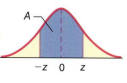

For the center case, look up the number $\dfrac{1 − A}{2}$ in the body of the table and use the corresponding $\pm z$ value.

COMMENT   When we use Table 3 of the Appendix to find a z value corresponding to a given area, we usually use the nearest area value rather than interpolating between values. However, when the area value given is exactly halfway between two area values of the table, we use the z value halfway between the z values of the corresponding table areas. Example 7 demonstrates this procedure. However, this interpolation convention is not always used, especially if the area is changing slowly, as it does in the tail ends of the distribution. *When the z value corresponding to an area is smaller than -2, the standard convention is to use the z value corresponding to the smaller area. Likewise, when the z value is larger than 2, the standard convention is to use the z value corresponding to the larger area.* We will see an example of this special case in Example 1 of Chapter 8.

EXAMPLE 7   FIND z

Find the z value such that 90% of the area under the standard normal curve lies between $-z$ and z.

SOLUTION: Sketch a picture showing the described area (see Figure 7-25).

FIGURE 7-25

Area between $-z$ and z Is 90%

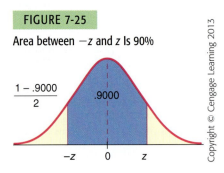

Copyright © Cengage Learning 2013

We find the corresponding area in the left tail.

$$\text{(Area left of } -z) = \frac{1 - 0.9000}{2}$$
$$= 0.0500$$

Looking in Table 7-5, we see that 0.0500 lies exactly between areas 0.0495 and 0.0505. The halfway value between $z = -1.65$ and $z = -1.64$ is $z = -1.645$. Therefore, we conclude that 90% of the area under the standard normal curve lies between the z values -1.645 and 1.645.

TABLE 7-5   **Excerpt from Table 3 of the Appendix**

z	. . .	.04	.05
:			
-1.6		.0505	.0495
		↑	
		0.0500	

Copyright © Cengage Learning 2013

GUIDED EXERCISE 7 | **Find z**

Find the z value such that 3% of the area under the standard normal curve lies to the right of z.

(a) Draw a sketch of the standard normal distribution showing the described area (Figure 7-26).

FIGURE 7-26 3% of Total Area Lies to the Right of z

97%

3%

0   z

(b) Find the area to the left of z.

Area to the left of $z = 1 - 0.0300 = 0.9700$.

(c) Look up the area in Table 7-6 and find the corresponding z.

The closest area is 0.9699. This area is to the left of $z = 1.88$.

TABLE 7-6 Excerpt from Table 3 of the Appendix

z	.00	.01	.02	.03	.04	.05	.06	.07	.08	.09
1.8	.9641	.9649	.9656	.9664	.9671	.9678	.9686	.9693	.9699	.9706
1.9	.9713	.9719	.9726	.9732	.9738	.9744	.9750	.9756	.9761	.9767

(d) Suppose the time to complete a test is normally distributed with $\mu = 40$ minutes and $\sigma = 5$ minutes. After how many minutes can we expect all but about 3% of the tests to be completed?

We are looking for an x value such that 3% of the normal distribution lies to the right of x. In part (c), we found that 3% of the standard normal curve lies to the right of $z = 1.88$. We convert $z = 1.88$ to an x value.

$$x = z\sigma + \mu$$
$$= 1.88(5) + 40 = 49.4 \text{ minutes}$$

All but about 3% of the tests will be complete after 50 minutes.

(e) Use Table 7-7 to find a z value such that 3% of the area under the standard normal curve lies to the left of z.

The closest area is 0.0301. This is the area to the left of $z = -1.88$.

TABLE 7-7 Excerpt from Table 3 of the Appendix

z	.00	.01	.02	.03	.04	.05	.06	.07	.08	.09
−1.9	.0287	.0281	.0274	.0268	.0262	.0256	.0250	.0244	.0239	.0233
−1.8	.0359	.0351	.0344	.0336	.0329	.0322	.0314	.0307	.0301	.0294

(f) Compare the z value of part (c) with the z value of part (e). Is there any relationship between the z values?

One z value is the negative of the other. This result is expected because of the symmetry of the normal distribution.

TECH NOTES   When we are given a z value and we find an area to the left of z, we are using a normal distribution function. When we are given an area to the left of z and we find the corresponding z, we are using an inverse normal distribution function. The TI-84Plus/TI-83Plus/TI-$n$spire calculators, Excel 2010, and Minitab all have inverse normal distribution functions for any normal distribution. For instance, to find an x value from a normal distribution with mean 40 and standard deviation 5 such that 97% of the area lies to the left of x, use the described instructions.

TI-84Plus/TI-83Plus/TI-*n*spire (with TI-84Plus keypad) Press the **DISTR** key and select 3:invNorm(area,μ,σ).

```
invNorm(.97,40,5)
          49.40396805
```

Copyright © Cengage Learning 2013

Excel 2010 Select **Insert Function** (f_x). In the dialogue box, select **Statistical** for the Category and then for the Function, select **NORM.INV**. Fill in the dialogue box.

f_x	=NORM.INV(0.97,40,5)
C	**D**
49.40395	

Copyright © Cengage Learning 2013

Minitab Use the menu selection **Calc ➤ Probability Distribution ➤ Normal.** Fill in the dialogue box, marking Inverse Cumulative.

```
Inverse Cumulative Distribution Function
Normal with mean = 40.000 and
  standard deviation = 5.00000
       P(X <= x)       x
          0.9700   49.4040
```

LOOKING FORWARD

In our work with confidence intervals (Chapter 8), we will use inverse probability distributions for the normal distribution and for the Student's t distribution (a similar distribution introduced in Chapter 8). Just as in Example 7, we'll use inverse probability distributions to identify values such that 90%, 95%, or 99% of the area under the distribution graph centered over the mean falls between the values.

**CRITICAL
THINKING**

Checking for Normality

How can we tell if data follow a normal distribution? There are several checks we can make. The following procedure lists some guidelines.

PROCEDURE

HOW TO DETERMINE WHETHER DATA HAVE A NORMAL DISTRIBUTION

The following guidelines represent some useful devices for determining whether or not data follow a normal distribution:

1. **Histogram:** Make a histogram. For a normal distribution, the histogram should be roughly bell-shaped.

2. **Outliers:** For a normal distribution, there should not be more than one outlier. One way to check for outliers is to use a box-and-whisker plot. Recall that outliers are those data values that are:

   above Q_3 by an amount greater than $1.5 \times$ interquartile range

   below Q_1 by an amount greater than $1.5 \times$ interquartile range

Pearson's index for skewness

3. **Skewness:** Normal distributions are symmetrical. One measure of skewness for sample data is given by *Pearson's index*:

$$\text{Person's index} = \frac{3(\overline{x} - \text{median})}{s}$$

An index value greater than 1 or less than -1 indicates skewness. Skewed distributions are not normal.

Normal quantile plot

4. **Normal quantile plot (or normal probability plot):** This plot is provided through statistical software on a computer or graphing calculator. The Using Technology feature, Application 1, at the end of this chapter gives a brief description of how such plots are constructed. The section also gives commands for producing such plots on the TI-84Plus/TI-83Plus/TI-*n*spire calculators, Minitab, or SPSS.

   Examine a normal quantile plot of the data.

   If the points lie close to a straight line, the data come from a distribution that is approximately normal.

   If the points do not lie close to a straight line or if they show a pattern that is not a straight line, the data are likely to come from a distribution that is not normal.

EXAMPLE 8   ASSESSING NORMALITY

Consider the following data, which are rounded to the nearest integer:

19	19	19	16	21	14	23	17	19	20	18	24	20	13	16
17	19	18	19	17	21	24	18	23	19	21	22	20	20	20
24	17	20	22	19	22	21	18	20	22	16	15	21	23	21
18	18	20	15	25										

(a) Look at the histogram and box-and-whisker plot generated by Minitab in Figure 7-27 and comment about normality of the data from these indicators.

FIGURE 7-27

Histogram and Box-and-Whisker Plot

Histogram

Box-and-whisker plot

SOLUTION: Note that the histogram is approximately normal. The box-and-whisker plot shows just one outlier. Both of these graphs indicate normality.

(b) Use Pearson's index to check for skewness.

SOLUTION: Summary statistics from Minitab:

Variable	N	N*	Mean	Se Mean	StDev	Minimum	Q1	Median	Q3
C2	50	0	19.460	0.382	2.705	13.000	18.000	19.500	21.000
Variable			Maximum						
C2			25.000						

We see that $\bar{x} = 19.46$, median = 19.5, and $s = 2.705$.

$$\text{Pearson's index} = \frac{3(19.46 - 19.5)}{2.705} \approx -0.04$$

Since the index is between -1 and 1, we detect no skewness. The data appear to be symmetrical.

(c) Look at the normal quantile plot in Figure 7-28 and comment on normality.

FIGURE 7-28

Normal Quantile Plot

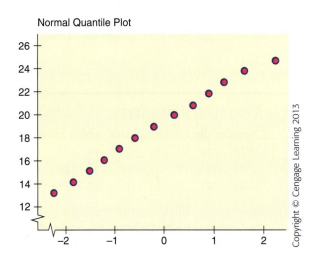

Normal Quantile Plot

SOLUTION: The points fall close to a straight line, so the data appear to come from a normal distribution.

(d) *Interpretation* Interpret the results.

SOLUTION: The histogram is roughly bell-shaped, there is only one outlier, Pearson's index does not indicate skewness, and the points on the normal quantile plot lie fairly close to a straight line. It appears that the data are likely from a distribution that is approximately normal.

VIEWPOINT | Want to Be an Archaeologist?

Each year about 4500 students work with professional archaeologists in scientific research at the Crow Canyon Archaeological Center, Cortez, Colorado. In fact, Crow Canyon was included in The Princeton Review Student Access Guide to America's Top 100 Internships. *The nonprofit, multidisciplinary program at Crow Canyon enables students and laypeople with little or no archaeology background to get started in archaeological research. The only requirement is that you be interested in Native American culture and history. By the way, a knowledge of introductory statistics could come in handy for this internship. For more information about the program, visit the Brase/Brase statistics site at* **http://www.cengagebrain.com** *and find the link to Crow Canyon.*

**SECTION 7.3
PROBLEMS**

 aplia   Access end-of-section exercises at www.aplia.com

SECTION 7.4

Sampling Distributions

FOCUS POINTS

• Review such commonly used terms as *random sample, relative frequency, parameter, statistic,* and *sampling distribution.*
• From raw data, construct a relative frequency distribution for $\bar{x}$ values and compare the result to a theoretical sampling distribution.

Let us begin with some common statistical terms. Most of these have been discussed before, but this is a good time to review them.

From a statistical point of view, a *population* can be thought of as a complete set of measurements (or counts), either existing or conceptual. We discussed populations at some length in Chapter 1. A *sample* is a subset of measurements from the population. For our purposes, the most important samples are *random samples*, which were discussed in Section 1.2.

When we compute a descriptive measure such as an average, it makes a difference whether it was computed from a population or from a sample.

Statistic
Parameter

A **statistic** is a numerical descriptive measure of a *sample*.
A **parameter** is a numerical descriptive measure of a *population*.

It is important to notice that for a given population, a specified parameter is a fixed quantity. On the other hand, the value of a statistic might vary depending on which sample has been selected.

Some commonly used statistics and corresponding parameters		
Measure	*Statistic*	*Parameter*
Mean	$\bar{x}$ (x bar)	μ (mu)
Variance	s^2	σ^2 (sigma squared)
Standard deviation	s	σ (sigma)
Proportion	$\hat{p}$ (p hat)	p

Often we do not have access to all the measurements of an entire population, because of constraints on time, money, or effort. So we must use measurements from a sample instead. In such cases, we will use a statistic (such as $\bar{x}$, s, or $\hat{p}$) to **Population parameter** make *inferences* about a corresponding *population parameter* (e.g., μ, σ, or p). The principal types of inferences we will make are the following:

Types of inferences

1. **Estimation:** In this type of inference, we estimate the *value* of a population parameter.
2. **Testing:** In this type of inference, we formulate a *decision* about the value of a population parameter.
3. **Regression:** In this type of inference, we make *predictions* or *forecasts* about the value of a statistical variable.

To evaluate the reliability of our inferences, we will need to know the probability distribution for the statistic we are using. Such a probability distribution is **Sampling distribution** called a *sampling distribution*. Perhaps Example 9 below will help clarify this discussion.

A **sampling distribution** is a probability distribution of a sample statistic based on all possible simple random samples of the *same* size from the same population.

EXAMPLE 9   SAMPLING DISTRIBUTION FOR $\bar{x}$

Pinedale, Wisconsin, is a rural community with a children's fishing pond. Posted rules state that all fish under 6 inches must be returned to the pond, only children under 12 years old may fish, and a limit of five fish may be kept per day. Susan is a college student who was hired by the community last summer to make sure the rules were obeyed and to see that the children were safe from accidents. The pond contains only rainbow trout and has been well stocked for many years. Each child has no difficulty catching his or her limit of five trout.

As a project for her biometrics class, Susan kept a record of the lengths (to the nearest inch) of all trout caught last summer. Hundreds of children visited the pond and caught their limit of five trout, so Susan has a lot of data. To make Table 7-8, Susan selected 100 children at random and listed the lengths of each of the five trout caught by a child in the sample. Then, for each child, she listed the mean length of the five trout that child caught.

Now let us turn our attention to the following question: What is the average (mean) length of a trout taken from the Pinedale children's pond last summer?

SOLUTION: We can get an idea of the average length by looking at the far right column of Table 7-8. But just looking at 100 of the $\bar{x}$ values doesn't tell us much.

TABLE 7-8   **Length Measurements of Trout Caught by a Random Sample of 100 Children at the Pinedale Children's Pond**

Sample	Length (to Nearest Inch)				$\bar{x}$ = Sample Mean	Sample	Length (to Nearest Inch)				$\bar{x}$ = Sample Mean
1	11	10	10	12	11	10.8	51	9	10	12	10	9	10.0
2	11	11	9	9	9	9.8	52	7	11	10	11	10	9.8
3	12	9	10	11	10	10.4	53	9	11	9	11	12	10.4
4	11	10	13	11	8	10.6	54	12	9	8	10	11	10.0
5	10	10	13	11	12	11.2	55	8	11	10	9	10	9.6
6	12	7	10	9	11	9.8	56	10	10	9	9	13	10.2
7	7	10	13	10	10	10.0	57	9	8	10	10	12	9.8
8	10	9	9	9	10	9.4	58	10	11	9	8	9	9.4
9	10	10	11	12	8	10.2	59	10	8	9	10	12	9.8
10	10	11	10	7	9	9.4	60	11	9	9	11	11	10.2
11	12	11	11	11	13	11.6	61	11	10	11	10	11	10.6
12	10	11	10	12	13	11.2	62	12	10	10	9	11	10.4
13	11	10	10	9	11	10.2	63	10	10	9	11	7	9.4
14	10	10	13	8	11	10.4	64	11	11	12	10	11	11.0
15	9	11	9	10	10	9.8	65	10	10	11	10	9	10.0
16	13	9	11	12	10	11.0	66	8	9	10	11	11	9.8
17	8	9	7	10	11	9.0	67	9	11	11	9	8	9.6
18	12	12	8	12	12	11.2	68	10	9	10	9	11	9.8
19	10	8	9	10	10	9.4	69	9	9	11	11	11	10.2
20	10	11	10	10	10	10.2	70	13	11	11	9	11	11.0
21	11	10	11	9	12	10.6	71	12	10	8	8	9	9.4
22	9	12	9	10	9	9.8	72	13	7	12	9	10	10.2
23	8	11	10	11	10	10.0	73	9	10	9	8	9	9.0
24	9	12	10	9	11	10.2	74	11	11	10	9	10	10.2
25	9	9	8	9	10	9.0	75	9	11	14	9	11	10.8
26	11	11	12	11	11	11.2	76	14	10	11	12	12	11.8
27	10	10	10	11	13	10.8	77	8	12	10	10	9	9.8
28	8	7	9	10	8	8.4	78	8	10	13	9	8	9.6
29	11	11	8	10	11	10.2	79	11	11	11	13	10	11.2
30	8	11	11	9	12	10.2	80	12	10	11	12	9	10.8
31	11	9	12	10	10	10.4	81	10	9	10	10	13	10.4
32	10	11	10	11	12	10.8	82	11	10	9	9	12	10.2
33	12	11	8	8	11	10.0	83	11	11	10	10	10	10.4
34	8	10	10	9	10	9.4	84	11	10	11	9	9	10.0
35	10	10	10	10	11	10.2	85	10	11	10	9	7	9.4
36	10	8	10	11	13	10.4	86	7	11	10	9	11	9.6
37	11	10	11	11	10	10.6	87	10	11	10	10	10	10.2
38	7	13	9	12	11	10.4	88	9	8	11	10	12	10.0
39	11	11	8	11	11	10.4	89	14	9	12	10	9	10.8
40	11	10	11	12	9	10.6	90	9	12	9	10	10	10.0
41	11	10	9	11	12	10.6	91	10	10	8	6	11	9.0
42	11	13	10	12	9	11.0	92	8	9	11	9	10	9.4
43	10	9	11	10	11	10.2	93	8	10	9	9	11	9.4
44	10	9	11	10	9	9.8	94	12	11	12	13	10	11.6
45	12	11	9	11	12	11.0	95	11	11	9	9	9	9.8
46	13	9	11	8	8	9.8	96	8	12	8	11	10	9.8
47	10	11	11	11	10	10.6	97	13	11	11	12	8	11.0
48	9	9	10	11	11	10.0	98	10	11	8	10	11	10.0
49	10	9	9	10	10	9.6	99	13	10	7	11	9	10.0
50	10	10	6	9	10	9.0	100	9	9	10	12	12	10.4

TABLE 7-9	Frequency Table for 100 Values of $\bar{x}$			
	Class Limits			
Class	Lower	Upper	f = Frequency	$f/100$ = Relative Frequency
1	8.39	8.76	1	0.01
2	8.77	9.14	5	0.05
3	9.15	9.52	10	0.10
4	9.53	9.90	19	0.19
5	9.91	10.28	27	0.27
6	10.29	10.66	18	0.18
7	10.67	11.04	12	0.12
8	11.05	11.42	5	0.05
9	11.43	11.80	3	0.03

Copyright © Cengage Learning 2013

Let's organize our $\bar{x}$ values into a frequency table. We used a class width of 0.38 to make Table 7-9.

Note: Techniques of Section 2.1 dictate a class width of 0.4. However, this choice results in the tenth class being beyond the data. Consequently, we shortened the class width slightly and also started the first class with a value slightly smaller than the smallest data value.

The far right column of Table 7-9 contains relative frequencies $f/100$. Recall that relative frequencies may be thought of as probabilities, so we effectively have a probability distribution. Because $\bar{x}$ represents the mean length of a trout (based on samples of five trout caught by each child), we estimate the probability of $\bar{x}$ falling into each class by using the relative frequencies. Figure 7-29 is a relative-frequency or probability distribution of the $\bar{x}$ values.

FIGURE 7-29

Estimates of Probabilities of $\bar{x}$ Values

Copyright © Cengage Learning 2013

The bars of Figure 7-29 represent our estimated probabilities of $\bar{x}$ values based on the data of Table 7-8. The bell-shaped curve represents the theoretical probability distribution that would be obtained if the number of children (i.e., number of $\bar{x}$ values) were much larger.

Figure 7-29 represents a *probability sampling distribution* for the sample mean $\bar{x}$ of trout lengths based on random samples of size 5. We see that the distribution is mound-shaped and even somewhat bell-shaped. Irregularities are due to the small number of samples used (only 100 sample means) and the rather small sample size (five trout per child). These irregularities would become less obvious and even disappear if the sample of children became much larger, if we used a larger number of classes in Figure 7-29, and if the number of trout in each sample became larger. In fact, the curve would eventually become a perfect bell-shaped curve. We will discuss this property at some length in the next section, which introduces the *central limit theorem*.

There are other sampling distributions besides the $\bar{x}$ distribution. Section 7.6 shows the sampling distribution for $\hat{p}$. In the chapters ahead, we will see that other statistics have different sampling distributions. However, the $\bar{x}$ sampling distribution is very important. It will serve us well in our inferential work in Chapters 8–10 on estimation and testing.

Let us summarize the information about sampling distributions in the following exercise.

GUIDED EXERCISE 8 | **Terminology**

(a) What is a population parameter? Give an example.

➡ A population parameter is a numerical descriptive measure of a population. Examples are μ, σ, and p. (There are many others.)

(b) What is a sample statistic? Give an example.

➡ A sample statistic or statistic is a numerical descriptive measure of a sample. Examples are $\bar{x}$, s, and $\hat{p}$.

(c) What is a sampling distribution?

➡ A sampling distribution is a probability distribution for the sample statistic we are using based on all possible samples of the same size.

(d) In Table 7-8, what makes up the members of the sample? What is the sample statistic corresponding to each sample? What is the sampling distribution? To which population parameter does this sampling distribution correspond?

➡ There are 100 samples, each of which comprises five trout lengths. In the first sample, the five trout have lengths 11, 10, 10, 12, and 11. The sample statistic is the sample mean $\bar{x} = 10.8$. The sampling distribution is shown in Figure 7-29. This sampling distribution relates to the population mean μ of all lengths of trout taken from the Pinedale children's pond (i.e., trout over 6 inches long).

(e) Where will sampling distributions be used in our study of statistics?

➡ Sampling distributions will be used for statistical inference. (Chapter 8 will concentrate on a method of inference called *estimation*. Chapters 9 and 10 will concentrate on a method of inference called *testing*.)

VIEWPOINT | "Chance Favors the Prepared Mind"

—Louis Pasteur

It has been said that a discovery is nothing more than an accident that meets a prepared mind. Sampling can be one of the best forms of preparation. In fact, sampling may be the primary way we humans venture into the unknown. Probability sampling distributions can provide new information for the sociologist, scientist, or economist. In addition, ordinary human sampling of life can help writers and artists develop preferences, styles, and insights. Ansel Adams became famous for photographing lyrical, unforgettable landscapes such as Moonrise, Hernandez, New Mexico. Adams claimed that he was a strong believer in the quote by Pasteur. In fact, he claimed that the Hernandez photograph was just such a favored chance happening that his prepared mind readily grasped. During his lifetime, Adams made over $25 million from sales and royalties on the Hernandez photograph.

SECTION 7.5

The Central Limit Theorem

FOCUS POINTS

- For a normal distribution, use μ and σ to construct the theoretical sampling distribution for the statistic $\bar{x}$.
- For large samples, use sample estimates to construct a good approximate sampling distribution for the statistic $\bar{x}$.
- Learn the statement and underlying meaning of the central limit theorem well enough to explain it to a friend who is intelligent but (unfortunately) doesn't know much about statistics.

The $\bar{x}$ Distribution, Given x Is Normal

In Section 7.4, we began a study of the distribution of $\bar{x}$ values, where $\bar{x}$ was the (sample) mean length of five trout caught by children at the Pinedale children's fishing pond. Let's consider this example again in the light of a very important theorem of mathematical statistics.

THEOREM 7.1   For a Normal Probability Distribution Let x be a random variable with a *normal distribution* whose mean is μ and whose standard deviation is σ. Let $\bar{x}$ be the sample mean corresponding to random samples of size n taken from the x distribution. Then the following are true:

(a) The $\bar{x}$ distribution is a *normal distribution*.
(b) The mean of the $\bar{x}$ distribution is μ.
(c) The standard deviation of the $\bar{x}$ distribution is $\sigma/\sqrt{n}$.

We conclude from Theorem 7.1 that when x has a normal distribution, the $\bar{x}$ distribution will be normal *for any sample size n*. Furthermore, we can convert the $\bar{x}$ distribution to the standard normal z distribution using the following formulas:

$$\mu_{\bar{x}} = \mu$$
$$\sigma_{\bar{x}} = \frac{\sigma}{\sqrt{n}}$$
$$z = \frac{\bar{x} - \mu_{\bar{x}}}{\sigma_{\bar{x}}} = \frac{\bar{x} - \mu}{\sigma/\sqrt{n}}$$

where n is the sample size,
$\quad \mu$ is the mean of the x distribution, and
$\quad \sigma$ is the standard deviation of the x distribution.

Theorem 7.1 is a wonderful theorem! It states that the $\bar{x}$ distribution will be normal, provided the x distribution is normal. The sample size n could be 2, 3, 4, or any other (fixed) sample size we wish. Furthermore, the mean of the $\bar{x}$ distribution is μ (same as for the x distribution), but the standard deviation is $\sigma/\sqrt{n}$ (which is, of course, smaller than σ). The next example illustrates Theorem 7.1.

EXAMPLE 10   PROBABILITY REGARDING x AND $\overline{x}$

Suppose a team of biologists has been studying the Pinedale children's fishing pond. Let x represent the length of a single trout taken at random from the pond. This group of biologists has determined that x has a normal distribution with mean $\mu = 10.2$ inches and standard deviation $\sigma = 1.4$ inches.

(a) What is the probability that a *single trout* taken at random from the pond is between 8 and 12 inches long?

SOLUTION: We use the methods of Section 7.3, with $\mu = 10.2$ and $\sigma = 1.4$, to get

$$z = \frac{x - \mu}{\sigma} = \frac{x - 10.2}{1.4}$$

Therefore,

$$P(8 < x < 12) = P\left(\frac{8 - 10.2}{1.4} < z < \frac{12 - 10.2}{1.4}\right)$$
$$= P(-1.57 < z < 1.29)$$
$$= 0.9015 - 0.0582 = 0.8433$$

Therefore, the probability is about 0.8433 that a *single* trout taken at random is between 8 and 12 inches long.

(b) What is the probability that the *mean length* $\overline{x}$ of five trout taken at random is between 8 and 12 inches?

SOLUTION: If we let $\mu_{\overline{x}}$ represent the mean of the distribution, then Theorem 7.1, part (b), tells us that

$$\mu_{\overline{x}} = \mu = 10.2$$

If $\sigma_{\overline{x}}$ represents the standard deviation of the $\sigma_{\overline{x}}$ distribution, then Theorem 7.1, part (c), tells us that

$$\sigma_{\overline{x}} = \sigma/\sqrt{n} = 1.4/\sqrt{5} \approx 0.63$$

To create a standard z variable from $\overline{x}$, we subtract $\mu_{\overline{x}}$ and divide by $\sigma_{\overline{x}}$:

$$z = \frac{\overline{x} - \mu_{\overline{x}}}{\sigma_{\overline{x}}} = \frac{\overline{x} - \mu}{\sigma/\sqrt{n}} \approx \frac{\overline{x} - 10.2}{0.63}$$

To standardize the interval $8 < \overline{x} < 12$, we use 8 and then 12 in place of $\overline{x}$ in the preceding formula for z.

$$8 < \overline{x} < 12$$
$$\frac{8 - 10.2}{0.63} < z < \frac{12 - 10.2}{0.63}$$
$$-3.49 < z < 2.86$$

Theorem 7.1, part (a), tells us that $\overline{x}$ has a normal distribution. Therefore,

$$P(8 < \overline{x} < 12) = P(-3.49 < z < 2.86) = 0.9979 - 0.0002 = 0.9977$$

The probability is about 0.9977 that the mean length based on a sample size of 5 is between 8 and 12 inches.

(c) Looking at the results of parts (a) and (b), we see that the probabilities (0.8433 and 0.9977) are quite different. Why is this the case?

SOLUTION: According to Theorem 7.1, both x and $\overline{x}$ have a normal distribution, and both have the same mean of 10.2 inches. The difference is in the standard deviations for x and $\bar{x}$. The standard deviation of the x distribution is $\sigma = 1.4$. The standard deviation of the $\bar{x}$ distribution is

$$\sigma_{\bar{x}} = \sigma/\sqrt{n} = 1.4/\sqrt{5} \approx 0.63$$

The standard deviation of the $\bar{x}$ distribution is less than half the standard deviation of the x distribution. Figure 7-30 shows the distributions of x and $\bar{x}$.

FIGURE 7-30

General Shapes of the x and $\bar{x}$ Distributions

(a) The x distribution with $\mu = 10.2$ and $\sigma = 1.4$

(b) The $\bar{x}$ distribution with $\mu_{\bar{x}} = 10.2$ and $\sigma_{\bar{x}} = 0.63$ for samples of size $n = 5$

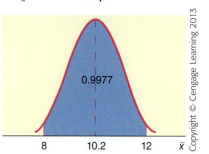

Looking at Figure 7-30(a) and (b), we see that both curves use the same scale on the horizontal axis. The means are the same, and the shaded area is above the interval from 8 to 12 on each graph. It becomes clear that the smaller standard deviation of the $\bar{x}$ distribution has the effect of gathering together much more of the total probability into the region over its mean. Therefore, the region from 8 to 12 has a much higher probability for the $\bar{x}$ distribution.

Standard error of the mean

Theorem 7.1 describes the distribution of a particular statistic: namely, the distribution of sample mean $\bar{x}$. The standard deviation of a statistic is referred to as the *standard error* of that statistic.

> The **standard error** is the standard deviation of a sampling distribution. For the $\bar{x}$ sampling distribution,
>
> $$\text{standard error} = \sigma_{\bar{x}} = \sigma/\sqrt{n}$$

Statistical software

The expression *standard error* appears commonly on printouts and refers to the standard deviation of the sampling distribution being used. (In Minitab, the expression SE MEAN refers to the standard error of the mean.)

The $\bar{x}$ Distribution, Given x Follows Any Distribution

Theorem 7.1 gives complete information about the $\bar{x}$ distribution, provided the original x distribution is known to be normal. What happens if we don't have information about the shape of the original x distribution? The *central limit theorem* tells us what to expect.

Central limit theorem

THEOREM 7.2   The Central Limit Theorem for Any Probability Distribution

If x possesses *any* distribution with mean μ and standard deviation σ, then the sample mean $\bar{x}$ based on a random sample of size n will have a distribution that approaches the distribution of a normal random variable with mean μ and standard deviation $\sigma/\sqrt{n}$ as n increases without limit.

The central limit theorem is indeed surprising! It says that x can have *any* distribution whatsoever, but that as the sample size gets larger and larger, the distribution of $\bar{x}$ will approach a *normal* distribution. From this relation, we begin to appreciate the scope and significance of the normal distribution.

In the central limit theorem, the degree to which the distribution of $\bar{x}$ values fits a normal distribution depends on both the selected value of n and the original distribution of x values. A natural question is: How large should the sample size be if we want to apply the central limit theorem? After a great deal of theoretical as well as empirical study, statisticians agree that if n is 30 or larger, the $\bar{x}$ distribution will appear to be normal and the central limit theorem will apply. However, this rule should not be applied blindly. If the x distribution is definitely not symmetrical about its mean, then the $\bar{x}$ distribution also will display a lack of symmetry. In such a case, a sample size larger than 30 may be required to get a reasonable approximation to the normal.

In practice, it is a good idea, when possible, to make a histogram of sample x values. If the histogram is approximately mound-shaped, and if it is more or less symmetrical, then we may be assured that, for all practical purposes, the $\bar{x}$ distribution will be well approximated by a normal distribution and the central limit theorem will apply when the sample size is 30 or larger. The main thing to remember is that in almost all practical applications, a sample size of 30 or more is adequate for the central limit theorem to hold. However, in a few rare applications, you may need a sample size larger than 30 to get reliable results.

Let's summarize this information for convenient reference: For almost all x distributions, if we use a random sample of size 30 or larger, the $\bar{x}$ distribution will be approximately normal. The larger the sample size becomes, the closer the $\bar{x}$ distribution gets to the normal. Furthermore, we may convert the $\bar{x}$ distribution to a standard normal distribution using the following formulas:

Large sample

> **Using the central limit theorem to convert the $\bar{x}$ distribution to the standard normal distribution**
>
> $$\mu_{\bar{x}} = \mu$$
>
> $$\sigma_{\bar{x}} = \frac{\sigma}{\sqrt{n}}$$
>
> $$z = \frac{\bar{x} - \mu_{\bar{x}}}{\sigma_{\bar{x}}} = \frac{\bar{x} - \mu}{\sigma/\sqrt{n}}$$
>
> where n is the sample size ($n \geq 30$),
> μ is the mean of the x distribution, and
> σ is the standard deviation of the x distribution.

Guided Exercise 9 shows how to standardize $\bar{x}$ when appropriate. Then, Example 11 demonstrates the use of the central limit theorem in a decision-making process.

GUIDED EXERCISE 9   Central limit theorem

(a) Suppose x has a *normal* distribution with mean $\mu = 18$ and standard deviation $\sigma = 13$. If you draw random samples of size 5 from the x distribution and $\bar{x}$ represents the sample mean, what can you say about the $\bar{x}$ distribution? How could you standardize the $\bar{x}$ distribution?

⟹ Since the x distribution is given to be *normal*, the $\bar{x}$ distribution also will be normal even though the sample size is much less than 30. The mean is $\mu_{\bar{x}} = \mu = 18$. The standard deviation is

$$\sigma_{\bar{x}} = \sigma/\sqrt{n} = 3/\sqrt{5} \approx 1.3$$

We could standardize $\bar{x}$ as follows:

$$z = \frac{\bar{x} - \mu}{\sigma/\sqrt{n}} \approx \frac{\bar{x} - 18}{1.3}$$

(b) Suppose you know that the x distribution has mean $\mu = 75$ and standard deviation $\sigma = 12$, but you have no information as to whether or not the x distribution is normal. If you draw samples of size 30 from the x distribution and $\bar{x}$ represents the sample mean, what can you say about the $\bar{x}$ distribution? How could you standardize the $\bar{x}$ distribution?

⟹ Since the sample size is large enough, the $\bar{x}$ distribution will be an approximately normal distribution. The mean of the $\bar{x}$ distribution is

$$\mu_{\bar{x}} = \mu = 75.$$

The standard deviation of the $\bar{x}$ distribution is

$$\sigma_{\bar{x}} = \sigma/\sqrt{n} = 12/\sqrt{30} \approx 2.2$$

We could standardize $\bar{x}$ as follows:

$$z = \frac{\bar{x} - \mu}{\sigma/\sqrt{n}} \approx \frac{\bar{x} - 75}{2.2}$$

(c) Suppose you did not know that x had a normal distribution. Would you be justified in saying that the $\bar{x}$ distribution is approximately normal if the sample size were $n = 8$?

⟹ No, the sample size should be 30 or larger if we don't know that x has a normal distribution.

EXAMPLE 11   CENTRAL LIMIT THEOREM

A certain strain of bacteria occurs in all raw milk. Let x be the bacteria count per milliliter of milk. The health department has found that if the milk is not contaminated, then x has a distribution that is more or less mound-shaped and symmetrical. The mean of the x distribution is $\mu = 2500$, and the standard deviation is $\sigma = 300$. In a large commercial dairy, the health inspector takes 42 random samples of the milk produced each day. At the end of the day, the bacteria count in each of the 42 samples is averaged to obtain the sample mean bacteria count $\bar{x}$.

(a) Assuming the milk is not contaminated, what is the distribution of $\bar{x}$?

SOLUTION: The sample size is $n = 42$. Since this value exceeds 30, the central limit theorem applies, and we know that $\bar{x}$ will be approximately normal, with mean and standard deviation

$$\mu_{\bar{x}} = \mu = 2500$$
$$\sigma_{\bar{x}} = \sigma/\sqrt{n} = 300/\sqrt{42} \approx 46.3$$

(b) Assuming the milk is not contaminated, what is the probability that the average bacteria count $\bar{x}$ for one day is between 2350 and 2650 bacteria per milliliter?

SOLUTION: We convert the interval

$$2350 \leq \bar{x} \leq 2650$$

to a corresponding interval on the standard z axis.

$$z = \frac{\bar{x} - \mu}{\sigma/\sqrt{n}} \approx \frac{\bar{x} - 2500}{46.3}$$

$$\bar{x} = 2350 \ \text{converts to} \ z = \frac{2350 - 2500}{46.3} \approx -3.24$$

$$\bar{x} = 2650 \ \text{converts to} \ z = \frac{2650 - 2500}{46.3} \approx 3.24$$

Therefore,

$$P(2350 \le \bar{x} \le 2650) = P(-3.24 \le z \le 3.24)$$
$$= 0.9994 - 0.0006$$
$$= 0.9988$$

The probability is 0.9988 that $\bar{x}$ is between 2350 and 2650.

(c) *Interpretation* At the end of each day, the inspector must decide to accept or reject the accumulated milk that has been held in cold storage awaiting shipment. Suppose the 42 samples taken by the inspector have a mean bacteria count $\bar{x}$ that is *not* between 2350 and 2650. If you were the inspector, what would be your comment on this situation?

SOLUTION: The probability that $\bar{x}$ is between 2350 and 2650 for milk that is not contaminated is very high. If the inspector finds that the average bacteria count for the 42 samples is not between 2350 and 2650, then it is reasonable to conclude that there is something wrong with the milk. If $\bar{x}$ is less than 2350, you might suspect someone added chemicals to the milk to artificially reduce the bacteria count. If $\bar{x}$ is above 2650, you might suspect some other kind of biological contamination.

PROCEDURE

How to find probabilities regarding $\bar{x}$

Given a probability distribution of x values where

n = sample size
μ = mean of the x distribution
σ = standard deviation of the x distribution

1. If the x distribution is *normal*, then the $\bar{x}$ distribution is *normal*.

2. Even if the x distribution is *not* normal, if the *sample size* $n \ge 30$, then, by the central limit theorem, the $\bar{x}$ distribution is *approximately normal*.

3. Convert $\bar{x}$ to z using the formula

$$z = \frac{\bar{x} - \mu_{\bar{x}}}{\sigma_{\bar{x}}} = \frac{\bar{x} - \mu}{\sigma/\sqrt{n}}$$

4. Use the standard normal distribution to find the corresponding probabilities of events regarding $\bar{x}$.

GUIDED EXERCISE 10 | ## Probability regarding x̄

In mountain country, major highways sometimes use tunnels instead of long, winding roads over high passes. However, too many vehicles in a tunnel at the same time can cause a hazardous situation. Traffic engineers are studying a long tunnel in Colorado. If x represents the time for a vehicle to go through the tunnel, it is known that the x distribution has mean $\mu = 12.1$ minutes and standard deviation $\sigma = 3.8$ minutes under ordinary traffic conditions. From a histogram of x values, it was found that the x distribution is mound-shaped with some symmetry about the mean.

Engineers have calculated that, *on average*, vehicles should spend from 11 to 13 minutes in the tunnel. If the time is less than 11 minutes, traffic is moving too fast for safe travel in the tunnel. If the time is more than 13 minutes, there is a problem of bad air quality (too much carbon monoxide and other pollutants).

Continued

Under ordinary conditions, there are about 50 vehicles in the tunnel at one time. What is the probability that the mean time for 50 vehicles in the tunnel will be from 11 to 13 minutes? We will answer this question in steps.

(a) Let $\bar{x}$ represent the sample mean based on samples of size 50. Describe the $\bar{x}$ distribution.

From the central limit theorem, we expect the $\bar{x}$ distribution to be approximately normal, with mean and standard deviation

$$\mu_{\bar{x}} = \mu = 12.1 \quad \text{and} \quad \sigma_{\bar{x}} = \frac{\sigma}{\sqrt{n}} = \frac{3.8}{\sqrt{50}} \approx 0.54$$

(b) Find $P(11 < \bar{x} < 13)$.

We convert the interval

$$11 < \bar{x} < 13$$

to a standard z interval and use the standard normal probability table to find our answer. Since

$$z = \frac{\bar{x} - \mu}{\sigma/\sqrt{n}} \approx \frac{\bar{x} - 12.1}{0.54},$$

$\bar{x} = 11$ converts to $z \approx \dfrac{11 - 12.1}{0.54} = -2.04$

and $\bar{x} = 13$ converts to $z \approx \dfrac{13 - 12.1}{0.54} = 1.67$

Therefore,

$$P(11 < \bar{x} < 13) = P(-2.04 < z < 1.67)$$
$$= 0.9525 - 0.0207$$
$$= 0.9318$$

(c) *Interpret* your answer to part (b).

It seems that about 93% of the time, there should be no safety hazard for average traffic flow.

CRITICAL THINKING

Bias and Variability

Whenever we use a sample statistic as an estimate of a population parameter, we need to consider both *bias* and *variability* of the statistic.

> A sample statistic is **unbiased** if the mean of its sampling distribution equals the value of the parameter being estimated.
>
> The spread of the sampling distribution indicates the **variability of the statistic**. The spread is affected by the sampling method and the sample size. Statistics from larger random samples have spreads that are smaller.

We see from the central limit theorem that the sample mean $\bar{x}$ is an unbiased estimator of the mean μ when $n \geq 30$. The variability of $\bar{x}$ decreases as the sample size increases.

In Section 7.6, we will see that the sample proportion $\hat{p}$ is an unbiased estimator of the population proportion of successes p in binomial experiments with sufficiently large numbers of trials n. Again, we will see that the variability of $\hat{p}$ decreases with increasing numbers of trials.

The sample variance s^2 is an unbiased estimator for the population variance σ^2.

LOOKING FORWARD

Sampling distributions for the mean $\overline{x}$ and for proportions $\hat{p}$ (Section 7.6) will form the basis of our work with estimation (Chapter 8) and hypothesis testing (Chapter 9). With these inferential statistics methods, we will be able to use information from a sample to make statements regarding the population. These statements will be made in terms of probabilities derived from the underlying sampling distributions.

VIEWPOINT   Chaos!

Is there a different side to random sampling? Can sampling be used as a weapon? According to The Wall Street Journal, *the answer could be yes! The acronym for Create Havoc around Our System is CHAOS. The Association of Flight Attendants (AFA) is a union that successfully used CHAOS against Alaska Airlines in 1994 as a negotiation tool. CHAOS involves a small sample of random strikes—a few flights at a time—instead of a mass walkout. The president of the AFA claims that by striking randomly, "We take control of the schedule." The entire schedule becomes unreliable, and that is something management cannot tolerate. In 1986, TWA flight attendants struck in a mass walkout, and all were permanently replaced! Using CHAOS, only a few jobs are put at risk, and these are usually not lost. It appears that random sampling can be used as a weapon.*

SECTION 7.5 PROBLEMS

 aplia™   Access end-of-section exercises at www.aplia.com

SECTION 7.6

Normal Approximation to Binomial Distribution and to p̂ Distribution

FOCUS POINTS

- State the assumptions needed to use the normal approximation to the binomial distribution.
- Compute μ and σ for the normal approximation.
- Use the continuity correction to convert a range of r values to a corresponding range of normal x values.
- Convert the x values to a range of standardized z scores and find desired probabilities.
- Describe the sampling distribution for proportions $\hat{p}$.

The probability that a new vaccine will protect adults from cholera is known to be 0.85. The vaccine is administered to 300 adults who must enter an area where the disease is prevalent. What is the probability that more than 280 of these adults will be protected from cholera by the vaccine?

This question falls into the category of a binomial experiment, with the number of trials n equal to 300, the probability of success p equal to 0.85, and the number of successes r greater than 280. It is possible to use the formula for the binomial distribution to compute the probability that r is greater than 280. However, this approach would involve a number of tedious and long calculations. There is an easier way to do this problem, for under the conditions stated below, the normal distribution can be used to approximate the binomial distribution:

> **Normal approximation to the binomial distribution**
>
> Consider a binomial distribution where
>
> n = number of trials
>
> r = number of successes
>
> p = probability of success on a single trial
>
> $q = 1 - p$ = probability of failure on a single trial
>
> If $np > 5$ and $nq > 5$, then r has a binomial distribution that is approximated by a normal distribution with
>
> $\mu = np$ and $\sigma = \sqrt{npq}$
>
> *Note:* As n increases, the approximation becomes better.

Criteria $np > 5$ and $nq > 5$

Example 12 demonstrates that as n increases, the normal approximation to the binomial distribution improves.

EXAMPLE 12   BINOMIAL DISTRIBUTION GRAPHS

Graph the binomial distributions for which $p = 0.25$, $q = 0.75$, and the number of trials is first $n = 3$, then $n = 10$, then $n = 25$, and finally $n = 50$.

SOLUTION: The authors used a computer program to obtain the binomial distributions for the given values of p, q, and n. The results have been organized and graphed in Figures 7-31, 7-32, 7-33, and 7-34.

FIGURE 7-31

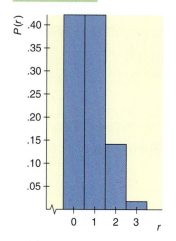

$n = 3$
$p = 0.25$
$q = 0.75$
$\mu = np = 0.75$
$\sigma = \sqrt{npq} = 0.75$

FIGURE 7-32

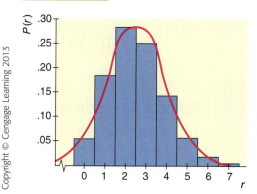

$n = 10$
$p = 0.25$
$q = 0.75$
$\mu = np = 2.5$
$\sigma = \sqrt{npq} = 1.37$

FIGURE 7-33

Good Normal Approximation;
$np > 5$ and $nq > 5$

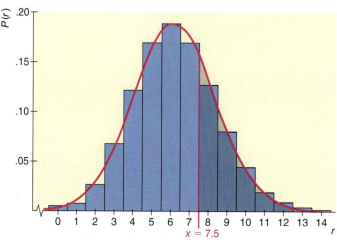

$n = 25$
$p = 0.25$
$q = 0.75$
$\mu = np = 6.25$
$\sigma = \sqrt{npq} = 2.17$

$x = 7.5$

FIGURE 7-34

Good Normal Approximation; $np > 5$ and $nq > 5$

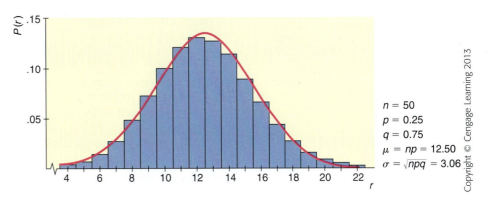

$n = 50$
$p = 0.25$
$q = 0.75$
$\mu = np = 12.50$
$\sigma = \sqrt{npq} = 3.06$

When $n = 3$, the outline of the histogram does not even begin to take the shape of a normal curve. But when $n = 10$, 25, or 50, it does begin to take a normal shape, indicated by the red curve. From a theoretical point of view, the histograms in Figures 7-32, 7-33, and 7-34 would have bars for all values of r from $r = 0$ to $r = n$. However, in the construction of these histograms, the bars of height less than 0.001 unit have been omitted—that is, in this example, probabilities less than 0.001 have been rounded to 0.

EXAMPLE 13   Normal approximation

The owner of a new apartment building must install 25 water heaters. From past experience in other apartment buildings, she knows that Quick Hot is a good brand. A Quick Hot heater is guaranteed for only 5 years, but from the owner's past experience, she knows that the probability it will last 10 years is 0.25.

(a) What is the probability that 8 or more of the 25 water heaters will last at least 10 years? Define success to mean a water heater that lasts at least 10 years.

SOLUTION: In this example, $n = 25$ and $p = 0.25$, so Figure 7-33 (on the preceding page) represents the probability distribution we will use. Let r be the binomial random variable corresponding to the number of successes out of $n = 25$ trials. We want to find $P(r \geq 8)$ by using the normal approximation. This probability is represented graphically (Figure 7-33) by the area of the bar over 8 plus the areas of all bars to the right of the bar over 8.

Let x be a normal random variable corresponding to a normal distribution, with $\mu = np = 25(0.25) = 6.25$ and $\sigma = \sqrt{npq} = \sqrt{25(0.25)(0.75)} \approx 2.17$. This normal curve is represented by the red line in Figure 7-33. The area under the normal curve from $x = 7.5$ to the right is approximately the same as the areas of the bars from the bar over $r = 8$ to the right. It is important to notice that we start with $x = 7.5$ because the bar over $r = 8$ really starts at $x = 7.5$.

The areas of the bars and the area under the corresponding red (normal) curve are approximately equal, so we conclude that $P(r \geq 8)$ is approximately equal to $P(x \geq 7.5)$.

When we convert $x = 7.5$ to standard units, we get

$$z = \frac{x - \mu}{\sigma} = \frac{7.5 - 6.25}{2.17} \approx 0.58 \quad \text{(Use } \mu = 6.25 \text{ and } \sigma = 2.17.\text{)}$$

The probability we want is

$$P(x \geq 7.5) = P(z \geq 0.58) = 1 - P(z \leq 0.58) = 1 - 0.7190 = 0.2810$$

(b) How does this result compare with the result we can obtain by using the formula for the binomial probability distribution with $n = 25$ and $p = 0.25$?

SOLUTION: Using the binomial distribution function on the TI-84Plus/TI-83Plus/TI-*n*spire calculators, the authors computed that $P(r \geq 8) \approx 0.2735$.

This means that the probability is approximately 0.27 that 8 or more water heaters will last at least 10 years.

(c) How do the results of parts (a) and (b) compare?

SOLUTION: The error of approximation is the difference between the approximate normal value (0.2810) and the binomial value (0.2735). The error is only $0.2810 - 0.2735 = 0.0075$, which is negligible for most practical purposes.

We knew in advance that the normal approximation to the binomial probability would be good, since $np = 25(0.25) = 6.25$ and $nq = 25(0.75) = 18.75$ are both greater than 5. These are the conditions that assure us that the normal approximation will be sufficiently close to the binomial probability for most practical purposes.

Continuity correction: Converting r values to x values

Remember that when we use the normal distribution to approximate the binomial, we are computing the areas under bars. The bar over the discrete variable r extends from $r - 0.5$ to $r + 0.5$. This means that the corresponding continuous normal variable x extends from $r - 0.5$ to $r + 0.5$. Adjusting the values of discrete random variables to obtain a corresponding range for a continuous random variable is called making a *continuity correction*.

PROCEDURE

HOW TO MAKE THE CONTINUITY CORRECTION

Convert the discrete random variable r (number of successes) to the continuous normal random variable x by doing the following:

1. If r is a **left point** of an interval, subtract 0.5 to obtain the corresponding normal variable x; that is, $x = r - 0.5$.

2. If r is a **right point** of an interval, add 0.5 to obtain the corresponding normal variable x; that is, $x = r + 0.5$.

For instance, $P(6 \leq r \leq 10)$, where r is a binomial random variable, is approximated by $P(5.5 \leq x \leq 10.5)$, where x is the corresponding normal random variable (see Figure 7-35).

FIGURE 7-35

$P(6 \leq r \leq 10)$ Is Approximately Equal to $P(5.5 \leq x \leq 10.5)$

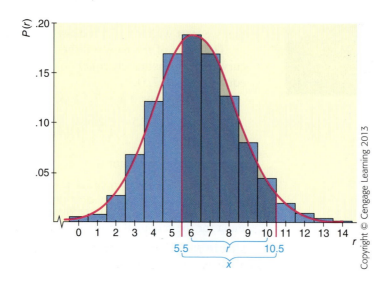

GUIDED EXERCISE 11 | **Continuity correction**

From many years of observation, a biologist knows that the probability is only 0.65 that any given Arctic tern will survive the migration from its summer nesting area to its winter feeding grounds. A random sample of 500 Arctic terns were banded at their summer nesting area. Use the normal approximation to the binomial and the following steps to find the probability that between 310 and 340 of the banded Arctic terns will survive the migration. Let r be the number of surviving terns.

Arctic tern

(a) To approximate $P(310 \leq r \leq 340)$, we use the normal curve with $\mu =$ _____ and $\sigma =$ _____.

→ We use the normal curve with
$\mu = np = 500(0.65) = 325$ and
$\sigma = \sqrt{npq} = \sqrt{500(0.65)(0.35)} \approx 10.67$

(b) $P(310 \leq r \leq 340)$ is approximately equal to $P($ _____ $\leq x \leq$ _____ $)$, where x is a variable from the normal distribution described in part (a).

→ Since 310 is the left endpoint, we subtract 0.5, and since 340 is the right endpoint, we add 0.5. Consequently,

$P(310 \leq r \leq 340) \approx P(309.5 \leq x \leq 340.5)$

(c) Convert the condition $309.5 \leq x \leq 340.5$ to a condition in standard units.

→ Since $\mu = 325$ and $\sigma \approx 10.67$, the condition $309.5 \leq x \leq 340.5$ becomes

$$\frac{309.5 - 325}{10.67} \leq z \leq \frac{340.5 - 325}{10.67}$$

or

$-1.45 \leq z \leq 1.45$

(d) $P(310 \leq r \leq 340) = P(309.5 \leq x \leq 340.5)$
$= P(-1.45 \leq z \leq 1.45)$
$=$ _____

→ $P(-1.45 \leq z \leq 1.45) = P(z \leq 1.45) - P(z \leq -1.45)$
$= 0.9265 - 0.0735$
$= 0.8530$

(e) Will the normal distribution make a good approximation to the binomial for this problem? Explain your answer.

→ Since $np = 500(0.65) = 35$ and $nq = 500(0.35) = 175$ are both greater than 5, the normal distribution will be a good approximation to the binomial.

Sampling Distributions for Proportions

In Sections 7.4 and 7.5, we studied the sampling distribution for the mean. Now we have the tools to look at sampling distributions for proportions. Suppose we repeat a binomial experiment with n trials again and again and, for each n trials, record the sample proportion of successes $\hat{p} = r/n$. The $\hat{p}$ values form a sampling distribution for proportions.

Sampling distribution for $\hat{p}$

Sampling distribution for the proportion $\hat{p} = \dfrac{r}{n}$

Given n = number of binomial trials (fixed constant)
   r = number of successes
   p = probability of success on each trial
   $q = 1 - p$ = probability of failure on each trial

If $np > 5$ and $nq > 5$, then the random variable $\hat{p} = r/n$ can be approximated by a normal random variable (x) with mean and standard deviation

$$\mu_{\hat{p}} = p \quad \text{and} \quad \sigma_{\hat{p}} = \sqrt{\frac{pq}{n}}$$

Standard error of a proportion

TERMINOLOGY   The *standard error* for the $\hat{p}$ distribution is the standard deviation $\sigma_{\hat{p}}$ of the $\hat{p}$ sampling distribution.

COMMENT   To obtain the information regarding the sampling distribution for the proportion $\hat{p} = r/n$, we consider the sampling distribution for r, the number of successes out of n binomial trials. Earlier we saw that when $np > 5$ and $nq > 5$, the r distribution is approximately normal, with mean $\mu_r = np$ and standard deviation $\sigma_r = \sqrt{npq}$. Notice that $\hat{p} = r/n$ is a linear function of r. This means that the $\hat{p}$ distribution is also approximately normal when np and nq are both greater than 5. In addition, $\mu_{\hat{p}} = \mu_r/n = np/n = p$ and $\sigma_{\hat{p}} = \sigma_r/n = \sqrt{npq}/n = \sqrt{pq/n}$. Although the values r/n are discrete for a fixed n, we do not use a continuity correction for the $\hat{p}$ distribution. This is the accepted standard practice for applications in inferential statistics, especially considering the requirements that $np > 5$ and $nq > 5$ are met.

We see from the sampling distribution for proportions that the mean of the $\hat{p}$ distribution is p, the population proportion of successes. This means that $\hat{p}$ is an *unbiased* estimator for p.

LOOKING FORWARD

We will use the sampling distribution for proportions in our work with estimation (Chapter 8) and hypothesis testing (Chapter 9).

EXAMPLE 14   SAMPLING DISTRIBUTION OF $\hat{p}$

The annual crime rate in the Capitol Hill neighborhood of Denver is 111 victims per 1000 residents. This means that 111 out of 1000 residents have been the victim of at least one crime (Source: *Neighborhood Facts*, Piton Foundation). For more information, visit the Brase/Brase statistics site at **http://www.cengagebrain.com** and find the link to the Piton Foundation. These crimes range from relatively minor crimes (stolen hubcaps or purse snatching) to major crimes (murder). The Arms is an apartment building in this neighborhood that has 50 year-round residents. Suppose we view each of the $n = 50$ residents as a binomial trial. The random variable r (which takes on values 0, 1, 2, . . . , 50) represents the number of victims of at least one crime in the next year.

(a) What is the population probability p that a resident in the Capital Hill neighborhood will be the victim of a crime next year? What is the probability q that a resident will not be a victim?

SOLUTION: Using the Piton Foundation report, we take

$$p = 111/1000 = 0.111 \ \text{ and } \ q = 1 - p = 0.889$$

(b) Consider the random variable

$$\hat{p} = \frac{r}{n} = \frac{r}{50}$$

Can we approximate the $\hat{p}$ distribution with a normal distribution? Explain.

SOLUTION: $np = 50(0.111) = 5.55$

$\qquad\qquad\quad nq = 50(0.889) = 44.45$

Since both np and nq are greater than 5, we can approximate the $\hat{p}$ distribution with a normal distribution.

(c) What are the mean and standard deviation for the $\hat{p}$ distribution?

SOLUTION: $\mu_{\hat{p}} = p = 0.111$

$$\sigma_{\hat{p}} = \sqrt{\frac{pq}{n}}$$

$$= \sqrt{\frac{(0.111)(0.889)}{50}} \approx 0.044$$

VIEWPOINT | Sunspots, Tree Rings, and Statistics

Ancient Chinese astronomers recorded extreme sunspot activity, with a peak around A.D. 1200. Mesa Verde tree rings in the period between 1276 and 1299 were unusually narrow, indicating a drought and/or a severe cold spell in the region at that time. A cooling trend could have narrowed the window of frost-free days below the approximately 80 days needed for cultivation of aboriginal corn and beans. Is this the reason the ancient Anasazi dwellings in Mesa Verde were abandoned? Is there a connection to the extreme sunspot activity? Much research and statistical work continue to be done on this topic.

Reference: *Prehistoric Astronomy in the Southwest*, by J. McKim Malville and C. Putnam, Department of Astronomy, University of Colorado.

SECTION 7.6 PROBLEMS   aplia   Access end-of-section exercises at www.aplia.com

Chapter Review

SUMMARY

In this chapter, we examined properties and applications of the normal probability distribution.

- A normal probability distribution is a distribution of a continuous random variable. Normal distributions are bell-shaped and symmetrical around the mean. The high point occurs over the mean, and most of the area occurs within 3 standard deviations of the mean. The mean and median are equal.

- The empirical rule for normal distributions gives areas within 1, 2, and 3 standard deviations of the mean. Approximately:

    68% of the data lie within the interval $\mu \pm \sigma$

    95% of the data lie within the interval $\mu \pm 2\sigma$

    99.7% of the data lie within the interval $\mu \pm 3\sigma$

- For symmetrical, bell-shaped distributions,

$$\text{standard deviation} \approx \frac{\text{range of data}}{4}$$

- A z score measures the number of standard deviations a raw score x lies from the mean.

$$z = \frac{x - \mu}{\sigma} \quad \text{and} \quad x = z\sigma + \mu$$

- For the standard normal distribution, $\mu = 0$ and $\sigma = 1$.

- Table 3 of the Appendix gives areas under a standard normal distribution that are to the left of a specified value of z.

- After raw scores x have been converted to z scores, the standard normal distribution table can be used to find probabilities associated with intervals of x values from any normal distribution.

- The inverse normal distribution is used to find z values associated with areas to the left of z. Table 3 of the Appendix can be used to find approximate z values associated with specific probabilities.

- Tools for assessing the normality of a data distribution include:

    Histogram of the data. A roughly bell-shaped histogram indicates normality.

    Presence of outliers. A limited number indicates normality.

    Skewness. For normality, Pearson's index is between −1 and 1.

    Normal quantile plot. For normality, points lie close to a straight line.

- Sampling distributions give us the basis for inferential statistics. By studying the distribution of a sample statistic, we can learn about the corresponding population parameter.

- For random samples of size n, the $\bar{x}$ distribution is the sampling distribution for the sample mean of an x distribution with population mean μ and population standard deviation σ. If the x distribution is normal, then the corresponding $\bar{x}$ distribution is normal.

    By the central limit theorem, when n is sufficiently large ($n \geq 30$), the $\bar{x}$ distribution is approximately normal even if the original x distribution is not normal.

    In both cases,

$$\mu_{\bar{x}} = \mu$$

$$\sigma_{\bar{x}} = \frac{\sigma}{\sqrt{n}}$$

- The binomial distribution can be approximated by a normal distribution with $\mu = np$ and $\sigma = \sqrt{npq}$, provided

    $np > 5$ and $nq > 5$, with $q = 1 - p$

    and a continuity correction is made.

- For n binomial trials with probability of success p on each trial, the $\hat{p}$ distribution is the sampling distribution of the sample proportion of successes. When $np > 5$ and $nq > 5$, the $\hat{p}$ distribution is approximately normal with

$$\mu_{\hat{p}} = p$$

$$\sigma_{\hat{p}} = \sqrt{\frac{pq}{n}}$$

Data from many applications follow distributions that are approximately normal. We will see normal distributions used extensively in later chapters.

IMPORTANT WORDS & SYMBOLS

VIEWPOINT | Nenana Ice Classic

The Nenana Ice Classic is a betting pool offering a large cash prize to the lucky winner who can guess the time, to the nearest minute, of the ice breakup on the Tanana River in the town of Nenana, Alaska. Official breakup time is defined as the time when the surging river dislodges a tripod on the ice. This breaks an attached line and stops a clock set to Yukon Standard Time. The event is so popular that the first state legislature of Alaska (1959) made the Nenana Ice Classic an official statewide lottery. Since 1918, the earliest breakup has been April 20, 1940, at 3:27 P.M., and the latest recorded breakup was May 20, 1964, at 11:41 A.M. Want to make a statistical guess predicting when the ice will break up? Breakup times from the years 1918 to 1996 are recorded in The Alaska Almanac, *published by Alaska Northwest Books, Anchorage.*

CHAPTER REVIEW PROBLEMS

1. | *Statistical Literacy* Describe a normal probability distribution.

2. | *Statistical Literacy* According to the empirical rule, approximately what percentage of the area under a normal distribution lies within 1 standard deviation of the mean? within 2 standard deviations? within 3 standard deviations?

3. | *Statistical Literacy* Random samples of size 9 are taken from a normal distribution with mean 10 and standard deviation 3. Describe the shape of the resulting $\bar{x}$ sampling distribution. What are the mean and standard deviation of the sampling distribution?

4. | *Statistical Literacy* Can a normal distribution always be used to approximate a binomial distribution? Explain.

5. | *Statistical Literacy* What characteristic of a normal quantile plot indicates that the data follow a distribution that is approximately normal?

6. | *Statistical Literacy* For a normal distribution, is it likely that a data value selected at random is more than 2 standard deviations above the mean?

7. | *Statistical Literacy* Give the formula for the *standard error* of the sample mean $\bar{x}$ distribution, based on samples of size n from a distribution with standard deviation σ.

8. | *Statistical Literacy* Give the formula for the *standard error* of the sample proportion $\hat{p}$ distribution, based on n binomial trials with probability of success p on each trial.

9. | *Critical Thinking* Let x be a random variable representing the amount of sleep each adult in New York City got last night. Consider a sampling distribution of sample means $\bar{x}$.
   (a) As the sample size becomes increasingly large, what distribution does the $\bar{x}$ distribution approach?
   (b) As the sample size becomes increasingly large, what value will the mean $\mu_{\bar{x}}$ of the $\bar{x}$ distribution approach?
   (c) What value will the standard deviation $\sigma_{\bar{x}}$ of the sampling distribution approach?
   (d) How do the two $\bar{x}$ distributions for sample size $n = 50$ and $n = 100$ compare?

10. | *Critical Thinking* If x has a normal distribution with mean $\mu = 15$ and standard deviation $\sigma = 3$, describe the distribution of $\bar{x}$ values for sample size n, where $n = 4$, $n = 16$, and $n = 100$. How do the $\bar{x}$ distributions compare for the various sample sizes?

11. | *Basic Computation: Probability* Given that x is a normal variable with mean $\mu = 47$ and standard deviation $\sigma = 6.2$, find:
   (a) $P(x \leq 60)$ (b) $P(x \geq 50)$ (c) $P(50 \leq x \leq 60)$

12. | *Basic Computation: Probability* Given that x is a normal variable with mean $\mu = 110$ and standard deviation $\sigma = 12$, find:
   (a) $P(x \leq 120)$ (b) $P(x \geq 80)$ (c) $P(108 \leq x \leq 117)$

13. | *Basic Computation: Inverse Normal* Find z such that 5% of the area under the standard normal curve lies to the right of z.

14. | *Basic Computation: Inverse Normal* Find z such that 99% of the area under the standard normal curve lies between $-z$ and z.

15. | *Medicine: Blood Type* Blood type AB is found in only 3% of the population (*Textbook of Medical Physiology*, by A. Guyton, M.D.). If 250 people are chosen at random, what is the probability that:
   (a) 5 or more will have this blood type?
   (b) between 5 and 10 will have this blood type?

16. | *Customer Complaints: Time* The customer service center in a large New York department store has determined that the amount of time spent with a customer about a complaint is normally distributed, with a mean of 9.3 minutes and a standard deviation of 2.5 minutes. What is the probability that for a randomly chosen customer with a complaint, the amount of time spent resolving the complaint will be:
   (a) less than 10 minutes?
   (b) longer than 5 minutes?
   (c) between 8 and 15 minutes?

17. | *Recycling: Aluminum Cans* One environmental group did a study of recycling habits in a California community. It found that 70% of the aluminum cans sold in the area were recycled.
   (a) If 400 cans are sold today, what is the probability that 300 or more will be recycled?
   (b) Of the 400 cans sold, what is the probability that between 260 and 300 will be recycled?

18. *Guarantee: CD Players* Future Electronics makes CD players. Its research department found that the life of the laser beam device is normally distributed, with mean 5000 hours and standard deviation 450 hours.
    (a) Find the probability that the laser beam device will wear out in 5000 hours or less.
    (b) *Inverse Normal Distribution* Future Electronics wants to place a guarantee on the players so that no more than 5% fail during the guarantee period. Because the laser pickup is the part most likely to wear out first, the guarantee period will be based on the life of the laser beam device. How many playing hours should the guarantee cover? (Round to the next playing hour.)

19. *Guarantee: Package Delivery* Express Courier Service has found that the delivery time for packages is normally distributed, with mean 14 hours and standard deviation 2 hours.
    (a) For a package selected at random, what is the probability that it will be delivered in 18 hours or less?
    (b) *Inverse Normal Distribution* What should be the guaranteed delivery time on all packages in order to be 95% sure that the package will be delivered before this time? *Hint:* Note that 5% of the packages will be delivered at a time beyond the guaranteed time period.

20. *Drugs: Effects* A new muscle relaxant is available. Researchers from the firm developing the relaxant have done studies that indicate that the time lapse between administration of the drug and beginning effects of the drug is normally distributed, with mean $\mu = 38$ minutes and standard deviation $\sigma = 5$ minutes.
    (a) The drug is administered to one patient selected at random. What is the probability that the time it takes to go into effect is 35 minutes or less?
    (b) The drug is administered to a random sample of 10 patients. What is the probability that the average time before it is effective for all 10 patients is 35 minutes or less?
    (c) Comment on the differences between the results in parts (a) and (b).

21. *Psychology: IQ Scores* Assume that IQ scores are normally distributed, with a standard deviation of 15 points and a mean of 100 points. If 100 people are chosen at random, what is the probability that the sample mean of IQ scores will not differ from the population mean by more than 2 points?

22. *Hatchery Fish: Length* A large tank of fish from a hatchery is being delivered to a lake. The hatchery claims that the mean length of fish in the tank is 15 inches and the standard deviation is 2 inches. A random sample of 36 fish is taken from the tank. Let $\bar{x}$ be the mean sample length of these fish. What is the probability that $\bar{x}$ is within 0.5 inch of the claimed population mean?

23. *Basic Computation: $\hat{p}$ Distribution* Suppose we have a binomial distribution with $n = 24$ trials and probability of success $p = 0.4$ on each trial. The sample proportion of successes is $\hat{p} = r/n$.
    (a) Is it appropriate to approximate the $\hat{p}$ distribution with a normal distribution? Explain.
    (b) What are the values of $\mu_{\hat{p}}$ and $\sigma_{\hat{p}}$?

24. *Green Behavior: Purchasing Habits* A recent Harris Poll on green behavior showed that 25% of adults often purchase used items instead of new ones. Consider a random sample of 75 adults. Let $\hat{p}$ be the sample proportion of adults who often purchase used instead of new items.
    (a) Is it appropriate to approximate the $\hat{p}$ distribution with a normal distribution? Explain.
    (b) What are the values of $\mu_{\hat{p}}$ and $\sigma_{\hat{p}}$?

DATA HIGHLIGHTS: GROUP PROJECTS

Wild iris

Break into small groups and discuss the following topics. Organize a brief outline in which you summarize the main points of your group discussion.

Iris setosa is a beautiful wildflower that is found in such diverse places as Alaska, the Gulf of St. Lawrence, much of North America, and even English meadows and parks. R. A. Fisher, with his colleague Dr. Edgar Anderson, studied these flowers extensively. Dr. Anderson described how he collected information on irises:

> I have studied such irises as I could get to see, in as great detail as possible, measuring iris standard after iris standard and iris fall after iris fall, sitting squat-legged with record book and ruler in mountain meadows, in cypress swamps, on lake beaches, and in English parks. [E. Anderson, "The Irises of the Gaspé Peninsula," *Bulletin of the American Iris Society*, Vol. 59, pp. 2–5, 1935.]

The data in Table 7-10 were collected by Dr. Anderson and were published by his friend and colleague R. A. Fisher in a paper entitled "The Use of Multiple Measurements in Taxonomic Problems" (*Annals of Eugenics*, part II, pp. 179–188, 1936). To find these data, visit the Brase/Brase statistics site at **http://www.cengagebrain.com** and find the link to DASL, the Carnegie Mellon University Data and Story Library. From the DASL site, look under Famous Datasets.

Let x be a random variable representing petal length. Using a TI-84Plus/TI-83Plus/TI-*n*spire calculator, it was found that the sample mean is $\bar{x} = 1.46$ centimeters (cm) and the sample standard deviation is $s = 0.17$ cm. Figure 7-36 shows a histogram for the given data generated on a TI-84Plus/TI-83Plus/TI-*n*spire calculator.

(a) Examine the histogram for petal lengths. Would you say that the distribution is approximately mound-shaped and symmetrical? Our sample has only 50 irises; if many thousands of irises had been used, do you think the distribution would look even more like a normal curve? Let x be the petal length of *Iris setosa*. Research has shown that x has an approximately normal distribution, with mean $\mu = 1.5$ cm and standard deviation $\sigma = 0.2$ cm.

(b) Use the empirical rule with $\mu = 1.5$ and $\sigma = 0.2$ to get an interval into which approximately 68% of the petal lengths will fall. Repeat this for 95% and 99.7%. Examine the raw data and compute the percentage of the raw data that actually fall into each of these intervals (the 68% interval, the 95% interval, and the 99.7% interval). Compare your computed percentages with those given by the empirical rule.

(c) Compute the probability that a petal length is between 1.3 and 1.6 cm. Compute the probability that a petal length is greater than 1.6 cm.

(d) Suppose that a random sample of 30 irises is obtained. Compute the probability that the average petal length for this sample is between 1.3 and 1.6 cm. Compute the probability that the average petal length is greater than 1.6 cm.

(e) Compare your answers for parts (c) and (d). Do you notice any differences? Why would these differences occur?

TABLE 7-10	Petal Length in Centimeters for *Iris setosa*			
1.4	1.4	1.3	1.5	1.4
1.7	1.4	1.5	1.4	1.5
1.5	1.6	1.4	1.1	1.2
1.5	1.3	1.4	1.7	1.5
1.7	1.5	1	1.7	1.9
1.6	1.6	1.5	1.4	1.6
1.6	1.5	1.5	1.4	1.5
1.2	1.3	1.4	1.3	1.5
1.3	1.3	1.3	1.6	1.9
1.4	1.6	1.4	1.5	1.4

Copyright © Cengage Learning 2013

FIGURE 7-36

Petal Length (cm) for *Iris setosa* (TI-84Plus/TI-83Plus/TI-*n*spire)

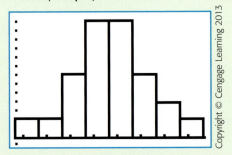

LINKING CONCEPTS: WRITING PROJECTS

Discuss each of the following topics in class or review the topics on your own. Then write a brief but complete essay in which you summarize the main points. Please include formulas and graphs as appropriate.

1. If you look up the word *empirical* in a dictionary, you will find that it means "relying on experiment and observation rather than on theory." Discuss the empirical rule in this context. The empirical rule certainly applies to the normal distribution, but does it also apply to a wide variety of other distributions that are not *exactly* (theoretically) normal? Discuss the terms *mound-shaped* and *symmetrical*. Draw several sketches of distributions that are mound-shaped *and* symmetrical. Draw sketches of distributions that are not mound-shaped or symmetrical. To which distributions will the empirical rule apply?

2. Why are standard z values so important? Is it true that z values have no units of measurement? Why would this be desirable for comparing data sets with *different* units of measurement? How can we assess differences in quality or performance by simply comparing z values under a standard normal curve? Examine the formula for computing standard z values. Notice that it involves *both* the mean and the standard deviation. Recall that in Chapter 3 we commented that the mean of a data collection is not entirely adequate to describe the data; you need the standard deviation as well. Discuss this topic again in light of what you now know about normal distributions and standard z values.

3. Most people would agree that increased information should give better predictions. Discuss how sampling distributions actually enable better predictions by providing more information. Examine Theorem 7.1 again. Suppose that x is a random variable with a *normal* distribution. Then $\bar{x}$, the sample mean based on random samples of size n, also will have a normal distribution for *any* value of $n = 1, 2, 3, \ldots$.

   What happens to the standard deviation of the $\bar{x}$ distribution as n (the sample size) increases? Consider the following table for different values of n:

n	1	2	3	4	10	50	100
$\sigma/\sqrt{n}$	1σ	0.71σ	0.58σ	0.50σ	0.32σ	0.14σ	0.10σ

   In this case, "increased information" means a larger sample size n. Give a brief explanation as to why a *large* standard deviation will usually result in poor statistical predictions, whereas a *small* standard deviation usually results in much better predictions. Since the standard deviation of the sampling distribution $\bar{x}$ is $\sigma/\sqrt{n}$, we can decrease the standard deviation by increasing n. In fact, if we look at the preceding table, we see that if we use a sample size of only $n = 4$, we cut the standard deviation of $\bar{x}$ by 50% of the standard deviation σ of x. If we were to use a sample size of $n = 100$, we would cut the standard deviation of $\bar{x}$ to 10% of the standard deviation σ of x.

   Give the preceding discussion some thought and explain why you should get much better predictions for μ by using $\bar{x}$ from a sample of size n rather than by just using x. Write a brief essay in which you explain why sampling distributions are an important tool in statistics.

4. In a way, the central limit theorem can be thought of as a kind of "grand central station." It is a connecting hub or center for a great deal of statistical work. We will use it extensively in Chapters 8–10. Put in a very elementary way, the central limit theorem states that as the sample size n increases, the distribution of the sample mean $\bar{x}$ will always approach a normal distribution, no matter where the original x variable came from. For most people, it is the complete generality of the central limit theorem that is so awe inspiring: It applies to practically everything. List and discuss at least three variables from everyday life for which you expect the variable x itself *not* to follow a normal or bell-shaped distribution. Then discuss what would happen to the sampling distribution $\bar{x}$ if the sample size were increased. Sketch diagrams of the $\bar{x}$ distributions as the sample size n increases.

USING TECHNOLOGY

Application 1

How can we determine if data originated from a normal distribution? We can look at a stem-and-leaf plot or histogram of the data to check for general symmetry, skewness, clusters of data, or outliers. However, a more sensitive way to check that a distribution is normal is to look at a special graph called a *normal quantile plot* (or a variation of this plot called a *normal probability plot* in some software packages). It really is not feasible to make a normal quantile plot by hand, but statistical software packages provide such plots. A simple version of the basic idea behind normal quantile plots involves the following process:

(a) Arrange the observed data values in order from smallest to largest, and determine the percentile occupied by each value. For instance, if there are 20 data values, the smallest datum is at the 5% point, the next smallest is at the 10% point, and so on.

(b) Find the z values that correspond to the percentile points. For instance, the z value that corresponds to the percentile 5% (i.e., percentage in the left tail of the distribution) is $z = -1.645$.

(c) Plot each data value x against the corresponding percentile z score. If the data are close to a normal distribution, the plotted points will lie close to a straight line. (If the data are close to a standard normal distribution, the points will lie close to the line $x = z$.)

The actual process that statistical software packages use to produce the z scores for the data is more complicated.

> **Interpreting normal quantile plots**
>
> If the points of a normal quantile plot lie close to a straight line, the plot indicates that the data follow a normal distribution. Systematic deviations from a straight line or bulges in the plot indicate that the data distribution is not normal. Individual points off the line may be outliers.

Consider Figure 7-37. This figure shows Minitab-generated quantile plots for two data sets. The black dots show the normal quantile plot for the salary data. The red dots show the normal quantile plot for a random sample of 42 data values drawn from a theoretical normal distribution with the same mean and standard deviation as the salary data ($\mu \approx 3421$, $\sigma \approx 709$).

FIGURE 7-37 Normal Quantile Plots

- Salary data for (city) government employees
- A random sample of 42 values from a theoretical normal distribution with the same mean and standard deviation as the salary data

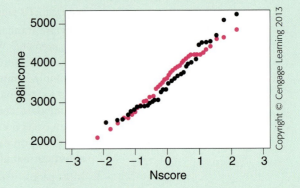

Copyright © Cengage Learning 2013

(a) Do the black dots lie close to a straight line? Do the salaries appear to follow a normal distribution? Are there any outliers on the low or high side? Would you say that any of the salaries are "out of line" for a normal distribution?

(b) Do the red dots lie close to a straight line? We know the red dots represent a sample drawn from a normal distribution. Is the normal quantile plot for the red dots consistent with this fact? Are there any outliers shown?

Technology Hints

TI-84Plus/TI-83Plus/TI-nspire

Enter the data. Press **STATPLOT** and select one of the plots. Highlight **ON**. Then highlight the sixth plot option. To get a plot similar to that of Figure 7-37, choose Y as the data axis.

Minitab

Minitab has several types of normal quantile plots that use different types of scales. To create a normal quantile plot similar to that of Figure 7-37, enter the data in column C1. Then use the menu choices **Calc ➤ Calculator**. In the dialogue box listing the functions, scroll to **Normal Scores**. Use **NSCOR(C1)** and store the results in column C2. Finally, use the menu choices **Graph ➤ Plot**. In the dialogue box, use C1 for variable y and C2 for variable x.

Enter the data. Use the menu choices **Analyze ➤ Descriptive Statistics ➤ Explore**. In the dialogue box, move your data variable to the dependent list. Click **Plots** .... Check "Normality plots with tests." The graph appears in the output window.

Application 2

As we have seen in this chapter, the value of a sample statistic such as $\bar{x}$ varies from one sample to another. The central limit theorem describes the distribution of the sample statistic $\bar{x}$ when samples are sufficiently large.

We can use technology tools to generate samples of the same size from the same population. Then we can look at the statistic $\bar{x}$ for each sample, and the resulting $\bar{x}$ distribution.

Project Illustrating the Central Limit Theorem

Step 1: Generate random samples of specified size n from a population.

The random-number table enables us to sample from the uniform distribution of digits 0 through 9. Use either the random-number table or a random-number generator to generate 30 samples of size 10.

Step 2: Compute the sample mean $\bar{x}$ of the digits in each sample.

Step 3: Compute the sample mean of the means (i.e., $\bar{x}_{\bar{x}}$) as well as the standard deviation $s_{\bar{x}}$ of the sample means.

The population mean of the uniform distribution of digits from 0 through 9 is 4.5. How does $\bar{x}_{\bar{x}}$ compare to this value?

Step 4: Compare the sample distribution of $\bar{x}$ values to a normal distribution having the mean and standard deviation computed in Step 3.

(a) Use the values of $\bar{x}_{\bar{x}}$ and $s_{\bar{x}}$ computed in Step 3 to create the intervals shown in column 1 of Table 7-11.

(b) Tally the sample means computed in Step 2 to determine how many fall into each interval of column 2. Then compute the percentage of data in each interval and record the results in column 3.

(c) The percentages listed in column 4 are those from a normal distribution (see Figure 7-3 showing the empirical rule). Compare the percentages in column 3 to those in column 4. How do the sample percentages compare with the hypothetical normal distribution?

Step 5: Create a histogram showing the sample means computed in Step 2.

Look at the histogram and compare it to a normal distribution with the mean and standard deviation of the $\bar{x}$s (as computed in Step 3).

Step 6: Compare the results of this project to the central limit theorem.

Increase the sample size of Step 1 to 20, 30, and 40 and repeat Steps 1–5.

TABLE 7-11   **Frequency Table of Sample Means**

1. Interval	2. Frequency	3. Percent	4. Hypothetical Normal Distribution
$\bar{x} - 3s$ to $\bar{x} - 2s$	Tally the sample	Compute	2 or 3%
$\bar{x} - 2s$ to $\bar{x} - s$	means computed	percents	13 or 14%
$\bar{x} - s$ to $\bar{x}$	in Step 2 and	from	About 34%
$\bar{x}$ to $\bar{x} + s$	place here.	column 2	About 34%
$\bar{x} + s$ to $\bar{x} + 2s$		and place	13 or 14%
$\bar{x} + 2s$ to $\bar{x} + 3s$		here.	2 or 3%

FIGURE 7-38

SPSS-Generated Histograms for Samples of Size 30 and Size 100

(a) $n = 30$

(b) $n = 100$

Technology Hints

The TI-84Plus/TI-83Plus/TI-nspire calculators, Excel 2010, Minitab, and SPSS all support the process of drawing random samples from a variety of distributions. Macros can be written in Excel 2010, Minitab, and the professional version of SPSS to repeat the six steps of the project. Figure 7-38 shows histograms generated by SPSS for random samples of size 30 and size 100. The samples are taken from a uniform probability distribution.

TI-84Plus/TI-83Plus/TI-*n*spire

You can generate random samples from uniform, normal, and binomial distributions. Press **MATH** and select **PRB**. Selection **5:randInt(lower, upper, sample size m)** generates *m* random integers from the specified interval. Selection **6:randNorm(μ, σ, sample size m)** generates *m* random numbers from a normal distribution with mean μ and standard deviation σ. Selection **7:randBin(number of trials n, p, sample size m)** generates *m* random values (number of successes out of *n* trials) for a binomial distribution with probability of success *p* on each trial. You can put these values in lists by using **Edit** under **Stat**. Highlight the list header, press **Enter**, and then select one of the options discussed.

Excel 2010

On the Home screen, click on the **Data** tab. In the **Analysis** group, select **Data Analysis**. In the dialogue box, select **Random Number Generation**. The next dialogue box provides choices for the population distribution, including uniform, binomial, and normal distributions. Fill in the required parameters and designate the location for the output.

Random Number Generation		? X
Number of Variables:	1	OK
Number of Random Numbers:	30	Cancel
Distribution:	Normal ▼	Help
Parameters		
Mean =	0	
Standard Deviation =	1	
Random Seed:		
Output options		
● Output Range:	$A$1:$A$30	
○ New Worksheet Ply:		
○ New Workbook		

Copyright © Cengage Learning 2013

258

Minitab

Use the menu selections **Calc ➤ Random Data.** Then select the population distribution. The choices include uniform, binomial, and normal distributions. Fill in the dialogue box, where the number of rows indicates the number of data in the sample.

SPSS

SPSS supports random samples from a variety of distributions, including binomial, normal, and uniform. In data view, generate a column of consecutive integers from 1 to n, where n is the sample size. In variable view, name the variables sample1, sample2, and so on, through sample30. These variables head the columns containing each of the 30 samples of size n. Then use the menu choices **Transform ➤ Compute.** In the dialogue box, use sample1 as the target variable for the first sample, and so forth.

In the function group, select **Random Numbers** and in the functions and special variables group, select **Rv.Uniform** for samples from a uniform distribution. Functions

Rv.Normal and **Rv.Binom** provide random samples from normal and binomial distributions, respectively. For each function, the necessary parameters are described.

Hubert Stadler/Corbis

8

Library of Congress

We dance round in a ring and suppose, But the Secret sits in the middle and knows.

—ROBERT FROST, "THE SECRET SITS"*

Rusty Jarrett/Getty Images for NASCAR/Getty Images

In Chapter 1, we said that statistics is the study of how to collect, organize, analyze, and interpret numerical data. That part of statistics concerned with analysis, interpretation, and forming conclusions about the source of the data is called *statistical inference*. Problems of statistical inference require us to draw a *random sample* of observations from a larger *population*. A sample usually contains incomplete information, so in a sense we must "dance round in a ring and suppose," to quote the words of the celebrated American poet Robert Lee Frost (1874–1963). Nevertheless, conclusions about the population can be obtained from sample data by the use of statistical estimates. This chapter introduces you to several widely used methods of estimation.

For online student resources, visit the Brase/Brase, *Understanding Basic Statistics,* 6th edition website at **http://www.cengagebrain.com**.

ESTIMATION

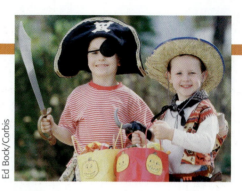

Ed Bock/Corbis

PREVIEW QUESTIONS

How do you estimate the expected value of a random variable? What assumptions are needed? How much confidence should be placed in such estimates? (SECTION 8.1)

At the beginning design stage of a statistical project, how large a sample size should you plan to get? (SECTION 8.1)

What famous statistician worked for the Guinness brewing company in Ireland? What has this got to do with constructing estimates from sample data? (SECTION 8.2)

How do you estimate the proportion p of successes in a binomial experiment? How does the normal approximation fit into this process? (SECTION 8.3)

FOCUS PROBLEM

Trick or Treat

About 28% of U.S. households turn out the lights and pretend not to be at home on Halloween (*Source: Are You Normal about Money?* by Bernice Kanner, Bloomberg Press).

Alice is a sociology student who is studying the affluent Cherry Creek neighborhood in Denver. As part of a larger survey, Alice interviewed a random sample of 35 households. One of the questions she asked was whether the resident turned out the lights and pretended not to be at home on Halloween. It was found that 11 of 35 residents actually did this practice.

Philip James/Corbis

(a) Compute a 90% confidence interval for p, the proportion of all households in Cherry Creek that pretend not to be home on Halloween.

(b) What assumptions are necessary to calculate the confidence interval of part (a)? Do you think these assumptions are met in this case? Explain.

(c) The national proportion is about 0.28. Is 0.28 in the confidence interval you computed? Based on your answer, does it seem that the Cherry Creek neighborhood is much different (either a higher or a lower proportion) from the population of all U.S. households? Explain.

(See Problem 18 of Section 8.3.)

SECTION 8.1

Estimating μ When σ Is Known

FOCUS POINTS

- Explain the meanings of confidence level, error of estimate, and critical value.
- Find the critical value corresponding to a given confidence level.
- Compute confidence intervals for μ when σ is known. Interpret the results.
- Compute the sample size to be used for estimating a mean μ.

Because of time and money constraints, difficulty in finding population members, and so forth, we usually do not have access to *all* measurements of an *entire* population. Instead we rely on information from a sample.

In this section, we develop techniques for estimating the population mean μ using sample data. We assume the population standard deviation σ is known.

Let's begin by listing some basic assumptions used in the development of our formulas for estimating μ when σ is known.

Assumptions about the random variable *x*

1. We have a *simple random sample* of size n drawn from a population of x values.
2. The value of σ, the population standard deviation of x, *is known*.
3. If the x *distribution is normal*, then our methods work for *any sample size n*.
4. If x has an unknown distribution, then we require a *sample size n $\geq$ 30*. However, if the x distribution is distinctly skewed and definitely not mound-shaped, a sample of size 50 or even 100 or higher may be necessary.

Point estimate

An estimate of a population parameter given by a single number is called a *point estimate* for that parameter. It will come as no great surprise that we use $\bar{x}$ (the sample mean) as the point estimate for μ (the population mean).

A **point estimate** of a population parameter is an estimate of the parameter using a single number.

$\bar{x}$ is the **point estimate** for μ.

Margin of error

Even with a large random sample, the value of $\bar{x}$ usually is not *exactly* equal to the population mean μ. The *margin of error* is the magnitude of the difference between the sample point estimate and the true population parameter value.

When using $\bar{x}$ as a point estimate for μ, the **margin of error** is the magnitude of $\bar{x} - \mu$, or $|\bar{x} - \mu|$.

Confidence level, c

We cannot say exactly how close $\bar{x}$ is to μ when μ is unknown. Therefore, the exact margin of error is unknown when the population parameter is unknown. Of course, μ is usually not known, or there would be no need to estimate it. In this section, we will use the language of probability to give us an idea of the size of the margin of error when we use $\bar{x}$ as a point estimate for μ.

FIGURE 8-1

Confidence Level *c* and Corresponding Critical Value z_c Shown on the Standard Normal Curve

Area = *c*

$-z_c$ 0 z_c

Finding the critical value, z_c

First, we need to learn about *confidence levels*. The reliability of an estimate will be measured by the confidence level.

Suppose we want a confidence level of *c* (see Figure 8-1). Theoretically, we can choose *c* to be any value between 0 and 1, but usually *c* is equal to a number such as 0.90, 0.95, or 0.99. In each case, the value z_c is the number such that the area under the standard normal curve falling between $-z_c$ and z_c is equal to *c*. The value z_c is called the *critical value* for a confidence level of *c*.

> For a confidence level *c*, the **critical value** z_c is the number such that the area under the standard normal curve between $-z_c$ and z_c equals *c*.

The area under the normal curve from $-z_c$ to z_c is the probability that the standardized normal variable *z* lies in that interval. This means that

$$P(-z_c < z < z_c) = c$$

EXAMPLE 1   FIND A CRITICAL VALUE

Let us use Table 3 of the Appendix to find a number $z_{0.99}$ such that 99% of the area under the standard normal curve lies between $-z_{0.99}$ and $z_{0.99}$. That is, we will find $z_{0.99}$ such that

$$P(-z_{0.99} < z < z_{0.99}) = 0.99$$

SOLUTION: In Section 7.3, we saw how to find the *z* value when we were given an area between $-z$ and *z*. The first thing we did was to find the corresponding area to the left of $-z$. If *A* is the area between $-z$ and *z*, then $(1 - A)/2$ is the area to the left of *z*. In our case, the area between $-z$ and *z* is 0.99. The corresponding area in the left tail is $(1 - 0.99)/2 = 0.005$ (see Figure 8-2).

Next, we use Table 3 of the Appendix to find the *z* value corresponding to a left-tail area of 0.0050. Table 8-1 shows an excerpt from Table 3 of the Appendix.

FIGURE 8-2

Area Between $-z$ and *z* Is 0.99

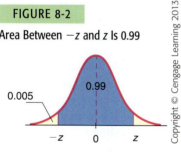

0.005

0.99

$-z$ 0   *z*

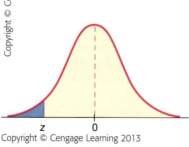

z   0

TABLE 8-1	Excerpt from Table 3 of the Appendix			
z	.00	... .07	.08	.09
−3.4	.0003	.0003	.0003	.0002
⋮				
−2.5	.0062	.0051	.0049	.0048
			↑ .0050	

From Table 8-1, we see that the desired area, 0.0050, is exactly halfway between the areas corresponding to $z = -2.58$ and $z = -2.57$. Because the two area values are so close together, we use the more conservative *z* value -2.58 rather than interpolate. In fact, $z_{0.99} \approx 2.576$. However, to two decimal places, we use $z_{0.99} = 2.58$ as the critical value for a confidence level of $c = 0.99$. We have

$$P(-2.58 < z < 2.58) \approx 0.99$$

The results of Example 1 will be used a great deal in our later work. For convenience, Table 8-2 gives some levels of confidence and corresponding critical values z_c. The same information is provided in Table 3(b) of the Appendix.

An estimate is not very valuable unless we have some kind of measure of how "good" it is. The language of probability can give us an idea of the size of the margin of error caused by using the sample mean $\bar{x}$ as an estimate for the population mean.

Remember that $\bar{x}$ is a random variable. Each time we draw a sample of size n from a population, we can get a different value for $\bar{x}$. According to the central limit theorem, if the sample size is large, then $\bar{x}$ has a distribution that is approximately normal with mean $\mu_{\bar{x}} = \mu$, the population mean we are trying to estimate. The standard deviation is $\sigma_{\bar{x}} = \sigma/\sqrt{n}$. If x has a normal distribution, these results are true *for any sample size*. (See Theorem 7.1.)

This information, together with our work on confidence levels, leads us (as shown in the optional derivation that follows) to the probability statement

$$P\left(-z_c\frac{\sigma}{\sqrt{n}} < \bar{x} - \mu < z_c\frac{\sigma}{\sqrt{n}}\right) = c \tag{1}$$

TABLE 8-2	Some Levels of Confidence and Their Corresponding Critical Values
Level of Confidence c	**Critical Value z_c**
0.70, or 70%	1.04
0.75, or 75%	1.15
0.80, or 80%	1.28
0.85, or 85%	1.44
0.90, or 90%	1.645
0.95, or 95%	1.96
0.98, or 98%	2.33
0.99, or 99%	2.58

Copyright © Cengage Learning 2013

FIGURE 8-3

Distribution of Sample Means $\bar{x}$

The probability is c that $\bar{x}$ is within $\pm z_c\frac{\sigma}{\sqrt{n}}$ of the true population mean μ.

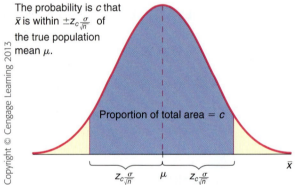

Proportion of total area $= c$

$z_c\frac{\sigma}{\sqrt{n}}$ μ $z_c\frac{\sigma}{\sqrt{n}}$

Copyright © Cengage Learning 2013

Equation (1) uses the language of probability to give us an idea of the size of the margin of error for the corresponding confidence level c. In words, Equation (1) states that the probability is c that our point estimate $\bar{x}$ is within a distance $\pm z_c(\sigma/\sqrt{n})$ of the population mean μ. This relationship is shown in Figure 8-3.

In the following optional discussion, we derive Equation (1). If you prefer, you may jump ahead to the discussion about the margin of error.

Optional derivation of Equation (1)

For a c confidence level, we know that

$$P(-z_c < z < z_c) = c \tag{2}$$

This statement gives us information about the size of z, but we want information about the size of $\bar{x} - \mu$. Is there a relationship between z and $\bar{x} - \mu$? The answer is yes since, by the central limit theorem, $\bar{x}$ has a distribution that is approximately normal, with mean μ and standard deviation $\sigma/\sqrt{n}$. We can convert $\bar{x}$ to a standard z score by using the formula

$$z = \frac{\bar{x} - \mu}{\sigma/\sqrt{n}} \tag{3}$$

Substituting this expression for z into Equation (2) gives

$$P\left(-z_c < \frac{\bar{x} - \mu}{\sigma/\sqrt{n}} < z_c\right) = c \tag{4}$$

Multiplying all parts of the inequality in (4) by $\sigma/\sqrt{n}$ gives us

$$P\left(-z_c\frac{\sigma}{\sqrt{n}} < \bar{x} - \mu < z_c\frac{\sigma}{\sqrt{n}}\right) = c \tag{1}$$

Equation (1) is precisely the equation we set out to derive.

The *margin of error* (or absolute error) using $\bar{x}$ as a point estimate for μ is $|\bar{x} - \mu|$. In most practical problems, μ is unknown, so the margin of error is also unknown. However, Equation (1) allows us to compute an *error tolerance E* that serves as a bound on the margin of error. Using a $c\%$ level of confidence, we can say that the point estimate $\bar{x}$ differs from the population mean μ by a *maximal margin of error*

Maximal margin of error, *E*

$$E = z_c\frac{\sigma}{\sqrt{n}} \tag{5}$$

Note: Formula (5) for E is based on the fact that the sampling distribution for $\bar{x}$ is exactly normal, with mean μ and standard deviation $\sigma/\sqrt{n}$. This occurs whenever the x distribution is normal with mean μ and standard deviation σ. If the x distribution is not normal, then according to the central limit theorem, large samples ($n \geq 30$) produce an $\bar{x}$ distribution that is approximately normal, with mean μ and standard deviation $\sigma/\sqrt{n}$.

Using Equations (1) and (5), we conclude that

$$P(-E < \bar{x} - \mu < E) = c \tag{6}$$

Equation (6) states that the probability is c that the difference between $\bar{x}$ and μ is no more than the maximal error tolerance E. If we use a little algebra on the inequality

$$-E < \bar{x} - \mu < E \tag{7}$$

we can rewrite it in the following mathematically equivalent way:

Confidence interval for μ when σ is known

$$\bar{x} - E < \mu < \bar{x} + E \tag{8}$$

Since formulas (7) and (8) are mathematically equivalent, their probabilities are the same. Therefore, from (6), (7), and (8), we obtain

$$P(\bar{x} - E < \mu < \bar{x} + E) = c \tag{9}$$

Equation (9) states that there is a chance c that the interval from $\bar{x} - E$ to $\bar{x} + E$ contains the population mean μ. We call this interval a *c confidence interval for μ*.

c confidence interval for μ

A ***c* confidence interval for μ** is an interval computed from sample data in such a way that c is the probability of generating an interval containing the actual value of μ. In other words, c is the proportion of confidence intervals, based on random samples of size n, that actually contain μ.

We may get a different confidence interval for each different sample that is taken. Some intervals will contain the population mean μ and others will not. However, in the long run, the proportion of confidence intervals that contain μ is c.

PROCEDURE

HOW TO FIND A CONFIDENCE INTERVAL FOR μ WHEN σ IS KNOWN

Requirements

Let x be a random variable appropriate to your application. Obtain a simple random sample (of size n) of x values from which you compute the sample mean $\bar{x}$. The value of σ is already known (perhaps from a previous study).

If you can assume that x has a normal distribution, then any sample size n will work. If you cannot assume this, then use a sample size of $n \geq 30$.

Confidence interval for μ when σ is known

$$\bar{x} - E < \mu < \bar{x} + E \tag{8}$$

where $\bar{x}$ = sample mean of a simple random sample

$$E = z_c \frac{\sigma}{\sqrt{n}}$$

c = confidence level $(0 < c < 1)$

z_c = critical value for confidence level c based on the standard normal distribution (see Table 3(b) of the Appendix for frequently used values).

EXAMPLE 2

CONFIDENCE INTERVAL FOR μ WITH σ KNOWN

Julia enjoys jogging. She has been jogging over a period of several years, during which time her physical condition has remained constantly good. Usually, she jogs 2 miles per day. The standard deviation of her times is $\sigma = 1.80$ minutes. During the past year, Julia has recorded her times to run 2 miles. She has a random sample of 90 of these times. For these 90 times, the mean was $\bar{x} = 15.60$ minutes. Let μ be the mean jogging time for the entire distribution of Julia's 2-mile running times (taken over the past year). Find a 0.95 confidence interval for μ.

SOLUTION: *Check Requirements* We have a simple random sample of running times, and the sample size $n = 90$ is large enough for the $\bar{x}$ distribution to be approximately normal. We also know σ. It is appropriate to use the normal distribution to compute a confidence interval for μ.

To compute E for the 95% confidence interval $\bar{x} - E$ to $\bar{x} + E$, we use the fact that $z_c = 1.96$ (see Table 8-2), together with the values $n = 90$ and $\sigma = 1.80$. Therefore,

$$E = z_c \frac{\sigma}{\sqrt{n}}$$

$$E = 1.96\left(\frac{1.80}{\sqrt{90}}\right)$$

$$E \approx 0.37$$

Using Equation (8), the given value of $\bar{x}$, and our computed value for E, we get the 95% confidence interval for μ.

$$\bar{x} - E < \mu < \bar{x} + E$$

$$15.60 - 0.37 < \mu < 15.60 + 0.37$$

$$15.23 < \mu < 15.97$$

Interpretation We conclude with 95% confidence that the interval from 15.23 minutes to 15.97 minutes is one that contains the population mean μ of jogging times for Julia.

CRITICAL THINKING

Interpreting Confidence Intervals

A few comments are in order about the general meaning of the term *confidence interval*.

- Since $\bar{x}$ is a random variable, the endpoints $\bar{x} \pm E$ are also random variables. Equation (9) states that we have a chance c of obtaining a sample such that the interval, once it is computed, will contain the parameter μ.

- After the confidence interval is numerically fixed for a specific sample, it either does or does not contain μ. So, the probability is 1 or 0 that the interval, when it is fixed, will contain μ. A nontrivial probability statement can be made only about variables, not constants.

- Equation (9), $P(\bar{x} - E < \mu < \bar{x} + E) = c$, really states that if we draw many random samples of size n and get lots of confidence intervals, then the proportion of all intervals that will turn out to contain the mean μ is c.

  For example, in Figure 8-4, the horizontal lines represent 0.90 confidence intervals for various samples of the same size from an x distribution. Some of these intervals contain μ and others do not. Since the intervals are 0.90 confidence intervals, about 90% of all such intervals should contain μ. For each sample, the interval goes from $\bar{x} - E$ to $\bar{x} + E$.

- Once we have a *specific* confidence interval for μ, such as $3 < \mu < 5$, all we can say is that we are $c\%$ confident that we have one of the intervals that actually contains μ. Another appropriate statement is that at the c confidence level, our interval contains μ.

FIGURE 8-4

0.90 Confidence Intervals for Samples of the Same Size

For each sample, the interval goes from $\bar{x} - E$ to $\bar{x} + E$.

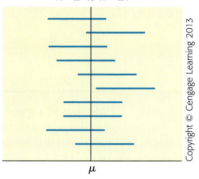

Copyright © Cengage Learning 2013

μ

COMMENT   Please see Using Technology at the end of this chapter for a computer demonstration of this discussion about confidence intervals.

GUIDED EXERCISE 1 | ## Confidence interval for μ with σ known

Walter usually meets Julia at the track. He prefers to jog 3 miles. From long experience, he knows that $\sigma = 2.40$ minutes for his jogging times. For a random sample of 90 jogging sessions, the mean time was $\bar{x} = 22.50$ minutes. Let μ be the mean jogging time for the entire distribution of Walter's 3-mile running times over the past several years. Find a 0.99 confidence interval for μ.

(a) *Check Requirements* Is the $\bar{x}$ distribution approximately normal? Do we know σ?

⟹ Yes; we know this from the central limit theorem. Yes, $\sigma = 2.40$ minutes.

(b) What is the value of $z_{0.99}$? (See Table 8-2.)

⟹ $z_{0.99} = 2.58$.

(c) What is the value of E?

⟹ $E = z_c \dfrac{\sigma}{\sqrt{n}} = 2.58 \left(\dfrac{2.40}{\sqrt{90}} \right) \approx 0.65.$

(d) What are the endpoints for a 0.99 confidence interval for μ?

⟹ The endpoints are given by
$$\bar{x} - E \approx 22.50 - 0.65 = 21.85 \text{ and}$$
$$\bar{x} + E \approx 22.50 + 0.65 = 23.15$$

(e) *Interpretation* Explain what the confidence interval tells us.

⟹ We are 99% certain that the interval from 21.85 to 23.15 is an interval that contains the population mean time μ.

When we use samples to estimate the mean of a population, we generate a small error. However, samples are useful even when it is possible to survey the entire population, because the use of a sample may yield savings of time or effort in collecting data.

LOOKING FORWARD

The basic structure for most confidence intervals for population parameters is

   sample statistic $- E <$ population parameter $<$ sample statistic $+ E$

where E is the maximal margin of error based on the sample statistic distribution and the level of confidence c. We will see this same format used for confidence intervals of the mean when σ is unknown (Section 8.2), for proportions (Section 8.3), for differences of means from independent samples (Section 10.2), for differences of proportions (Section 10.3), and for parameters of linear regression (Section 11.4). This structure for confidence intervals is so basic that some software packages, such as Excel, simply give the value of E for a confidence interval and expect the user to finish the computation.

TECH NOTES   The TI-84Plus/TI-83Plus/TI-*n*spire calculators, Excel 2010, and Minitab all support confidence intervals for μ from large samples. The level of support varies according to the technology. When a confidence interval is given, the standard mathematical notation (lower value, upper value) is used. For instance, the notation (15.23, 15.97) means the interval from 15.23 to 15.97.

TI-84Plus/TI-83Plus/TI-*n*spire (with TI-84Plus keypad) These calculators give the most extensive support. The user can opt to enter raw data or just summary statistics. In each case, the value of σ must be specified. Press the **STAT** key, then select **TESTS**, and use **7:ZInterval**. The TI-84Plus/TI-83Plus/TI-*n*spire output shows the results for Example 2.

```
ZInterval
 Inpt:Data Stats
 σ:1.8
 x̄:15.6
 n:90
 C-Level:95
 Calculate
```

```
ZInterval
 (15.228, 15.972)
 x̄=15.6
 n=90
```

Copyright © Cengage Learning 2013

Copyright © Cengage Learning 2013

Excel 2010 Excel gives only the value of the maximal error of estimate E. On the **Home** screen click the **Insert Function** (f_x). In the dialogue box, select **Statistical** for the category, and then select **CONFIDENCE.NORM**. In the resulting dialogue box, the value of **alpha** is ($1 -$ confidence level). For example, alpha is 0.05 for a 95% confidence interval. The values of σ and n are also required. The Excel output shows the value of E for Example 2.

f_x	=CONFIDENCE.NORM(0.05,1.8,90)	
C	**D**	**E**
0.371876		

Copyright © Cengage Learning 2013

An alternate approach incorporating raw data (using the Student's t distribution presented in the next section) uses a selection from the Data Analysis package. Click the **Data** tab on the home ribbon. From the Analysis group, select **Data Analysis**. In the dialogue box, select **Descriptive Statistics**. Check the box by **Confidence Level for Mean** and enter the confidence level. Again, the value of E for the interval is given.

Minitab Raw data are required. Use the menu choices **Stat ➤ Basic Statistics ➤ 1-SampleZ**.

Sample Size for Estimating the Mean μ

In the design stages of statistical research projects, it is a good idea to decide in advance on the confidence level you wish to use and to select the *maximal* margin of error E you want for your project. How you choose to make these decisions depends on the requirements of the project and the practical nature of the problem.

   Whatever specifications you make, the next step is to determine the sample size. Solving the formula that gives the maximal margin of error E for n enables us to determine the minimal sample size.

PROCEDURE

How to find the sample size n for estimating μ when σ is known

Requirements

The distribution of sample means $\bar{x}$ is approximately normal.

Formula for sample size

$$n = \left(\frac{z_c \sigma}{E}\right)^2 \tag{10}$$

where E = specified maximal margin of error

   σ = population standard deviation

   z_c = critical value from the normal distribution for the desired confidence level c. Commonly used values of z_c can be found in Table 3(b) of the Appendix.

   If n is not a whole number, increase n to the next higher whole number. Note that n is the minimal sample size for a specified confidence level and maximal error of estimate E.

COMMENT If you have a *preliminary study* involving a sample size of 30 or larger, then for most practical purposes it is safe to approximate σ with the sample standard deviation s in the formula for sample size.

EXAMPLE 3

Sample size for estimating μ

A wildlife study is designed to find the mean weight of salmon caught by an Alaskan fishing company. A preliminary study of a random sample of 50 salmon showed $s \approx 2.15$ pounds. How large a sample should be taken to be 99% confident that the sample mean $\bar{x}$ is within 0.20 pound of the true mean weight μ?

Photo Researchers

Salmon moving upstream

SOLUTION: In this problem, $z_{0.99} = 2.58$ (see Table 8-2) and $E = 0.20$. The preliminary study of 50 fish is large enough to permit a good approximation of σ by $s = 2.15$. Therefore, Equation (10) becomes

$$n = \left(\frac{z_c \sigma}{E}\right)^2 \approx \left(\frac{(2.58)(2.15)}{0.20}\right)^2 = 769.2$$

Note: In determining sample size, any fractional value of n is always rounded to the *next higher whole number*. We conclude that a sample size of 770 will be large enough to satisfy the specifications. Of course, a sample size larger than 770 also works.

Music and Techno Theft

Performing rights organizations ASCAP (American Society of Composers, Authors, and Publishers) and BMI (Broadcast Music, Inc.) collect royalties for songwriters and music publishers. Radio, television, cable, nightclubs, restaurants, elevators, and even beauty salons play music that is copyrighted by a composer or publisher. The royalty payment for this music turns out to be more than a billion dollars a year (Source: The Wall Street Journal). How do ASCAP and BMI know who is playing what music? The answer is, they don't! Instead of tracking exactly what gets played, they use random sampling and confidence intervals. For example, each radio station (there are more than 10,000 in the United States) has randomly chosen days of programming analyzed every year. The results are used to assess royalty fees. In fact, Deloitte & Touche (a financial services company) administers the sampling process.

Although the system is not perfect, it helps bring order into an otherwise chaotic accounting system. Such methods of "copyright policing" help prevent techno theft, ensuring that many songwriters and recording artists get a reasonable return for their creative work.

**SECTION 8.1
PROBLEMS**

aplia   Access end-of-section exercises at www.aplia.com

SECTION 8.2

Estimating μ When σ Is Unknown

FOCUS POINTS
- Learn about degrees of freedom and Student's t distributions.
- Find critical values using degrees of freedom and confidence levels.
- Compute confidence intervals for μ when σ is unknown. What does this information tell you?

In order to use the normal distribution to find confidence intervals for a population mean μ, we need to know the value of σ, the population standard deviation. However, much of the time, when μ is unknown, σ is unknown as well. In such cases, we use the sample standard deviation s to approximate σ. When we use s to approximate σ, the sampling distribution for $\bar{x}$ follows a new distribution called a *Student's t distribution*.

Student's t distribution

Student's t Distributions

Student's t distributions were discovered in 1908 by W. S. Gosset. He was employed as a statistician by the Guinness brewing company, a company that discouraged publication of research by its employees. As a result, Gosset published his research under the pseudonym "Student." Gosset was the first to recognize the importance of developing statistical methods for obtaining reliable information from samples of populations with unknown σ. Gosset used the variable t when he introduced the distribution in 1908. To this day and in his honor, it is still called a Student's t distribution. It might be more fitting to call this distribution *Gosset's t distribution*; however, in the literature of mathematical statistics, it is known as a *Student's t distribution*.

The variable t is defined as follows. A Student's t distribution depends on sample size n.

Assume that x has a normal distribution with mean μ. For samples of size n with sample mean $\overline{x}$ and sample standard deviation s, the t **variable**

$$t = \frac{\overline{x} - \mu}{\frac{s}{\sqrt{n}}} \tag{11}$$

Degrees of freedom, *d.f.*

has a **Student's t distribution** with **degrees of freedom $d.f. = n - 1$.**

If many random samples of size n are drawn, then we get many t values from Equation (11). These t values can be organized into a frequency table, and a histogram can be drawn, thereby giving us an idea of the shape of the t distribution (for a given n).

Fortunately, all this work is not necessary because mathematical theorems can be used to obtain a formula for the t distribution. However, it is important to observe that these theorems say that the shape of the t distribution depends only on n, provided the basic variable x has a normal distribution. So, *when we use a t distribution, we will assume that the x distribution is normal.*

Table 4 of the Appendix gives values of the variable t corresponding to what we call the number of *degrees of freedom*, abbreviated $d.f.$ For the methods used in this section, the number of degrees of freedom is given by the formula

$$d.f. = n - 1 \tag{12}$$

where $d.f.$ stands for the degrees of freedom and n is the sample size. Each choice for $d.f.$ gives a different t distribution.

The graph of a t distribution is always symmetric about its mean, which (as for the z distribution) is 0. The main observable difference between a t distribution and the standard normal z distribution is that a t distribution has somewhat thicker tails.

Figure 8-5 shows a standard normal z distribution and Student's t distribution with $d.f. = 3$ and $d.f. = 5$.

FIGURE 8-5

A Standard Normal Distribution and Student's t Distribution with $d.f. = 3$ and $d.f. = 5$

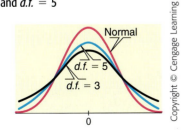

Copyright © Cengage Learning 2013

Properties of a Student's t distribution

1. The distribution is *symmetric* about the mean 0.

2. The distribution depends on the *degrees of freedom, d.f.* ($d.f. = n - 1$ for μ confidence intervals).

3. The distribution is *bell-shaped*, but has thicker tails than the standard normal distribution.

4. As the degrees of freedom increase, the t distribution *approaches* the standard normal distribution.

5. The area under the entire curve is 1.

Using Table 4 to Find Critical Values for Confidence Intervals

Table 4 of the Appendix gives various t values for different degrees of freedom $d.f.$ We will use this table to find *critical values* t_c for a c confidence level. In other words, we want to find t_c such that an area equal to c under the t distribution for a given number of degrees of freedom falls between $-t_c$ and t_c. In the language of probability, we want to find t_c such that

$$P(-t_c < t < t_c) = c$$

This probability corresponds to the shaded area in Figure 8-6.

Table 4 of the Appendix has been arranged so that c is one of the column headings, and the degrees of freedom $d.f.$ are the row headings. To find t_c for any specific c, we find the column headed by that c value and read down until we reach the row headed by the appropriate number of degrees of freedom $d.f.$ (You will notice two other column headings: one-tail area and two-tail area. We will use these later, but for the time being, just ignore them.)

Critical values, t_c

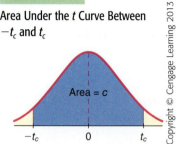

FIGURE 8-6

Area Under the t Curve Between $-t_c$ and t_c

Copyright © Cengage Learning 2013

Convention for using a Student's t distribution table

If the degrees of freedom $d.f.$ you need are not in the table, use the closest $d.f.$ in the table that is *smaller*. This procedure results in a critical value t_c that is more conservative, in the sense that it is larger. The resulting confidence interval will be longer and have a probability that is slightly higher than c.

EXAMPLE 4   STUDENT'S t DISTRIBUTION

Use Table 8-3 (an excerpt from Table 4 of the Appendix) to find the critical value t_c for a 0.99 confidence level for a t distribution with sample size $n = 5$.

SOLUTION:

(a) First, we find the column with c heading 0.990.

(b) Next, we compute the number of degrees of freedom:
$d.f. = n - 1 = 5 - 1 = 4.$

(c) We read down the column under the heading $c = 0.99$ until we reach the row headed by 4 (under $d.f.$). The entry is 4.604. Therefore, $t_{0.99} = 4.604$.

TABLE 8-3   Student's t Distribution Critical Values
(Excerpt from Table 4 of the Appendix)

one-tail area		—	—	—	—
two-tail area		—	—	—	—
d.f. \ **c**		**... 0.900**	**0.950**	**0.980**	**0.990 ...**
⋮					
3		... 2.353	3.182	4.541	5.841 ...
4		... 2.132	2.776	3.747	4.604 ...
⋮					
7		... 1.895	2.365	2.998	3.449 ...
8		... 1.860	2.306	2.896	3.355 ...

Copyright © Cengage Learning 2013

GUIDED EXERCISE 2 | *Student's t distribution table*

Use Table 4 of the Appendix (or Table 8-3, showing an excerpt from the table) to find t_c for a 0.90 confidence level for a t distribution with sample size $n = 9$.

GUIDED EXERCISE 2

(a) We find the column headed by $c =$ _____.   ➡ $c = 0.900$.

(b) The degrees of freedom are given by   ➡ $d.f. = n - 1 = 9 - 1 = 8$.
    $d.f. = n - 1 =$ _____.

(c) Read down the column found in part (a) until   ➡ $t_{0.90} = 1.860$ for a sample of size $n = 9$.
    you reach the entry in the row headed by
    $d.f. = 8$. The value of $t_{0.90}$ is _____ for
    a sample of size 9.

(d) Find t_c for a 0.95 confidence level for a t ➡ $t_{0.95} = 2.306$ for a sample of size $n = 9$.
    distribution with sample size $n = 9$.

LOOKING FORWARD

Student's t distributions will be used again in Chapter 9 when testing μ and in Chapter 10 when testing differences of means. The distributions are also used for confidence intervals and testing of parameters of linear regression (Section 11.4).

Confidence Intervals for μ When σ Is Unknown

Maximal margin of error, E

In Section 8.1, we found bounds $\pm E$ on the margin of error for a c confidence level. Using the same basic approach, we arrive at the conclusion that

$$E = t_c \frac{s}{\sqrt{n}}$$

is the maximal margin of error for a c confidence level when σ is unknown (i.e., $|\overline{x} - \mu| < E$ with probability c). The analogue of Equation (1) in Section 8.1 is

$$P\left(-t_c \frac{s}{\sqrt{n}} < \overline{x} - \mu < t_c \frac{s}{\sqrt{n}}\right) = c \tag{13}$$

COMMENT   Comparing Equation (13) with Equation (1) in Section 8.1, it becomes evident that we are using the same basic method on the t distribution that we used on the z distribution.

Likewise, for samples from normal populations with unknown σ, Equation (9) of Section 8.1 becomes

$$P(\overline{x} - E < \mu < \overline{x} + E) = c \tag{14}$$

where $E = t_c(s/\sqrt{n})$. Let us organize what we have been doing in a convenient summary.

PROCEDURE

Confidence interval for μ when σ is unknown

HOW TO FIND A CONFIDENCE INTERVAL FOR μ WHEN σ IS UNKNOWN

Requirements
Let x be a random variable appropriate to your application. Obtain a simple random sample (of size n) of x values from which you compute the sample mean $\overline{x}$ and the sample standard deviation s.

Continued

If you can assume that x has a normal distribution or simply a mound-shaped, symmetric distribution, then any sample size n will work. If you cannot assume this, then use a sample size of $n \geq 30$.

Confidence interval for μ when σ is unknown

$$\bar{x} - E < \mu < \bar{x} + E \tag{8}$$

where $\bar{x}$ = sample mean of a simple random sample

$$E = t_c \frac{s}{\sqrt{n}}$$

c = confidence level $(0 < c < 1)$

t_c = critical value for confidence level c and degrees of freedom
$d.f. = n - 1$ (See Table 4 of the Appendix.)

Problem 23 presents an alternate method for computing confidence intervals when σ is unknown but $n \geq 30$. The alternate method approximates σ by the sample standard deviation s and utilizes the normal distribution.

COMMENT   In our applications of Student's t distributions, we have made the basic assumption that x has a normal distribution. However, the same methods apply even if x is only approximately normal. In fact, the main requirement for using a Student's t distribution is that the distribution of x values be reasonably symmetric and mound-shaped. If this is the case, then the methods we employ with the t distribution can be considered valid for most practical applications.

EXAMPLE 5

CONFIDENCE INTERVAL FOR μ WITH σ UNKNOWN

Suppose an archaeologist discovers seven fossil skeletons from a previously unknown species of miniature horse. Reconstructions of the skeletons of these seven miniature horses show the shoulder heights (in centimeters) to be

45.3   47.1   44.2   46.8   46.5   45.5   47.6

For these sample data, the mean is $\bar{x} \approx 46.14$ and the sample standard deviation is $s \approx 1.19$. Let μ be the mean shoulder height (in centimeters) for this entire species of miniature horse, and assume that the population of shoulder heights is approximately normal.

Find a 99% confidence interval for μ, the mean shoulder height of the entire population of such horses.

SOLUTION: *Check Requirements* We assume that the shoulder heights of the reconstructed skeletons form a random sample of shoulder heights for all the miniature horses of the unknown species. The x distribution is assumed to be approximately normal. Since σ is unknown, it is appropriate to use a Student's t distribution and sample information to compute a confidence interval for μ.

In this case, $n = 7$, so $d.f. = n - 1 = 7 - 1 = 6$. For $c = 0.990$, Table 4 of the Appendix gives $t_{0.99} = 3.707$ (for $d.f. = 6$). The sample standard deviation is $s = 1.19$.

$$E = t_c \frac{s}{\sqrt{n}} = (3.707) \frac{1.19}{\sqrt{7}} \approx 1.67$$

The 99% confidence interval is

$$\bar{x} - E < \mu < \bar{x} + E$$
$$46.14 - 1.67 < \mu < 46.14 + 1.67$$
$$44.5 < \mu < 47.8$$

Interpretation   The archaeologist can be 99% confident that the interval from 44.5 cm to 47.8 cm is an interval that contains the population mean μ for shoulder height of this species of miniature horse.

GUIDED EXERCISE 3 | ### Confidence interval for μ with σ unknown

A company has a new process for manufacturing large artificial sapphires. In a trial run, 37 sapphires are produced. The distribution of weights is mound-shaped and symmetric. The mean weight for these 37 gems is $\bar{x} = 6.75$ carats, and the sample standard deviation is $s = 0.33$ carat. Let μ be the mean weight for the distribution of all sapphires produced by the new process.

(a) *Check Requirements* Is it appropriate to use a Student's t distribution to compute a confidence interval for μ?

⇒ Yes, we assume that the 37 sapphires constitute a simple random sample of all sapphires produced under the new process. The requirement that the x distribution be approximately normal can be dropped since the sample size is large enough to ensure that the $\bar{x}$ distribution is approximately normal.

(b) What is *d.f.* for this setting?

⇒ $d.f. = n - 1$, where n is the sample size. Since $n = 37$, $d.f. = 37 - 1 = 36$.

(c) Use Table 4 of the Appendix to find $t_{0.95}$. Note that *d.f.* = 36 is not in the table. Use the *d.f.* closest to 36 that is *smaller* than 36.

⇒ $d.f. = 35$ is the closest *d.f.* in the table that is *smaller* than 36. Using *d.f.* = 35 and $c = 0.95$, we find $t_{0.95} = 2.030$.

(d) Find E.

⇒ $E = t_{0.95}\dfrac{s}{\sqrt{n}}$

$\approx 2.030 \dfrac{0.33}{\sqrt{37}} \approx 0.11$ carat

(e) Find a 95% confidence interval for μ.

⇒ $\bar{x} - E < \mu < \bar{x} + E$

$6.75 - 0.11 < \mu < 6.75 + 0.11$

$6.64 \text{ carats} < \mu < 6.86 \text{ carats}$

(f) *Interpretation* What does the confidence interval tell us in the context of the problem?

⇒ The company can be 95% confident that the interval from 6.64 to 6.86 contains the population mean weight of sapphires produced by the new process.

We have several formulas for confidence intervals for the population mean μ. How do we choose an appropriate one? We need to look at the sample size, the distribution of the original population, and whether or not the population standard deviation σ is known.

Summary: Confidence intervals for the mean

Assume that you have a random sample of size n from an x distribution and that you have computed $\bar{x}$ and s. A confidence interval for μ is

$\bar{x} - E < \mu < \bar{x} + E$

where E is the margin of error. How do you find E? It depends on how much you know about the x distribution.

Continued

Situation I (most common)
You don't know the population standard deviation σ. In this situation, you use the *t* distribution with margin of error

$$E = t_c \frac{s}{\sqrt{n}}$$

where degrees of freedom

$$d.f. = n - 1$$

Although a *t* distribution can be used in many situations, you need to observe some guidelines. If *n* is less than 30, *x* should have a distribution that is mound-shaped and approximately symmetric. It's even better if the *x* distribution is normal. If *n* is 30 or more, the central limit theorem (Chapter 7) implies that these restrictions can be relaxed.

Situation II (almost never happens!)
You actually know the population value of σ. In addition, you know that *x* has a normal distribution. If you don't know that the *x* distribution is normal, then your sample size *n* must be 30 or larger. In this situation, you use the standard normal *z* distribution with margin of error

$$E = z_c \frac{\sigma}{\sqrt{n}}$$

Which distribution should you use for $\bar{x}$?

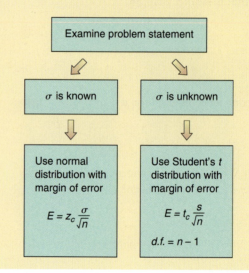

COMMENT   To find confidence intervals for μ based on small samples, we need to know that the population distribution is approximately normal. What if this is not the case? A procedure called *bootstrap* utilizes computer power to generate an approximation for the $\bar{x}$ sampling distribution. Essentially, the bootstrap method treats the sample as if it were the population. Then, using repetition, it takes many samples (often thousands) from the original sample. This process is called *resampling*. The sample mean $\bar{x}$ is computed for each resample and a distribution of sample means is created. For example, a 95% confidence interval reflects the range for the middle 95% of the bootstrap $\bar{x}$ distribution. If you read Using Technology at the end of this chapter, you will find one (of many) bootstrap methods (Reference: *An Introduction to the Bootstrap* by B. Efron and R. Tibshirani).

TECH NOTES   The TI-84Plus/TI-83Plus/TI-*n*spire calculators, Excel 2010, and Minitab support confidence intervals using the Student's *t* distribution.

TI-84Plus/TI-83Plus/TI-*n*spire (with TI-84Plus keypad) Press the **STAT** key, select **TESTS**, and choose the option **8:TInterval**. You may use either raw data in a list or summary statistics.

Excel 2010 Excel gives only the value of the maximal margin of error *E*. You can easily construct the confidence interval by computing $\bar{x} - E$ and $\bar{x} + E$. For raw data, on the Home screen, click the **Data** tab. In the Analysis group, click **Data Analysis**. In the dialogue box, select **Descriptive Statistics**. In the dialogue box, check Summary Statistics and check Confidence Level for Mean. Then set the desired confidence level. Under these choices, Excel uses the Student's *t* distribution.

An alternate approach utilizing only the sample standard deviation and sample size is to use the **Insert Function** (f_x) button on the Home screen. In the dialogue box, select **Statistical** for the category and then **CONFIDENCE.T** for the function. In the next dialogue box, the value of **alpha** is (1 − the confidence level). For example, alpha is 0.05 for a 95% confidence interval. Enter the standard deviation and the sample size. The output is the value of the maximal margin of error *E*.

Minitab Use the menu choices **Stat ➤ Basic Statistics ➤ 1-Sample *t***. In the dialogue box, indicate the column that contains the raw data. The Minitab output shows the confidence interval for Example 5.

```
T Confidence Intervals
Variable      N      Mean      StDev      SE Mean       99.0 % CI
C1            7      46.143    1.190      0.450         (44.475, 47.810)
```

VIEWPOINT | Earthquakes!

California, Washington, Nevada, and even Yellowstone National Park all have earthquakes. Some earthquakes are severe! Earthquakes often bring fear and anxiety to people living near the quake. Is San Francisco due for a really big quake like the 1906 major earthquake? How big are the sizes of recent earthquakes compared with really big earthquakes? What is the duration of an earthquake? How long is the time span between major earthquakes? One way to answer questions such as these is to use existing data to estimate confidence intervals for the average size, duration, and time interval between quakes. Recent data sets for computing such confidence intervals can be found at the National Earthquake Information Center at the U.S. Geological Survey website. To access the site, visit the Brase/Brase statistics site at **http://www.cengagebrain.com** *and find the link to the National Earthquake Information Center.*

SECTION 8.2 PROBLEMS

Access end-of-section exercises at www.aplia.com

Estimating *p* in the Binomial Distribution

FOCUS POINTS
- Compute the maximal margin of error for proportions using a given level of confidence.
- Compute confidence intervals for *p* and interpret the results.
- Interpret poll results.
- Compute the sample size to be used for estimating a proportion *p* when we have an estimate for *p*.
- Compute the sample size to be used for estimating a proportion *p* when we have no estimate for *p*.

Basic criteria

The binomial distribution is completely determined by the number of trials n and the probability p of success on a single trial. For most experiments, the number of trials is chosen in advance. Then the distribution is completely determined by p. In this section, we will consider the problem of estimating p under the assumption that n has already been selected.

We are employing what are called *large-sample methods*. We will assume that the normal curve is a good approximation to the binomial distribution, and when necessary, we will use sample estimates for the standard deviation. Empirical studies have shown that these methods are quite good, provided *both*

Basic criteria

$$np > 5 \quad \text{and} \quad nq > 5, \quad \text{where } q = 1 - p$$

Point estimate for p

Let r be the number of successes out of n trials in a binomial experiment. We will take the sample proportion of successes $\hat{p}$ (read "p hat") $= r/n$ as our *point estimate for p*, the population proportion of successes.

> The **point estimates for *p* and *q*** are
>
> $$\hat{p} = \frac{r}{n}$$
> $$\hat{q} = 1 - \hat{p}$$
>
> where n = number of trials and r = number of successes.

For example, suppose that 800 students are selected at random from a student body of 20,000 and that they are each given a shot to prevent a certain type of flu. These 800 students are then exposed to the flu, and 600 of them do not get the flu. What is the probability p that the shot will be successful for any single student selected at random from the entire population of 20,000 students? We estimate p for the entire student body by computing r/n from the sample of 800 students. The value $\hat{p} = r/n$ is 600/800, or 0.75. The value $\hat{p} = 0.75$ is then the *point estimate* for p.

The difference between the actual value of p and the estimate $\hat{p}$ is the size of our error caused by using $\hat{p}$ as a point estimate for p. The magnitude of $\hat{p} - p$ is

Margin of error

called the *margin of error* for using $\hat{p} = r/n$ as a point estimate for p. In absolute value notation, the margin of error is $|\hat{p} - p|$.

To compute the bounds for the margin of error, we need some information about the distribution of $\hat{p} = r/n$ values for different samples of the same size n. It turns out that, for large samples, the distribution of $\hat{p}$ values is well approximated by a *normal curve* with

$$\text{mean } \mu = p \quad \text{and} \quad \text{standard error } \sigma = \sqrt{pq/n}$$

Since the distribution of $\hat{p} = r/n$ is approximately normal, we use features of the standard normal distribution to find the bounds for the difference $\hat{p} - p$. Recall that z_c is the number such that an area equal to c under the standard normal curve falls between $-z_c$ and z_c. Then, in the language of probability,

$$P\left(-z_c\sqrt{\frac{pq}{n}} < \hat{p} - p < z_c\sqrt{\frac{pq}{n}}\right) = c \tag{15}$$

Equation (15) says that the chance is c that the numerical difference between $\hat{p}$ and p is between $-z_c\sqrt{pq/n}$ and $z_c\sqrt{pq/n}$. With the c confidence level, our estimate $\hat{p}$ differs from p by no more than

Maximal margin of error, E

$$E = z_c\sqrt{pq/n}$$

As in Sections 8.1 and 8.2 we call E the *maximal margin of error*.

Optional derivation of equation (15)

First, we need to show that $\hat{p} = r/n$ has a distribution that is approximately normal, with $\mu = p$ and $\sigma = \sqrt{pq/n}$. From Section 7.6, we know that for sufficiently large n, the binomial distribution can be approximated by a normal distribution with mean $\mu = np$ and standard deviation $\sigma = \sqrt{npq}$. If r is the number of successes out of n trials of a binomial experiment, then r is a binomial random variable with a binomial distribution. When we convert r to standard z units, we obtain

$$z = \frac{r-\mu}{\sigma} = \frac{r-np}{\sqrt{npq}}$$

For sufficiently large n, r will be approximately normally distributed, so z will be too.

If we divide both the numerator and the denominator of the last expression by n, the value of z will not change.

$$z = \frac{\dfrac{r-np}{n}}{\dfrac{\sqrt{npq}}{n}} \quad \text{Simplified, we find } z = \frac{\dfrac{r}{n}-p}{\sqrt{\dfrac{pq}{n}}}. \tag{16}$$

Equation (16) tells us that the $\hat{p} = r/n$ distribution is approximated by a normal curve with $\mu = p$ and $\sigma = \sqrt{pq/n}$.

The probability is c that z lies in the interval between $-z_c$ and z_c because an area equal to c under the standard normal curve lies between $-z_c$ and z_c. Using the language of probability, we write

$$P(-z_c < z < z_c) = c$$

From Equation (16), we know that

$$z = \frac{\hat{p}-p}{\sqrt{\dfrac{pq}{n}}}$$

If we put this expression for z into the preceding equation, we obtain

$$P\left(-z_c < \frac{\hat{p}-p}{\sqrt{\dfrac{pq}{n}}} < z_c\right) = c$$

If we multiply all parts of the inequality by $\sqrt{pq/n}$, we obtain the equivalent statement

$$P\left(-z_c\sqrt{\frac{pq}{n}} < \hat{p} - p < z_c\sqrt{\frac{pq}{n}}\right) = c \tag{15}$$

Confidence interval for p

To find a c confidence interval for p, we will use E in place of the expression $z_c\sqrt{pq/n}$ in Equation (15). Then we get

$$P(-E < \hat{p} - p < E) = c \qquad (17)$$

Some algebraic manipulation produces the mathematically equivalent statement

$$P(\hat{p} - E < p < \hat{p} + E) = c \qquad (18)$$

Equation (18) states that the probability is c that p lies in the interval from $\hat{p} - E$ to $\hat{p} + E$. Therefore, the interval from $\hat{p} - E$ to $\hat{p} + E$ is the c confidence interval for p that we wanted to find.

There is one technical difficulty in computing the c confidence interval for p. The expression $E = z_c\sqrt{pq/n}$ requires that we know the values of p and q. In most situations, we will not know the actual values of p or q, so we will use our point estimates

$$p \approx \hat{p} \quad \text{and} \quad q = 1 - p \approx 1 - \hat{p}$$

to estimate E. These estimates are reliable for most practical purposes, since we are dealing with large-sample theory ($np > 5$ and $nq > 5$).

For convenient reference, we'll summarize the information about c confidence intervals for p, the probability of success in a binomial distribution.

PROCEDURE

Problem 28 of this section shows the method for computing a "plus four confidence interval for p." This is an alternate method that generally results in a slightly smaller confidence interval than the ones computed with standard methods presented in this section and used in most statistical technology packages.

HOW TO FIND A CONFIDENCE INTERVAL FOR A PROPORTION p

Requirements
Consider a binomial experiment with n trials, where p represents the population probability of success on a single trial and $q = 1 - p$ represents the population probability of failure. Let r be a random variable that represents the number of successes out of the n binomial trials.

The point estimates for p and q are

$$\hat{p} = \frac{r}{n} \quad \text{and} \quad \hat{q} = 1 - \hat{p}$$

The number of trials n should be sufficiently large so that both $n\hat{p} > 5$ and $n\hat{q} > 5$.

Confidence interval for p

$$\hat{p} - E < p < \hat{p} + E$$

where $E \approx z_c\sqrt{\dfrac{\hat{p}\hat{q}}{n}} = z_c\sqrt{\dfrac{\hat{p}(1 - \hat{p})}{n}}$

$\qquad c$ = confidence level ($0 < c < 1$)

$\qquad z_c$ = critical value for confidence level c based on the standard normal distribution (see Table 3(b) of the Appendix for frequently used values.)

COMMENT   Problem 6 of this section asks you to show that the two conditions $n\hat{p} > 5$ and $n\hat{q} > 5$ are equivalent to the two conditions that the number of successes $r > 5$ and the number of failures $n - r > 5$.

EXAMPLE 6   CONFIDENCE INTERVAL FOR p

Let's return to our flu shot experiment described at the beginning of this section. Suppose that 800 students were selected at random from a student body of 20,000 and given shots to prevent a certain type of flu. All 800 students were exposed to the flu, and 600 of them did not get the flu. Let p represent the probability that the shot will be successful for any single student selected at random from the entire population of 20,000. Let q be the probability that the shot is not successful.

(a) What is the number of trials n? What is the value of r?

SOLUTION: Since each of the 800 students receiving the shot may be thought of as a trial, then $n = 800$, and $r = 600$ is the number of successful trials.

(b) What are the point estimates for p and q?

SOLUTION: We estimate p by the sample point estimate

$$\hat{p} = \frac{r}{n} = \frac{600}{800} = 0.75$$

We estimate q by

$$\hat{q} = 1 - \hat{p} = 1 - 0.75 = 0.25$$

(c) *Check Requirements* Would it seem that the number of trials is large enough to justify a normal approximation to the binomial?

SOLUTION: Since $n = 800, p \approx 0.75$, and $q \approx 0.25$, then

$$np \approx (800)(0.75) = 600 > 5 \quad \text{and} \quad np \approx (800)(0.25) = 200 > 5$$

A normal approximation is certainly justified.

(d) Find a 99% confidence interval for p.

SOLUTION:

$$z_{0.99} = 2.58 \text{ (see Table 8-2 or Table 3(b) of the Appendix)}$$

$$E \approx z_{0.99}\sqrt{\frac{\hat{p}(1 - \hat{p})}{n}} \approx 2.58\sqrt{\frac{(0.75)(0.25)}{800}} \approx 0.0395$$

The 99% confidence interval is then

$$\hat{p} - E < p < \hat{p} + E$$
$$0.75 - 0.0395 < p < 0.75 + 0.0395$$
$$0.71 < p < 0.79$$

Interpretation   We are 99% confident that the probability that a flu shot will be effective for a student selected at random is between 0.71 and 0.79.

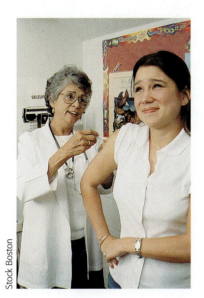

Stock Boston

GUIDED EXERCISE 4 | *Confidence interval for p*

A random sample of 188 books purchased at a local bookstore showed that 66 of the books were murder mysteries. Let p represent the proportion of books sold by this store that are murder mysteries.

(a) What is a point estimate for p?

$$\hat{p} = \frac{r}{n} = \frac{66}{188} = 0.35$$

Continued

GUIDED EXERCISE 4   *continued*

(b) Find a 90% confidence interval for p.

$$E = z_c \sqrt{\frac{\hat{p}(1 - \hat{p})}{n}}$$

$$= 1.645 \sqrt{\frac{(0.35)(1 - 0.35)}{188}} \approx 0.0572$$

The confidence interval is

$$\hat{p} - E < p < \hat{p} + E$$
$$0.35 - 0.0572 < p < 0.35 + 0.0572$$
$$0.29 < p < 0.41$$

(c) *Interpretation* What does the confidence interval you just computed mean in the context of this application?

If we had computed the interval for many different sets of 188 books, we would have found that about 90% of the intervals actually contained p, the population proportion of mysteries. Consequently, we can be 90% confident that our interval is one of the intervals that contains the unknown value p.

(d) *Check Requirements* To compute the confidence interval, we used a normal approximation. Does this seem justified?

$n = 188; p \approx 0.35; q \approx 0.65$

Since $np \approx 65.8 > 5$ and $np \approx 122.2 > 5$, the approximation is justified.

It is interesting to note that our sample point estimate $\hat{p} = r/n$ and the confidence interval for the population proportion p do not depend on the size of the population. In our bookstore example, it made no difference how many books the store sold. On the other hand, the size of the sample does affect the accuracy of a statistical estimate. At the end of this section, we will study the effect of sample size on the reliability of our estimate.

TECH NOTES   The TI-84Plus/TI-83Plus/TI-*n*spire calculators and Minitab provide confidence intervals for proportions.

TI-84Plus/TI-83Plus/TI-*n*spire (with TI-84Plus keypad) Press the **STAT** key, select **TESTS**, and choose option **A:1-PropZInt**. The letter x represents the number of successes r. The TI-84Plus/TI-83Plus/TI-*n*spire output shows the results for Guided Exercise 4.

```
1-PropZInt
 (.29381,.40832)
 p̂=.3510638298
 n=188
```

Minitab Use the menu selections **Stat ➤ Basic Statistics ➤ 1 Proportion**. In the dialogue box, select Summarized Data and fill in the number of trials and the number of successes. Under Options, select a confidence interval. Minitab uses the binomial distribution directly unless Normal is checked. The Minitab output shows the results for Guided Exercise 4. Information from Chapter 9 material is also shown.

Test and Confidence Interval for One Proportion (Using Binomial)
Test of p = 0.5 vs p not = 0.5

Sample	X	N	Sample p	90.0 % CI	Exact P-Value
1	66	188	0.351064	(0.293222, 0.412466)	0.000

```
Test and Confidence Interval for One Proportion (Using Normal)
Test of p = 0.5 vs p not = 0.5
Sample   X    N    Sample p       90.0 % CI      Z-Value   P-Value
1       66   188   0.351064   (0.293805, 0.408323)   -4.08    0.000
```

Interpreting Results from a Poll

Newspapers frequently report the results of an opinion poll. In articles that give more information, a statement about the margin of error accompanies the poll results. In most polls, the margin of error is given for a *95% confidence interval.*

Margin of error for polls

General interpretation of poll results

1. When a poll states the results of a survey, the proportion reported to respond in the designated manner is $\hat{p}$, the sample estimate of the population proportion.

2. The *margin of error* is the maximal error E of a 95% confidence interval for p.

3. A 95% confidence interval for the population proportion p is
   (poll report $\hat{p}$ − margin of error E) $< p <$
   (poll report $\hat{p}$ + margin of error E)

COMMENT   Leslie Kish, a statistician at the University of Michigan, was the first to apply the term *margin of error.* He was a pioneer in the study of population sampling techniques. His book *Survey Sampling* is still widely used all around the world.

Some articles clarify the meaning of the margin of error further by saying that it is an error due to sampling. For instance, the following comments accompany results of a political poll reported in an issue of *The Wall Street Journal*:

How the Poll Was Conducted

The *Wall Street Journal*/NBC News poll was based on nationwide telephone interviews of 1508 adults conducted last Friday through Tuesday by the polling organizations of Peter Hart and Robert Teeter.

The sample was drawn from 315 randomly selected geographic points in the continental United States. Each region was represented in proportion to its population. Households were selected by a method that gave all telephone numbers . . . an equal chance of being included.

One adult, 18 years or older, was selected from each household by a procedure to provide the correct number of male and female respondents.

Chances are 19 of 20 that if all adults with telephones in the United States had been surveyed, the findings would differ from these poll results by no more than 2.6 percentage points in either direction.

GUIDED EXERCISE 5 | *Reading a poll*

Read the last paragraph of the article excerpt "How the Poll Was Conducted."

(a) What confidence level corresponds to the phrase   ⟹ $\dfrac{19}{20} = 0.95$
"chances are 19 of 20 that if . . ."?

A 95% confidence interval is being discussed.

Continued

GUIDED EXERCISE 5   *continued*

(b) The complete article indicates that everyone in the sample was asked the question "Which party, the Democratic Party or the Republican Party, do you think would do a better job handling . . . education?" Possible responses were "Democrats," "neither," "both," or "Republicans." The poll reported that 32% of the respondents said "Democrats." Does 32% represent the sample statistic $\hat{p}$ or the population parameter p for the proportion of adults responding "Democrats"?

32% represents a sample statistic $\hat{p}$ because 32% represents the percentage of the adults in the *sample* who responded "Democrats."

(c) Continue reading the last paragraph of the article. It goes on to state, "If all adults with telephones in the U.S. had been surveyed, the findings would differ from these poll results by no more than 2.6 percentage points in either direction." Use this information, together with parts (a) and (b), to find a 95% confidence interval for the proportion p of the specified population who responded "Democrats" to the question.

The value 2.6 percentage points represents the margin of error. Since the margin of error is for a 95% confidence interval, the confidence interval is

$$32\% - 2.6\% < p < 32\% + 2.6\%$$
$$29.4\% < p < 34.6\%$$

The poll indicates that at the time of the poll, between 29.4% and 34.6% of the specified population thought Democrats would do a better job handling education.

Sample Size for Estimating p

Suppose you want to specify the maximal margin of error in advance for a confidence interval for p at a given confidence level c. What sample size do you need? The answer depends on whether or not you have a preliminary estimate for the population probability of success p in a binomial distribution.

PROCEDURE

HOW TO FIND THE SAMPLE SIZE n FOR ESTIMATING A PROPORTION p

$$n = p(1 - p)\left(\frac{z_c}{E}\right)^2 \text{ if you have a preliminary estimate for } p \qquad (19)$$

$$n = \frac{1}{4}\left(\frac{z_c}{E}\right)^2 \text{ if you do } not \text{ have a preliminary estimate for } p \qquad (20)$$

where E = specified maximal error of estimate

   z_c = critical value from the normal distribution for the desired confidence level c. Commonly used values of z_c can be found in Table 3(b) of the Appendix.

If n is not a whole number, increase n to the next higher whole number. Also, if necessary, increase the sample size n to ensure that both $np > 5$ and $nq > 5$. Note that n is the minimal sample size for a specified confidence level and maximal error of estimate.

COMMENT   To obtain Equation (19), simply solve the formula that gives the maximal error of estimate E of p for the sample size n. When you don't have an estimate for p, a little algebra can be used to show that the maximum value of $p(1 - q)$ is 1/4. See Problem 27.

EXAMPLE 7   SAMPLE SIZE FOR ESTIMATING p

A company is in the business of selling wholesale popcorn to grocery stores. The company buys directly from farmers. A buyer for the company is examining a large amount of corn from a certain farmer. Before the purchase is made, the buyer wants to estimate p, the probability that a kernel will pop.

Suppose a random sample of n kernels is taken and r of these kernels pop. The buyer wants to be 95% sure that the point estimate $\hat{p} = r/n$ for p will be in error either way by less than 0.01.

(a) If no preliminary study is made to estimate p, how large a sample should the buyer use?

SOLUTION: In this case, we use Equation (20) with $z_{0.95} = 1.96$ (see Table 8-2) and $E = 0.01$.

$$n = \frac{1}{4}\left(\frac{z_c}{E}\right)^2 = \frac{1}{4}\left(\frac{1.96}{0.01}\right)^2 = 0.25(38{,}416) = 9604$$

The buyer would need a sample of $n = 9604$ kernels.

(b) A preliminary study showed that p was approximately 0.86. If the buyer uses the results of the preliminary study, how large a sample should he use?

SOLUTION: In this case, we use Equation (19) with $p \approx 0.86$. Again, from Table 8-2, $z_{0.95} = 1.96$, and from the problem, $E = 0.01$.

$$n = p(1 - p)\left(\frac{z_c}{E}\right)^2 = (0.86)(0.14)\left(\frac{1.96}{0.01}\right)^2 = 4625.29$$

The sample size should be at least $n = 4626$. This sample is less than half the sample size necessary without the preliminary study.

VIEWPOINT   "Band-Aid Surgery"

Faster recovery time and less pain: sounds great! An alternate surgical technique called laparoscopic *("Band-Aid") surgery involves small incisions through which tiny video cameras and long surgical instruments are maneuvered. Instead of a 10-inch incision, surgeons might use four little slits of about $\frac{1}{2}$ inch in length. However, not every such surgery is successful. An article in the Health section of* The Wall Street Journal *recommends using a surgeon who has done at least 50 such surgeries. Then the prospective patient should ask about the* rate of conversion—that *is, the proportion* p *of times the surgeon has been forced by complications to switch in midoperation to conventional surgery. A confidence interval for the proportion* p *would be useful patient information!*

aplia™   Access end-of-section exercises at www.aplia.com

PROCEDURE

HOW TO COMPUTE A PLUS FOUR CONFIDENCE INTERVAL FOR p

Requirements

Consider a binomial experiment with n trials, where p represents the population probability of success and $q = 1 - p$ represents the population probability of failure. Let r be a random variable that represents the number of successes out of the n binomial trials.

The plus four point estimates for p and q are

$$\tilde{p} = \frac{r + 2}{n + 4} \quad \text{and} \quad \tilde{q} = 1 - \tilde{p}$$

The number of trials n should be at least 10.

Approximate confidence interval for p

$$\tilde{p} - E < p < \tilde{p} + E$$

where $\quad E \approx z_c \sqrt{\dfrac{\tilde{p}\tilde{q}}{n + 4}} = z_c \sqrt{\dfrac{\tilde{p}(1 - \tilde{p})}{n + 4}}$

c = confidence level $(0 < c < 1)$

z_c = critical value for confidence level c based on the standard normal distribution

---

Chapter Review

SUMMARY

How do you get information about a population by looking at a random sample? One way is to use point estimates and confidence intervals.

- Point estimates and their corresponding parameters are

    $\bar{x}$ for μ $\hat{p}$ for p

- Confidence intervals are of the form

    (point estimate $- E$) < parameter < (point estimate $+ E$)

- E is the maximal margin of error. Specific values of E depend on the parameter, the level of confidence, whether population standard deviations are known, the sample size, and the shapes of the original population distributions.

For μ: $E = z_c \dfrac{\sigma}{\sqrt{n}}$ when σ is known;

$E = t_c \dfrac{s}{\sqrt{n}}$ with $d.f. = n - 1$ when σ is unknown

For p: $E = z_c \sqrt{\dfrac{\hat{p}(1 - \hat{p})}{n}}$ when $n\hat{p} > 5$ and $n\hat{q} > 5$.

- Confidence intervals have an associated probability c called the confidence level. For a given sample size, the proportion of all corresponding confidence intervals that contain the parameter in question is c.

IMPORTANT WORDS & SYMBOLS

Section 8.1
Point estimate for μ 262
Confidence level, c 262
Critical value, z_c 263
Maximal margin of error, E 265
c confidence interval for μ 265
Sample size for estimating μ 269

Section 8.2
Student's t distribution 270
Degrees of freedom, *d.f.* 271
Critical values, t_c 272

Section 8.3
Point estimate for p, $\hat{p}$ 278
Confidence interval for p 280
Margin of error for polls 283
Sample size for estimating p 284

VIEWPOINT | All Systems Go?

On January 28, 1986, the space shuttle Challenger *caught fire and blew up only seconds after launch. A great deal of good engineering had gone into the design of the* Challenger. *However, when a system has several confidence levels operating at once, it can happen in rare cases that risks increase rather than cancel out. (See Chapter Review Problem 13.) Diane Vaughan is a professor of sociology at Boston College and author of the book* The Challenger Launch Decision *(University of Chicago Press). Her book contains an excellent discussion of risks, the normalization of deviance, and cost–safety tradeoffs. Vaughan's book is described as "a remarkable and important analysis of how social structures can induce consequential errors in a decision process" (Robert K. Merton, Columbia University).*

CHAPTER REVIEW PROBLEMS

1. *Statistical Literacy* In your own words, carefully explain the meanings of the following terms: *point estimate, critical value, maximal margin of error, confidence level,* and *confidence interval.*

2. *Critical Thinking* Suppose you are told that a 95% confidence interval for the average price of a gallon of regular gasoline in your state is from $3.15 to $3.45. Use the fact that the confidence interval for the mean has the form $\bar{x} - E$ to $\bar{x} + E$ to compute the sample mean and the maximal margin of error E.

3. *Critical Thinking* If you have a 99% confidence interval for μ based on a simple random sample:
   (a) is it correct to say that the *probability* that μ is in the specified interval is 99%? Explain.
   (b) is it correct to say that in the long run, if you computed many, many confidence intervals using the prescribed method, about 99% of such intervals would contain μ? Explain.

   For Problems 4–12, categorize each problem according to the parameter being estimated: proportion p, or mean μ. Then solve the problem.

4. *Auto Insurance: Claims* Anystate Auto Insurance Company took a random sample of 370 insurance claims paid out during a 1-year period. The average claim paid was $1570. Assume σ = $250. Find 0.90 and 0.99 confidence intervals for the mean claim payment.

5. *Psychology: Closure* Three experiments investigating the relationship between need for cognitive closure and persuasion were reported in "Motivated Resistance and Openness to Persuasion in the Presence or Absence of Prior Information" by A. W. Kruglanski (*Journal of Personality and Social Psychology*, Vol. 65, No. 5, pp. 861–874). Part of the study involved administering a "need for closure scale" to a group of students enrolled in an introductory psychology course. The "need for closure scale" has scores ranging from 101 to 201. For the 73 students in the highest quartile of the distribution, the mean score was $\bar{x}$ = 178.70. Assume a population standard deviation of σ = 7.81. These students were all classified as high on their need for closure. Assume that the 73 students represent a random sample of all students who are classified as high on their need for closure. Find a 95% confidence interval for the population mean score μ on the "need for closure scale" for all students with a high need for closure.

6. *Psychology: Closure* How large a sample is needed in Problem 5 if we wish to be 99% confident that the sample mean score is within 2 points of the population mean score for students who are high on the need for closure?

7. *Archaeology: Excavations* The Wind Mountain archaeological site is located in southwestern New Mexico. Wind Mountain was home to a culture of prehistoric Native Americans called Anasazi. A random sample of excavations at Wind Mountain gave the following depths (in centimeters) from present-day surface grade to the location of significant archaeological artifacts (Source: *Mimbres Mogollon Archaeology*, by A. Woosley and A. McIntyre, University of New Mexico Press):

85	45	120	80	75	55	65	60
65	95	90	70	75	65	68	

(a) Use a calculator with mean and sample standard deviation keys to verify that $\bar{x} \approx$ 74.2 cm and $s \approx$ 18.3 cm.

(b) Compute a 95% confidence interval for the mean depth μ at which archaeological artifacts from the Wind Mountain excavation site can be found.

8. *Archaeology: Pottery* Shards of clay vessels were put together to reconstruct rim diameters of the original ceramic vessels found at the Wind Mountain archaeological site (see source in Problem 7). A random sample of ceramic vessels gave the following rim diameters (in centimeters):

15.9	13.4	22.1	12.7	13.1	19.6	11.7	13.5	17.7	18.1

(a) Use a calculator with mean and sample standard deviation keys to verify that $\bar{x} \approx$ 15.8 cm and $s \approx$ 3.5 cm.

(b) Compute an 80% confidence interval for the population mean μ of rim diameters for such ceramic vessels found at the Wind Mountain archaeological site.

9. *Telephone Interviews: Survey* The National Study of the Changing Workforce conducted an extensive survey of 2958 wage and salaried workers on issues ranging from relationships with their bosses to household chores. The data were gathered through hour-long telephone interviews with a nationally representative sample (*The Wall Street Journal*). In response to the question "What does success mean to you?" 1538 responded "Personal satisfaction from doing a good job." Let p be the population proportion of all wage and salaried workers who would respond the same way to the stated question. Find a 90% confidence interval for p.

10. *Telephone Interviews: Survey* How large a sample is needed in Problem 9 if we wish to be 95% confident that the sample percentage of those equating success with personal satisfaction is within 1% of the population percentage? *Hint:* Use $p \approx 0.52$ as a preliminary estimate.

11. *Archaeology: Pottery* Three-circle, red-on-white is one distinctive pattern painted on ceramic vessels of the Anasazi period found at the Wind Mountain archaeological site (see source for Problem 7). At one excavation, a sample of 167 potsherds indicated that 68 were of the three-circle, red-on-white pattern.
    (a) Find a point estimate $\hat{p}$ for the proportion of all ceramic potsherds at this site that are of the three-circle, red-on-white pattern.
    (b) Compute a 95% confidence interval for the population proportion p of all ceramic potsherds with this distinctive pattern found at the site.

12. *Archaeology: Pottery* Consider the three-circle, red-on-white pattern discussed in Problem 11. How many ceramic potsherds must be found and identified if we are to be 95% confident that the sample proportion $\hat{p}$ of such potsherds is within 6% of the population proportion of three-circle, red-on-white patterns found at this excavation site? *Hint:* Use the results of Problem 11 as a preliminary estimate.

13. *Expand Your Knowledge: Two Confidence Intervals* What happens if we want several confidence intervals to hold at the same time (concurrently)? Do we still have the same level of confidence we had for *each* individual interval?
    (a) Suppose we have two independent random variables x_1 and x_2 with respective population means μ_1 and μ_2. Let us say that we use sample data to construct two 80% confidence intervals.

Confidence Interval	Confidence Level
$A_1 < \mu_1 < B_1$	0.80
$A_2 < \mu_2 < B_2$	0.80

Now, what is the probability that *both* intervals hold at the same time? Use methods of Section 5.2 to show that

$$P(A_1 < \mu_1 < B_1 \ and \ A_2 < \mu_2 < B_2) = 0.64$$

Hint: You are combining independent events. If the confidence is 64% that both intervals hold concurrently, explain why the risk that at least one interval does not hold (i.e., fails) must be 36%.

    (b) Suppose we want *both* intervals to hold with 90% confidence (i.e., only 10% risk level). How much confidence c should each interval have to achieve this combined level of confidence? (Assume that each interval has the same confidence level c.)

Hint: $P(A_1 < \mu_1 < B_1 \ and \ A_2 < \mu_2 < B_2) = 0.90$
$P(A_1 < \mu_1 < B_1) \times P(A_2 < \mu_2 < B_2) = 0.90$
$c \times c = 0.90$

Now solve for c.

    (c) If we want *both* intervals to hold at the 90% level of confidence, then the individual intervals must hold at a *higher* level of confidence. Write a brief but detailed explanation of how this could be of importance in a large, complex engineering design such as a rocket booster or a spacecraft.

DATA HIGHLIGHTS: GROUP PROJECTS

Judy Griesedieck/Encyclopedia/Corbis

Digging clams

Break into small groups and discuss the following topics. Organize a brief outline in which you summarize the main points of your group discussion.

1. Garrison Bay is a small bay in Washington state. A popular recreational activity in the bay is clam digging. For several years, this harvest has been monitored and the size distribution of clams recorded. Data for lengths and widths of littleneck clams (*Protothaca staminea*) were recorded by a method of systematic sampling in a study done by S. Scherba and V. F. Gallucci ("The Application of Systematic Sampling to a Study of Infaunal Variation in a Soft Substrate Intertidal Environment," *Fishery Bulletin*, Vol. 74, pp. 937–948). The data in Tables 8-4 and 8-5 give lengths and widths for 35 littleneck clams.

    (a) Use a calculator to compute the sample mean and sample standard deviation for the lengths and widths. Compute the coefficient of variation for each.

    (b) Compute a 95% confidence interval for the population mean length of all Garrison Bay littleneck clams.

    (c) How many more littleneck clams would be needed in a sample if you wanted to be 95% sure that the sample mean length is within a maximal margin of error of 10 mm of the population mean length?

    (d) Compute a 95% confidence interval for the population mean width of all Garrison Bay littleneck clams.

    (e) How many more littleneck clams would be needed in a sample if you wanted to be 95% sure that the sample mean width is within a maximal margin of error of 10 mm of the population mean width?

    (f) The *same* 35 clams were used for measures of length and width. Are the sample measurements length and width independent or dependent? Why?

TABLE 8-4 Lengths of Littleneck Clams (mm)

530	517	505	512	487	481	485	479	452	468
459	449	472	471	455	394	475	335	508	486
474	465	420	402	410	393	389	330	305	169
91	537	519	509	511					

TABLE 8-5 Widths of Littleneck Clams (mm)

494	477	471	413	407	427	408	430	395	417
394	397	402	401	385	338	422	288	464	436
414	402	383	340	349	333	356	268	264	141
77	498	456	433	447					

2. Examine Figure 8-7, "Fall Back" on the next page.

    (a) Of the 1024 adults surveyed, 66% were reported to favor daylight saving time. How many people in the sample preferred daylight saving time? Using the statistic $\hat{p} = 0.66$ and sample size $n = 1024$, find a 95% confidence interval for the proportion of people p who favor daylight saving time. How could you report this information in terms of a margin of error?

    (b) Look at Figure 8-7 to find the sample statistic $\hat{p}$ for the proportion of people preferring standard time. Find a 95% confidence interval for the population proportion p of people who favor standard time. Report the same information in terms of a margin of error.

FIGURE 8-7

Fall Back
Each fall, we roll the clocks back to standard time. However, not everyone likes going back to standard time. Percentage of adults who prefer

Standard time 28%
No preference 6%
66% Daylight saving time

Source: Hilton Time Survey of 1024 adults

FIGURE 8-8   Coupons: Limited Use

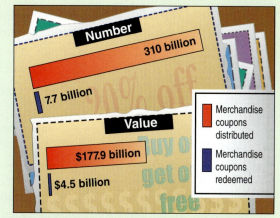

Number
310 billion
7.7 billion

Value
$177.9 billion
$4.5 billion

Merchandise coupons distributed
Merchandise coupons redeemed

Source: NCH Promotional Services

3. Examine Figure 8-8, "Coupons: Limited Use."

(a) Use Figure 8-8 to estimate the percentage of merchandise coupons that were redeemed. Also estimate the percentage dollar value of the coupons that were redeemed. Are these numbers approximately equal?

(b) Suppose you are a marketing executive working for a national chain of toy stores. You wish to estimate the percentage of coupons that will be redeemed for the toy stores. How many coupons should you check to be 95% sure that the percentage of coupons redeemed is within 1% of the population proportion of all coupons redeemed for the toy store?

(c) Use the results of part (a) as a preliminary estimate for p, the percentage of coupons that are redeemed, and redo part (b).

(d) Suppose you sent out 937 coupons and found that 27 were redeemed. Explain why you could be 95% confident that the proportion of such coupons redeemed in the future would be between 1.9% and 3.9%.

(e) Suppose the dollar value of a collection of coupons was $10,000. Use the data in Figure 8-8 to find the expected value and standard deviation of the dollar value of the redeemed coupons. What is the probability that between $225 and $275 (out of the $10,000) was redeemed?

LINKING CONCEPTS: WRITING PROJECTS

Discuss each of the following topics in class or review the topics on your own. Then write a brief but complete essay in which you summarize the main points. Please include formulas and graphs as appropriate.

1. In this chapter, we have studied confidence intervals. Carefully read the following statements about confidence intervals:

(a) Once the endpoints of the confidence interval are numerically fixed, the parameter in question (either μ or p) does or does not fall inside the "fixed" interval.

(b) A given fixed interval either does or does not contain the parameter μ or p; therefore, the probability is 1 or 0 that the parameter is in the interval.

Next, read the following statements. Then discuss all four statements in the context of what we actually mean by a confidence interval.

(c) Nontrivial probability statements can be made only about variables, not about constants.

(d) The confidence level c represents the proportion of all (fixed) intervals that would contain the parameter if we repeated the process many, many times.

2. Throughout Chapter 8, we have used the normal distribution, the central limit theorem, and the Student's t distribution.

   (a) Give a brief outline describing how confidence intervals for means use the normal distribution or Student's t distribution in their basic construction.

   (b) Give a brief outline describing how the normal approximation to the binomial distribution is used in the construction of confidence intervals for a proportion p.

   (c) Give a brief outline describing how the sample size for a predetermined error tolerance and level of confidence is determined from the normal distribution.

3. When the results of a survey or a poll are published, the sample size is usually given, as well as the margin of error. For example, suppose the *Honolulu Star-Bulletin* reported that it surveyed 385 Honolulu residents and 78% said they favor mandatory jail sentences for people convicted of driving under the influence of drugs or alcohol (with margin of error of 3 percentage points in either direction). Usually the confidence level of the interval is not given, but it is standard practice to use the margin of error for a 95% confidence interval when no other confidence level is given.

   (a) The paper reported a point estimate of 78%, with margin of error of $\pm 3\%$. Write this information in the form of a confidence interval for p, the population proportion of residents favoring mandatory jail sentences for people convicted of driving under the influence. What is the assumed confidence level?

   (b) The margin of error is simply the error due to using a sample instead of the entire population. It does not take into account the bias that might be introduced by the wording of the question, by the truthfulness of the respondents, or by other factors. Suppose the question was asked in this fashion: "Considering the devastating injuries suffered by innocent victims in auto accidents caused by drunken or drugged drivers, do you favor a mandatory jail sentence for those convicted of driving under the influence of drugs or alcohol?" Do you think the wording of the question would influence the respondents? Do you think the population proportion of those favoring mandatory jail sentences would be accurately represented by a confidence interval based on responses to such a question? Explain your answer.

   If the question had been "Considering existing overcrowding of our prisons, do you favor a mandatory jail sentence for people convicted of driving under the influence of drugs or alcohol?" do you think the population proportion of those favoring mandatory sentences would be accurately represented by a confidence interval based on responses to such a question? Explain.

USING TECHNOLOGY

Application 1

Finding a Confidence Interval for a Population Mean μ

Cryptanalysis, the science of breaking codes, makes extensive use of language patterns. The frequency of various letter combinations is an important part of the study. A letter combination consisting of a single letter is a monograph, while combinations consisting of two letters are called digraphs, and those with three letters are called trigraphs. In the English language, the most frequent digraph is the letter combination TH.

The *characteristic rate* of a letter combination is a measurement of its rate of occurrence. To compute the characteristic rate, count the number of occurrences of a given letter combination and divide by the number of letters in the text. For instance, to estimate the characteristic rate of the digraph TH, you could select a newspaper text and pick a random starting place. From that place, mark off 2000 letters and count the number of times that TH occurs. Then divide the number of occurrences by 2000.

The characteristic rate of a digraph can vary slightly depending on the style of the author, so to estimate an overall characteristic frequency, you want to consider several samples of newspaper text by different authors. Suppose you did this with a random sample of 15 articles and found the characteristic rates of the digraph TH in the articles. The results follow.

0.0275	0.0230	0.0300	0.0255
0.0280	0.0295	0.0265	0.0265
0.0240	0.0315	0.0250	0.0265
0.0290	0.0295	0.0275	

(a) Find a 95% confidence interval for the mean characteristic rate of the digraph TH.

(b) Repeat part (a) for a 90% confidence interval.

(c) Repeat part (a) for an 80% confidence interval.

(d) Repeat part (a) for a 70% confidence interval.

(e) Repeat part (a) for a 60% confidence interval.

(f) For each confidence interval in parts (a)–(e), compute the length of the given interval. Do you notice a

relation between the confidence level and the length of the interval?

A good reference for cryptanalysis is a book by Sinkov:

Sinkov, Abraham. *Elementary Cryptanalysis.* New York: Random House.

In the book, other common digraphs and trigraphs are given.

Application 2

Confidence Interval Demonstration

When we generate different random samples of the same size from a population, we discover that $\bar{x}$ varies from sample to sample. Likewise, different samples produce different confidence intervals for μ. The endpoints $\bar{x} \pm E$ of a confidence interval are statistical variables. A 90% confidence interval tells us that if we obtain lots of confidence intervals (for the same sample size), then the proportion of all intervals that will turn out to contain μ is 90%.

(a) Use the technology of your choice to generate 10 large random samples from a population with a known mean μ.

(b) Construct a 90% confidence interval for the mean for each sample.

(c) Examine the confidence intervals and note the percentage of the intervals that contain the population mean μ. We have 10 confidence intervals. Will exactly 90% of 10 intervals always contain μ? Explain. What if we have 1000 intervals?

Technology Hints for Confidence Interval Demonstration

TI-84Plus/TI-83Plus/TI-*n*spire

The TI-84Plus/TI-83Plus/TI-*n*spire (with TI-84Plus keypad) calculators generate random samples from uniform, normal, and binomial distributions. Press the **MATH** key and select **PRB**. Choice **5:randInt(lower, upper, sample size *n*)** generates random samples of size *n* from the integers between the specified lower and upper values. Choice **6:randNorm(μ, σ, sample size *n*)** generates random samples of size *n* from a normal distribution with specified mean and standard deviation.

Choice 7:randBin(number of trials, *p*, sample size *n*) generates samples of the specified size from the designated binomial distribution. Under **STAT**, select **EDIT** and highlight the list name, such as L1. At the = sign, use the **MATH** key to access the desired population distribution. Finally, use **Zinterval** under the **TESTS** option of the **STAT** key to generate 90% confidence intervals.

Excel 2010

On the **Home** screen, click the **Data** tab. Then in the Analysis Group, click **Data Analysis**. In the resulting dialogue box, select **Random Number Generator**. In that dialogue box, the number of variables refers to the number of samples. The number of random numbers refers to the number of data values in each sample. Select the population distribution (uniform, normal, binomial, etc.). When you click **OK** the data appear in columns on a spreadsheet, with each sample appearing in a separate column. Click on **Insert Function** $\boxed{f_x}$. In the dialogue box, select **Statistical** for the category and then select **Confidence.t**. In the dialogue box for Confidence, **alpha** = 1 − *c*, so for a 90% confidence interval, enter 0.10 for alpha. Then enter the **population standard deviation S**, and the **sample size**. The resulting output gives the value of the maximal margin of error *E* for the confidence interval for the mean μ. Note that if you use the population standard deviation σ in the function, the value of *E* will be the same for all samples of the same size. Next, find the sample mean $\overline{x}$ for each sample (use **Insert Function** $\boxed{f_x}$ with **Statistical** for category in the dialogue box and select **Average**). Finally, construct the endpoints $\overline{x} \pm E$ of the confidence interval for each sample.

Minitab

Minitab provides options for sampling from a variety of distributions. To generate random samples from a specific distribution, use the menu selection **Calc ➤ Random Data** and then select the population distribution. In the dialogue box, the *number of rows of data* represents the *sample size*. The *number of samples* corresponds to the number of columns selected for data storage. For example, C1–C10 in data storage produces 10 different random samples of the specified size. Use the menu selection **Stat ➤ Basic Statistics ➤ 1 sample z** to generate confidence intervals for the mean μ from each sample. In the variables box, list all the columns containing your samples. For instance, using C1–C10 in the variables list will produce confidence intervals for each of the 10 samples stored in columns C1 through C10.

The Minitab display shows 90% confidence intervals for 10 different random samples of size 50 taken from a normal distribution with $\mu = 30$ and $\sigma = 4$. Notice that only 8 out of 10 of the intervals contain $\mu = 30$. As the number of 90% confidence intervals increases, the percentage containing μ will tend toward 90%.

Minitab Display

Z Confidence Intervals (Samples from a Normal Population with $\mu = 30$ and $\sigma = 4$)

The assumed sigma = 4.00

Variable	N	Mean	StDev	SE Mean	90.0 % CI
C1	50	30.265	4.300	0.566	(29.334, 31.195)
C2	50	31.040	3.957	0.566	(30.109, 31.971)
C3	50	29.940	4.195	0.566	(29.010, 30.871)
C4	50	30.753	3.842	0.566	(29.823, 31.684)
C5	50	30.047	4.174	0.566	(29.116, 30.977)
C6	50	29.254	4.423	0.566	(28.324, 30.185)
C7	50	29.062	4.532	0.566	(28.131, 29.992)
C8	50	29.344	4.487	0.566	(28.414, 30.275)
C9	50	30.062	4.199	0.566	(29.131, 30.992)
C10	50	29.989	3.451	0.566	(29.058, 30.919)

SPSS uses a Student's t distribution to generate confidence intervals for the mean and difference of means (Section 10.2). Use the menu choices **Analyze ➤ Compare Means** and then **One-Sample T Test** or **Independent-Sample T Tests** for confidence intervals for a single mean or difference of means, respectively. In the dialogue box, use 0 for the test value. Click **Options . . .** to provide the confidence level.

To generate 10 random samples of size $n = 30$ from a normal distribution with $\mu = 30$ and $\sigma = 4$, first enter consecutive integers from 1 to 30 in a column of the data editor. Then, under variable view, enter the variable names Sample1 through Sample10. Use the menu choices **Transform ➤ Compute Variable**. In the dialogue box, use Sample1 for the target variable. In the function group select **Random Numbers**. Then select the function **Rv.Normal**. Use 30 for the mean and 4 for the standard deviation. Continue until you have 10 samples. To sample from other distributions, use appropriate functions in the Compute dialogue box.

The SPSS display shows 90% confidence intervals for 10 different random samples of size 30 taken from a normal distribution with $\mu = 30$ and $\sigma = 4$. Notice that, as expected, 9 of the 10 intervals contain the population mean $\mu = 30$.

SPSS Display

90% t-confidence intervals for random samples of size n = 30 from a normal distribution with $\mu = 30$ and $\sigma = 4$.

	t	df	Sig(2-tail)	Mean	Lower	Upper
SAMPLE1	42.304	29	.000	29.7149	28.5214	30.9084
SAMPLE2	43.374	29	.000	30.1552	28.9739	31.3365
SAMPLE3	53.606	29	.000	31.2743	30.2830	32.2656
SAMPLE4	35.648	29	.000	30.1490	28.7120	31.5860
SAMPLE5	47.964	29	.000	31.0161	29.9173	32.1148
SAMPLE6	34.718	29	.000	30.3519	28.8665	31.8374
SAMPLE7	34.698	29	.000	30.7665	29.2599	32.2731
SAMPLE8	39.731	29	.000	30.2388	28.9456	31.5320
SAMPLE9	44.206	29	.000	29.7256	28.5831	30.8681
SAMPLE10	49.981	29	.000	29.7273	28.7167	30.7379

Application 3
Bootstrap Demonstration

Bootstrap can be used to construct confidence intervals for μ when traditional methods cannot be used. For example, if the sample size is small and the sample shows extreme outliers or extreme lack of symmetry, use of the Student's t distribution is inappropriate. Bootstrap makes no assumptions about the population.

Consider the following random sample of size 20:

12	15	21	2	6	3	15	51	22	18
37	12	25	19	33	15	14	17	12	27

A stem-and-leaf display shows that the data are skewed with one outlier.

```
0 | 2   represents 2
0 | 236
1 | 2224555789
2 | 1257
3 | 37
4 |
5 | 1
```

We can use Minitab to model the bootstrap method for constructing confidence intervals for μ. (The Professional edition of Minitab is required because of spreadsheet size and other limitations of the Student edition.) This demonstration uses only 1000 samples. Bootstrap uses many thousands.

Step 1: Create 1000 new samples, each of size 20, by sampling *with replacement* from the original data. To do this in Minitab, we enter the original 20 data values in column C1. Then, in column C2, we place equal probabilities of 0.05 beside each of the original data values. Use the menu choices **Calc ➤ Random Data ➤ Discrete**. In the dialogue box, fill in 1000 as the number of rows, store the data in columns C11–C30, and use column C1 for values and column C2 for probabilities.

Step 2: Find the sample mean of each of the 1000 samples. To do this in Minitab, use the menu choices **Calc ➤ Row Statistics**. In the dialogue box, select **mean**. Use columns C11–C30 as the input variables and store the results in column C31.

Step 3: Order the 1000 means from smallest to largest. In Minitab, use the menu choices **Manip ➤ Sort**. In the dialogue box, indicate C31 as the column to be sorted. Store the results in column C32. Sort by values in column C31.

Step 4: Create a 95% confidence interval by finding the boundaries for the middle 95% of the data. In other words, you need to find the values of the 2.5 percentile ($P_{2.5}$) and the 97.5 percentile ($P_{97.5}$). Since there are 1000 data values, the 2.5 percentile is the data value in position 25, while the 97.5 percentile is the data value in position 975. The confidence interval is $P_{2.5} < \mu < P_{97.5}$.

Demonstration Results

Figure 8-9 shows a histogram of the 1000 $\bar{x}$ values from one bootstrap simulation. Three bootstrap simulations produced the following 95% confidence intervals:

13.90 to 23.90
14.00 to 24.15
14.05 to 23.80

Using the t distribution on the sample data, Minitab produced the interval 13.33 to 24.27. The results of the bootstrap simulations and the t distribution method are quite close.

FIGURE 8-9 Bootstrap Simulation, $\bar{x}$ Distribution

9.1 Introduction to Statistical Tests

9.2 Testing the Mean μ

9.3 Testing a Proportion p

Mary Evans Picture Library/Arthur Rackham/
The Image Works

Sam Abell/National
Geographic/Getty Images

"Would you tell me, please, which way I ought to go from here?"

"That depends a good deal on where you want to get to," said the Cat.

"I don't much care where—" said Alice.

"Then it doesn't matter which way you go," said the Cat.

—LEWIS CARROLL
Alice's Adventures in Wonderland

Charles Lutwidge Dodgson (1832–1898) was an English mathematician who loved to write children's stories in his free time. The dialogue between Alice and the Cheshire Cat occurs in the masterpiece *Alice's Adventures in Wonderland*, written by Dodgson under the pen name Lewis Carroll. These lines relate to our study of hypothesis testing. Statistical tests cannot answer all of life's questions. They cannot always tell us "where to go," but after this decision is made on other grounds, they can help us find the best way to get there.

HYPOTHESIS TESTING

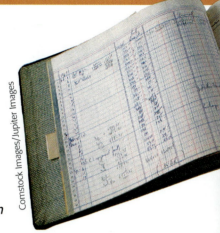

PREVIEW QUESTIONS

Many of life's questions require a yes or no answer. When you must act on incomplete (sample) information, how do you decide whether to accept or reject a proposal? (SECTION 9.1)

What is the P-value of a statistical test? What does this measurement have to do with performance reliability? (SECTION 9.1)

How do you construct statistical tests for μ? Does it make a difference whether σ is known or unknown? (SECTION 9.2)

How do you construct statistical tests for the proportion p of successes in a binomial experiment? (SECTION 9.3)

FOCUS PROBLEM

Benford's Law: The Importance of Being Number 1

Benford's Law states that in a wide variety of circumstances, numbers have "1" as their first nonzero digit disproportionately often. Benford's Law applies to such diverse topics as the drainage areas of rivers; properties of chemicals; populations of towns; figures in newspapers, magazines, and government reports; and the half-lives of radioactive atoms!

Specifically, such diverse measurements begin with "1" about 30% of the time, with "2" about 18% of time, and with "3" about 12.5% of the time. Larger digits occur less often. For example, less than 5% of the numbers in circumstances such as these begin with the digit 9. This is in dramatic contrast to a random sampling situation, in which each of the digits 1 through 9 has an equal chance of appearing.

The first nonzero digits of numbers taken from large bodies of numerical records such as tax returns, population studies, government records, and so forth, show the probabilities of occurrence as displayed in the table on the next page.

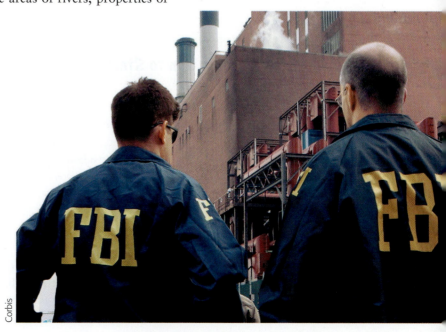

First nonzero digit	1	2	3	4	5	6	7	8	9
Probability	0.301	0.176	0.125	0.097	0.079	0.067	0.058	0.051	0.046

More than 100 years ago, the astronomer Simon Newcomb noticed that books of logarithm tables were much dirtier near the fronts of the tables. It seemed that people were more frequently looking up numbers with a low first digit. This was regarded as an odd phenomenon and a strange curiosity. The phenomenon was rediscovered in 1938 by physicist Frank Benford (hence the name "Benford's Law").

More recently, Ted Hill, a mathematician at the Georgia Institute of Technology, studied situations that might demonstrate Benford's Law. Professor Hill showed that such probability distributions are likely to occur when we have a "distribution of distributions." Put another way, large random collections of random samples tend to follow Benford's Law. This seems to be especially true for samples taken from large government data banks, accounting reports for large corporations, large collections of astronomical observations, and so forth. For more information, see *American Scientist*, Vol. 86, pp. 358–363, and *Chance*, American Statistical Association, Vol. 12, No. 3, pp. 27–31.

Can Benford's Law be applied to help solve a real-world problem? Well, one application might be accounting fraud! Suppose the first nonzero digits of the entries in the accounting records of a large corporation (such as Enron or WorldCom) do not follow Benford's Law. Should this set off an accounting alarm for the FBI or the stockholders? How "significant" would this be? Such questions are the subject of statistics.

In Section 9.3, you will see how to use sample data to test whether the proportion of first nonzero digits of the entries in a large accounting report follows Benford's Law. Problems 7 and 8 of Section 9.3 relate to Benford's Law and accounting discrepancies. In one problem, you are asked to use sample data to determine if accounting books have been "cooked" by "pumping numbers up" to make the company look more attractive or perhaps to provide a cover for money laundering. In the other problem, you are asked to determine if accounting books have been "cooked" by artificially lowered numbers, perhaps to hide profits from the Internal Revenue Service or to divert company profits to unscrupulous employees. (See Problems 7 and 8 of Section 9.3.)

SECTION 9.1   Introduction to Statistical Tests

FOCUS POINTS

- Understand the rationale for statistical tests.
- Identify the null and alternate hypotheses in a statistical test.
- Identify right-tailed, left-tailed, and two-tailed tests.
- Use a test statistic to compute a *P*-value.
- Recognize types of errors, level of significance, and power of a test.
- Understand the meaning and risks of rejecting or not rejecting the null hypothesis.

In Chapter 1, we emphasized the fact that one of a statistician's most important jobs is to draw inferences about populations based on samples taken from the populations. Most statistical inference centers around the parameters of a population (often the mean or the probability of success in a binomial trial). Methods for drawing inferences about parameters are of two types: Either we make decisions concerning the value of the parameter, or we actually estimate the value of the parameter. When we estimate the value (or location) of a parameter, we are using methods of estimation such as those studied in Chapter 8. Decisions concerning the value of a parameter are obtained by *hypothesis testing*, the topic we shall study in this chapter.

Students often ask which method should be used on a particular problem—that is, should the parameter be estimated, or should we test a *hypothesis* involving the parameter? The answer lies in the practical nature of the problem and the questions posed about it. Some people prefer to test theories concerning the parameters. Others prefer to express their inferences as estimates. Both estimation and hypothesis testing are found extensively in the literature of statistical applications.

Stating Hypotheses

Our first step is to establish a working hypothesis about the population parameter in question. This hypothesis is called the *null hypothesis*, denoted by the symbol H_0. The value specified in the null hypothesis is often a historical value, a claim, or a production specification. For instance, if the average height of a professional male basketball player was 6.5 feet 10 years ago, we might use a null hypothesis H_0: $\mu = 6.5$ feet for a study involving the average height of this year's professional male basketball players. If television networks claim that the average length of time devoted to commercials in a 60-minute program is 12 minutes, we would use H_0: $\mu = 12$ minutes as our null hypothesis in a study regarding the average length of time devoted to commercials. Finally, if a repair shop claims that it should take an average of 25 minutes to install a new muffler on a passenger automobile, we would use H_0: $\mu = 25$ minutes as the null hypothesis for a study of how well the repair shop is conforming to specified average times for a muffler installation.

Any hypothesis that differs from the null hypothesis is called an *alternate hypothesis*. An alternate hypothesis is constructed in such a way that it is the hypothesis to be accepted when the null hypothesis must be rejected. The alternate hypothesis is denoted by the symbol H_1. For instance, if we believe the average height of professional male basketball players is taller than it was 10 years ago, we would use an alternate hypothesis H_1: $\mu > 6.5$ feet with the null hypothesis H_0: $\mu = 6.5$ feet.

Null hypothesis H_0: This is the statement that is under investigation or being tested. Usually the null hypothesis represents a statement of "no effect," "no difference," or, put another way, "things haven't changed."
Alternate hypothesis H_1: This is the statement you will adopt in the situation in which the evidence (data) is so strong that you reject H_0. A statistical test is designed to assess the strength of the evidence (data) against the null hypothesis.

EXAMPLE 1   NULL AND ALTERNATE HYPOTHESES

A car manufacturer advertises that its new subcompact models get 47 miles per gallon (mpg). Let μ be the mean of the mileage distribution for these cars. You assume that the manufacturer will not underrate the car, but you suspect that the mileage might be overrated.

(a) What shall we use for H_0?

SOLUTION: We want to see if the manufacturer's claim that $\mu = 47$ mpg can be rejected. Therefore, our null hypothesis is simply that $\mu = 47$ mpg. We denote the null hypothesis as

H_0: $\mu = 47$ mpg

(margin notes)

Hypothesis testing

Hypothesis

Null hypothesis, H_0

Alternate hypothesis, H_1

(b) What shall we use for H_1?

> **SOLUTION:** From experience with this manufacturer, we have every reason to believe that the advertised mileage is too high. If μ is not 47 mpg, we are sure it is less than 47 mpg. Therefore, the alternate hypothesis is
>
> H_1: $\mu < 47$ mpg

GUIDED EXERCISE 1 | **Null and alternate hypotheses**

A company manufactures ball bearings for precision machines. The average diameter of a certain type of ball bearing should be 6.0 mm. To check that the average diameter is correct, the company formulates a statistical test.

(a) What should be used for H_0? (*Hint:* What is the company trying to test?)

➡ If μ is the mean diameter of the ball bearings, the company wants to test whether $\mu = 6.0$ mm. Therefore, H_0: $\mu = 6.0$ mm.

(b) What should be used for H_1? (*Hint:* An error either way, too small or too large, would be serious.)

➡ An error either way could occur, and it would be serious. Therefore, H_1: $\mu \neq 6.0$ mm (μ is either smaller than or larger than 6.0 mm).

COMMENT: NOTATION REGARDING THE NULL HYPOTHESIS In statistical testing, the null hypothesis H_0 always contains the equals symbol. However, in the null hypothesis, some statistical software packages and texts also include the inequality symbol that is the opposite of the symbol shown in the alternate hypothesis. For instance, if the alternate hypothesis is "μ is less than 3" ($\mu < 3$), then the corresponding null hypothesis is sometimes written as "μ is greater than or equal to 3" ($\mu \geq 3$). The mathematical construction of a statistical test uses the null hypothesis to assign a specific number (rather than a range of numbers) to the parameter μ in question. The null hypothesis establishes a single fixed value for μ, so we are working with a single distribution having a specific mean. In this case, H_0 assigns $\mu = 3$. So when H_1: $\mu < 3$ is the alternate hypothesis, we follow the commonly used convention of writing the null hypothesis simply as H_0: $\mu = 3$.

Types of Tests

The null hypothesis H_0 always states that the parameter of interest *equals* a specified value. The alternate hypothesis H_1 states that the parameter is *less than*, *greater than*, or simply *not equal to* the same value. We categorize a statistical test as *left-tailed*, *right-tailed*, or *two-tailed* according to the alternate hypothesis.

Types of statistical tests

A statistical test is:

Left-tailed test   **left-tailed** if H_1 states that the parameter is less than the value claimed in H_0

Right-tailed test   **right-tailed** if H_1 states that the parameter is greater than the value claimed in H_0

Two-tailed test   **two-tailed** if H_1 states that the parameter is different from (or not equal to) the value claimed in H_0

TABLE 9-1	**Null and Alternate Hypotheses for Tests of the Mean** μ

Null Hypothesis	Alternate Hypotheses and Types of Test		
Claim about μ or historical value of μ $H_0: \mu = k$	You believe that μ is less than the value stated in H_0. $H_1: \mu < k$ Left-tailed test	You believe that μ is more than the value stated in H_0. $H_1: \mu > k$ Right-tailed test	You believe that μ is different from the value stated in H_0. $H_1: \mu \neq k$ Two-tailed test

In this introduction to statistical tests, we discuss tests involving a population mean μ. However, you should keep an open mind and be aware that the methods outlined apply to testing other parameters as well (e.g., p, σ, $\mu_1 - \mu_2$, $p_1 - p_2$, and so on). Table 9-1 shows how tests of the mean μ are categorized.

Hypothesis Tests of μ, Given That x Is Normal and σ Is Known

Once you have selected the null and alternate hypotheses, how do you decide which hypothesis is likely to be valid? Data from a simple random sample and the sample test statistic, together with the corresponding sampling distribution of the test statistic, will help you decide. Example 2 leads you through the decision process.

Parameter

First, a quick review of Section 7.4 is in order. Recall that a population *parameter* is a numerical descriptive measurement of the entire population. Examples of population parameters are μ, p, and σ. It is important to remember that for a given population, the parameters are *fixed* values. They do not vary! The null hypothesis H_0 makes a statement about a population parameter.

Statistic

A *statistic* is a numerical descriptive measurement of a sample. Examples of statistics are $\bar{x}$, $\hat{p}$, and s. Statistics usually *vary* from one sample to the next. The probability distribution of the statistic we are using is called a *sampling distribution*.

Sample test statistic for μ, given that x is normal and σ is known

For hypothesis testing, we take a *simple* random sample and compute a *sample test statistic* corresponding to the parameter in H_0. Based on the sampling distribution of the statistic, we can assess how compatible the sample test statistic is with H_0.

In this section, we use hypothesis tests about the mean to introduce the concepts and vocabulary of hypothesis testing. In particular, let's suppose that x has *a normal distribution* with mean μ and standard deviation σ. Then, Theorem 7.1 tells us that $\bar{x}$ has a *normal distribution* with mean μ and standard deviation $\sigma/\sqrt{n}$.

PROCEDURE

Requirements The x distribution is *normal* with known standard deviation σ.

Then $\bar{x}$ has a normal distribution. The standardized test statistic is

$$\text{test statistic} = z = \frac{\bar{x} - \mu}{\sigma/\sqrt{n}}$$

where $\bar{x}$ = mean of a simple random sample

μ = value stated in H_0

n = sample size

EXAMPLE 2   STATISTICAL TESTING PREVIEW

Rosie is an aging sheep dog in Montana who gets regular checkups from her owner, the local veterinarian. Let x be a random variable that represents Rosie's resting heart rate (in beats per minute). From past experience, the vet knows that

x has a normal distribution with $\sigma = 12$. The vet checked *The Merck Veterinary Manual* and found that for dogs of this breed, $\mu = 115$ beats per minute.

Over the past six weeks, Rosie's heart rate (beats/min) measured

93	109	110	89	112	117

The sample mean is $\bar{x} = 105.0$. The vet is concerned that Rosie's heart rate may be slowing. Do the data indicate that this is the case?

SOLUTION:

(a) Establish the null and alternate hypotheses.

If "nothing has changed" from Rosie's earlier life, then her heart rate should be nearly average. This point of view is represented by the null hypothesis

$$H_0: \mu = 115$$

However, the vet is concerned about Rosie's heart rate slowing. This point of view is represented by the alternate hypothesis

$$H_1: \mu < 115$$

(b) Are the observed sample data compatible with the null hypothesis?

Are the six observations of Rosie's heart rate compatible with the null hypothesis $H_0: \mu = 115$? To answer this question, we need to know the *probability* of obtaining a sample mean of 105.0 or less from a population with true mean $\mu = 115$. If this probability is small, we conclude that $H_0: \mu = 115$ is not the case. Rather, $H_1: \mu < 115$ and Rosie's heart rate is slowing.

(c) How do we compute the probability in part (b)?

Well, you probably guessed it! We use the sampling distribution for $\bar{x}$ and compute $P(\bar{x} < 105.0)$. Figure 9-1 shows the $\bar{x}$ distribution and the corresponding standard normal distribution with the desired probability shaded.

Check Requirements Since x has a normal distribution, $\bar{x}$ will also have a normal distribution for any sample size n and given σ (see Theorem 7.1).

Note that using $\mu = 115$ from H_0, $\sigma = 12$, and $n = 6$, the sample mean $\bar{x} = 105.0$ converts to

$$\text{test statistic} = z = \frac{\bar{x} - \mu}{\sigma/\sqrt{n}} = \frac{105.0 - 115}{12/\sqrt{6}} \approx -2.04$$

Using the standard normal distribution table, we find that

$$P(\bar{x} < 105.0) = P(z < -2.04) = 0.0207$$

P-value

The area in the left tail that is more extreme than $\bar{x} = 105.0$ is called the *P-value* of the test. In this example, P-value = 0.0207. We will learn more about P-values later.

FIGURE 9-1

Sampling Distribution for $\bar{x}$ and Corresponding z Distribution

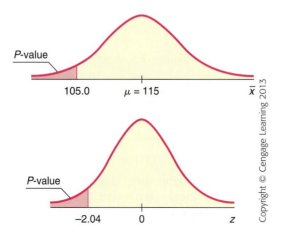

P-value

105.0 $\mu = 115$ $\bar{x}$

P-value

−2.04   0 z

(d) *Interpretation* What conclusion can be drawn about Rosie's average heart rate?

    If $H_0: \mu = 115$ is in fact true, the probability of getting a sample mean of $\bar{x} \leq 105.0$ is only about 2%. Because this probability is small, we reject $H_0: \mu = 115$ and conclude that $H_1: \mu < 115$. Rosie's average heart rate seems to be slowing.

(e) Have we proved $H_0: \mu = 115$ to be false and $H_1: \mu < 115$ to be true?

    No! The sample data do not *prove* H_0 to be false and H_1 to be true! We say that H_0 has been "discredited" by a small *P*-value of 0.0207. Therefore, we abandon the claim $H_0: \mu = 115$ and adopt the claim $H_1: \mu < 115$.

The *P*-value of a Statistical Test

Rosie the sheep dog has helped us to "sniff out" an important statistical concept.

> **P-value**
>
> Assuming H_0 is true, the *probability* that the test statistic will take on values as extreme as or more extreme than the observed test statistic (computed from sample data) is called the **P-value** of the test. The smaller the *P*-value computed from sample data, the stronger the evidence against H_0.

    The *P*-value, sometimes called the *probability of chance*, can be thought of as the probability that the results of a statistical experiment are due only to chance. The lower the *P*-value, the greater the likelihood of obtaining the same (or very similar) results in a repetition of the statistical experiment. Thus, a low *P*-value is a good indication that your results are not due to random chance alone.

    The *P*-value associated with the observed test statistic takes on different values depending on the alternate hypothesis and the type of test. Let's look at *P*-values and types of tests when the test involves the mean and standard normal distribution. Notice that in Example 2, part (c), we computed a *P*-value for a left-tailed test. Guided Exercise 3 asks you to compute a *P*-value for a two-tailed test.

P-values and types of tests

Let $z_{\bar{x}}$ represent the standardized sample test statistic for testing a mean μ using the standard normal distribution. That is, $z_{\bar{x}} = (\bar{x} - \mu)/(\sigma/\sqrt{n})$.

I. Left-tailed Test
$H_0: \mu = k$ $H_1: \mu < k$

Copyright © Cengage Learning 2013

P-value $= P(z < z_{\bar{x}})$

This is the probability of getting a test statistic as low as or lower than $z_{\bar{x}}$.

Continued

II. Right-tailed Test

$H_0: \mu = k$ $H_1: \mu > k$

P-value $= P(z > z_{\bar{x}})$

This is the probability of getting a test statistic as high as or higher than $z_{\bar{x}}$.

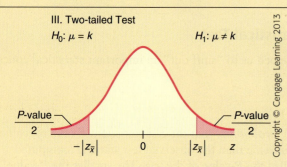

III. Two-tailed Test

$H_0: \mu = k$ $H_1: \mu \neq k$

$$\frac{P\text{-value}}{2} = P(z > |z_{\bar{x}}|); \text{ therefore,}$$

$$P\text{-value} = 2P(z > |z_{\bar{x}}|)$$

This is the probability of getting a test statistic either lower than $-|z_{\bar{x}}|$ or higher than $|z_{\bar{x}}|$.

Copyright © Cengage Learning 2013

Types of Errors

Type I error
Type II error

If we *reject the null hypothesis when it is, in fact, true*, we have made an error that is called a *type I error*. On the other hand, if we *accept the null hypothesis when it is in fact false*, we have made an error that is called a *type II error*. Table 9-2 indicates how these errors occur.

For tests of hypotheses to be well constructed, they must be designed to minimize possible errors of decision. (Usually, we do not know if an error has been made, and therefore, we can talk only about the probability of making an error.) Usually, for a given sample size, an attempt to reduce the probability of one type of error results in an increase in the probability of the other type of error. In practical applications, one type of error may be more serious than another. In such a case, careful attention is given to the more serious error. If we increase the sample size, it is possible to reduce both types of errors, but increasing the sample size may not be possible.

Good statistical practice requires that we announce in advance how much evidence against H_0 will be required to reject H_0. The probability with which we are willing to risk a type I error is called the *level of significance* of a test. The level of significance is denoted by the Greek letter α (pronounced "alpha").

Level of significance, α

The **level of significance** α is the probability of rejecting H_0 when it is true. This is the probability of a type I error.

TABLE 9-2   **Type I and Type II Errors**

	Our Decision	
Truth of H_0	And if we do not reject H_0	And if we reject H_0
If H_0 is true	Correct decision; no error	Type I error
If H_0 is false	Type II error	Correct decision; no error

Copyright © Cengage Learning 2013

TABLE 9-3	**Probabilities Associated with a Statistical Test**	
	Our Decision	
Truth of H_0	**And if we accept H_0 as true**	**And if we reject H_0 as false**
If H_0 is true	Correct decision, with corresponding probability $1 - \alpha$	Type I error, with corresponding probability α, called the *level of significance of the test*
If H_0 is false	Type II error, with corresponding probability β	Correct decision, with corresponding probability $1 - \beta$, called the *power of the test*

Copyright © Cengage Learning 2013

Probability of a type II error, β

The *probability of making a type II error* is denoted by the Greek letter β (pronounced "beta").

Methods of hypothesis testing require us to choose α and β values to be as small as possible. In elementary statistical applications, we usually choose α first.

Power of a test, $(1 - \beta)$

The quantity $1 - \beta$ is called the *power of a test* and represents the probability of rejecting H_0 when it is, in fact, false.

For a given level of significance, how much power can we expect from a test? The actual value of the power is usually difficult (and sometimes impossible) to obtain, since it requires us to know the H_1 distribution. However, we can make the following general comments:

1. The power of a statistical test increases as the level of significance α increases. A test performed at the $\alpha = 0.05$ level has more power than one performed at $\alpha = 0.01$. This means that the less stringent we make our significance level α, the more likely we will be to reject the null hypothesis when it is false.

2. Using a larger value of α will increase the power, but it also will increase the probability of a type I error. Despite this fact, most business executives, administrators, social scientists, and scientists use *small* α values. This choice reflects the conservative nature of administrators and scientists, who are usually more willing to make an error by failing to reject a claim (i.e., H_0) than to make an error by accepting another claim (i.e., H_1) that is false. Table 9-3 summarizes the probabilities of errors associated with a statistical test.

COMMENT   Since the calculation of the probability of a type II error is treated in advanced statistics courses, we will restrict our attention to the probability of a type I error.

GUIDED EXERCISE 2 | **Types of errors**

Let's reconsider Guided Exercise 1, which dealt with the manufacturing specifications for the diameter of ball bearings. The hypotheses were

H_0: $\mu = 6.0$ mm (manufacturer's specification) H_1: $\mu \neq 6.0$ mm (cause for adjusting process)

(a) Suppose the manufacturer requires a 1% level of significance. Describe a type I error, its consequence, and its probability.   A type I error is caused when sample evidence indicates that we should reject H_0 when in fact the average diameter of the ball bearings being produced is 6.0 mm. A type I error will cause a needless

Continued

adjustment and delay of the manufacturing process. The probability of such an error is 1% because $\alpha = 0.01$.

(b) Discuss a type II error and its consequences.

➡️ A type II error occurs if the sample evidence tells us not to reject the null hypothesis H_0: $\mu = 6.0$ mm when in fact the average diameter of the ball bearing is either too large or too small to meet specifications. Such an error would mean that the production process would not be adjusted even though it really needed to be adjusted. This could possibly result in a large production of ball bearings that do not meet specifications.

Concluding a Statistical Test

Usually, α is specified in advance before any samples are drawn, so that results will not influence the choice for the level of significance. To conclude a statistical test, we compare our α value with the *P*-value computed using sample data and the sampling distribution.

PROCEDURE

> ### HOW TO CONCLUDE A TEST USING THE *P*-VALUE AND LEVEL OF SIGNIFICANCE α
>
> Statistical significance
>
> If *P*-value $\leq \alpha$, we **reject** the null hypothesis and say the data are **statistically significant** at the level α.
>
> If *P*-value $> \alpha$, we **do not reject** the null hypothesis.

In what sense are we using the word *significant*? *Webster's Dictionary* gives two interpretations of *significance*: (1) having or signifying *meaning*, or (2) being important or momentous.

In statistical work, significance does not necessarily imply momentous importance. For us, "significant" at the α level has a special meaning. It says that at the α level of risk, the evidence (sample data) against the null hypothesis H_0 is sufficient to discredit H_0, so we adopt the alternate hypothesis H_1.

In any case, we do not claim that we have "proved" or "disproved" the null hypothesis H_0. We can say that the probability of a type I error (rejecting H_0 when it is in fact true) is α.

> **Basic components of a statistical test**
>
> A statistical test can be thought of as a package of five basic ingredients.
>
> 1. **Null hypothesis H_0, alternate hypothesis H_1, and preset level of significance α**
>
>   If the evidence (sample data) against H_0 is strong enough, we reject H_0 and adopt H_1. The level of significance α is the probability of rejecting H_0 when it is in fact true.
>
> 2. **Test statistic and sampling distribution**
>
>   These are mathematical tools used to measure the compatibility of sample data and the null hypothesis.

3. **P-value**

This is the probability of obtaining a test statistic from the sampling distribution that is as extreme as, or more extreme (as specified by H_1) than, the sample test statistic computed from the data under the assumption that H_0 is true.

4. **Test conclusion**

If P-value $\leq \alpha$, we reject H_0 and say that the data are significant at level α. If P-value $> \alpha$, we do not reject H_0.

5. **Interpretation of the test results**

Give a simple explanation of your conclusions in the context of the application.

GUIDED EXERCISE 3 | ## Constructing a statistical test for μ (normal distribution)

The Environmental Protection Agency has been studying Miller Creek regarding ammonia nitrogen concentration. For many years, the concentration has been 2.3 mg/L. However, a new golf course and new housing developments are raising concern that the concentration may have changed because of lawn fertilizer. Any change (either an increase or a decrease) in the ammonia nitrogen concentration can affect plant and animal life in and around the creek (Reference: *EPA Report 832-R-93-005*). Let x be a random variable representing ammonia nitrogen concentration (in mg/L). Based on recent studies of Miller Creek, we may assume that x has a normal distribution with $\sigma = 0.30$. Recently, a random sample of eight water tests from the creek gave the following x values:

2.1   2.5   2.2   2.8   3.0   2.2   2.4   2.9

The sample mean is $\bar{x} \approx 2.51$.

Let us construct a statistical test to examine the claim that the concentration of ammonia nitrogen has changed from 2.3 mg/L. Use level of significance $\alpha = 0.01$.

(a) What is the null hypothesis? What is the alternate hypothesis? What is the level of significance α?

$\implies$ H_0: $\mu = 2.3$
H_1: $\mu \neq 2.3$
$\alpha = 0.01$

(b) Is this a right-tailed, left-tailed, or two-tailed test? $\implies$ Since H_1: $\mu \neq 2.3$, this is a two-tailed test.

(c) *Check Requirements* What sampling distribution shall we use? Note that the value of μ is given in the null hypothesis, H_0.

$\implies$ Since the x distribution is normal and σ is known, we use the standard normal distribution with

$$z = \frac{\bar{x} - \mu}{\dfrac{\sigma}{\sqrt{n}}} = \frac{\bar{x} - 2.3}{\dfrac{0.3}{\sqrt{8}}}$$

(d) What is the sample test statistic? Convert the sample mean $\bar{x}$ to a standard z value.

$\implies$ The sample of eight measurements has mean $\bar{x} = 2.51$. Converting this measurement to z, we have

$$\text{test statistic} = z = \frac{2.51 - 2.3}{\dfrac{0.3}{\sqrt{8}}} \approx 1.98$$

Continued

(e) Draw a sketch showing the *P*-value area on the standard normal distribution. Find the *P*-value.

→ *P*-value = $2P(z > 1.98) = 2(0.0239) = 0.0478$.

FIGURE 9-2 *P*-value

$\dfrac{P\text{-value}}{2}$ $\dfrac{P\text{-value}}{2}$

−1.98   0   1.98   *z*

$\bar{x} = 2.51$
Sample Test Statistic

(f) Compare the level of significance α and the *P*-value. What is your conclusion?

→ Since *P*-value $0.0478 \geq 0.01$, we see that *P*-value $> \alpha$. We fail to reject H_0.

(g) *Interpret* your results in the context of this problem.

→ The sample data are not significant at the $\alpha = 1\%$ level. At this point in time, there is not enough evidence to conclude that the ammonia nitrogen concentration has changed in Miller Creek.

Interpretation of level of significance

In most statistical applications, the level of significance is specified to be $\alpha = 0.05$ or $\alpha = 0.01$, although other values can be used. If $\alpha = 0.05$, then we say we are using a 5% level of significance. This means that in 100 similar situations, H_0 will be rejected 5 times, on average, when it should not have been rejected.

Meaning of accepting H_0

Using Technology at the end of this chapter shows a simulation of this phenomenon.

When we accept (or fail to reject) the null hypothesis, we should understand that we are *not proving the null hypothesis*. We are saying only that the sample evidence (data) is not strong enough to justify rejection of the null hypothesis. The word *accept* sometimes has a stronger meaning in common English usage than we are willing to give it in our application of statistics. Therefore, we often use the expression *fail to reject H_0* instead of *accept H_0*. "*Fail to reject* the null hypothesis" simply means that the evidence in favor of rejection was not strong enough (see Table 9-4). Often, in the case that H_0 cannot be rejected, a confidence interval is used to estimate the parameter in question. The confidence interval gives the statistician a range of possible values for the parameter.

TABLE 9-4   **Meanings of the Terms *Fail to Reject H_0* and *Reject H_0***

Term	Meaning
Fail to reject H_0	There is not enough evidence in the data (and the test being used) to justify rejection of H_0. This means that we retain H_0 with the understanding that we have not proved it to be true beyond all doubt.
Reject H_0	There is enough evidence in the data (and the test employed) to justify rejection of H_0. This means that we choose the alternate hypothesis H_1 with the understanding that we have not proved H_1 to be true beyond all doubt.

Fail to reject H_0

Reject H_0

COMMENT   Some comments about *P*-values and level of significance α should be made. The level of significance α should be a fixed, prespecified value. Usually, α is chosen before any samples are drawn. The level of significance α is the probability of a type I error. So α is the probability of rejecting H_0 when in fact H_0 is true.

Interpreting the *P*-value of a test statistic

The *P*-value should *not* be interpreted as the probability of a type I error. The level of significance (in theory) is set in advance before any samples are drawn. The *P*-value cannot be set in advance, since it is determined from the random sample. The *P*-value, together with α, should be regarded as tools used to conclude the test. If *P*-value ≤ α, then reject H_0, and if *P*-value > α, then do not reject H_0.

In most computer applications and journal articles, only the *P*-value is given. It is understood that the person using this information will supply an appropriate level of significance α. From a historical point of view, the English statistician F. Y. Edgeworth (1845–1926) was one of the first to use the term *significant* to imply that the sample data indicate a "meaningful" difference from a previously held view.

In this book, we are using the most popular method of testing, which is called the *P-value method*. At the end of the next section, you will learn about another (equivalent) method of testing called the *critical region method*. An extensive discussion regarding the *P*-value method of testing versus the critical region method can be found in *The American Statistician*, Vol. 57, No. 3, pp. 171–178, American Statistical Association.

VIEWPOINT | Lovers, Take Heed!!!

If you are going to whisper sweet nothings to your sweetheart, be sure to whisper them in the left ear. Professor T. C. Sim of Sam Houston State University (Huntsville, Texas) found that emotionally loaded words have a higher recall rate when spoken into a person's left ear than into the right. Professor Sim presented his findings at the European Congress of Psychology. He told the congress that his findings are consistent with the hypothesis that the brain's right hemisphere has more influence in the processing of emotional stimuli. (The left ear is controlled by the right side of the brain.) Sim's research involved statistical tests like the ones you will study in this chapter.

SECTION 9.1 PROBLEMS

 aplia   Access end-of-section exercises at www.aplia.com

SECTION 9.2

Testing the Mean μ

FOCUS POINTS
- Review the general procedure for testing using *P*-values.
- Test μ when σ is known using the normal distribution.
- Test μ when σ is unknown using a Student's *t* distribution.
- Understand the "traditional" method of testing that uses critical regions and critical values instead of *P*-values.

In this section, we continue our study of testing the mean μ. The method we are using is called the *P*-value method. It was used extensively by the famous statistician R. A. Fisher and is the most popular method of testing in use today. At the end of this section, we present another method of testing called the *critical region method* (or *traditional method*). The critical region method was used extensively by the statisticians J. Neyman and E. Pearson. In recent years, the use of this method has been declining. It is important to realize that for a fixed, preset level of significance α, both methods are logically equivalent.

In Section 9.1, we discussed the vocabulary and method of hypothesis testing using *P*-values. Let's quickly review the basic process.

1. We first state a proposed value for a population parameter in the null hypothesis H_0. The alternate hypothesis H_1 states alternative values of the parameter, either $<$, $>$, or $\neq$ the value proposed in H_0. We also set the level of significance α. This is the risk we are willing to take of committing a type I error. That is, α is the probability of rejecting H_0 when it is in fact true.

2. We use a corresponding sample statistic from a simple random sample to challenge the statement made in H_0. We convert the sample statistic to a test statistic, which is the corresponding value of the appropriate sampling distribution.

3. We use the sampling distribution of the test statistic and the type of test to compute the *P*-value of this statistic. Under the assumption that the null hypothesis is true, the *P*-value is the probability of getting a sample statistic as extreme as or more extreme than the observed statistic from our random sample.

4. Next, we conclude the test. If the *P*-value is very small, we have evidence to reject H_0 and adopt H_1. What do we mean by "very small"? We compare the *P*-value to the preset level of significance α. If the *P*-value $\leq \alpha$, then we say that we have evidence to reject H_0 and adopt H_1. Otherwise, we say that the sample evidence is insufficient to reject H_0.

5. Finally, we *interpret* the results in the context of the application.

Knowing the sampling distribution of the sample test statistic is an essential part of the hypothesis testing process. For tests of μ, we use one of two sampling distributions for $\bar{x}$: the standard normal distribution or a Student's *t* distribution. As discussed in Chapters 7 and 8, the appropriate distribution depends upon our knowledge of the population standard deviation σ, the nature of the *x* distribution, and the sample size.

Part I: Testing μ When σ Is Known

In most real-world situations, σ is simply not known. However, in some cases a preliminary study or other information can be used to get a realistic and accurate value for σ.

PROCEDURE

HOW TO TEST μ WHEN σ IS KNOWN

Requirements

Let *x* be a random variable appropriate to your application. Obtain a simple random sample (of size *n*) of *x* values from which you compute the sample mean $\bar{x}$. The value of σ is already known (perhaps from a previous study). If you can assume that *x* has a normal distribution, then any sample size *n* will work. If you cannot assume this, then use a sample size $n \geq 30$.

Procedure

1. In the context of the application, state the *null and alternate hypotheses* and set the *level of significance α*.

2. Use the known σ, the sample size *n*, the value of $\bar{x}$ from the sample, and μ from the null hypothesis to compute the standardized *sample test statistic*.

$$z = \frac{\bar{x} - \mu}{\frac{\sigma}{\sqrt{n}}}$$

3. Use the standard normal distribution and the type of test, one-tailed or two-tailed, to find the *P-value* corresponding to the test statistic.

4. *Conclude* the test. If *P*-value $\leq \alpha$, then reject H_0. If *P*-value $> \alpha$, then do not reject H_0.

5. *Interpret your conclusion* in the context of the application.

In Section 9.1, we examined *P*-value tests for normal distributions with relatively small sample sizes ($n < 30$). The next example does not assume a normal distribution, but has a large sample size ($n \geq 30$).

EXAMPLE 3

TESTING μ WITH σ KNOWN

Sunspots have been observed for many centuries. Records of sunspots from ancient Persian and Chinese astronomers go back thousands of years. Some archaeologists think sunspot activity may somehow be related to prolonged periods of drought in the southwestern United States. Let x be a random variable representing the average number of sunspots observed in a four-week period. A random sample of 40 such periods from Spanish colonial times gave the following data (Reference: M. Waldmeier, *Sun Spot Activity*, International Astronomical Union Bulletin):

12.5	14.1	37.6	48.3	67.3	70.0	43.8	56.5	59.7	24.0
12.0	27.4	53.5	73.9	104.0	54.6	4.4	177.3	70.1	54.0
28.0	13.0	6.5	134.7	114.0	72.7	81.2	24.1	20.4	13.3
9.4	25.7	47.8	50.0	45.3	61.0	39.0	12.0	7.2	11.3

The sample mean is $\bar{x} \approx 47.0$. Previous studies of sunspot activity during this period indicate that $\sigma = 35$. It is thought that for thousands of years, the mean number of sunspots per four-week period was about $\mu = 41$. Sunspot activity above this level may (or may not) be linked to gradual climate change. Do the data indicate that the mean sunspot activity during the Spanish colonial period was higher than 41? Use $\alpha = 0.05$.

SOLUTION:

(a) Establish the null and alternate hypotheses.
Since we want to know whether the average sunspot activity during the Spanish colonial period was higher than the long-term average of $\mu = 41$,

$H_0: \mu = 41$ and $H_1: \mu > 41$

(b) *Check Requirements* What distribution do we use for the sample test statistic?
Compute the test statistic from the sample data. Since $n \geq 30$ and we know σ, we use the standard normal distribution. Using $\bar{x} = 47$ from the sample, $\sigma = 35$, $\mu = 41$ from H_0, and $n = 40$,

$$z = \frac{\bar{x} - \mu}{\sigma/\sqrt{n}} \approx \frac{47 - 41}{35/\sqrt{40}} \approx 1.08$$

(c) Find the *P*-value of the test statistic.
Figure 9-3 shows the *P*-value. Since we have a right-tailed test, the *P*-value is the area to the right of $z = 1.08$ shown in Figure 9-3. Using Table 3 of the Appendix, we find that

P-value $= P(z > 1.08) \approx 0.1401$

FIGURE 9-3

P-value Area

Copyright © Cengage Learning 2013

(d) Conclude the test.
   Since the *P*-value of 0.1401 > 0.05 for α, we do not reject H_0.

(e) *Interpretation* Interpret the results in the context of the problem.
   At the 5% level of significance, the evidence is not sufficient to reject H_0. Based on the sample data, we do not think the average sunspot activity during the Spanish colonial period was higher than the long-term mean.

Part II: Testing μ When σ Is Unknown

In many real-world situations, you have only a random sample of data values. In addition, you may have some limited information about the probability distribution of your data values. Can you still test μ under these circumstances? In most cases, the answer is yes!

PROCEDURE

HOW TO TEST μ WHEN σ IS UNKNOWN

Requirements

Let x be a random variable appropriate to your application. Obtain a simple random sample (of size n) of x values from which you compute the sample mean $\bar{x}$ and the sample standard deviation s. If you can assume that x has a normal distribution or simply a mound-shaped and symmetrical distribution, then any sample size n will work. If you cannot assume this, use a sample size $n \geq 30$.

Procedure

1. In the context of the application, state the *null and alternate hypotheses* and set the *level of significance* α.

2. Use $\bar{x}$, s, and n from the sample, with μ from H_0, to compute the *sample test statistic*.

d.f. for testing μ when σ is unknown

$$t = \frac{\bar{x} - \mu}{\dfrac{s}{\sqrt{n}}} \text{ with degrees of freedom } d.f. = n - 1$$

3. Use the Student's t distribution and the type of test, one-tailed or two-tailed, to find (or estimate) the *P-value* corresponding to the test statistic.

4. *Conclude* the test. If *P*-value $\leq \alpha$, then reject H_0. If *P*-value $> \alpha$, then do not reject H_0.

5. *Interpret your conclusion* in the context of the application.

Copyright © Cengage Learning 2013

Using the Student's *t* table to estimate *P*-values

In Section 8.2, we used Table 4 of the Appendix, "Critical Values for Student's *t* Distribution," to find critical values t_c for confidence intervals. The critical values are in the body of the table. We find *P*-values in the *rows* headed by "one-tail area" and "two-tail area," depending on whether we have a one-tailed or two-tailed test. If the test statistic *t* for the sample statistic $\bar{x}$ is negative, look up the *P*-value for the corresponding *positive* value of *t* (i.e., look up the *P*-value for $|t|$).

Note: In Table 4, areas are given in *one tail* beyond positive *t* on the right or negative *t* on the left, and in *two tails* beyond $\pm t$. Notice that in each column, the two-tail area = 2(one-tail area). Consequently, we use *one-tail areas* as endpoints of the interval containing the *P*-value for *one-tailed tests*. We use *two-tail areas* as endpoints of the interval containing the *P*-value for *two-tailed tests*. (See Figure 9-4.)

Example 4 and Guided Exercise 4 show how to use Table 4 of the Appendix to find an interval containing the *P*-value corresponding to a test statistic *t*.

FIGURE 9-4

P-value for One-Tailed Tests and for Two-Tailed Tests

| Use *one-tail area* to estimate *P*-value for left-tailed tests | Use *one-tail area* to estimate *P*-value for right-tailed tests | Use *two-tail area* to estimate *P*-value for two-tailed tests |

EXAMPLE 4

TESTING μ WITH σ UNKNOWN

The drug 6-mP (6-mercaptopurine) is used to treat leukemia. The following data represent the remission times (in weeks) for a random sample of 21 patients using 6-mP (Reference: E. A. Gehan, University of Texas M. D. Anderson Cancer Center):

| 10 | 7 | 32 | 23 | 22 | 6 | 16 | 34 | 32 | 25 | 11 |
| 20 | 19 | 6 | 17 | 35 | 6 | 13 | 9 | 6 | 10 | |

The sample mean is $\bar{x} \approx 17.1$ weeks, with sample standard deviation $s \approx 10.0$ weeks. Let *x* be a random variable representing the remission time (in weeks) for all patients using 6-mP. Assume the *x* distribution is mound-shaped and symmetrical. A previously used drug treatment had a mean remission time of $\mu = 12.5$ weeks. Do the data indicate that the mean remission time using the drug 6-mP is different (either way) from 12.5 weeks? Use $\alpha = 0.01$.

SOLUTION:

(a) Establish the null and alternate hypotheses.
    Since we want to determine if the drug 6-mP provides a mean remission time that is different from that provided by a previously used drug having $\mu = 12.5$ weeks,

    H_0: $\mu = 12.5$ weeks and H_1: $\mu \neq 12.5$ weeks

(b) *Check Requirements* What distribution do we use for the sample test statistic *t*? Compute the sample test statistic from the sample data.
    The *x* distribution is assumed to be mound-shaped and symmetrical. Because we don't know σ, we use a Student's *t* distribution with *d.f.* = 20. Using $\bar{x} \approx 17.1$ and $s \approx 10.0$ from the sample data, $\mu = 12.5$ from H_0, and $n = 21$,

$$t \approx \frac{\bar{x} - \mu}{s/\sqrt{n}} \approx \frac{17.1 - 12.5}{10.0/\sqrt{21}} \approx 2.108$$

(c) Find the *P*-value or the interval containing the *P*-value.

Figure 9-5 shows the *P*-value. Using Table 4 of the Appendix, we find an interval containing the *P*-value. Since this is a two-tailed test, we use entries from the row headed by *two-tail area*. Look up the *t* value in the row headed by $d.f. = n - 1 = 21 - 1 = 20$. The sample statistic $t = 2.108$ falls between 2.086 and 2.528. The *P*-value for the sample *t* falls between the corresponding two-tail areas 0.050 and 0.020. (See Table 9-5.)

$$0.020 < P\text{-value} < 0.050$$

FIGURE 9-5

P-value

$\bar{x} = 17.1$
Sample Test Statistic

TABLE 9-5	Excerpt from Student's t Distribution (Table 4 of the Appendix)		
one-tail area		...	...
✓ two-tail area		0.050	0.020
$d.f. = 20$		2.086	2.528

↑
Sample $t = 2.108$

(d) Conclude the test.

The following diagram shows the interval that contains the single *P*-value corresponding to the test statistic. Note that there is just one *P*-value corresponding to the test statistic. Table 4 of the Appendix does not give that specific value, but it does give a range that contains that specific *P*-value. As the diagram shows, the entire range is greater than α. This means the specific *P*-value is greater than α, so we cannot reject H_0.

α
0.01   0.020   0.050

Note: Using the raw data, computer software gives *P*-value ≈ 0.048. This value is in the interval we estimated. It is larger than the α value of 0.01, so we do not reject H_0.

(e) *Interpretation* Interpret the results in the context of the problem.

At the 1% level of significance, the evidence is not sufficient to reject H_0. Based on the sample data, we cannot say that the drug 6-mP provides a different average remission time than the previous drug.

GUIDED EXERCISE 4 | Testing μ with σ unknown

Archaeologists become excited when they find an anomaly in discovered artifacts. The anomaly may (or may not) indicate a new trading region or a new method of craftsmanship. Suppose the lengths of projectile points (arrowheads) at a certain archaeological site have mean length $\mu = 2.6$ cm. A random sample of 61 recently discovered projectile points in an adjacent cliff dwelling gave the following lengths (in cm) (Reference: A. Woosley and A. McIntyre, *Mimbres Mogollon Archaeology*, University of New Mexico Press):

3.1	4.1	1.8	2.1	2.2	1.3	1.7	3.0	3.7	2.3	2.6	2.2	2.8	3.0
3.2	3.3	2.4	2.8	2.8	2.9	2.9	2.2	2.4	2.1	3.4	3.1	1.6	3.1
3.5	2.3	3.1	2.7	2.1	2.0	4.8	1.9	3.9	2.0	5.2	2.2	2.6	1.9
4.0	3.0	3.4	4.2	2.4	3.5	3.1	3.7	3.7	2.9	2.6	3.6	3.9	3.5
1.9	4.0	4.0	4.6	1.9									

Continued

GUIDED EXERCISE 4 *continued*

The sample mean is $\bar{x} \approx 2.92$ cm and the sample standard deviation is $s \approx 0.85$ cm, where x is a random variable that represents the lengths (in cm) of all projectile points found at the adjacent cliff-dwelling site. Do these data indicate that the mean length of projectile points in the adjacent cliff dwelling is longer than 2.6 cm? Use a 1% level of significance.

(a) State H_0, H_1, and α.

 H_0: $\mu = 2.6$ cm; H_1: $\mu > 2.6$ cm; $\alpha = 0.01$.

(b) *Check Requirements* What sampling distribution should you use? What is the value of the sample test statistic t?

Because $n \geq 30$ and σ is unknown, use the Student's t distribution with $d.f. = n - 1 = 61 - 1 = 60$. Using $\bar{x} \approx 2.92$, $s \approx 0.85$, $\mu = 2.6$ from H_0, and $n = 61$,

$$t = \frac{\bar{x} - \mu}{s/\sqrt{n}} \approx \frac{2.92 - 2.6}{0.85/\sqrt{61}} \approx 2.940$$

(c) When you use Table 4 of the Appendix to find an interval containing the *P*-value, do you use one-tail or two-tail areas? Why? Sketch a figure showing the *P*-value. Find an interval containing the *P*-value.

This is a right-tailed test, so use a one-tail area.

FIGURE 9-6 *P*-value

TABLE 9-6   Excerpt from Student's *t* Table

✓ one-tail area	...	0.005	0.0005
two-tail area	...	0.010	0.0010
$d.f. = 60$	...	2.660	3.460

Sample $t = 2.940$

Using $d.f. = 60$, we find that the sample $t = 2.940$ is between the critical values 2.660 and 3.460. The sample *P*-value is then between the one-tail areas 0.005 and 0.0005.

$0.0005 < P\text{-value} < 0.005$

(d) Do we reject or fail to reject H_0?

Since the interval containing the *P*-value lies to the left of $\alpha = 0.01$, we reject H_0.

Note: Using the raw data, computer software gives *P*-value ≈ 0.0022. This value is in our estimated range and is less than $\alpha = 0.01$, so we reject H_0.

(e) *Interpretation* Interpret your results in the context of the application.

At the 1% level of significance, sample evidence is sufficiently strong to reject H_0 and conclude that the average projectile point length at the adjacent cliff dwelling site is longer than 2.6 cm.

TECH NOTES   The TI-84Plus/TI-83Plus/TI-*n*spire calculators, Excel 2010, and Minitab all support testing of μ using the standard normal distribution. The TI-84Plus/TI-83Plus/TI-*n*spire calculators and Minitab support testing of μ using a Student's *t* distribution. All the technologies return a *P*-value for the test.

TI-84Plus/TI-83Plus/TI-*nspire* (with TI-84Plus keypad) You can select to enter raw data (**Data**) or summary statistics (**Stats**). Enter the value of μ_0 used in the null hypothesis $H_0: \mu = \mu_0$. Select the symbol used in the alternate hypothesis ($\neq \mu_0, < \mu_0, > \mu_0$). To test μ using the standard normal distribution, press **Stat**, select **Tests**, and use option **1:Z-Test.** The value for σ is required. To test μ using a Student's t distribution, use option **2:T-Test.** Using data from Example 4 regarding remission times, we have the following displays. The P-value is given as "p."

```
T-Test
 Inpt:Data Stats
 μ₀ :12.5
 List:L₁
 Freq:1
 μ: ≠μ₀ <μ₀ >μ₀
 Calculate Draw
```
Copyright © Cengage Learning 2013

```
T-Tests
 μ≠12.5
 t=2.105902924
 p=.0480466063
 x̄=17.0952381
 Sx=9.999523798
 n=21
```
Copyright © Cengage Learning 2013

Excel 2010 In Excel, the **Z.TEST** function finds the P-values for a right-tailed test. Click the ribbon choice **Insert Function** $\boxed{f_x}$. In the dialogue box, select **Statistical** for the category and **Z.TEST** for the function. In the next dialogue box, give the cell range containing your data for the array. Use the value of μ stated in H_0 for x. Provide σ. Otherwise, Excel uses the sample standard deviation computed from the data.

Minitab Enter the raw data from a sample. Use the menu selections **Stat ➤ Basic Stat ➤ 1-Sample z** for tests using the standard normal distribution. For tests of μ using a Student's t distribution, select **1-Sample t.**

Part III: Testing μ Using Critical Regions (Traditional Method)

The most popular method of statistical testing is the P-value method. For that reason, the P-value method is emphasized in this book. Another method of testing is called the *critical region method* or *traditional method.*

Critical region method

For a fixed, preset value of the level of significance α, both methods are logically equivalent. Because of this, we treat the traditional method as an "optional" topic and consider only the case of testing μ when σ is known.

Consider the null hypothesis $H_0: \mu = k$. We use information from a random sample, together with the sampling distribution for $\overline{x}$ and the level of significance α, to determine whether or not we should reject the null hypothesis. The essential question is "How much can $\overline{x}$ vary from $\mu = k$ before we suspect that $H_0: \mu = k$ is false and reject it?"

The answer to the question regarding the relative sizes of $\overline{x}$ and μ, as stated in the null hypothesis, depends on the sampling distribution of $\overline{x}$, the alternate hypothesis H_1, and the level of significance α. If the sample test statistic $\overline{x}$ is sufficiently different from the claim about μ made in the null hypothesis, we reject the null hypothesis.

The values of $\overline{x}$ for which we reject H_0 are called the *critical region* of the $\overline{x}$ distribution. Depending on the alternate hypothesis, the critical region is located on the left side, the right side, or both sides of the $\overline{x}$ distribution. Figure 9-7 shows the relationship of the critical region to the alternate hypothesis and the level of significance α.

Another method for concluding two-tailed tests involves the use of confidence intervals. Problems 25 and 26 at the end of this section discuss the confidence interval method.

Notice that the total area in the critical region is preset to be the level of significance α. This is *not* the P-value discussed earlier! In fact, you cannot set the P-value in advance because it is determined from a random sample. Recall that the level of significance α should (in theory) be a fixed, preset number assigned before drawing any samples.

FIGURE 9-7

Critical Regions for H_0: $\mu = k$

(a) H_1: $\mu < k$
     Left-tailed

Area $= \alpha$

k

Critical region

(b) H_1: $\mu > k$
     Right-tailed

Area $= \alpha$

k

Critical region

(c) H_1: $\mu \neq k$
     Two-tailed

Area $= \alpha/2$ Area $= \alpha/2$

k

Critical regions

The most commonly used levels of significance are $\alpha = 0.05$ and $\alpha = 0.01$. Critical regions of a standard normal distribution are shown for these levels of significance in Figure 9-8. *Critical values* are the boundaries of the critical region. Critical values designated as z_0 for the standard normal distribution are shown in Figure 9-8. For easy reference, they are also included in Table 3 of the Appendix, "Hypothesis Testing Critical Values z_0."

The procedure for hypothesis testing using critical regions follows the same first two steps as the procedure using P-values. However, instead of finding a P-value for the sample test statistic, we check if the sample test statistic falls in the critical region. If it does, we reject H_0. Otherwise, we do not reject H_0.

FIGURE 9-8

Critical Values z_0 for Tests Involving a Mean (Large Samples)

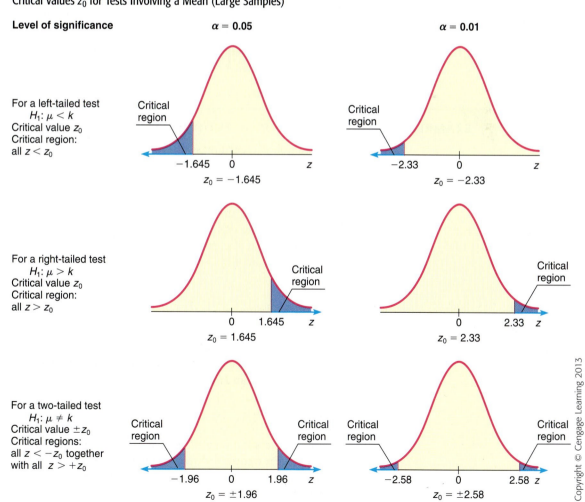

Level of significance $\alpha = 0.05$ $\alpha = 0.01$

For a left-tailed test
  H_1: $\mu < k$
Critical value z_0
Critical region:
all $z < z_0$

Critical region

-1.645 0 z

$z_0 = -1.645$

Critical region

-2.33 0 z

$z_0 = -2.33$

For a right-tailed test
  H_1: $\mu > k$
Critical value z_0
Critical region:
all $z > z_0$

Critical region

0 1.645 z

$z_0 = 1.645$

Critical region

0 2.33 z

$z_0 = 2.33$

For a two-tailed test
  H_1: $\mu \neq k$
Critical value $\pm z_0$
Critical regions:
all $z < -z_0$ together
with all $z > +z_0$

Critical region   Critical region

-1.96 0 1.96 z

$z_0 = \pm 1.96$

Critical region   Critical region

-2.58 0 2.58 z

$z_0 = \pm 2.58$

Critical values

PROCEDURE

HOW TO TEST μ WHEN σ IS KNOWN (CRITICAL REGION METHOD)

Requirements

Let x be a random variable appropriate to your application. Obtain a simple random sample (of size n) of x values from which you compute the sample mean $\bar{x}$. The value of σ is already known (perhaps from a previous study). If you can assume that x has a normal distribution, then any sample size n will work. If you cannot assume this, use a sample size $n \geq 30$. Then $\bar{x}$ follows a distribution that is normal or approximately normal.

Procedure

1. In the context of the application, state the *null and alternate hypotheses* and set the *level of significance* α. We use the most popular choices, $\alpha = 0.05$ or $\alpha = 0.01$.

2. Use the known σ, the sample size n, the value of $\bar{x}$ from the sample, and μ from the null hypothesis to compute the standardized sample *test statistic*.

$$z = \frac{\bar{x} - \mu}{\dfrac{\sigma}{\sqrt{n}}}$$

3. Show the *critical region* and *critical value(s)* on a graph of the sampling distribution. The level of significance α and the alternate hypothesis determine the locations of critical regions and critical values.

4. *Conclude* the test. If the test statistic z computed in Step 2 is in the critical region, then reject H_0. If the test statistic z is not in the critical region, then do not reject H_0.

5. *Interpret your conclusion* in the context of the application.

EXAMPLE 5

CRITICAL REGION METHOD OF TESTING μ

Consider Example 3 regarding sunspots. Let x be a random variable representing the number of sunspots observed in a four-week period. A random sample of 40 such periods from Spanish colonial times gave the number of sunspots per period. The raw data are given in Example 3. The sample mean is $\bar{x} \approx 47.0$. Previous studies indicate that for this period, $\sigma = 35$. It is thought that for thousands of years, the mean number of sunspots per four-week period was about $\mu = 41$. Do the data indicate that the mean sunspot activity during the Spanish colonial period was higher than 41? Use $\alpha = 0.05$.

SOLUTION:

(a) Set the null and alternate hypotheses.
   As in Example 3, we use H_0: $\mu = 41$ and H_1: $\mu > 41$.

(b) Compute the sample test statistic.
   As in Example 3, we use the standard normal distribution, with $\bar{x} = 47$, $\sigma = 35$, $\mu = 41$ from H_0, and $n = 40$.

$$z = \frac{\bar{x} - \mu}{\sigma/\sqrt{n}} \approx \frac{47 - 41}{35/\sqrt{40}} \approx 1.08$$

(c) Determine the critical region and critical value based on H_1 and $\alpha = 0.05$.
   Since we have a right-tailed test, the critical region is the rightmost 5% of the standard normal distribution. According to Figure 9-8, the critical value is $z_0 = 1.645$.

(d) Conclude the test.

We conclude the test by showing the critical region, critical value, and sample test statistic $z = 1.08$ on the standard normal curve. For a right-tailed test with $\alpha = 0.05$, the critical value is $z_0 = 1.645$. Figure 9-9 shows the critical region. As we can see, the sample test statistic does not fall in the critical region. Therefore, we fail to reject H_0.

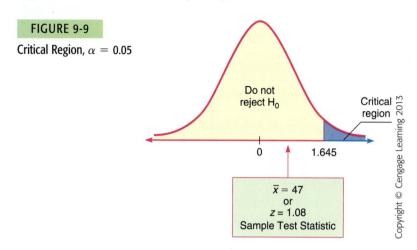

FIGURE 9-9

Critical Region, $\alpha = 0.05$

(e) *Interpretation* Interpret the results in the context of the application.

At the 5% level of significance, the sample evidence is insufficient to justify rejecting H_0. It seems that the average sunspot activity during the Spanish colonial period was the same as the historical average.

(f) How do the results of the critical region method compare to the results of the P-value method for a 5% level of significance?

The results, as expected, are the same. In both cases, we fail to reject H_0.

The critical region method of testing as outlined applies to tests of other parameters. As with the P-value method, you need to know the sampling distribution of the sample test statistic. Critical values for distributions are usually found in tables rather than in computer software outputs. For example, Table 4 of the Appendix provides critical values for Student's t distributions.

The critical region method of hypothesis testing is very general. The following procedure box outlines the process of concluding a hypothesis test using the critical region method:

PROCEDURE

HOW TO CONCLUDE TESTS USING THE CRITICAL REGION METHOD

1. Compute the sample test statistic using an appropriate sampling distribution.

2. Using the same sampling distribution, find the critical value(s) as determined by the level of significance α and the nature of the test: right-tailed, left-tailed, or two-tailed.

3. Compare the sample test statistic to the critical value(s).

   (a) For a right-tailed test:

      i. If sample test statistic $\geq$ critical value, reject H_0.

      ii. If sample test statistic $<$ critical value, fail to reject H_0.

Continued

(b) For a left-tailed test:

    i. If sample test statistic $\leq$ critical value, reject H_0.

    ii. If sample test statistic $>$ critical value, fail to reject H_0.

(c) For a two-tailed test:

    i. If sample test statistic lies at or beyond critical values, reject H_0.

    ii. If sample test statistic lies between critical values, fail to reject H_0.

VIEWPOINT | Predator or Prey?

Consider animals such as the arctic fox, gray wolf, desert lion, and South American jaguar. Each animal is a predator. What are the total sleep time (hours per day), maximum life span (years), and overall danger index from other animals? Now consider prey such as rabbits, deer, wild horses, and the Brazilian tapir (a wild pig). Is there a statistically significant difference in average sleep time, life span, and danger index? What about other variables, such as the ratio of brain weight to body weight or the sleep exposure index (sleeping in a well-protected den or out in the open)? How did prehistoric humans fit into this picture? Scientists have collected a lot of data, and a great deal of statistical work has been done regarding such questions. For more information, see the website at **http://lib.stat.cmu.edu/** *and follow the links to Datasets and then Sleep.*

SECTION 9.2 PROBLEMS

 aplia   Access end-of-section exercises at www.aplia.com

SECTION 9.3

Testing a Proportion p

FOCUS POINTS

- Identify the components needed for testing a proportion.
- Compute the sample test statistic.
- Find the *P*-value and conclude the test.

Many situations arise that call for tests of proportions or percentages rather than means. For instance, a college registrar may want to determine if the proportion of students wanting 3-week intensive courses has increased.

How can we make such a test? In this section, we will study tests involving proportions (i.e., percentages or proportions). Such tests are similar to those in Sections 9.1 and 9.2. The main difference is that we are working with a distribution of proportions.

Tests for a single proportion

Throughout this section, we will assume that the situations we are dealing with satisfy the conditions underlying the binomial distribution. In particular, we will let r be a binomial random variable. This means that r is the number of successes out of n independent binomial trials (for the definition of a binomial trial, see Section 6.2). We will use $\hat{p} = r/n$ as our estimate for p, the population probability of success on each trial. The letter q again represents the population probability of failure on each trial, and so $q = 1 - p$. We also assume that the samples are large (i.e., $np > 5$ and $nq > 5$).

Criteria for using normal approximation to binomial, $np > 5$ and $nq > 5$

For large samples, $np > 5$ and $nq > 5$, the distribution of $\hat{p} = r/n$ values is well approximated by a *normal curve* with mean μ and standard deviation σ as follows:

$$\mu = p$$

$$\sigma = \sqrt{\frac{pq}{n}}$$

Hypothesis for testing p

The null and alternate hypotheses for tests of proportions are

Left-Tailed Test	Right-Tailed Test	Two-Tailed Test
H_0: $p = k$	H_0: $p = k$	H_0: $p = k$
H_1: $p < k$	H_1: $p > k$	H_1: $p \neq k$

depending on what is asked for in the problem. Notice that since p is a probability, the value k must be between 0 and 1.

Sample test statistic $\hat{p}$

For tests of proportions, we need to convert the sample test statistic $\hat{p}$ to a z value. Then we can find a P-value appropriate for the test. The $\hat{p}$ distribution is approximately normal, with mean p and standard deviation $\sqrt{pq/n}$. Therefore, the conversion of $\hat{p}$ to z follows the formula

$$z = \frac{\hat{p} - p}{\sqrt{\dfrac{pq}{n}}}$$

where $\hat{p} = r/n$ is the sample test statistic
n = number of trials
p = proportion specified in H_0
$q = 1 - p$

Using this mathematical information about the sampling distribution for $\hat{p}$, the basic procedure is similar to tests you have conducted before.

PROCEDURE

How to Test a Proportion p

Requirements

Consider a binomial experiment with n trials, where p represents the population probability of success and $q = 1 - p$ represents the population probability of failure. Let r be a random variable that represents the number of successes out of the n binomial trials. The number of trials n should be sufficiently large so that both $np > 5$ and $nq > 5$ (use p from the null hypothesis). In this case, $\hat{p} = r/n$ can be approximated by the normal distribution.

Procedure

1. In the context of the application, state the *null and alternate hypotheses* and set the *level of significance* α.

2. Compute the standardized *sample test statistic*

$$z = \frac{\hat{p} - p}{\sqrt{\dfrac{pq}{n}}}$$

where p is the value specified in H_0 and $q = 1 - p$.

3. Use the standard normal distribution and the type of test, one-tailed or two-tailed, to find the *P-value* corresponding to the test statistic.

4. *Conclude* the test. If P-value $\leq \alpha$, then reject H_0. If P-value $> \alpha$, then do not reject H_0.

5. *Interpret your conclusion* in the context of the application.

EXAMPLE 6

TESTING p

A team of eye surgeons has developed a new technique for a risky eye operation to restore the sight of people blinded from a certain disease. Under the old method, it is known that only 30% of the patients who undergo this operation recover their eyesight.

Suppose that surgeons in various hospitals have performed a total of 225 operations using the new method and that 88 have been successful (i.e., the patients fully recovered their sight). Can we justify the claim that the new method is better than the old one? (Use a 1% level of significance.)

SOLUTION:

(a) Establish H_0 and H_1 and note the level of significance.

The level of significance is $\alpha = 0.01$. Let p be the probability that a patient fully recovers his or her eyesight. The null hypothesis is that p is still 0.30, even for the new method. The alternate hypothesis is that the new method has improved the chances of a patient recovering his or her eyesight. Therefore,

$$H_0: p = 0.30 \quad \text{and} \quad H_1: p > 0.30$$

(b) *Check Requirements* Is the sample sufficiently large to justify use of the normal distribution for $\hat{p}$? Find the sample test statistic $\hat{p}$ and convert it to a z value, if appropriate.

Using p from H_0 we note that $np = 225(0.3) = 67.5$ is greater than 5 and that $nq = 225(0.7) = 157.5$ is also greater than 5, so we can use the normal distribution for the sample statistic $\hat{p}$.

$$\hat{p} = \frac{r}{n} = \frac{88}{225} \approx 0.39$$

The z value corresponding to $\hat{p}$ is

$$z = \frac{\hat{p} - p}{\sqrt{\dfrac{pq}{n}}} \approx \frac{0.39 - 0.30}{\sqrt{\dfrac{0.30(0.70)}{225}}} \approx 2.95$$

In the formula, the value for p is from the null hypothesis. H_0 specifies that $p = 0.30$, so $q = 1 - 0.30 = 0.70$.

(c) Find the P-value of the test statistic.

Figure 9-10 shows the P-value. Since we have a right-tailed test, the P-value is the area to the right of $z = 2.95$. Using the normal distribution (Table 3 of the Appendix), we find that P-value $= P(z > 2.95) \approx 0.0016$.

FIGURE 9-10

P-value Area

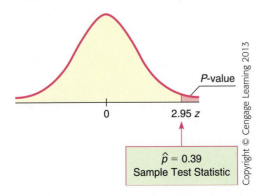

$\hat{p} = 0.39$
Sample Test Statistic

(d) Conclude the test.
   Since the P-value of $0.0016 \leq 0.01$ for α, we reject H_0.

(e) *Interpretation* Interpret the results in the context of the problem.
   At the 1% level of significance, the evidence shows that the population probability of success for the new surgery technique is higher than that of the old technique.

GUIDED EXERCISE 5 | ## Testing p

A botanist has produced a new variety of hybrid wheat that is better able to withstand drought than other varieties. The botanist knows that for the parent plants, the proportion of seeds germinating is 80%. The proportion of seeds germinating for the hybrid variety is unknown, but the botanist claims that it is 80%. To test this claim, 400 seeds from the hybrid plant are tested, and it is found that 312 germinate. Use a 5% level of significance to test the claim that the proportion germinating for the hybrid is 80%.

(a) Let p be the proportion of hybrid seeds that will germinate. Notice that we have no prior knowledge about the germination proportion for the hybrid plant. State H_0 and H_1. What is the required level of significance?

➡ $H_0: p = 0.80$; $H_1: p \neq 0.80$; $\alpha = 0.05$.

(b) *Check Requirements* Using the value of p in H_0, are both $np > 5$ and $nq > 5$? Can we use the normal distribution for $\hat{p}$?

➡ From H_0, $p = 0.80$ and $q = 1 - p = 0.20$.
$np = 400(0.8) = 320 > 5$
$nq = 400(0.2) = 80 > 5$
So we can use the normal distribution for $\hat{p}$.

Calculate the sample test statistic $\hat{p}$.

The number of trials is $n = 400$, and the number of successes is $r = 312$. Thus,

$$\hat{p} = \frac{r}{n} = \frac{312}{400} = 0.78$$

(c) Next, we convert the sample test statistic $\hat{p} = 0.78$ to a z value. Based on our choice for H_0, what value should we use for p in our formula? Since $q = 1 - p$, what value should we use for q? Using these values for p and q, convert $\hat{p}$ to a z value.

➡ According to H_0, $p = 0.80$. Then $q = 1 - p = 0.20$. Using these values in the following formula gives

$$z = \frac{\hat{p} - p}{\sqrt{\dfrac{pq}{n}}} = \frac{0.78 - 0.80}{\sqrt{\dfrac{0.80(0.20)}{400}}} = -1.00$$

CALCULATOR NOTE   If you evaluate the denominator separately, be sure to carry at least four digits after the decimal.

Continued

(d) Is the test right-tailed, left-tailed, or two-tailed? Find the *P*-value of the sample test statistic and sketch a standard normal curve showing the *P*-value.

For a two-tailed test, using the normal distribution (Table 3 of the Appendix), we find that

$$P\text{-value} = 2P(z < -1.00) = 2(0.1587) = 0.3174$$

FIGURE 9-11 *P*-value

(e) Do we reject or fail to reject H_0?

We fail to reject H_0, because *P*-value of $0.3174 > 0.05$ for α.

(f) *Interpretation* Interpret your conclusion in the context of the application.

At the 5% level of significance, there is insufficient evidence to conclude that the germination rate is not 80%.

CRITICAL THINKING

Issues Related to Hypothesis Testing

Through our work with hypothesis tests of μ and p, we've gained experience in setting up, performing, and interpreting results of such tests.

We know that different random samples from the same population are very likely to have sample statistics $\bar{x}$ or $\hat{p}$ that differ from their corresponding parameters μ or p. Some values of a statistic from a random sample will be close to the corresponding parameter. Others may be farther away simply because we happened to draw a random sample of more extreme data values.

> The central question in hypothesis testing is whether or not you think the value of the sample test statistic is too far away from the value of the population parameter proposed in H_0 to occur by chance alone.

This is where the *P*-value of the sample test statistic comes into play. The *P*-value of the sample test statistic tells you the probability that you would get a sample statistic as far away, or farther from, the value of the parameter as stated in the null hypothesis H_0.

If the *P*-value is very small, you reject H_0. But what does "very small" mean? It is customary to define "very small" as smaller than the preset level of significance α.

When you reject H_0, are you absolutely certain that you are making a correct decision? The answer is no! You are simply willing to take a chance that you are making a mistake (a type I error). The level of significance α describes the chance of making a mistake if you reject H_0 when it is in fact true.

Several issues come to mind:

1. What if the *P*-value is so close to α that we "barely" reject or fail to reject H_0? In such cases, researchers might attempt to clarify the results by
   - increasing the sample size or
   - controlling the experiment to reduce the standard deviation.

   Both actions tend to increase the magnitude of the *z* or *t* value of the sample test statistic, resulting in a smaller corresponding *P*-value.

2. How reliable are the study and the measurements in the sample?
   - When reading results of a statistical study, be aware of the source of the data and the reliability of the organization doing the study.
   - Is the study sponsored by an organization that might profit or benefit from the stated conclusions? If so, look at the study carefully to ensure that the measurements, sampling technique, and handling of data are proper and meet professional standards.

TECH NOTES   The TI-84Plus/TI-83Plus/TI-*n*spire calculators and Minitab support tests of proportions. The output for both technologies includes the sample proportion $\hat{p}$ and the *P*-value of $\hat{p}$. Minitab also includes the *z* value corresponding to $\hat{p}$.

TI-84Plus/TI-83Plus/TI-*n*spire (with TI-84Plus keypad) Press **STAT**, select **TESTS**, and use option **5:1-PropZTest**. The value of p_0 is from the null hypothesis $H_0: p = p_0$. The number of successes is the value for *x*.

Minitab Menu selections: **Stat ▶ Basic Statistics ▶ 1 Proportion.** Under options, set the test proportion as the value in H_0. Choose to use the normal distribution.

Critical region method

Since the $\hat{p}$ sampling distribution is approximately normal, we use Table 3 in the Appendix, "Areas of a Standard Normal Distribution," to find critical values.

EXAMPLE 7   CRITICAL REGION METHOD FOR TESTING p

Let's solve Guided Exercise 5 using the critical region approach. In that problem, 312 of 400 seeds from a hybrid wheat variety germinated. For the parent plants, the proportion of germinating seeds is 80%. Use a 5% level of significance to test the claim that the population proportion of germinating seeds from the hybrid wheat is different from that of the parent plants.

SOLUTION:

(a) As in Guided Exercise 5, we have $\alpha = 0.05$, $H_0: p = 0.80$, and $H_1: p \neq 0.80$. The next step is to find $\hat{p}$ and the corresponding sample test statistic *z*. This was done in Guided Exercise 5, where we found that $\hat{p} = 0.78$, with corresponding $z = -1.00$.

(b) Now we find the critical value z_0 for a two-tailed test using $\alpha = 0.05$. This means that we want the total area 0.05 divided between two tails, one to the right of z_0 and one to the left of $-z_0$. As shown in Figure 9-8 of Section 9.2, the critical value(s) are ± 1.96 (See also Table 3, part (c), of the Appendix for critical values of the z distribution.)

(c) Figure 9-12 shows the critical regions and the location of the sample test statistic.

FIGURE 9-12

Critical Regions, $\alpha = 0.05$

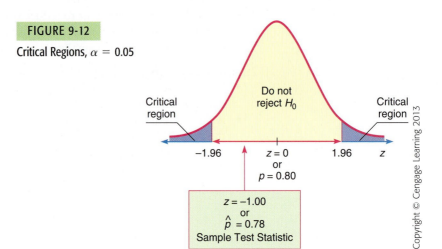

(d) Finally, we conclude the test and compare the results to Guided Exercise 5. Since the sample test statistic does not fall in the critical region, we fail to reject H_0 and conclude that, at the 5% level of significance, the evidence is not strong enough to reject the botanist's claim. This result, as expected, is consistent with the conclusion obtained by using the P-value method.

VIEWPOINT | Who Did What?

Art, music, literature, and science share a common need to classify things: Who painted that picture? Who composed that music? Who wrote that document? Who should get that patent? In statistics, such questions are called classification problems. *For example, the* Federalist Papers *were published anonymously in 1787–1788 by Alexander Hamilton, John Jay, and James Madison. But who wrote what? That question is addressed by F. Mosteller (Harvard University) and D. Wallace (University of Chicago) in the book* Statistics: A Guide to the Unknown, *edited by J. M. Tanur. Other scholars have studied authorship regarding Plato's* Republic *and* Dialogues, *including the* Symposium. *For more information on this topic, see the source in Problems 15 and 16 of this exercise set.*

SECTION 9.3 PROBLEMS

aplia™   Access end-of-section exercises at www.aplia.com

Chapter Review

SUMMARY

Hypothesis testing is a major component of inferential statistics. In hypothesis testing, we propose a specific value for the population parameter in question. Then we use sample data from a random sample and probability to determine whether or not to reject this specific value for the parameter.

Basic components of a hypothesis test are:

- The *null hypothesis* H_0 states that a parameter equals a specific value.

- The *alternate hypothesis* H_1 states that the parameter is greater than, less than, or simply not equal to the value specified in H_0.

- The *level of significance* α of the test is the probability of rejecting H_0 when it is true.

- The *sample test statistic* corresponding to the parameter in H_0 is computed from a random sample and appropriate sampling distribution.

- Assuming H_0 is true, the probability that a sample test statistic will take on a value as extreme as, or more extreme than, the observed sample test statistic is the *P-value* of the test. The P-value is computed by using the sample test statistic, the corresponding sampling distribution, H_0, and H_1.

- If *P*-value $\leq \alpha$, we reject H_0. If *P*-value $> \alpha$, we fail to reject H_0.

- We say that sample data are *significant* if we can reject H_0.

An alternative way to conclude a test of hypotheses is to use critical regions based on the alternate hypothesis and α. Critical values z_0 are found in Table 3(c) of the Appendix. Critical values t_0 are found in Table 4 of the Appendix. If the sample test statistic falls beyond the critical values—that is, in the critical region—we reject H_0.

The methods of hypothesis testing are very general, and we will see them used again in later chapters. In this chapter, we looked at tests involving:

- parameter μ. Use standard normal or Student's t distribution. See procedure displays in Section 9.2.

- parameter p. Use standard normal distribution. See procedure displays in Section 9.3.

FINDING THE *P*-VALUE CORRESPONDING TO A SAMPLE TEST STATISTIC

Use the appropriate sampling distribution as described in procedure displays for each of the various tests.

Left-Tailed Test
P-value = area to the left of the sample test statistic.

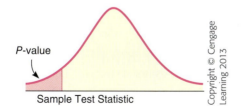

Right-Tailed Test
P- value = area to the right of the sample test statistic.

Two-Tailed Test

Sample test statistic lies to the *left* of center
P-value = twice the area to the left of the sample test statistic

Sample test statistic lies to the *right* of center
P-value = twice area to the right of the sample test statistic

Sampling Distributions for Inferences Regarding μ or p

Parameter	Condition	Sampling Distribution
μ	σ is known and x has a normal distribution or $n \geq 30$	Normal distribution
μ	σ is not known and x has a normal or mound-shaped, symmetrical distribution or $n \geq 30$	Student's t distribution with $d.f. = n - 1$
p	$np > 5$ and $n(1 - p) > 5$	Normal distribution

IMPORTANT WORDS & SYMBOLS

Section 9.1
Hypothesis testing 301
Hypothesis 301
Null hypothesis, H_0 301
Alternate hypothesis, H_1 301
Left-tailed test 302
Right-tailed test 302
Two-tailed test 302
Sample test statistic 303
P-value 304
Type I error 306
Type II error 306
Level of significance of a test and probability of a type I error, α 306

Probability of a type II error, β 307
Power of a test, $(1 - \beta)$ 307
Statistical significance 308
Fail to reject H_0 310
Reject H_0 310

Section 9.2
$d.f.$ for testing μ when σ is unknown 314
Critical region method 318
Critical value 319

Section 9.3
Criteria for using normal approximation to binomial, $np > 5$ and $nq > 5$ 323

VIEWPOINT | Will It Rain?

*Do cloud-seed experiments ever work? If you seed the clouds, will it rain? If it does rain, who will benefit? Who will be displeased by the rain? If you seed the clouds and nothing happens, will taxpayers (who support the effort) complain or rejoice? Maybe this should be studied over a remote island—such as Tasmania (near Australia). Using what you already know about statistical testing, you can conduct your own tests, given the appropriate data. Remember, there are sociological questions (pleased/displeased with result) as well as technical questions (number of inches of rain produced). For data regarding cloud-seeding experiments over Tasmania, visit the Brase/Brase statistics site at **http://www.cengagebrain.com** and find the link to DASL, the Carnegie Mellon University Data and Story Library. From the DASL site, look under Datasets for Cloud.*

CHAPTER REVIEW PROBLEMS

1. *Statistical Literacy* When testing μ, how do we decide whether to use the standard normal distribution or a Student's t distribution?

2. *Statistical Literacy* What do we mean when we say a test is *significant*? Does this necessarily mean the results are important?

3. *Critical Thinking* All other conditions being equal, does a larger sample size increase or decrease the corresponding magnitude of the z or t value of the sample test statistic?

4. *Critical Thinking* All other conditions being equal, does a z or t value with larger magnitude have a larger or smaller corresponding P-value?

Before you solve each problem below, first categorize it by answering the following question: Are we testing a single mean or a single proportion? Assume underlying population distributions are mound-shaped and symmetrical for problems with small samples that involve testing a mean. Then provide the following information for Problems 5–12:

(a) What is the level of significance? State the null and alternate hypotheses.

(b) *Check Requirements* What sampling distribution will you use? What assumptions are you making? What is the value of the sample test statistic?

(c) Find (or estimate) the P-value. Sketch the sampling distribution and show the area corresponding to the P-value.

(d) Based on your answers in parts (a)–(c), will you reject or fail to reject the null hypothesis? Are the data statistically significant at level α?

(e) *Interpret* your conclusion in the context of the application.
   Note: For degrees of freedom d.f. not in the Student's t table, use the closest d.f. that is *smaller*. In some situations, this choice of d.f. may increase the P-value by a small amount and therefore produce a slightly more "conservative" answer. Answers may vary due to rounding.

5. *Vehicles: Mileage* Based on information in the *Statistical Abstract of the United States* (116th edition), the average annual miles driven per vehicle in the United States is 11.1 thousand miles, with $\sigma \approx 600$ miles. Suppose that a random sample of 36 vehicles owned by residents of Chicago showed that the average mileage driven last year was 10.8 thousand miles. Does this indicate that the average miles driven per vehicle in Chicago is different from (higher or lower than) the national average? Use a 0.05 level of significance.

6. *Student Life: Employment* Professor Jennings claims that only 35% of the students at Flora College work while attending school. Dean Renata thinks that the professor has underestimated the number of students with part-time or full-time jobs. A random sample of 81 students shows that 39 have jobs. Do the data indicate that more than 35% of the students have jobs? (Use a 5% level of significance.)

7. *Toys: Electric Trains* The Toylot Company makes an electric train with a motor that it claims will draw an average of only 0.8 ampere (A) under a normal load. A sample of nine motors was tested, and it was found that the mean current was $\bar{x} = 1.4$ A, with a sample standard deviation of $s = 0.41$ A. Do the data indicate that the Toylot claim of 0.8 A is too low? (Use a 1% level of significance.)

8. *Matches: Number per Box* The Nero Match Company sells matchboxes that are supposed to have an average of 40 matches per box, with $\sigma = 9$. A random sample of 94 Nero matchboxes shows the average number of matches per box to be 43.1. Using a 1% level of significance, can you say that the average number of matches per box is more than 40?

9. *Medicine: Plasma Compress* A hospital reported that the normal death rate for patients with extensive burns (more than 40% of skin area) has been significantly reduced by the use of new fluid plasma compresses. Before the new treatment, the mortality rate for extensively burned patients was about 60%. Using the new compresses, the hospital found that only 40 of 90 patients with extensive burns died. Use a 1% level of significance to test the claim that the mortality rate has dropped.

10. *Civil Service: College Degrees* The Congressional Budget Office reports that 36% of federal civilian employees have a bachelor's degree or higher (*The Wall Street Journal*). A random sample of 120 employees in the private sector showed that 33 have a bachelor's degree or higher. Does this indicate that the percentage of employees holding bachelor's degrees or higher in the private sector is less than that in the federal civilian sector? Use $\alpha = 0.05$.

11. *Vending Machines: Coffee* A machine in the student lounge dispenses coffee. The average cup of coffee is supposed to contain 7.0 ounces. Eight cups of coffee from this machine show the average content to be 7.3 ounces with a standard deviation of 0.5 ounce. Do you think that the machine has slipped out of adjustment and that the average amount of coffee per cup is different from 7 ounces? Use a 5% level of significance.

12. *Sports Car: Fuel Injection* The manufacturer of a sports car claims that the fuel injection system lasts 48 months before it needs to be replaced. A consumer group tests this claim by surveying a random sample of 10 owners who had the fuel injection system replaced. The ages of the cars at the time of replacement were (in months):

29	42	49	48	53	46	30	51	42	52

    i. Use your calculator to verify that the mean age of a car when the fuel injection system fails is $\bar{x} = 44.2$ months, with standard deviation $s \approx 8.61$ months.
    ii. Test the claim that the fuel injection system lasts less than an average of 48 months before needing replacement. Use a 5% level of significance.

DATA HIGHLIGHTS: GROUP PROJECTS

Break into small groups and discuss the following topic. Organize a brief outline in which you summarize the main points of your group discussion.

"With Sampling, There Is Too a Free Lunch"—this is a headline that appeared in *The Wall Street Journal*. The article is about food-product samples available at grocery stores. Giving out food samples is expensive and labor intensive. It clogs supermarket aisles. It is risky. What if a customer tries an item and spits it out on the floor or says the product is awful? It creates litter. Some customers drop toothpicks or small paper cups on the floor or spill the product. However, the budget that companies are willing to spend to have their products sampled is growing. The director of communications for Bigg's "hypermarket" (a combination grocery and general-merchandise store) says that more than 60% of customers sample products and about 37% of those who sample buy the product.

(a) Let's test the hypothesis that 60% of customers sample a particular product. What is the null hypothesis? Do you believe that the percentage of customers who sample products is less than, more than, or just different from 60%? What will you use for the alternate hypothesis?

(b) Choose a level of significance α.

(c) Go to a grocery store when special products are being sampled (not just the usual in-house store samples often available at the deli or bakery). Count the number of customers going by the display when a sample is available and the number of customers who try the sample. Be sure the number of customers n is large enough to use the normal distribution to approximate the binomial.

(d) Using your sample data, conclude the hypothesis test. What is your conclusion?

(e) Do you think different food products might have a higher or lower percentage of customers trying them? For instance, do a higher percentage of customers try samples of pizza than samples of yogurt? How could you use statistics to justify your answer?

(f) Do you want to include young children in your sample? Do they pick up items to include in the customer's basket, or do they just munch the samples?

LINKING CONCEPTS: WRITING PROJECTS

Discuss each of the following topics in class or review the topics on your own. Then write a brief but complete essay in which you summarize the main points. Please include formulas and graphs as appropriate.

The most important questions in life usually cannot be answered with absolute certainty. Many important questions are answered by giving an estimate and a measure of confidence in the estimate. This was the focus of Chapter 8. However, sometimes important questions must be answered in a more straightforward manner by a simple yes or no. Hypothesis testing is the statistical process of answering questions with a straightforward yes or no *and* providing an estimate of the risk in accepting the answer.

1. Review and discuss type I and type II errors associated with hypothesis testing.

2. Review and discuss the level of significance and power of a statistical test.

3. The following statements are very important. Give them some careful thought and discuss them.

    (a) When we fail to reject the null hypothesis, we do not claim that it is absolutely true. We simply claim that at the given level of significance, the data were not sufficient to reject the null hypothesis.

    (b) When we accept the alternate hypothesis, we do not claim that the null hypothesis is absolutely false. We do claim that at the given level of significance, the data presented enough evidence to reject the null hypothesis.

4. In the text, it is said that a statistical test is a package of five basic ingredients. List these ingredients, discuss them in class, and write a short description of how these ingredients relate to the above discussion questions.

5. As access to computers becomes more and more prevalent, we see the *P*-value reported in hypothesis testing more frequently. Review the use of the *P*-value in hypothesis testing. What is the difference between the level of significance of a test and the *P*-value? Considering both the *P*-value and level of significance, under what conditions do we reject or fail to reject the null hypothesis?

USING TECHNOLOGY

Simulation

Recall that the level of significance α is the probability of mistakenly rejecting a true null hypothesis. If $\alpha = 0.05$, then we expect to mistakenly reject a true null hypothesis about 5% of the time. The following simulation conducted with Minitab demonstrates this phenomenon.

We draw 40 random samples of size 50 from a population that is normally distributed with mean $\mu = 30$ and standard deviation $\sigma = 2.5$. The display shows the results of a hypothesis test with

$$H_0: \mu = 30 \quad H_1: \mu > 30$$

for each of the 40 samples labeled C1 through C40. Because each of the 40 samples is drawn from a population with mean $\mu = 30$, the null hypothesis $H_0: \mu = 30$ is true for the test based on each sample. However, as the display shows, for some samples we reject the true null hypothesis.

(a) How many of the 40 samples have a sample mean $\bar{x}$ above $\mu = 30$? below $\mu = 30$?

(b) Look at the P-value of the sample statistic $\bar{x}$ in each of the 40 samples. How many P-values are less than or equal to $\alpha = 0.05$? What percentage of the P-values are less than or equal to α? What percentage of the samples have us reject H_0 when in fact each of the samples was drawn from a normal distribution with $\mu = 30$, as hypothesized in the null hypothesis?

(c) If you have access to computer or calculator technology that creates random samples from a normal distribution with a specified mean and standard deviation, repeat this simulation. Do you expect to get the same results? Why or why not?

Minitab Display: Random samples of size 50 from a normal population with $\mu = 30$ and $\sigma = 2.5$

```
Z-Test
Test of mu = 30.000 vs. mu > 30.000
The assumed sigma = 2.50
Variable   N     Mean    StDev   SE Mean      Z       P
C1        50   30.002   2.776    0.354     0.01    0.50
C2        50   30.120   2.511    0.354     0.34    0.37
C3        50   30.032   2.721    0.354     0.09    0.46
C4        50   30.504   2.138    0.354     1.43    0.077
C5        50   29.901   2.496    0.354    -0.28    0.61
C6        50   30.059   2.836    0.354     0.17    0.43
C7        50   30.443   2.519    0.354     1.25    0.11
C8        50   29.775   2.530    0.354    -0.64    0.74
C9        50   30.188   2.204    0.354     0.53    0.30
C10       50   29.907   2.302    0.354    -0.26    0.60
C11       50   30.036   2.762    0.354     0.10    0.46
C12       50   30.656   2.399    0.354     1.86    0.032
C13       50   30.158   2.884    0.354     0.45    0.33
C14       50   29.830   3.129    0.354    -0.48    0.68
C15       50   30.308   2.241    0.354     0.87    0.19
C16       50   29.751   2.165    0.354    -0.70    0.76
C17       50   29.833   2.358    0.354    -0.47    0.68
C18       50   29.741   2.836    0.354    -0.73    0.77
C19       50   30.441   2.194    0.354     1.25    0.11
C20       50   29.820   2.156    0.354    -0.51    0.69
C21       50   29.611   2.360    0.354    -1.10    0.86
C22       50   30.569   2.659    0.354     1.61    0.054
C23       50   30.294   2.302    0.354     0.83    0.20
C24       50   29.978   2.298    0.354    -0.06    0.53
C25       50   29.836   2.438    0.354    -0.46    0.68
C26       50   30.102   2.322    0.354     0.29    0.39
C27       50   30.066   2.266    0.354     0.19    0.43
C28       50   29.071   2.219    0.354    -2.63    1.00
C29       50   30.597   2.426    0.354     1.69    0.046
C30       50   30.092   2.296    0.354     0.26    0.40
C31       50   29.803   2.495    0.354    -0.56    0.71
C32       50   29.546   2.335    0.354    -1.28    0.90
C33       50   29.702   1.902    0.354    -0.84    0.80
C34       50   29.233   2.657    0.354    -2.17    0.98
C35       50   30.097   2.472    0.354     0.28    0.39
C36       50   29.733   2.588    0.354    -0.76    0.78
C37       50   30.379   2.976    0.354     1.07    0.14
C38       50   29.424   2.827    0.354    -1.63    0.95
C39       50   30.288   2.396    0.354     0.81    0.21
C40       50   30.195   3.051    0.354     0.55    0.29
```

Technology Hints

Press **STAT** and select **EDIT**. Highlight the list name, such as L1. Then press **MATH**, select **PRB**, and highlight **6:randNorm(μ, σ, sample size)**. Press Enter. Fill in the values of $\mu = 30$, $\sigma = 2.5$, and sample size = 50. Press Enter. Now list L1 contains a random sample from the normal distribution specified.

To test the hypothesis H_0: $\mu = 30$ against H_1: $\mu > 30$, press **STAT**, select **TESTS**, and use option **1:Z-Test.** Fill in the value 30 for μ_0, 2.5 for σ, and $> \mu_0$. The output provides the value of the sample statistic $\bar{x}$, its corresponding z value, and the P-value of the sample statistic.

Excel 2010

To draw random samples from a normal distribution with $\mu = 30$ and $\sigma = 2.5$, click the **Data** tab on the home ribbon and select **Data Analysis** in the Analysis Group. In the dialogue box, select **Random Number Generator.** In the first dialogue box, select **Normal** for the Distribution. In the resulting dialogue box shown below, the number of variables is the number of samples. Fill in the rest of the dialogue box as shown.

To conduct a hypothesis test of H_0: $\mu = 30$ against H_1: $\mu > 30$, click **Insert Function** f_x on the home screen. For the category, select **Statistical** and then select the function **Z.TEST.** Fill in the dialogue box with the array containing the random numbers, 30 for X, the value in H_0, and 2.5 for sigma.

Minitab

To generate random samples from a normal distribution, use the menu choices **Calc ➤ Random Data ➤ Normal.** In the dialogue box, the number of rows refers to the sample size. Use 50 rows. Then designate the columns for the samples. Using C1–C40 will generate 40 random samples and put the samples in columns C1 through C40.

To test the hypothesis H_0: $\mu = 30$ against H_1: $\mu > 30$, use the menu choices **Stat ➤ Basic Statistics ➤ 1-SampleZ.** Use columns C1–C40 as the variables. Fill in 30 for the test mean, use "greater than" for the alternate hypothesis, and use 2.5 for sigma.

SPSS

SPSS uses a Student's t distribution to test the mean and difference of means. SPSS uses the sample standard deviation s even if the population σ is known. Use the menu choices **Analyze ➤ Compare Means** and then **One-Sample T Test** or **Independent-Sample T Tests** for tests of a single mean or a difference of means (Section 10.1), respectively. In the dialogue box, fill in the test value of the null hypothesis.

To generate 40 random samples of size $n = 50$ from a normal distribution with $\mu = 30$ and $\sigma = 2.5$, first enter consecutive integers from 1 to 50 in a column of the data editor. Then, under variable view, enter the variable names Sample1 through Sample40. Use the menu choices **Transform ➤ Compute Variable.** In the dialogue box, use Sample1 for the target variable, select **Random Numbers** for the Function Group and then select the function **Rv.Normal(mean, stddev).** Use 30 for the mean and 2.5 for the standard deviation. Continue until you have 40 samples. To sample from other distributions, use appropriate functions in the Compute dialogue box.

The SPSS display shows the test results ($H_0: \mu = 30$; $H_1: \mu \neq 30$) for a sample of size $n = 50$ drawn from a normal distribution with $\mu = 30$ and $\sigma = 2.5$. The P-value is given as the significance for a two-tailed test.

For a one-tailed test, divide the significance by 2. In the display, the significance is 0.360 for a two-tailed test. So for a one-tailed test, the P-value is 0.360/2, or 0.180.

SPSS Display

T-Test

One-Sample Statistics

	N	Mean	Std. Deviation	Std. Error Mean
SAMPLE1	50	30.3228	2.46936	.34922

One-Sample Test

	Test Value = 30					
					90% Confidence Interval of the Difference	
	t	df	Sig. (2-tailed)	Mean Difference	Lower	Upper
SAMPLE1	.924	49	.360	.3228	−.2627	.9083

Cumulative Review Problems

CHAPTERS 7–9

Answers may vary due to rounding.

1. *Statistical Literacy* What are the values of the mean and standard deviation of a standard normal distribution?

2. *Statistical Literacy* According to the empirical rule, about what percentage of a normal distribution falls within 2 standard deviations of the mean?

3. *Statistical Literacy* Please give a careful but brief answer to each of the following questions.
   (a) What is a population? How do you get a simple random sample? Give examples.
   (b) What is a sample statistic? What is a sampling distribution? Give examples.
   (c) Give a careful and complete statement of the central limit theorem.
   (d) List at least three areas of everyday life to which the above concepts can be applied. Be specific.

4. *Normal Distribution* Oxygen demand is a term biologists use to describe the oxygen needed by fish and other aquatic organisms for survival. The Environmental Protection Agency conducted a study of a wetland area in Marin County, California. In this wetland environment, the mean oxygen demand was $\mu = 9.9$ mg/L, with 95% of the data ranging from 6.5 mg/L to 13.3 mg/L (Reference: *EPA Report 832-R-93-005*). Let x be a random variable that represents oxygen demand in this wetland environment. Assume x has a probability distribution that is approximately normal.

   (a) Use the 95% data range to estimate the standard deviation for oxygen demand. *Hint:* See Problem 31 of Section 7.3.
   (b) An oxygen demand below 8 mg/L indicates that some organisms in the wetland environment may be dying. What is the probability that the oxygen demand will fall below 8 mg/L?
   (c) A high oxygen demand can also indicate trouble. An oxygen demand above 12 mg/L may indicate an overabundance of organisms that endanger some types of plant life. What is the probability that the oxygen demand will exceed 12 mg/L?

5. *Normal Approximation to the Binomial* The majority of house burglars simply walk into a house that is unlocked! In fact, about 57% of all house burglars gain entrance through an unlocked window or door (Reference: *The Book of Risks* by Larry Laudan). Suppose that $n = 129$ house burglaries will occur tomorrow in Los Angeles. Let r be a binomial random variable that represents the number of burglaries that required no forced entrance.
   (a) We want to approximate the binomial random variable r by a normal random variable x. Is this appropriate? What requirements must be satisfied before we can do this? Do you think these requirements are satisfied in this case? Explain.
   (b) Compute μ and σ for the normal approximation.
   (c) What is the probability that at least 65 of the burglaries required no forced entry?

Patrol car at scene of house robbery

6. *Sampling Distribution x* Workers at a large toxic cleanup project are concerned that their white blood cell counts may have been reduced. Let x be a random variable that represents white blood cell count per cubic millimeter of whole blood in a healthy adult. Then $\mu = 7500$ and $\sigma \approx 1750$ (Reference: *Diagnostic Tests with Nursing Applications*, S. Loeb). A random sample of $n = 50$ workers from the toxic cleanup site were given a blood test that showed $\bar{x} = 6820$. What is the probability that, for healthy adults, $\bar{x}$ will be this low or lower?
   (a) How does the central limit theorem apply? Explain.
   (b) Compute $P(\bar{x} \le 6820)$.
   (c) *Interpretation* Based on your answer to part (b), would you recommend that additional facts be obtained, or would you recommend that the workers' claims be dismissed? Explain.

7. *Sampling Distribution $\hat{p}$* Do you have a great deal of confidence in the advice given to you by your medical doctor? About 45% of all adult Americans claim they do have a great deal of confidence in their MDs (Reference: National Opinion Research Center, University of Chicago). Suppose a random sample of $n = 32$ adults in a health insurance program are asked about their confidence in the medical advice their doctors give.

(a) Is the normal approximation to the proportion $\hat{p} = r/n$ valid?

(b) Find the values of $\mu_{\hat{p}}$ and $\sigma_{\hat{p}}$.

In Problems 8–12, please use the following steps (i)–(v) for all hypothesis tests:

(i) What is the level of significance? State the null and alternate hypotheses.

(ii) *Check Requirements* What sampling distribution will you use? What assumptions are you making? What is the value of the sample test statistic?

(iii) Find (or estimate) the P-value. Sketch the sampling distribution and show the area corresponding to the P-value.

(iv) Based on your answers in parts (i)–(iii), will you reject or fail to reject the null hypothesis? Are the data statistically significant at level a?

(v) *Interpret* your conclusion in the context of the application.

Note: For degrees of freedom d.f. not in the Student's t table, use the closest d.f. that is *smaller*. In some situations, this choice of d.f. may increase the P-value a small amount and thereby produce a slightly more "conservative" answer.

8. *Testing and Estimating μ with σ Known* Let x be a random variable that represents micrograms of lead per liter of water (μg/L). An industrial plant discharges water into a creek. The Environmental Protection Agency has studied the discharged water and found x to have a normal distribution, with $\sigma = 0.7$ μg/L (Reference: *EPA Wetlands Case Studies*).

(a) The industrial plant says that the population mean value of x is $\mu = 2.0$ μg/L. However, a random sample of $n = 10$ water samples showed that $\bar{x} = 2.56$ μg/L. Does this indicate that the lead-

concentration population mean is higher than the industrial plant claims? Use $\alpha = 1\%$.

(b) Find a 95% confidence interval for μ using the sample data and the EPA value for σ.

(c) How large a sample should be taken to be 95% confident that the sample mean $\bar{x}$ is within a margin of error $E = 0.2\mu$g/L of the population mean?

9. *Testing and Estimating μ with σ Unknown* Carboxyhemoglobin is formed when hemoglobin is exposed to carbon monoxide. Heavy smokers tend to have a high percentage of carboxyhemoglobin in their blood (Reference: *A Manual of Laboratory and Diagnostic Tests* by F. Fischbach). Let x be a random variable representing percentage of carboxyhemoglobin in the blood. For a person who is a regular heavy smoker, x has a distribution that is approximately normal. A random sample of $n = 12$ blood tests given to a heavy smoker gave the following results (percentage of carboxyhemoglobin in the blood):

9.1	9.5	10.2	9.8	11.3	12.2
11.6	10.3	8.9	9.7	13.4	9.9

(a) Use a calculator to verify that $\bar{x} \approx 10.49$ and $s \approx 1.36$.

(b) A long-term population mean $\mu = 10\%$ is considered a health risk. However, a long-term population mean above 10% is considered a clinical alert that the person may be asymptomatic. Do the data indicate that the population mean percentage is higher than 10% for this patient? Use $\alpha = 0.05$.

(c) Use the given data to find a 99% confidence interval for μ for this patient.

10. *Testing and Estimating a Proportion p* Although older Americans are most afraid of crime, it is young people who are more likely to be the actual victims of crime. It seems that older people are more cautious about the people with whom they associate. A national survey showed that 10% of all people ages 16–19 have been victims of crime (Reference: *Bureau of Justice Statistics*). At Jefferson High School, a random sample of $n = 68$ students (ages 16–19) showed that $r = 10$ had been victims of a crime.

(a) Do these data indicate that the population proportion of students in this school (ages 16–19) who have been victims of a crime is different (either way) from the national rate for this age group? Use $\alpha = 0.05$. Do you think the conditions $np > 5$ and $nq > 5$ are satisfied in this setting? Why is this important?

(b) Find a 90% confidence interval for the proportion of students in this school (ages 16–19) who have been victims of a crime.

(c) How large a sample size should be used to be 95% sure that the sample proportion $\hat{p}$ is within a margin of error $E = 0.05$ of the population proportion of all students in this school (ages 16–19) who have been victims of a crime? *Hint*: Use sample data $\hat{p}$ as a preliminary estimate for p.

11. *Essay and Project* In Chapters 8 and 9 you studied estimation and hypothesis testing.

(a) Write a brief essay in which you discuss using information from samples to infer information about populations. Be sure to include methods of estimation and hypothesis testing in your discussion. What two sampling distributions are used in estimation and hypothesis testing of population means and proportions? What are the criteria for determining the appropriate sampling distribution? What is the level of significance of a test? What is the *P*-value? How is the *P*-value related to the alternate hypothesis? How is the null hypothesis related to the sample test statistic? Explain.

(b) Suppose you want to study the length of time devoted to commercial breaks for two different types of television programs. Identify the types of programs you want to study (e.g., sitcoms, sports events, movies, news, children's programs, etc.). Write a brief outline for your study. Discuss how to obtain random samples. How large should the sample be for a specified margin of error? Describe the protocol you will follow to measure the times of the commercial breaks. Determine whether you are going to compare the average time devoted to commercials or the proportion of time devoted to commercials. What assumptions will you make regarding population distributions? What graphics might be appropriate? What methods of estimation will you use? What methods of testing will you use?

12. *Critical Thinking* Explain hypothesis testing to a friend, using the following scenario as a model. Describe the hypotheses, the sample statistic, the *P*-value, the meanings of type I and type II errors, and the level of significance. Discuss the significance of the results. Formulas are not required.

A team of research doctors designed a new knee surgery technique utilizing much smaller incisions than the standard method. They believe recovery times are shorter when the new method is used. Under the old method, the average recovery time for full use of the knee is 4.5 months. A random sample of 38 surgeries using the new method showed the average recovery time to be 3.6 months, with sample standard deviation of 1.7 months. The *P*-value for the test is 0.0011. The research team states that the results are statistically significant at the 1% level of significance.

10

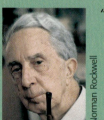

"So what!"

—ANONYMOUS

Norman Rockwell

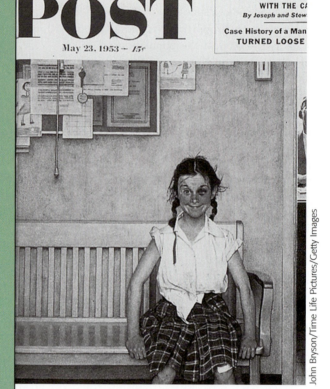

John Bryson/Time Life Pictures/Getty Images

Girl with Black Eye *by Norman Rockwell (1894–1978)*

Norman Rockwell painted everyday people and situations. In the cover for the *Saturday Evening Post* (May 23, 1953) shown above, a young lady is about to have a conference with her school principal about a schoolyard fight. So what!

Philologists (people who study cultural linguistics) tell us that this expression "So what!" is a shortened version of "So what is the difference?" They also tell us that there are similar popular or slang expressions about differences in all languages and cultures. It is human nature to challenge the claim that something is better, worse, or just simply different. In this chapter, we will focus on this very human theme by studying a variety of topics regarding questions of whether or not differences exist between two populations.

For online student resources, visit the Brase/Brase, *Understanding Basic Statistics*, 6th edition website at **http://www.cengagebrain.com**.

INFERENCES ABOUT DIFFERENCES

Bobby Deal/RealDealPhoto, 2010/Used under license from Shutterstock.com

PREVIEW QUESTIONS

What are the statistical advantages of paired data values? How do we construct statistical tests? (SECTION 10.1)

How do we compare means from two independent populations when we know σ for each population? (SECTION 10.2)

What if we want to compare means from two independent populations, but we do not know σ for each population? (SECTION 10.2)

How do we use sample data to compare proportions from two independent populations? (SECTION 10.3)

FOCUS PROBLEM

The Trouble with Wood Ducks

The National Wildlife Federation published an article entitled "The Trouble with Wood Ducks" (*National Wildlife,* Vol. 31, No. 5). In this article, wood ducks are described as beautiful birds living in forested areas such as the Pacific Northwest and Southeast United States. Because of overhunting and habitat destruction, these birds were in danger of extinction. A federal ban on hunting wood ducks in 1918 helped save the species from extinction. Wood ducks like to nest in tree cavities. However, many such trees were disappearing due to heavy timber cutting. For a period of time, it seemed that nesting boxes were the solution to disappearing trees. At first, the wood duck population grew, but after a few seasons, the population declined sharply. Good biology research combined with good statistics provided an answer to this disturbing phenomenon.

Cornell University professors of ecology Paul Sherman and Brad Semel found that the nesting boxes were placed too close to each other. Female wood ducks prefer a secluded nest that is a considerable distance from the next wood duck nest. In fact, female wood duck behavior changed when the nests were too close to each other. Some females would lay their eggs in another female's nest. The result was too many eggs in one nest. The biologists found that if there were too many eggs in a nest, the proportion of eggs that hatched was considerably reduced. In the long run, this meant a decline in the population of wood ducks.

In their study, Sherman and Semel used two placements of nesting boxes. Group I boxes were well separated from each

Photo Researchers

other and well hidden by available brush. Group II boxes were highly visible and grouped closely together.

In group I boxes, there were a total of 474 eggs, of which a field count showed that about 270 hatched. In group II boxes, there were a total of 805 eggs, of which a field count showed that, again, about 270 hatched.

(a) Find a 95% confidence interval for $p_1 - p_2$. Does the interval indicate that the proportion of eggs hatched from group I nesting boxes is higher than, lower than, or not different from the proportion of eggs hatched from group II nesting boxes?

(b) Use a 5% level of significance to test the hypothesis that the proportion of hatches from group I placements is greater than that from group II placements. Do we reject or fail to reject H_0? Is this result consistent with the information from the 95% confidence interval?

(c) What conclusions about placement of nesting boxes can be drawn? In the article, additional concerns are raised about the higher cost of placing and maintaining group I nest boxes. Also at issue is the cost efficiency per successful wood duck hatch. Data in the article do not include information that would help us answer questions of *cost* efficiency. However, the data presented do help us answer questions about the proportions of successful hatches in the two nesting box configurations.

(See Problem 22 of Section 10.3.)

SECTION 10.1

Tests Involving Paired Differences (Dependent Samples)

FOCUS POINTS
- Identify paired data and dependent samples.
- Explain the advantages of paired data tests.
- Compute differences and the sample test statistic.
- Estimate the *P*-value and conclude the test.

Paired data
Creating data pairs

Many statistical applications use *paired data* samples to draw conclusions about the difference between two population means. *Data pairs* occur very naturally in "before and after" situations, where the *same* object or item is measured both before and after a treatment. Applied problems in social science, natural science, and business administration frequently involve a study of matching pairs. Psychological studies of identical twins; biological studies of plant growth on plots of land matched for soil type, moisture, and sun; and business studies on sales of matched inventories are examples of paired data studies.

When working with paired data, it is very important to have a definite and uniform method of creating data pairs that clearly utilizes a natural matching of characteristics. The next example and exercise demonstrate this feature.

Dependent samples

> Two samples are **dependent** if each data value in one sample can be paired in a meaningful way with a corresponding data value in the other sample.

EXAMPLE 1   PAIRED DATA

A shoe manufacturer claims that among the general population of adults in the United States, the average length of the left foot is longer than that of the right. To compare the average length of the left foot with that of the right, we can take a random sample of 15 U.S. adults and measure the length of the left foot and then the length of the right foot for each person in the sample. Is there a natural way of pairing the measurements? How many pairs will we have?

SOLUTION: In this case, we can pair each left foot measurement with the same person's right foot measurement. The person serves as the "matching link" between the two distributions. We will have 15 pairs of measurements.

GUIDED EXERCISE 1 | **Paired data**

A psychologist has developed a series of exercises called the Instrumental Enrichment (IE) program, which he claims is useful in overcoming cognitive deficiencies in mentally retarded children. To test the program, extensive statistical tests are being conducted. In one experiment, a random sample of 10-year-old students with IQ scores below 80 was selected. An IQ test was given to these students before they spent 2 years in an IE program, and an IQ test was given to the same students after the program.

(a) On what basis can you pair the IQ scores?

⟹ Take the "before and after" IQ scores of each individual student.

(b) If there were 20 students in the sample, how many data pairs would you have?

⟹ Twenty data pairs. Note that there would be 40 IQ scores, but only 20 pairs.

COMMENT   To compare two populations, we cannot always employ paired data tests, but when we can, what are the advantages? Using matched or paired data often can reduce the danger of introducing extraneous or uncontrollable factors into our sample measurements because the matched or paired data have essentially the *same* characteristics except for the *one* characteristic that is being measured. Furthermore, it can be shown that pairing data has the theoretical effect of reducing measurement variability (i.e., variance), which increases the accuracy of statistical conclusions.

When we wish to compare the means of two samples, the first item to be determined is whether or not there is a natural pairing between the data in the two samples. Again, data pairs are created from "before and after" situations, or from matching data by using studies of the same object, or by a process of taking measurements of closely matched items.

Testing the differences, d

When testing *paired* data, we take the difference d of the data pairs *first* and look at the mean difference $\bar{d}$. Then we use a test on $\bar{d}$. Theorem 10.1 provides the basis for our work with paired data.

THEOREM 10.1   Consider a random sample of n data pairs. Suppose the differences d between the first and second members of each data pair are (approximately) normally distributed, with population mean μ_d. Then the t values

μ_d, mean of differences from data pairs

$$t = \frac{\bar{d} - \mu_d}{s_d/\sqrt{n}}$$

where $\bar{d}$ is the sample mean of the d values, n is the number of data pairs, and

s_d, standard deviation of differences from data pairs

$$s_d = \sqrt{\frac{\Sigma(d - \bar{d})^2}{n - 1}}$$

is the sample standard deviation of the d values, follow a Student's t distribution with degrees of freedom $d.f. = n - 1$.

Hypotheses for testing the mean of paired differences

When testing the mean of the differences of paired data values, the null hypothesis is that there is no difference among the pairs. That is, the mean of the differences μ_d is zero:

$$H_0: \mu_d = 0$$

The alternate hypothesis depends on the problem and can be

$H_1: \mu_d < 0$	$H_1: \mu_d > 0$	$H_1: \mu_d \neq 0$
(left-tailed)	(right-tailed)	(two-tailed)

Sample test statistic

For paired difference tests, we make our decision regarding H_0 according to the evidence of the sample mean $\overline{d}$ of the differences of measurements. By Theorem 10.1, we convert the sample test statistic $\overline{d}$ to a t value using the formula

$$t = \frac{\overline{d} - \mu_d}{s_d/\sqrt{n}} \text{ with } d.f. = n - 1$$

where s_d = sample standard deviation of the differences d

n = number of data pairs

$\mu_d = 0$, as specified in H_0

P-values from Table 4 of the Appendix

To find the P-value (or an interval containing the P-value) corresponding to the test statistic t computed from $\overline{d}$, we use the Student's t distribution table (Table 4 of the Appendix). Recall from Section 9.2 that we find the test statistic t (or, if t is negative, $|t|$) in the row headed by $d.f. = n - 1$, where n is the number of data pairs. The P-value for the test statistic is the column entry in the *one-tail area* row for one-tailed tests (right or left). For two-tailed tests, the P-value is the column entry in the *two-tail area* row. Usually the exact test statistic t is not in the table, so we obtain an interval that contains the P-value by using adjacent entries in the table. Table 10-1 gives the basic structure for using the Student's t distribution table to find the P-value or an interval containing the P-value.

TABLE 10-1   Using Student's *t* Distribution Table for *P*-values

For one-tailed tests:	one-tail area	P-value	P-value
For two-tailed tests:	two-tail area	P-value	P-value
Use row header	d.f. = n − 1		Find t value ↑

Copyright © Cengage Learning 2013

With the preceding information, you are now ready to test paired differences. First let's summarize the procedure.

PROCEDURE

HOW TO TEST PAIRED DIFFERENCES USING THE STUDENT'S t DISTRIBUTION

Requirements Obtain a simple random sample of n matched data pairs A, B. Let d be a random variable representing the difference between the values in a matched data pair. Compute the sample mean $\overline{d}$ and sample standard deviation s_d. If you can assume that d has a normal distribution or simply has a mound-shaped, symmetrical distribution, then any sample size n will work. If you cannot assume this, then use a sample size $n \geq 30$.

Continued

Procedure

1. Use the *null hypothesis* of *no difference*, $H_0: \mu_d = 0$. In the context of the application, choose the *alternate hypothesis* to be $H_1: \mu_d > 0$, $H_1: \mu_d < 0$, or $H_1: \mu_d \neq 0$. Set the *level of significance* α.

2. Use $\bar{d}, s_d$, the sample size n, and $\mu_d = 0$ from the null hypothesis to compute the sample *test statistic*

   $$t = \frac{\bar{d} - 0}{\frac{s_d}{\sqrt{n}}} = \frac{\bar{d}\sqrt{n}}{s_d}$$

   with degrees of freedom $d.f. = n - 1$.

3. Use the Student's t distribution and the type of test, one-tailed or two-tailed, to find (or estimate) the *P-value* corresponding to the test statistic.

4. *Conclude* the test. If P-value $\leq \alpha$, then reject H_0. If P-value $> \alpha$, then do not reject H_0.

5. *Interpret your conclusion* in the context of the application.

EXAMPLE 2   PAIRED DIFFERENCE TEST

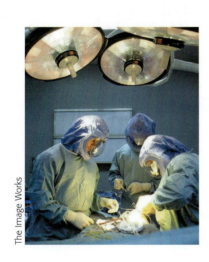

A team of heart surgeons at Saint Ann's Hospital knows that many patients who undergo corrective heart surgery have a dangerous buildup of anxiety before their scheduled operations. The staff psychiatrist at the hospital has started a new counseling program intended to reduce this anxiety. A test of anxiety is given to patients who know they must undergo heart surgery. Then each patient participates in a series of counseling sessions with the staff psychiatrist. At the end of the counseling sessions, each patient is retested to determine anxiety level. Table 10-2 on the next page indicates the results for a random sample of nine patients. Higher scores mean higher levels of anxiety. Assume the distribution of differences is mound-shaped and symmetric.

From the given data, can we conclude that the counseling sessions reduce anxiety? Use a 0.01 level of significance.

SOLUTION: Before we answer this question, let us notice two important points: (1) we have a *random sample* of nine patients, and (2) we have a *pair* of measurements taken on the same patient before and after counseling sessions. In our problem, the sample size is $n = 9$ pairs (i.e., patients), and the d values are found in the fourth column of Table 10-2.

(a) Note the level of significance and set the hypotheses.
    In the problem statement, $\alpha = 0.01$. We want to test the claim that the counseling sessions reduce anxiety. This means that the anxiety level before counseling is expected to be higher than the anxiety level after counseling. In symbols, $d = B - A$ should tend to be positive, and the population mean of differences μ_d also should be positive. Therefore, we have

   $$H_0: \mu_d = 0 \qquad \text{and} \qquad H_1: \mu_d > 0$$

(b) *Check Requirements* Is it appropriate to use a Student's t distribution for the sample test statistic? Explain. What degrees of freedom are used?
    We have a random sample of paired differences d. Under the assumption that the d distribution is mound-shaped and symmetrical, we use a Student's t distribution with $d.f. = n - 1 = 9 - 1 = 8$.

(c) Find the sample test statistic $\bar{d}$ and convert it to a corresponding test statistic t.
    First we need to compute $\bar{d}$ and s_d. Using formulas or a calculator and the d values shown in Table 10-2, we find that

   $$\bar{d} \approx 33.33 \qquad \text{and} \qquad s_d \approx 22.92$$

TABLE 10-2

Patient	B Score before Counseling	A Score after Counseling	$d = B - A$ Difference
Jan	121	76	45
Tom	93	93	0
Diane	105	64	41
Barbara	115	117	−2
Mike	130	82	48
Bill	98	80	18
Frank	142	79	63
Carol	118	67	51
Alice	125	89	36

Using these values together with $n = 9$ and $\mu_d = 0$, we have

$$t = \frac{\bar{d} - 0}{s_d/\sqrt{n}} \approx \frac{33.33}{22.92/\sqrt{9}} \approx 4.363$$

(d) Find the *P*-value for the test statistic and sketch the *P*-value on the *t* distribution. Since we have a right-tailed test, the *P*-value is the area to the right of $t = 4.363$, as shown in Figure 10-1. In Table 4 of the Appendix, we find an interval containing the *P*-value. Use entries from the row headed by $d.f. = n - 1 = 9 - 1 = 8$. The test statistic $t = 4.363$ falls between 3.355 and 5.041. The *P*-value for the sample *t* falls between the corresponding one-tail areas 0.005 and 0.0005. (See Table 10-3, an excerpt from Table 4 of the Appendix.)

$$0.0005 < P\text{-value} < 0.005$$

FIGURE 10-1

P-value

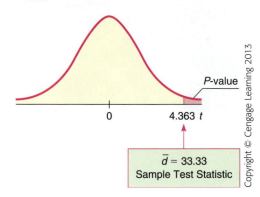

$\bar{d} = 33.33$
Sample Test Statistic

TABLE 10-3   **Excerpt from Student's *t* Distribution Table (Table 4 of the Appendix)**

✓ one-tail area	0.005	0.0005
two-tail area	0.010	0.0010
$d.f. = 8$	3.355	5.041

↑
Sample *t* = 4.363

(e) Conclude the test.

Since the interval containing the *P*-value lies to the left of $\alpha = 0.01$, we reject H_0. *Note:* Using the raw data and software, *P*-value ≈ 0.0012.

(f) *Interpretation* Interpret the results.
At the 1% level of significance, we conclude that the counseling sessions reduce the average anxiety level of patients about to undergo corrective heart surgery.

The problem we have just solved is a paired difference problem of the "before and after" type. The next guided exercise demonstrates a paired difference problem of the "matched pair" type.

GUIDED EXERCISE 2 | ## Paired difference test

Do educational toys make a difference in the age at which a child learns to read? To study this question, researchers designed an experiment in which one group of preschool children spent 2 hours each day (for 6 months) in a room well supplied with "educational" toys such as alphabet blocks, puzzles, ABC readers, coloring books featuring letters, and so forth. A control group of children spent 2 hours a day for 6 months in a "noneducational" toy room. It was anticipated that IQ differences and home environment might be uncontrollable factors unless identical twins could be used. Therefore, six pairs of identical twins of preschool age were randomly selected. From each pair, one member was randomly selected to participate in the experimental (i.e., educational toy room) group and the other in the control (i.e., noneducational toy room) group. For each twin, the data item recorded is the age in months at which the child began reading at the primary level (Table 10-4). Assume the distribution of differences is mound-shaped and symmetric.

TABLE 10-4 **Reading Ages for Identical Twins (in Months)**

Twin Pair	Experimental Group B = Reading Age	Control Group A = Reading Age	Difference $d = B - A$
1	58	60	
2	61	64	
3	53	52	
4	60	65	
5	71	75	
6	62	63	

Copyright © Cengage Learning 2013

(a) Compute the entries in the $d = B - A$ column of Table 10-4. Using formulas for the mean and sample standard deviation or a calculator with mean and sample standard deviation keys, compute $\bar{d}$ and s_d.

➡️

Pair	$d = B - A$
1	−2
2	−3
3	1
4	−5
5	−4
6	−1

$\bar{d} \approx -2.33$

$s_d \approx 2.16$

(b) What is the null hypothesis?

➡️ $H_0: \mu_d = 0$.

(c) To test the claim that the experimental group learned to read at a *different age* (either younger or older), what should be the alternate hypothesis?

➡️ $H_1: \mu_d \neq 0$.

(d) **Check Requirements** What distribution does the sample test statistic t follow? Find the degrees of freedom.

➡️ Because the d distribution is mound-shaped and symmetrical and we have a random sample of $n = 6$ paired differences, the sample test statistic follows a Student's t distribution with $d.f. = n - 1 = 6 - 1 = 5$.

(e) Convert the sample test statistic $\bar{d}$ to a t value.

➡️ Using $\mu_d = 0$ from H_0, $\bar{d} = -2.33$, $n = 6$, and $s_d = 2.16$, we get

$$t = \frac{\bar{d} - \mu_d}{s_d/\sqrt{n}} \approx \frac{-2.33 - 0}{2.16/\sqrt{6}} \approx -2.642$$

Continued

GUIDED EXERCISE 2 *continued*

(f) When we use Table 4 of the Appendix to find an interval containing the P-value, do we use a one-tail or two-tail area? Why? Sketch a figure showing the P-value. Find an interval containing the P-value.

TABLE 10-5 Excerpt from Student's *t* Table

one-tail area	0.025	0.010
✓ two-tail area	0.050	0.020
d.f. = 5	2.571	3.365

Sample *t* = 2.642

Copyright © Cengage Learning 2013

 This is a two-tailed test, so we use a two-tail area.

FIGURE 10-2 *P*-value

$\dfrac{P\text{-value}}{2}$ $\dfrac{P\text{-value}}{2}$

−2.642   0   2.642 *t*

$\bar{d} = -2.33$
Sample Test Statistic

Copyright © Cengage Learning 2013

The sample *t* is between 2.571 and 3.365.

$0.020 < P\text{-value} < 0.050$

Note: Using the raw data and software, $P\text{-value} \approx 0.0457$.

(g) Using $\alpha = 0.05$, do we reject or fail to reject H_0?.

Since the interval containing the P-value has values that are all smaller than 0.05, we reject H_0.

α

0.020   0.050

(h) *Interpretation* What do the results mean in the context of this application?

At the 5% level of significance, the experiment indicates that educational toys make a difference in the age at which a child learns to read.

TECH NOTES   Both Excel 2010 and Minitab support paired difference tests directly. On the TI-84Plus and TI-83Plus/TI-*n*spire calculators, construct a column of differences and then do a *t* test on the data in that column. For each technology, be sure to relate the alternate hypothesis to the "before and after" assignments. All the displays show the results for the data of Guided Exercise 2.

TI-84Plus/TI-83Plus/TI-*n*spire (with TI-84Plus keypad) Enter the "before" data in column L1 and the "after" data in column L2. Highlight L3, type L1 − L2, and press Enter. The column L3 now contains the $B - A$ differences. To conduct the test, press **STAT**, select **TESTS**, and use option **2:T-Test**. Note that the letter *x* is used in place of *d*.

```
L1        L2        L3       3
58        60        -2
61        64        -3
53        52        1
60        65        -5
71        75        -4
62        63        -1

L3(7) =
```

```
T-Test
μ≠0
t=-2.645751311
p=.0456591238
x̄=-2.333333333
Sx=2.160246899
n=6
```

Excel 2010 Enter the data in two columns. Use the menu choices **Tools ➤ Data Analysis ➤ t-Test: Paired Two-Sample for Means.** Fill in the dialogue box with the hypothesized mean difference of 0. Set alpha. Another way to conduct a paired-difference *t* test is to press **Insert Function** f_x and select **Statistical** for the category and then **T.TEST** for the function. In the dialogue box, use **1** (paired) for the type of test.

	B	C	D
	t-Test: Paired Two Sample for Means		
		Variable 1	*Variable 2*
	Mean	60.83333	63.16666667
	Variance	34.96667	55.76666667
	Observations	6	6
	Pearson Correlation	0.974519	
	Hypothesized Mean Difference	0	
	df	5	
	t Stat	−2.64575	
	P(T<=t) one-tail	0.02283	
	t Critical one-tail	2.015049	
P-value →	P(T<=t) two-tail	0.045659	
	t Critical two-tail	2.570578	

Minitab Enter the data in two columns. Use the menu selection **Stat ➤ Basic Statistics ➤ Paired _t_.** Under Options, set the null and alternate hypotheses.

Problem 18 shows how to find confidence intervals for μ_d.

```
Paired T-Test and Confidence Interval
Paired T for B - A
                 N      Mean     St Dev SE   Mean
B                6      60.83       5.91      2.41
A                6      63.17       7.47      3.05
Difference       6      -2.333      2.160     0.882
95% CI for mean difference: (-4.601, -0.066)
T-Test of mean difference = 0 (vs not = 0):
      T-Value = -2.65 P-Value = 0.046
```

EXAMPLE 3   CRITICAL REGION METHOD

Let's revisit Guided Exercise 2 regarding educational toys and reading age and conclude the test using the critical region method. Recall that there were six pairs of twins. One twin of each set was given educational toys and the other was not. The difference *d* in reading ages for each pair of twins was measured, and $\alpha = 0.05$.

SOLUTION: From Guided Exercise 2, we have

$$H_0: \mu_d = 0 \quad \text{and} \quad H_1: \mu_d \neq 0$$

We computed the sample test statistic $\bar{d} \approx -2.33$ with corresponding $t \approx -2.642$.

(a) Find the critical values for $\alpha = 0.05$.
Since the number of pairs is $n = 6$, $d.f. = n - 1 = 5$. In the Student's *t* distribution table (Table 4 of the Appendix), look in the row headed by 5. To find the column, locate $\alpha = 0.05$ in the *two-tail area* row, since we have a two-tailed test. The critical values are $\pm t_0 = \pm 2.571$.

(b) Sketch the critical regions and place the *t* value of the sample test statistic $\bar{d}$ on the sketch. Conclude the test. Compare the result to the result given by the *P*-value method of Guided Exercise 2.
Since the sample test statistic falls in the critical region (see Figure 10-3), we reject H_0 at the 5% level of significance. At this level, educational toys seem to make a difference in reading age. Notice that this conclusion is consistent with the conclusion obtained using the *P*-value.

FIGURE 10-3

Critical Region with $\alpha = 0.05, d.f. = 5$

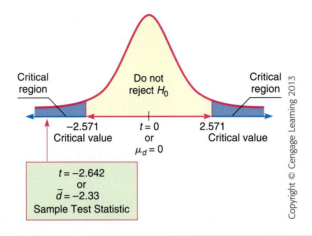

VIEWPOINT   DUI

DUI usually means "driving under the influence" of alcohol, but driving under the influence of sleep loss can be just as dangerous. Researchers in Australia have found that after staying awake for 24 hours straight, a person will be about as impaired as if he or she had had enough alcohol to be legally drunk in most U.S. states (Source: Rocky Mountain News). Using driver simulation exams and statistical tests (paired difference tests) found in this section, it is possible to show that the null hypothesis H_0: $\mu_d = 0$ cannot be rejected. Or, put another way, the average level of impairment for a given individual from alcohol (at the DUI level) is about the same as the average level of impairment from sleep loss (24 hours without sleep).

SECTION 10.1 PROBLEMS

aplia

Access end-of-section exercises at www.aplia.com

SECTION 10.2

Inferences about the Difference of Two Means $\mu_1 - \mu_2$

FOCUS POINTS

- Identify independent samples and sampling distributions.
- Compute the sample test statistic and *P*-value for testing $\mu_1 - \mu_2$.
- Find confidence intervals for $\mu_1 - \mu_2$.

Independent Samples

Many practical applications of statistics involve a comparison of two population means or two population proportions. In Section 10.1, we considered tests of difference of means for *dependent samples*. With dependent samples, we could pair the data and then consider the differences of the data measurements *d*. In this section, we will turn our attention to inferences regarding differences of means from *independent samples*.

Independent samples

> Two samples are **independent** if the selection of sample data from one population is completely unrelated to the selection of sample data from the other population.

Independent samples occur very naturally when we draw *two random samples*, one from the first population and one from the second population.

GUIDED EXERCISE 3	Distinguish between independent and dependent samples

For each experiment, categorize the sampling as independent or dependent, and explain your choice.

(a) In many medical experiments, a sample of subjects is randomly divided into two groups. One group is given a specific treatment, and the other group is given a placebo. After a certain period of time, both groups are measured for the same condition. Do the measurements from these two groups constitute independent or dependent samples?

➡ Since the subjects are *randomly assigned* to the two treatment groups (one receives a treatment, the other a placebo), the resulting measurements would form independent samples.

(b) In an accountability study, a group of students in an English composition course is given a pretest. After the course, the same students are given a posttest covering similar material. Are the two groups of scores independent or dependent?

➡ Since the pretest scores and the posttest scores are from the same students, the samples are dependent. Each student has both a pretest score and a posttest score, so there is a natural pairing of data values.

Because *both* samples are random samples, there is no pairing of measurements between the two populations.

Hypothesis Tests and Confidence Intervals for $\mu_1 - \mu_2$ (σ_1 and σ_2 Known)

Properties of the $\bar{x}_1 - \bar{x}_2$ sampling distribution with σ_1 and σ_2 known

In this section, we will use probability distributions that arise from a difference of means. How do we obtain such distributions? Suppose that we have two statistical variables x_1 and x_2, each with its own distribution. We take *independent* random samples of size n_1 from the x_1 distribution and of size n_2 from the x_2 distribution. Then we compute the respective means $\bar{x}_1$ and $\bar{x}_2$. Now consider the difference $\bar{x}_1 - \bar{x}_2$. This expression represents a difference of means. If we repeat this sampling process over and over, we will create lots of $\bar{x}_1 - \bar{x}_2$ values. Figure 10-4 on the next page illustrates the sampling distribution of $\bar{x}_1 - \bar{x}_2$.

The values of $\bar{x}_1 - \bar{x}_2$ that come from repeated (independent) sampling of populations 1 and 2 can be arranged in a relative-frequency table and a relative-frequency histogram (see Section 2.1). This would give us an experimental idea of the theoretical probability distribution of $\bar{x}_1 - \bar{x}_2$.

Fortunately, it is not necessary to carry out this lengthy process for each example. The results have been worked out mathematically. The next theorem presents the main results.

THEOREM 10.2   Let x_1 and x_2 have normal distributions with means μ_1 and μ_2 and standard deviations σ_1 and σ_2, respectively. If we take independent random samples of size n_1 from the x_1 distribution and of size n_2 from the x_2 distribution, then the variable $\bar{x}_1 - \bar{x}_2$ has

1. a normal distribution

2. mean $\mu_1 - \mu_2$

3. standard deviation $\sqrt{\dfrac{\sigma_1^2}{n_1} + \dfrac{\sigma_2^2}{n_2}}$

COMMENT   The theorem requires that x_1 and x_2 have *normal* distributions. However, if *both* n_1 and n_2 are 30 or larger, then the central limit theorem (Section 7.5) assures us that $\bar{x}_1$ and $\bar{x}_2$ are approximately normally distributed. In this case, the conclusions of the theorem are again valid even if the original x_1 and x_2 distributions are not exactly normal.

FIGURE 10-4

Sampling Distribution of $\bar{x}_1 - \bar{x}_2$

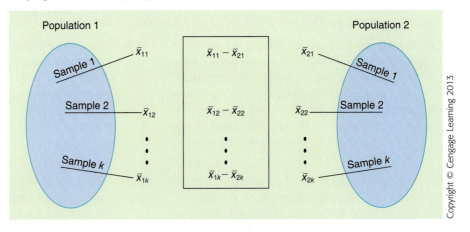

Theorem 10.2 gives us the basis for hypothesis tests and confidence intervals for $\mu_1 - \mu_2$ when σ_1 and σ_2 are both known.

When testing the difference of means, it is customary to use the null hypothesis

Hypotheses for testing difference of means

$$H_0: \mu_1 - \mu_2 = 0 \text{ or, equivalently, } H_0: \mu_1 = \mu_2$$

As mentioned in Section 9.1, the null hypothesis is set up to see if it can be rejected. When testing the difference of means, we first set up the hypothesis H_0 that there is no difference. The alternate hypothesis could then be any of the ones listed in Table 10-6. The alternate hypothesis and consequent type of test used depend on the particular problem. Note that μ_1 is always listed first.

Using Theorem 10.2 and the central limit theorem (Section 7.5), we can summarize the procedure for testing $\mu_1 - \mu_2$ and finding a confidence interval when both σ_1 and σ_2 are known.

TABLE 10-6   Alternate Hypotheses and Types of Test: Difference of Two Means

H_1			Type of Test
$H_1: \mu_1 - \mu_2 < 0$	or equivalently	$H_1: \mu_1 < \mu_2$	Left-tailed test
$H_1: \mu_1 - \mu_2 > 0$	or equivalently	$H_1: \mu_1 > \mu_2$	Right-tailed test
$H_1: \mu_1 - \mu_2 \neq 0$	or equivalently	$H_1: \mu_1 \neq \mu_2$	Two-tailed test

PROCEDURE

HOW TO TEST $\mu_1 - \mu_2$ AND FIND A CONFIDENCE INTERVAL WHEN BOTH σ_1 AND σ_2 ARE KNOWN

Requirements Let σ_1 and σ_2 be the population standard deviations of populations 1 and 2. Obtain two independent random samples from populations 1 and 2, where

$\bar{x}_1$ and $\bar{x}_2$ are sample means from populations 1 and 2

n_1 and n_2 are sample sizes from populations 1 and 2

If you can assume that both population distributions 1 and 2 are normal, any sample sizes n_1 and n_2 will work. If you cannot assume this, then use sample sizes $n_1 \geq 30$ and $n_2 \geq 30$.

Hypothesis test for $\mu_1 - \mu_2$
(σ_1 and σ_2 known)

Procedure for Testing $\mu_1 - \mu_2$

1. In the context of the application, state the *null and alternate hypotheses* and set the *level of significance* α. It is customary to use $H_0: \mu_1 - \mu_2 = 0$.

2. Use $\mu_1 - \mu_2 = 0$ from the null hypothesis together with $\bar{x}_1$, $\bar{x}_2$, σ_1, σ_2, n_1, and n_2 to compute the sample *test statistic*.

$$z = \frac{(\bar{x}_1 - \bar{x}_2) - (\mu_1 - \mu_2)}{\sqrt{\dfrac{\sigma_1^2}{n_1} + \dfrac{\sigma_2^2}{n_2}}} = \frac{\bar{x}_1 - \bar{x}_2}{\sqrt{\dfrac{\sigma_1^2}{n_1} + \dfrac{\sigma_2^2}{n_2}}}$$

3. Use the standard normal distribution and the type of test, one-tailed or two-tailed, to find the *P-value* corresponding to the sample test statistic.

4. *Conclude the test.* If P-value $\le \alpha$, then reject H_0. If P-value $> \alpha$, then do not reject H_0.

5. *Interpret your conclusion* in the context of the application.

Confidence interval for $\mu_1 - \mu_2$
(σ_1 and σ_2 known)

Procedure for Confidence Interval for $\mu_1 - \mu_2$

$$(\bar{x}_1 - \bar{x}_2) - E < \mu_1 - \mu_2 < (\bar{x}_1 - \bar{x}_2) + E$$

where $E = z_c \sqrt{\dfrac{\sigma_1^2}{n_1} + \dfrac{\sigma_2^2}{n_2}}$

c = confidence level ($0 < c < 1$)

z_c = critical value for confidence level c based on the standard normal distribution (see Table 3(b) of the Appendix for commonly used values)

EXAMPLE 4   TESTING THE DIFFERENCE OF MEANS AND FINDING A CONFIDENCE INTERVAL (σ_1 AND σ_2 KNOWN)

A consumer group is testing camp stoves. To test the heating capacity of a stove, it measures the time required to bring 2 quarts of water from 50°F to boiling (at sea level). Two competing models are under consideration. Ten stoves of the first model and 12 stoves of the second model are selected at random and tested. The following results are obtained:

Model 1: Mean time $\bar{x}_1 = 11.4$ min; $\sigma_1 = 2.5$ min; $n_1 = 10$

Model 2: Mean time $\bar{x}_2 = 9.9$ min; $\sigma_2 = 3.0$ min; $n_2 = 12$

Assume that the time required to bring water to a boil is normally distributed for each stove.

Hypothesis test: Is there any difference (either way) between the performances of these two models? Use a 5% level of significance.

SOLUTION:

(a) State the null and alternate hypotheses and note the value of α.

Let μ_1 and μ_2 be the means of the distributions of times for models 1 and 2, respectively. We set up the null hypothesis to state that there is no difference:

$H_0: \mu_1 = \mu_2$ or $H_0: \mu_1 - \mu_2 = 0$

The alternate hypothesis states that there is a difference:

$H_1: \mu_1 \ne \mu_2$ or $H_1: \mu_1 - \mu_2 \ne 0$

The level of significance is $\alpha = 0.05$.

(b) *Check Requirements* What distribution does the sample test statistic follow?
The sample test statistic z follows a standard normal distribution because the original distributions from which samples are drawn are normal. In addition, the population standard deviations of the original distributions are known and the samples are independent.

(c) Compute the sample test statistic $\bar{x}_1 - \bar{x}_2$ and then convert it to a z value.

We are given the values $\bar{x}_1 = 11.4$ and $\bar{x}_2 = 9.9$. Therefore, the sample test statistic is $\bar{x}_1 - \bar{x}_2 = 11.4 - 9.9 = 1.5$. To convert this to a z value, we use the values $\sigma_1 = 2.5$, $\sigma_2 = 3.0$, $n_1 = 10$, and $n_2 = 12$. From the null hypothesis, $\mu_1 - \mu_2 = 0$.

$$z = \frac{(\bar{x}_1 - \bar{x}_2) - (\mu_1 - \mu_2)}{\sqrt{\dfrac{\sigma_1^2}{n_1} + \dfrac{\sigma_2^2}{n_2}}} = \frac{1.5}{\sqrt{\dfrac{2.5^2}{10} + \dfrac{3.0^2}{12}}} \approx 1.28$$

(d) Find the *P*-value and sketch the area on the standard normal curve.
Figure 10-5 shows the *P*-value. Use the standard normal distribution (Table 3 of the Appendix) and the fact that we have a two-tailed test. *P*-value $\approx 2(0.1003) = 0.2006$.

FIGURE 10-5

P-value

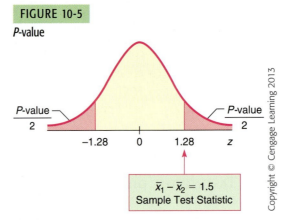

(e) Conclude the test.
The *P*-value is 0.2006 and $\alpha = 0.05$. Since *P*-value $> \alpha$, do not reject H_0.

(f) *Interpretation* Interpret the results.
At the 5% level of significance, the sample data do not indicate any difference in the population mean times for boiling water for the two stove models.

Confidence interval: Find a 95% confidence interval for the population difference $\mu_1 - \mu_2$ of mean times to boil water for the two stoves. Interpret the results.

SOLUTION: Note that both σ_1 and σ_2 are known and that the times for each stove to bring 2 quarts of water to a boil are normally distributed. Therefore, by Theorem 10.2, $\bar{x}_1 - \bar{x}_2$ is normal. We use the normal distribution to find the critical value $z_{0.95}$. From Table 3(b) of the Appendix, we see that $z_{0.95} = 1.96$. The confidence interval is

$$(\bar{x}_1 - \bar{x}_2) - E < \mu_1 - \mu_2 < (\bar{x}_1 - \bar{x}_2) + E$$

where $\bar{x}_1 - \bar{x}_2 = 11.4 - 9.9 = 1.5$ min and

$$E = z_c \sqrt{\frac{\sigma_1^2}{n_1} + \frac{\sigma_2^2}{n_2}} = 1.96\sqrt{\frac{2.5^2}{10} + \frac{3.0^2}{12}} \approx 2.30 \text{ min}$$

The confidence interval is

$$-0.8 \text{ min} < \mu_1 - \mu_2 < 3.8 \text{ min}$$

Notice that the confidence interval contains both negative and positive values. At the 95% confidence level, we cannot conclude that μ_1 is either less than or greater than μ_2. This result is consistent with the conclusion of the hypothesis test—that is, at the 5% level, the evidence is not sufficient to conclude that there is a difference in the average times for boiling 2 quarts of water between the two types of stoves.

Meaning of confidence interval for $\mu_1 - \mu_2$

In Example 4, we saw that the confidence interval for $\mu_1 - \mu_2$ contained both positive and negative numbers. At the 95% level of confidence, we could not conclude that there was a difference in value between μ_1 and μ_2. There are two other cases: either all the values in the confidence interval are positive, or they are all negative. The next procedure summarizes the interpretation of all three cases.

PROCEDURE

HOW TO INTERPRET CONFIDENCE INTERVALS FOR DIFFERENCES

Suppose that we construct a $c\%$ confidence interval for $\mu_1 - \mu_2$. Then three cases arise:

1. The $c\%$ confidence interval contains only *negative values*. In this case, we conclude that $\mu_1 - \mu_2 < 0$, and we are therefore $c\%$ confident that $\mu_1 < \mu_2$.

2. The $c\%$ confidence interval contains only *positive values*. In this case, we conclude that $\mu_1 - \mu_2 > 0$, and we can be $c\%$ confident that $\mu_1 > \mu_2$.

3. The $c\%$ confidence interval contains *both positive and negative values*. In this case, we cannot at the $c\%$ confidence level conclude that either μ_1 or μ_2 is larger. However, if we *reduce* the confidence level c to a *smaller value*, then the confidence interval will, in general, be shorter (explain why). A shorter confidence interval *might* put us back into case 1 or case 2 above (again, explain why).

GUIDED EXERCISE 4 | **Interpret a confidence interval**

(a) A study reported a 90% confidence interval for the difference of means to be

$$10 < \mu_1 - \mu_2 < 20$$

For this interval, what can you conclude about the respective values of μ_1 and μ_2?

⟹ At a 90% level of confidence, we can say that the difference $\mu_1 - \mu_2$ is positive, so $\mu_1 - \mu_2 > 0$ and $\mu_1 > \mu_2$.

(b) A study reported a 95% confidence interval for the difference of means to be

$$-0.32 < \mu_1 - \mu_2 < 0.16$$

From this interval, what can you conclude about the respective values of μ_1 and μ_2?

⟹ At the 95% confidence level, we see that the difference of means ranges from negative to positive values. We cannot tell from this interval if μ_1 is greater than μ_2 or μ_1 is less than μ_2.

COMMENT   In the case of large samples ($n_1 \geq 30$ and $n_2 \geq 30$), it is not unusual to see σ_1 and σ_2 approximated by s_1 and s_2. Then Theorem 10.2 is used as a basis for approximating confidence intervals for $\mu_1 - \mu_2$. In other words, when samples are large, sample estimates for σ_1 and σ_2 can be used together with the standard normal distribution to test $\mu_1 - \mu_2$ and find confidence intervals. However, in this text, we follow the more common convention of using a Student's t distribution whenever σ_1 and σ_2 are unknown.

Hypothesis Tests and Confidence Intervals for $\mu_1 - \mu_2$ When σ_1 and σ_2 Are Unknown

When σ_1 and σ_2 are unknown, we turn to a Student's t distribution. As before, when we use a Student's t distribution, we require that our populations be normal or approximately normal (mound-shaped and symmetrical) when the sample sizes n_1 and n_2 are less than 30. We also replace σ_1 by s_1 and σ_2 by s_2. Then we consider the approximate t value, attributed to Welch (*Biometrika*, Vol. 29, pp. 350–362):

$$t \approx \frac{(\bar{x}_1 - \bar{x}_2) - (\mu_1 - \mu_2)}{\sqrt{\frac{s_1^2}{n_1} + \frac{s_2^2}{n_2}}}$$

d.f. for $\mu_1 - \mu_2$ distribution when σ_1 and σ_2 are unknown

Unfortunately, this approximation is *not* exactly a Student's t distribution. However, it will be a good approximation, provided we adjust the degrees of freedom by one of the following methods.

1. The adjustment for the degrees of freedom is calculated from sample data. The formula is called *Satterthwaite's approximation*. It is rather complicated. Satterthwaite's approximation is used in statistical software packages such as Minitab and in the TI-84Plus/TI-83Plus/TI-*n*spire calculators. See Problem 23 for the formula.

2. An alternative method, which is much simpler, is to approximate the degrees of freedom using the *smaller* of $n_1 - 1$ and $n_2 - 1$.

For hypothesis tests and confidence intervals, we take the degrees of freedom *d.f.* to be the smaller of $n_1 - 1$ and $n_2 - 1$. This commonly used choice for the degrees of freedom is more conservative than Satterthwaite's approximation in the sense that it produces a slightly larger P-value for testing or a slightly larger margin of error for confidence intervals.

Applying methods similar to those used for hypothesis tests and confidence intervals for μ when σ is unknown, and using the Welch approximation for t, we obtain the following results:

PROCEDURE

HOW TO TEST $\mu_1 - \mu_2$ AND FIND A CONFIDENCE INTERVAL WHEN σ_1 AND σ_2 ARE UNKNOWN

Requirements
Obtain two independent random samples from populations 1 and 2, where

$\bar{x}_1$ and $\bar{x}_2$ are sample means from populations 1 and 2

s_1 and s_2 are sample standard deviations from populations 1 and 2

n_1 and n_2 are sample sizes from populations 1 and 2

If you can assume that both population distributions 1 and 2 are normal or at least mound-shaped and symmetrical, then any sample sizes n_1 and n_2 will work. If you cannot assume this, then use sample sizes $n_1 \geq 30$ and $n_2 \geq 30$.

Hypothesis test for $\mu_1 - \mu_2$
(σ_1 and σ_2 unknown)

Problem 23 discusses Satterthwaite's approximation for the degrees of freedom.

Procedure for Testing $\mu_1 - \mu_2$

1. In the context of the application, state the *null and alternate hypotheses* and set the *level of significance* α. It is customary to use $H_0: \mu_1 - \mu_2 = 0$.

2. Use $\mu_1 - \mu_2 = 0$ from the null hypothesis together with $\bar{x}_1$, $\bar{x}_2$, s_1, s_2, n_1, and n_2 to compute the sample *test statistic*.

$$t = \frac{(\bar{x}_1 - \bar{x}_2) - (\mu_1 - \mu_2)}{\sqrt{\dfrac{s_1^2}{n_1} + \dfrac{s_2^2}{n_2}}} = \frac{\bar{x}_1 - \bar{x}_2}{\sqrt{\dfrac{s_1^2}{n_1} + \dfrac{s_2^2}{n_2}}}$$

The sample test statistic distribution is approximately that of a Student's *t* with *degrees of freedom d.f. = smaller of $n_1 - 1$ and $n_2 - 1$.*

Note that statistical software gives a slightly more accurate and larger *d.f.* based on Satterthwaite's approximation (see Problem 23).

3. Use a Student's *t* distribution and the type of test, one-tailed or two-tailed, to find the *P-value* corresponding to the sample test statistic.

4. *Conclude the test.* If P-value ≤ α, then reject H_0. If P-value > α, then do not reject H_0.

5. *Interpret your conclusion* in the context of the application.

Procedure for Confidence Interval for $\mu_1 - \mu_2$

$$(\bar{x}_1 - \bar{x}_2) - E < \mu_1 - \mu_2 < (\bar{x}_1 - \bar{x}_2) + E$$

where $E = t_c \sqrt{\dfrac{s_1^2}{n_1} + \dfrac{s_2^2}{n_2}}$

c = confidence level ($0 < c < 1$)

t_c = critical value for confidence level c (see Table 4 of the Appendix.)

$d.f.$ = smaller of $n_1 - 1$ and $n_2 - 1$. Note that statistical software gives a slightly more accurate and larger *d.f.* based on Satterthwaite's approximation.

Confidence interval for $\mu_1 - \mu_2$
(σ_1 and σ_2 unknown)

EXAMPLE 5   TESTING THE DIFFERENCE OF MEANS AND FINDING A CONFIDENCE INTERVAL (σ_1 AND σ_2 UNKNOWN)

Two competing headache remedies claim to give fast-acting relief. An experiment was performed to compare the mean lengths of time required for bodily absorption of brand A and brand B headache remedies.

Twelve people were randomly selected and given an oral dose of brand A. Another 12 were randomly selected and given an equal dose of brand B. The lengths of time in minutes for the drugs to reach a specified level in the blood were recorded. The means, standard deviations, and sizes of the two samples follow.

Brand A: $\bar{x}_1 = 21.8$ min; $s_1 = 8.7$ min; $n_1 = 12$

Brand B: $\bar{x}_2 = 18.9$ min; $s_2 = 7.5$ min; $n_2 = 12$

Past experience with the drug composition of the two remedies permits researchers to assume that both distributions are approximately normal.

Hypothesis test: Use a 5% level of significance to test the claim that there is no difference in the mean time required for bodily absorption. Also, find or estimate the *P-value* of the sample test statistic.

SOLUTION:

(a) $\alpha = 0.05$. The null hypothesis is

$$H_0: \mu_1 = \mu_2 \qquad \text{or} \qquad H_0: \mu_1 - \mu_2 = 0$$

Since we have no prior knowledge about which brand is faster, the alternate hypothesis is

$$H_1: \mu_1 \neq \mu_2 \qquad \text{or} \qquad H_1: \mu_1 - \mu_2 \neq 0$$

(b) *Check Requirements* What distribution does the sample test statistic follow? We use a Student's t distribution with $d.f.$ = smaller sample size $- 1 = 12 - 1 = 11$. Note that both samples are of size 12. Degrees of freedom can be computed also by Satterthwaite's approximation. A Student's t distribution is appropriate because the original populations are approximately normal, the population standard deviations are not known, and the samples are independent.

(c) Compute the sample test statistic.
We're given $\bar{x}_1 = 21.8$ and $\bar{x}_2 = 18.9$, so the sample difference is $\bar{x}_1 - \bar{x}_2 = 21.8 - 18.9 = 2.9$. Using $s_1 = 8.7$, $s_2 = 7.5$, $n_1 = 12$, $n_2 = 12$, and $\mu_1 - \mu_2 = 0$ from H_0, we compute the sample test statistic.

$$t = \frac{(\bar{x}_1 - \bar{x}_2) - (\mu_1 - \mu_2)}{\sqrt{\dfrac{s_1^2}{n_1} + \dfrac{s_2^2}{n_2}}} = \frac{2.9}{\sqrt{\dfrac{8.7^2}{12} + \dfrac{7.5^2}{12}}} \approx 0.875$$

(d) Estimate the P-value and sketch the area on a t graph.
Figure 10-6 shows the P-value. The degrees of freedom are $d.f. = 11$ (since both samples are of size 12). Because the test is a two-tailed test, the P-value is the area to the right of 0.875 together with the area to the left of -0.875. In the Student's t distribution table (Table 4 of the Appendix), we find an interval containing the P-value. Find 0.875 in the row headed by $d.f. = 11$. The test statistic 0.875 falls between the entries 0.697 and 1.214. Because this is a two-tailed test, we use the corresponding P-values 0.500 and 0.250 from the *two-tail area* row (see Table 10-7, an excerpt from Table 4). The P-value for the sample t is in the interval

$$0.250 < P\text{-value} < 0.500$$

(e) Conclude the test.

$$
\begin{array}{c}
\alpha \\
\underset{0.05}{|} \qquad \underset{0.250}{(\;\rule[0.3ex]{2.5em}{0.6ex}\;)} \underset{0.500}{}
\end{array}
$$

Since the interval containing the P-value lies to the right of $\alpha = 0.05$, we fail to reject H_0.
Note: Using Satterthwaite's approximation for the degrees of freedom, $d.f. \approx 21.53$, the P-value ≈ 0.3915. This value is in the interval we computed.

FIGURE 10-6

P-value

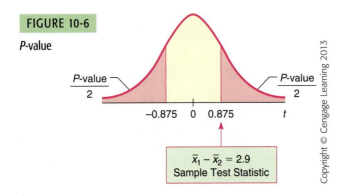

$$\bar{x}_1 - \bar{x}_2 = 2.9$$
Sample Test Statistic

TABLE 10-7   **Excerpt from Table 4 of the Appendix**

one-tail area	0.250	0.125
✓ two-tail area	0.500	0.250
$d.f. = 11$	0.697	1.214

Sample t = 0.875

(f) *Interpretation* Interpret the results.

At the 5% level of significance, there is insufficient evidence to conclude that there is a difference in mean times for the remedies to reach the specified level in the bloodstream.

Confidence interval: Find a 95% confidence interval for the population difference $\mu_1 - \mu_2$ of mean times for the competing drugs to reach a specified level in the blood. Interpret the results.

SOLUTION: Because neither σ_1 nor σ_2 is known and because the distributions of times for each drug to enter the bloodstream are approximately normal, it is appropriate to use a Student's t distribution. We approximate the degrees of freedom by using the smaller of $n_1 - 1$ and $n_2 - 1$. Since n_1 and n_2 are both 12, $d.f. = 12 - 1 = 11$. From Table 4 of the Appendix, we find that $t_{0.95} = 2.201$. The confidence interval is

$$(\bar{x}_1 - \bar{x}_2) - E < \mu_1 - \mu_2 < (\bar{x}_1 - \bar{x}_2) + E$$

with $\bar{x}_1 - \bar{x}_2 = 21.8 - 18.9 = 2.9$ min and

$$E = t_c \sqrt{\frac{s_1^2}{n_1} + \frac{s_2^2}{n_2}} = 2.201 \sqrt{\frac{8.7^2}{12} + \frac{7.5^2}{12}} \approx 7.3 \text{ min}$$

The confidence interval is

$$-4.4 \text{ min} < \mu_1 - \mu_2 < 10.2 \text{ min}$$

Note: Using Satterthwaite's formula, $d.f. = 21.53$ and the 95% confidence interval is from -3.99 to 9.79 minutes.

Interpretation Since the confidence interval contains both negative and positive values, we conclude that at the 95% confidence level, there is no difference in the population mean times for the two drugs to reach the bloodstream. This result is consistent with the two-tailed hypothesis test at the 5% level of significance.

GUIDED EXERCISE 5	**Testing the difference of means and finding a confidence interval (σ_1 and σ_2 unknown)**

Suppose the experiment to measure the time in minutes for the headache remedies to enter the bloodstream (Example 5) yielded sample means, sample standard deviations, and sample sizes as follows:

Brand A: $\bar{x}_1 = 20.1$ min; $s_1 = 8.7$ min; $n_1 = 12$

Brand B: $\bar{x}_2 = 11.2$ min; $s_2 = 7.5$ min; $n_2 = 8$

Brand B claims to be faster.

Hypothesis test: Is this claim justified at the 5% level of significance? (Use the following steps to obtain the answer.)

(a) What is α? State H_0 and H_1.

  $\alpha = 0.05$.

H_0: $\mu_1 = \mu_2$ or H_0: $\mu_1 - \mu_2 = 0$.

H_1: $\mu_1 > \mu_2$ (or H_1: $\mu_1 - \mu_2 > 0$). This says that the mean time for brand B is less than the mean time for brand A.

(b) *Check Requirements* What distribution does the sample test statistic follow? Explain.

Student's t distribution with $d.f. = n_2 - 1 = 8 - 1 = 7$ since $n_2 < n_1$. From Example 5, we know that the original distributions are approximately normal. Both σ_1 and σ_2 are unknown and the samples are independent.

Continued

GUIDED EXERCISE 5 *continued*

(c) Compute the sample test statistic
$\bar{x}_1 - \bar{x}_2$ and convert it to a t value.

➡ $\bar{x}_1 - \bar{x}_2 = 20.1 - 11.2 = 8.9$. Using $s_1 = 8.7$, $s_2 = 7.5$, $n_1 = 12$, $n_2 = 8$, and $\mu_1 - \mu_2 = 0$ from H_0, we have

$$t = \frac{\bar{x}_1 - \bar{x}_2}{\sqrt{\dfrac{s_1^2}{n_1} + \dfrac{s_2^2}{n_2}}} = \frac{8.9}{\sqrt{\dfrac{8.7^2}{12} + \dfrac{7.5^2}{8}}} \approx 2.437$$

(d) What degrees of freedom do you use? To find
an interval containing the P-value, do you use
one-tail or two-tail areas in Table 4 of the
Appendix? Sketch a figure showing the P-value.
Find an interval for the P-value.

➡ Since $n_2 < n_1$, $d.f. = n_2 - 1 = 8 - 1 = 7$. Use *one-tail area* in Table 4 of the Appendix.

TABLE 10-8 Excerpt from Student's t Table

✓ one-tail area	0.025	0.010
two-tail area	0.050	0.020
$d.f. = 7$	2.365	2.998

↑
Sample t = 2.437

Copyright © Cengage Learning 2013

FIGURE 10-7 P-value

P-value

0   2.437 *t*

$\bar{x}_1 - \bar{x}_2 = 8.9$
Sample Test Statistic

Copyright © Cengage Learning 2013

The sample t is between 2.365 and 2.998.

$0.010 < P\text{-value} < 0.025$

(e) Do we reject or fail to reject H_0?

➡ Since the interval containing the P-value has values that are all less than 0.05, we reject H_0.

α

0.010   0.025   0.05

Note: On the calculator with Satterthwaite's approximation for $d.f.$, we have $d.f. = 16.66$ and $P\text{-value} \approx 0.013$.

(f) *Interpret* the results in the context of the
application.

➡ At the 5% level of significance, there is evidence that the mean time for brand B to enter the bloodstream is less than the mean time for brand A.

Confidence interval: Find a 90% confidence interval for the difference of population means $\mu_1 - \mu_2$ of times for the two competing headache remedies to enter the bloodstream.

(g) Find the degrees of freedom and the critical
value $t_{0.90}$.

➡ Since $n_2 < n_1$, $d.f. = n_2 - 1 = 8 - 1 = 7$.
From Table 4 of the Appendix, $t_{0.90} = 1.895$.

(h) Find the maximal error of estimate E for a 90%
confidence interval.

➡ $E = t_c \sqrt{\dfrac{s_1^2}{n_1} + \dfrac{s_2^2}{n_2}} = 1.895 \sqrt{\dfrac{8.7^2}{12} + \dfrac{7.5^2}{8}} \approx 6.9$ min

Continued

GUIDED EXERCISE 5 *continued*

(i) Find the sample difference $\bar{x}_1 - \bar{x}_2$ and a 90% confidence interval for $\mu_1 - \mu_2$.

$\bar{x}_1 - \bar{x}_2 = 20.1 - 11.2 = 8.9$ min

$(\bar{x}_1 - \bar{x}_2) - E < \mu_1 - \mu_2 < (\bar{x}_1 - \bar{x}_2) + E$

$8.9 - 6.9 < \mu_1 - \mu_2 < 8.9 + 6.9$

$2\text{ min} < \mu_1 - \mu_2 < 15.8\text{ min}$

With Satterthwaite's formula, *d.f.* = 16.66 and the 90% confidence interval is 2.54 to 15.26 min.

(j) *Interpret* the results in the context of the application.

The confidence interval contains all positive numbers. This indicates that at the 90% confidence level, $\mu_1 > \mu_2$. It seems to take longer for brand A to enter the bloodstream.

COMMENT In Guided Exercise 5, we used a right-tailed test at the 5% level of significance and a 90% confidence interval. With both techniques, we concluded that the time for brand A to enter the bloodstream was longer than the time for brand B. This agreement of results is not an accident. In general, when dealing with the difference of means (and difference of proportions, as discussed in the next section), we will arrive at the same conclusion using either of the following two combinations:

1. Two-tailed hypothesis test at level of significance α and $(1 - \alpha)$ confidence interval

2. One-tailed hypothesis test at level of significance α and $(1 - 2\alpha)$ confidence interval

Alternate method using pooled standard deviation

Problem 24 discusses the use of the pooled standard deviation.

Pooled standard deviation

There is another method for testing $\mu_1 - \mu_2$ and finding confidence intervals when σ_1 and σ_2 are unknown. Suppose that the sample values s_1 and s_2 are sufficiently close and that there is reason to believe $\sigma_1 = \sigma_2$ (or that the standard deviations are approximately equal). This situation can happen when you make a slight change or alteration to a known process or method of production. The standard deviation may not change much, but the outputs or means could be very different. When there is reason to believe that $\sigma_1 = \sigma_2$, it is best to use a *pooled standard deviation*. The sample test statistic $\bar{x}_1 - \bar{x}_2$ has a corresponding t variable with an *exact* Student's t distribution and degrees of freedom *d.f.* = $n_1 + n_2 - 2$. Problem 24 at the end of this section provides the details for using the pooled standard deviation in hypothesis tests of and confidence intervals for $\mu_1 - \mu_2$.

Summary

Depending on the information available, there are several methods for testing and constructing confidence intervals for the difference of means $\mu_1 - \mu_2$ from two independent random samples. To use the normal probability distribution, you need to know σ_1 and σ_2. In addition, you need to know that the original population distributions are both normal or that n_1 and n_2 are both of size at least 30. To use a Student's t distribution, you need to know that the original population distributions are both normal or at least mound-shaped, or that n_1 and n_2 are both at least 30. The decision chart on the next page summarizes the appropriate formulas.

Continued

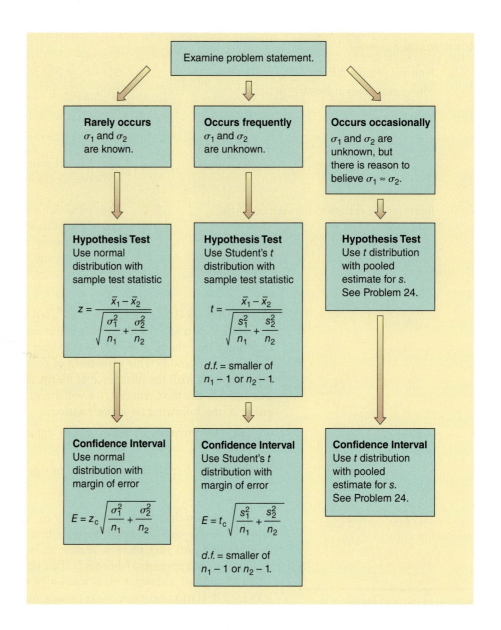

The TI-84Plus/TI-83Plus/TI-*n*spire calculators, Excel 2010, and Minitab all support testing the difference of means for independent samples. The TI-84Plus/TI-83Plus/TI-*n*spire calculators and Minitab also supply confidence intervals for the difference of means.

TI-84Plus/TI-83Plus/TI-*n*spire (with TI-84Plus keypad) Enter either summary statistics or raw data. Press **STAT** and select **TESTS**. Options **3:2-SampZTest** and **4:2-SampTTest** test the difference of means using the normal distribution and a Student's *t* distribution, respectively. Choice **9:2-SampZInt** finds a confidence interval for a difference of means when σ_1 and σ_2 are known. Choice **0:2-SampTInt** finds a confidence interval for a difference of means when σ_1 and σ_2 are unknown. In general, use **No** for Pooled. Then Satterthwaite's approximation for the degrees of freedom is used. However, if $\sigma_1 \approx \sigma_2$, use **Yes** for Pooled.

Excel 2010 Enter the data in two columns. Use the menu choices **Tools ➤ Data Analysis.** The choice **z-Test Two Sample Means** tests the difference of means using the normal distribution. The choice **t-Test: Two-Sample Assuming Unequal Variances** tests the difference of means using a Student's t distribution with Satterthwaite's approximation for the degrees of freedom. The choice **t-Test: Two-Sample Assuming Equal Variances** conducts a test using the pooled standard deviation. Another way to conduct a paired-difference-of-means test is to press **Insert Function** (f_x) and select **Statistical** for the category and then **T.TEST** for the function. In the dialogue box for type of test, use **2** for the t test with Satterthwaite's approximation for degrees of freedom or **3** for tests with pooled standard deviation.

Minitab For differences of means, enter the data into two columns. Use the menu choices **STAT ➤ Basic Statistics ➤ 2 sample t**. Minitab always uses a Student's t distribution for $\mu_1 - \mu_2$. In the dialogue box, leave the box **Assume equal variances** unchecked to use Satterthwaite's approximation for the degrees of freedom. Check the box **Assume equal variances** to use the pooled standard deviation.

Testing $\mu_1 - \mu_2$ Using Critical Regions (Optional)

For a set level of significance, the traditional critical region method yields the same results as the P-value method of hypothesis testing. Critical values z_0 for tests using the standard normal distribution can be found in Table 3(c) of the Appendix. Critical values t_0 for tests using a Student's t distribution are found in Table 4 of the Appendix. Use the row headed by the appropriate degrees of freedom and the column that includes the value of α (level of significance) in the *one-tail area* row for one-tailed tests or the *two-tail area* row for two-tailed tests.

EXAMPLE 6   CRITICAL REGION METHOD

Use the critical region method to solve the application in Example 5 (test $\mu_1 - \mu_2$ when σ_1 and σ_2 are unknown).

SOLUTION: Example 5 involves testing the difference in average times for two headache remedies to reach the bloodstream. For brand A, $\bar{x}_1 = 21.8$ min, $s_1 = 8.7$ min, and $n_1 = 12$; for brand B, $\bar{x}_2 = 18.9$ min, $s_2 = 7.5$ min, and $n_2 = 12$. The level of significance α is 0.05. The Student's t distribution is appropriate because both populations are approximately normal.

(a) To use the critical region method to test for a difference in average times, we use the same hypotheses and the same sample test statistic as in Example 5.

$$H_0: \mu_1 = \mu_2; H_1: \mu_1 \neq \mu_2$$

sample test statistic: $\bar{x}_1 - \bar{x}_2 = 2.9$ min; sample $t = 0.875$

(b) Instead of finding the P-value of the sample test statistic, we use α and H_1 to find the critical values in Table 4 of the Appendix. We have $d.f. = 11$ (since both samples are of size 12). We find $\alpha = 0.05$ in the *two-tail area* row, since we have a two-tailed test. The critical values are $\pm t_0 = \pm 2.201$.

Next compare the sample test statistic $t = 0.875$ to the critical values. Figure 10-8 shows the critical regions and the sample test statistic. We see that the sample test statistic falls in the "do not reject H_0" region. At the 5% level of significance, the sample evidence does not show a difference in times for the drugs to reach the bloodstream. The result is consistent with the result obtained by the P-value method of Example 5.

FIGURE 10-8

Critical Regions $\alpha = 0.05; d.f. = 11$

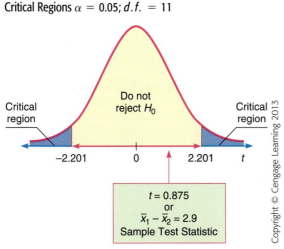

Critical region

Do not reject H_0

Critical region

−2.201   0   2.201 t

$t = 0.875$
or
$\bar{x}_1 - \bar{x}_2 = 2.9$
Sample Test Statistic

VIEWPOINT   Temper! Temper!

In her book Red Ink Behaviors, *Jean Hollands discusses inappropriate, problem behaviors of professional employees in the corporate business world. Temper tantrums, flaming e-mails, omitting essential information, sabotaging fellow workers, and the arrogant opinion that others are "dumb and dispensable" create personnel problems that cost companies a lot in the form of wasted time, reduced productivity, and lost revenues. A study of major industries in the Silicon Valley area gave Hollands data for estimating just how much time and money are wasted by such "red ink behaviors." For more information, see Problems 21 and 22 in this section.*

SECTION 10.2 PROBLEMS

aplia   Access end-of-section exercises at www.aplia.com

SECTION 10.3

Inferences about the Difference of Two Proportions $p_1 - p_2$

FOCUS POINTS
- Compute the sample test statistic and *P*-value for testing $p_1 - p_2$.
- Find confidence intervals for $p_1 - p_2$.

Testing a Difference of Proportions $p_1 - p_2$

Is the population proportion of people favoring more wilderness areas different between men and women, between younger people and older people, between ranchers and energy employees, etc.? To answer questions of this type, we explore the sampling distribution $\hat{p}_1 - \hat{p}_2$ from two independent binomial experiments.

THEOREM 10.3   Suppose we have two independent binomial experiments—that is, outcomes from one binomial experiment are in no way paired with outcomes from the other. We use the notation

Binomial Experiment 1	*Binomial Experiment 2*
n_1 = number of trials	n_2 = number of trials
r_1 = number of successes	r_2 = number of successes
p_1 = population probability of success on a single trial	p_2 = population probability of success on a single trial

For *large* values of n_1 and n_2, the distribution of sample differences

$$\hat{p}_1 - \hat{p}_2 = \frac{r_1}{n_1} - \frac{r_2}{n_2}$$

is closely approximated by a *normal distribution* with mean μ and standard deviation σ as shown:

$$\mu = p_1 - p_2 \qquad \sigma = \sqrt{\frac{p_1 q_1}{n_1} + \frac{p_2 q_2}{n_2}}$$

where $q_1 = 1 - p_1$ and $q_2 = 1 - p_2$.

Pooled estimate of proportion, $\bar{p}$

For most practical problems involving a comparison of two binomial populations, the experimenters will want to test the null hypothesis $p_1 = p_2$. Consequently, this is the type of test we shall consider. Since the values of p_1 and p_2 are unknown, and since specific values are not assumed under the null hypothesis $p_1 = p_2$, the best estimate for the common value is the total number of successes $(r_1 + r_2)$ divided by the total number of trials $(n_1 + n_2)$. If we denote this *pooled estimate of proportion* by $\bar{p}$ (read "p bar"), then

$$\bar{p} = \frac{r_1 + r_2}{n_1 + n_2}$$

This formula gives the best sample estimate $\bar{p}$ for p_1 and p_2 *under the assumption that $p_1 = p_2$.* Also, $\bar{q} = 1 - \bar{p}$.

Criteria for using the normal approximation to the binomial

COMMENT   For most practical applications, the sample sizes n_1 and n_2 are considered large samples if each of the four quantities

$$n_1\bar{p} \qquad n_1\bar{q} \qquad n_2\bar{p} \qquad n_2\bar{q}$$

is larger than 5 (see Section 7.6).

Theorem 10.3 leads to the following procedure for testing $p_1 - p_2$:

PROCEDURE

How to test a difference of proportions $p_1 - p_2$

Procedure Consider two independent binomial experiments.

Binomial Experiment 1	**Binomial Experiment 2**
n_1 = number of trials	n_2 = number of trials
r_1 = number of successes out of n_1 trials	r_2 = number of successes out of n_2 trials
$\hat{p}_1 = \dfrac{r_1}{n_1}$	$\hat{p}_2 = \dfrac{r_2}{n_2}$
p_1 = population probability of success on a single trial	p_2 = population probability of success on a single trial

1. Use the *null hypothesis* of no difference, $H_0: p_1 - p_2 = 0$. In the context of the application, choose the *alternate hypothesis*. Set the *level of significance* α.

2. The null hypothesis claims that $p_1 = p_2$; therefore, *pooled best estimates* for the population probabilities of success and failure are

$$\bar{p} = \frac{r_1 + r_2}{n_1 + n_2} \qquad \text{and} \qquad \bar{q} = 1 - \bar{p}$$

Requirements The number of trials should be sufficiently large so that all four quantities $n_1\bar{p}$, $n_1\bar{q}$, $n_2\bar{p}$, and $n_2\bar{q}$ are each larger than 5. In this case you compute the sample *test statistic*

$$z = \frac{\hat{p}_1 - \hat{p}_2}{\sqrt{\dfrac{\bar{p}\bar{q}}{n_1} + \dfrac{\bar{p}\bar{q}}{n_2}}}$$

Continued

3. Use the standard normal distribution and type of test, one-tailed or two-tailed, to find the *P-value* corresponding to the sample test statistic.

4. *Conclude the test.* If *P*-value $\leq \alpha$, then reject H_0. If *P*-value $> \alpha$, then do not reject H_0.

5. *Interpret your conclusion* in the context of the application.

EXAMPLE 7

TESTING THE DIFFERENCE OF PROPORTIONS

The Macek County clerk wishes to improve voter registration. One method under consideration is to send reminders in the mail to all citizens in the county who are eligible to register. As part of a pilot study to determine if this method will actually improve voter registration, a random sample of 1250 potential voters was taken. This sample was then randomly divided into two groups.

> *Group 1:* There were 625 people in this group. No reminders to register were sent to them. The number of potential voters from this group who registered was 295.

> *Group 2:* This group also contained 625 people. Reminders were sent in the mail to each member in the group, and the number who registered to vote was 350.

The county clerk claims that the proportion of people who registered was significantly greater in group 2. On the basis of this claim, the clerk recommends that the project be funded for the entire population of Macek County. Use a 5% level of significance to test the claim that the proportion of potential voters who registered was greater in group 2, the group that received reminders.

The Image Works

SOLUTION:

(a) Note that $\alpha = 0.05$. Let p_1 be the proportion of voters who registered from group 1, and let p_2 be the proportion who registered from group 2. The null hypothesis is that there is no difference in proportions, so

$$H_0: p_1 = p_2 \quad \text{or} \quad H_0: p_1 - p_2 = 0$$

The alternate hypothesis is that the proportion of voters who registered is greater for the group that received reminders.

$$H_1: p_1 < p_2 \quad \text{or} \quad H_1: p_1 - p_2 < 0$$

(b) Calculate the *pooled estimates* $\overline{p}$ and $\overline{q}$.
Under the null hypothesis that $p_1 = p_2$ we find

$$\overline{p} = \frac{r_1 + r_2}{n_1 + n_2} = \frac{295 + 350}{625 + 625} \approx 0.516 \quad \text{and} \quad \overline{q} = 1 - \overline{p} \approx 0.484$$

(c) *Check Requirements* What distribution does the sample test statistic follow? The sample test statistic follows a standard normal distribution because the number of binomial trials is large enough that each of the products $n_1\overline{p}$, $n_1\overline{q}$, $n_2\overline{p}$, and $n_2\overline{q}$ exceeds 5.

(d) Compute the sample statistic $\hat{p}_1 - \hat{p}_2$, and convert it to a z value.

CALCULATOR NOTE Carry the values for $\hat{p}_1$, $\hat{p}_2$, and the pooled estimates $\overline{p}$ and $\overline{q}$ to at least three places after the decimal. Then round the z value of the corresponding test statistic to two places after the decimal.

For the first group, the number of successes is $r_1 = 295$ out of $n_1 = 625$ trials. For the second group, there are $r_2 = 350$ successes out of $n_2 = 625$ trials. Since

$$\hat{p}_1 = \frac{r_1}{n_1} = \frac{295}{625} = 0.472 \qquad \text{and} \qquad \hat{p}_2 = \frac{r_2}{n_2} = \frac{350}{625} = 0.560$$

then

$$\hat{p}_1 - \hat{p}_2 = 0.472 - 0.560 = -0.088$$

Using these values and the values of $\bar{p}$ and $\bar{q}$ found in part (b), we find that

$$z = \frac{\hat{p}_1 - \hat{p}_2}{\sqrt{\dfrac{\bar{p}\bar{q}}{n_1} + \dfrac{\bar{p}\bar{q}}{n_2}}} = \frac{-0.088}{\sqrt{\dfrac{(0.516)(0.484)}{625} + \dfrac{(0.516)(0.484)}{625}}} \approx -3.11$$

(e) Find the P-value and sketch the area on the standard normal curve.
Figure 10-9 shows the P-value. This is a left-tailed test, so the P-value is the area to the left of -3.11. Using the standard normal distribution (Table 3 of the Appendix), we find $P\text{-value} = P(z < -3.11) \approx 0.0009$.

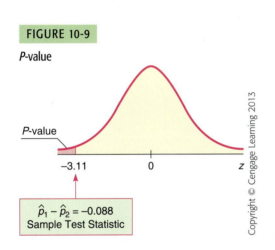

FIGURE 10-9

P-value

P-value

-3.11 0   z

$\hat{p}_1 - \hat{p}_2 = -0.088$
Sample Test Statistic

Copyright © Cengage Learning 2013

FIGURE 10-10

Critical Region, $\alpha = 0.05$

Critical region

-1.645 0   z

$z = -3.11$
Sample Test Statistic

Copyright © Cengage Learning 2013

(f) Conclude the test.
Since P-value of $0.0009 \le 0.05$ for α, we reject H_0.

(g) *Interpretation* Interpret the results in the context of the application.
At the 5% level of significance, the data indicate that the population proportion of potential voters who registered was greater in group 2, the group that received reminders.

(h) **Critical region method (optional):** Use the critical region method to conclude the test at the 5% level of significance. Compare your results with the P-value method.
In part (d), we found that the sample test statistic is $z = -3.11$. In Table 3(c) of the Appendix, we see that the critical value $z_0 = -1.645$ for a left-tailed test with $\alpha = 0.05$. In Figure 10-10, we see that the sample test statistic falls in the critical region, so we reject H_0. This conclusion is consistent with the conclusion obtained using the P-value method. There is sufficient evidence to conclude that at the 5% level of significance, the population of potential voters who registered was greater in the group receiving reminders.

GUIDED EXERCISE 6 | *Testing the difference of proportions*

In Example 7 about voter registration, suppose that a random sample of 1100 potential voters was randomly divided into two groups.

    Group 1: 500 potential voters; no registration reminders sent; 248 registered to vote

    Group 2: 600 potential voters; registration reminders sent; 332 registered to vote

Do these data support the claim that the proportion of voters who registered was greater in the group that received reminders than in the group that did not? Use a 1% level of significance.

(a) What is α? State H_0 and H_1.

$\alpha = 0.01$. As before, $H_0: p_1 = p_2$ and $H_1: p_1 < p_2$.

(b) Under the null hypothesis $p_1 = p_2$, calculate the *pooled estimates* $\overline{p}$ and $\overline{q}$.

$n_1 = 500,\ r_1 = 248;\ n_2 = 600,\ r_2 = 332$

$$\overline{p} = \frac{r_1 + r_2}{n_1 + n_2} = \frac{248 + 332}{500 + 600} \approx 0.527$$

$$\overline{q} = 1 - \overline{p} \approx 1 - 0.527 \approx 0.473$$

(c) *Check Requirements* What distribution does the sample test statistic follow?

The sample test statistic z follows a standard normal distribution because the number of binomial trials is large enough that each of the products $n_1\overline{p}$, $n_1\overline{q}$, $n_2\overline{p}$, and $n_2\overline{q}$ exceeds 5.

(d) What is the value of the sample statistic $\hat{p}_1 - \hat{p}_2$?

$$\hat{p}_1 = \frac{r_1}{n_1} = \frac{248}{500} = 0.496 \quad \hat{p}_2 = \frac{r_2}{n_2} = \frac{332}{600} \approx 0.553$$

$$\hat{p}_1 - \hat{p}_2 = -0.057$$

(e) Convert the sample statistic $\hat{p}_1 - \hat{p}_2 = -0.057$ to a z value.

$$z = \frac{\hat{p}_1 - \hat{p}_2}{\sqrt{\dfrac{\overline{p}\,\overline{q}}{n_1} + \dfrac{\overline{p}\,\overline{q}}{n_2}}} = \frac{-0.057}{\sqrt{\dfrac{(0.527)(0.473)}{500} + \dfrac{(0.527)(0.473)}{600}}} \approx -1.89$$

(f) Find the *P*-value and sketch the area on the standard normal curve.

Figure 10-11 on the next page shows the *P*-value. It is the area to the left of $z = -1.89$. Using Table 3 of the Appendix, we find *P*-value $= P(z < -1.89) = 0.0294$.

FIGURE 10-11 *P*-value

Copyright © Cengage Learning 2013

(g) Conclude the test.

Since *P*-value of 0.0294 > 0.01 for α, we cannot reject H_0.

(h) *Interpret* the results in the context of the application.

At the 1% level of significance, the data do not support the claim that the reminders increased the proportion of registered voters.

Estimating a Difference of Proportions $p_1 - p_2$

We conclude this section with a discussion of confidence intervals for $p_1 - p_2$, the difference of two proportions from two independent binomial probability distributions. Theorem 10.3 gives us the basis for constructing such confidence intervals. Using the notation of Theorem 10.3, recall that for sufficiently large sample sizes n_1 and n_2, the sampling distribution $\hat{p}_1 - \hat{p}_2$ of differences of proportions from two independent binomial probability distributions is approximately normal, with mean $\mu = p_1 - p_2$ and standard deviation $\sigma = \sqrt{p_1 q_1 / n_1 - p_2 q_2 / n_2}$.

There is one technical difficulty in computing the standard deviation σ of the $\hat{p}_1 - \hat{p}_2$ distribution: We don't know the values of p_1 or p_2. However, if all four quantities $n_1 \hat{p}_1, n_1 \hat{q}_1, n_2 \hat{p}_2$, and $n_2 \hat{q}_2$ are greater than 5, then a good approximation for the standard deviation σ of the $\hat{p}_1 - \hat{p}_2$ distribution is given by

$$\sigma \approx \hat{\sigma} = \sqrt{\frac{\hat{p}_1 \hat{q}_1}{n_1} + \frac{\hat{p}_2 \hat{q}_2}{n_2}}$$

This leads to the following procedure for confidence intervals:

PROCEDURE

How to find a confidence interval for $p_1 - p_2$

Requirements Consider two independent binomial experiments.

Binomial Experiment 1	*Binomial Experiment 2*
n_1 = number of trials	n_2 = number of trials
r_1 = number of successes out of n_1 trials	r_2 = number of successes out of n_2 trials
$\hat{p}_1 = \dfrac{r_1}{n_1};\ \hat{q}_1 = 1 - \hat{p}_1$	$\hat{p}_2 = \dfrac{r_2}{n_2};\ \hat{q}_2 = 1 - \hat{p}_2$
p_1 = population probability of success	p_2 = population probability of success

The number of trials should be sufficiently large so that all four of the following inequalities are true:

$$n_1 \hat{p}_1 > 5;\quad n_1 \hat{q}_1 > 5;\quad n_2 \hat{p}_2 > 5;\quad n_2 \hat{q}_2 > 5$$

Procedures for Confidence Interval for $p_1 - p_2$

$$(\hat{p}_1 - \hat{p}_2) - E \le p_1 - p_2 \le (\hat{p}_1 - \hat{p}_2) + E$$

where

$$E \approx z_c \hat{\sigma} = z_c \sqrt{\frac{\hat{p}_1 \hat{q}_1}{n_1} + \frac{\hat{p}_2 \hat{q}_2}{n_2}}$$

c = confidence level, $0 < c < 1$

z_c = critical value for confidence level c based on the standard normal distribution (see Table 3(b) of the Appendix for commonly used values)

EXAMPLE 8   Confidence interval for $p_1 - p_2$

In his book *Secrets of Sleep*, Professor Borbély describes research on dreams in the sleep laboratory at the University of Zurich Medical School. During normal sleep, there is a phase known as *REM* (rapid eye movement). For most people,

REM sleep occurs about every 90 minutes or so, and it is thought that dreams occur just before or during the REM phase. Using electronic equipment in the sleep laboratory, it is possible to detect the REM phase in a sleeping person. If a person is wakened immediately after the REM phase, he or she usually can describe a dream that has just taken place. Based on a study of over 650 people in the Zurich sleep laboratory, it was found that about one-third of all dream reports contain feelings of fear, anxiety, or aggression. There is a conjecture that if a person is in a good mood when going to sleep, the proportion of "bad" dreams (fear, anxiety, aggression) might be reduced.

Suppose that two groups of subjects were randomly chosen for a sleep study. In group I, before going to sleep, the subjects spent 1 hour watching a comedy movie. In this group, there were a total of $n_1 = 175$ dreams recorded, of which $r_1 = 49$ were dreams with feelings of anxiety, fear, or aggression. In group II, the subjects did not watch a movie but simply went to sleep. In this group, there were a total of $n_2 = 180$ dreams recorded, of which $r_2 = 63$ were dreams with feelings of anxiety, fear, or aggression.

(a) *Check Requirements* Why could groups I and II be considered independent binomial distributions? Why do we have a "large-sample" situation?

> **SOLUTION:** Since the two groups were chosen randomly, it is reasonable to assume that neither group's response would be related to the other's. In both groups, each recorded dream could be thought of as a trial, with success being a dream with feelings of fear, anxiety, or aggression.

$$\hat{p}_1 = \frac{r_1}{n_1} = \frac{49}{175} = 0.28 \quad \text{and} \quad \hat{q}_1 = 1 - \hat{p}_1 = 0.72$$

$$\hat{p}_2 = \frac{r_2}{n_2} = \frac{63}{180} = 0.35 \quad \text{and} \quad \hat{q}_2 = 1 - \hat{p}_2 = 0.65$$

Since

$$n_1\hat{p}_1 = 49 > 5 \quad n_1\hat{q}_1 = 126 > 5$$

$$n_2\hat{p}_2 = 63 > 5 \quad n_2\hat{q}_2 = 117 > 5$$

then large-sample theory is appropriate.

(b) What is $p_1 - p_2$? Compute a 95% confidence interval for $p_1 - p_2$.

> **SOLUTION:** p_1 is the population proportion of successes (bad dreams) for all people who watch comedy movies before bed. Thus, p_1 can be thought of as the percentage of bad dreams for all people who are in a "good mood" when they go to bed. Likewise, p_2 is the percentage of bad dreams for the population of all people who just go to bed (no movie). The difference $p_1 - p_2$ is the population difference.

To find a confidence interval for $p_1 - p_2$, we need the values of z_c, $\hat{\sigma}$, and E. From Table 3(b) of the Appendix, we see that $z_{0.95} = 1.96$, so

$$\hat{\sigma} = \sqrt{\frac{\hat{p}_1 \hat{q}_1}{n_1} + \frac{\hat{p}_2 \hat{q}_2}{n_2}} = \sqrt{\frac{(0.28)(0.72)}{175} + \frac{(0.35)(0.65)}{180}}$$

$$\approx \sqrt{0.0024} \approx 0.0492$$

$$E = z_c \hat{\sigma} = 1.96(0.0492) \approx 0.096$$

$$(\hat{p}_1 - \hat{p}_2) - E < p_1 - p_2 < (\hat{p}_1 - \hat{p}_2) + E$$

$$(0.28 - 0.35) - 0.096 < p_1 - p_2 < (0.28 - 0.35) + 0.096$$

$$-0.166 < p_1 - p_2 < 0.026$$

(c) *Interpretation* Explain the meaning of the confidence interval that you constructed in part (b).

SOLUTION: We are 95% sure that the interval between -16.6% and 2.6% is one that contains the percentage difference of "bad" dreams for group I and group II. Since the interval -0.166 to 0.026 is not all negative (or all positive), we cannot say that $p_1 - p_2 < 0$ (or $p_1 - p_2 > 0$). Thus, at the 95% confidence level, we *cannot* conclude that $p_1 < p_2$ or $p_1 > p_2$. The comedy movies before bed help some people reduce the percentage of "bad" dreams, but at the 95% confidence level, we cannot say that the *population difference* is reduced.

TECH NOTES   The TI-84Plus/TI-83Plus/TI-*n*spire calculators and Minitab support both testing and confidence intervals for the difference of proportions.

TI-84Plus/TI-83Plus/TI-*n*spire (with TI-84Plus calculator) Use the **STAT** key and highlight **TESTS**. The menu choices **B:2-PropZInt** and **6:2-PropZTest** provide confidence intervals and testing, respectively. Note that the symbol x is used to designate the number of successes r.

Minitab Use the menu choices **STAT ➤ Basic Statistics ➤ 2 proportions**. In the dialogue box, under Options, select the null and alternate hypotheses and set the confidence level.

VIEWPOINT | What's the difference?

Will two 15-minute piano lessons a week significantly improve a child's analytical reasoning skills? Why piano? Why not computer keyboard instruction or maybe voice lessons? Professor Frances Rauscher, University of Wisconsin, and Professor Gordon Shaw, University of California, Irvine, claim that there is a difference! How could this be measured? A large number of piano students were given complicated tests of mental ability. Independent control groups of other students were given the same tests. Techniques involving the study of differences of means and proportions were used to draw the conclusion that students taking piano lessons did better on tests measuring analytical reasoning skills (reported in The Denver Post).

SECTION 10.3 PROBLEMS   aplia   Access end-of-section exercises at www.aplia.com

Chapter Review

In this chapter, we continued the discussion of hypothesis testing and confidence intervals. In particular, we looked at inferences involving paired differences, differences of means, and differences of proportions.

- The first task is to determine whether two samples are independent or dependent. Two samples are dependent if each data value in one population can be paired in a meaningful way with a corresponding data value in the other population. Two samples are independent if the selection of data drawn from one population is completely unrelated to the selection of data drawn from the other population.

- Section 10.1 presented the paired difference test for difference of means $\bar{d}$ from dependent samples. The Student's t distribution is used. See procedure displays in Section 10.1 for specific formulas and methods.

- Section 10.2 presented inferences regarding the parameter $\mu_1 - \mu_2$ from independent populations. Both hypothesis tests and confidence intervals were discussed. Depending on our knowledge of population standard deviations, either the normal distribution or the Student's t distribution is used. See procedure displays in Section 10.2.

- Section 10.3 presented inferences regarding the parameter $p_1 - p_2$ from large independent populations. Both hypothesis tests and confidence intervals were discussed. For testing procedures, the pooled estimate for the proportion $\bar{p}$ of successes is used. For sufficiently large numbers of trials, the normal distribution is employed for both confidence intervals and testing. See procedure displays in Section 10.3.

IMPORTANT WORDS & SYMBOLS

Section 10.1
Dependent samples   342
Data pairs   342
μ_d, mean of differences from data pairs   343
σ_d, standard deviation of differences from data pairs   343

Section 10.2
Independent samples   350
d.f. for $\mu_1 - \mu_2$ distribution when σ_1 and σ_2 are unknown   356
Pooled standard deviation   361

Section 10.3
Pooled estimate of proportion, $\bar{p}$ 365

VIEWPOINT | Who Watches Cable TV?

Consider the following claim: Average cable TV viewers are, generally speaking, as affluent as newspaper readers and better off than radio or magazine audiences. How do we know that this claim is true? One way to answer such a question is to construct several individual tests of difference of means. One test would compare cable TV viewers with newspaper readers, a second test would compare cable TV viewers with radio audiences, and a third test would compare cable TV viewers with magazine audiences. Another way to handle all three tests at once would be to use techniques of ANOVA (analysis of variance)––a subject often introduced in a second course in statistics. For more information and data, see American Demographics *(Vol. 17, No. 6).*

CHAPTER REVIEW PROBLEMS

1. | *Statistical Literacy* When are two random samples independent?

2. | *Statistical Literacy* When are two random samples dependent?

For each hypothesis test in Problems 3–10, please provide the following information:

(i)   What is the level of significance? State the null and alternate hypotheses.

(ii) *Check Requirements* What sampling distribution will you use? What asumptions are you making? What is the value of the sample test statistic?

(iii) Find (or estimate) the *P*-value. Sketch the sampling distribution and show the area corresponding to the *P*-value.

(iv) Based on your answers in parts (i) – (iii), will you reject or fail to reject the null hypothesis? Are the data statistically significant at level α?

(v) *Interpret* your conclusion in the context of the application.

Note: For degrees of freedom *d.f.* not in the Student's *t* table, use the closest *d.f.* that is smaller. In some cases, this choice will increase the *P*-value by a small amount or increase the length of a confidence interval, thereby making the answer slightly more "conservative." Answers may vary due to rounding.

3. *Agriculture: Bell Peppers* The following data represent soil water content (percentage of water by volume) for independent random samples of soil taken from two experimental fields growing bell peppers (Reference: *Journal of Agricultural, Biological, and Environmental Statistics*). *Note:* These data are also available for download at http://www.cengagebrain.com.

Soil water content from field I: x_1; $n_1 = 72$

15.1	11.2	10.3	10.8	16.6	8.3	9.1	12.3	9.1	14.3
10.7	16.1	10.2	15.2	8.9	9.5	9.6	11.3	14.0	11.3
15.6	11.2	13.8	9.0	8.4	8.2	12.0	13.9	11.6	16.0
9.6	11.4	8.4	8.0	14.1	10.9	13.2	13.8	14.6	10.2
11.5	13.1	14.7	12.5	10.2	11.8	11.0	12.7	10.3	10.8
11.0	12.6	10.8	9.6	11.5	10.6	11.7	10.1	9.7	9.7
11.2	9.8	10.3	11.9	9.7	11.3	10.4	12.0	11.0	10.7
8.8	11.1								

Soil water content from field II: x_2; $n_2 = 80$

12.1	10.2	13.6	8.1	13.5	7.8	11.8	7.7	8.1	9.2
14.1	8.9	13.9	7.5	12.6	7.3	14.9	12.2	7.6	8.9
13.9	8.4	13.4	7.1	12.4	7.6	9.9	26.0	7.3	7.4
14.3	8.4	13.2	7.3	11.3	7.5	9.7	12.3	6.9	7.6
13.8	7.5	13.3	8.0	11.3	6.8	7.4	11.7	11.8	7.7
12.6	7.7	13.2	13.9	10.4	12.8	7.6	10.7	10.7	10.9
12.5	11.3	10.7	13.2	8.9	12.9	7.7	9.7	9.7	11.4
11.9	13.4	9.2	13.4	8.8	11.9	7.1	8.5	14.0	14.2

(a) Use a calculator with mean and standard deviation keys to verify that $\bar{x}_1 \approx 11.42$, $s_1 \approx 2.08$, $\bar{x}_2 \approx 10.65$, and $s_2 \approx 3.03$.

(b) Let μ_1 be the population mean for x_1 and let μ_2 be the population mean for x_2. Find a 95% confidence interval for $\mu_1 - \mu_2$.

(c) Examine the confidence interval and explain what it means in the context of this problem. Does the interval consist of numbers that are all positive? all negative? of different signs? At the 95% level of confidence, is the population mean soil water content of the first field higher than that of the second field?

(d) Which distribution (standard normal or Student's *t*) did you use? Why? Do you need information about the soil water content distributions?

(e) Use $\alpha = 0.01$ to test the claim that the population mean soil water content of the first field is higher than that of the second.

4. *Stocks: Retail and Utility* How profitable are different sectors of the stock market? One way to answer such a question is to examine profit as a percentage of stockholder equity. A random sample of 32 retail stocks such as Toys "Я" Us, Best Buy, and Gap was studied for x_1, profit as a percentage of stockholder equity. The result was $\bar{x}_1 = 13.7$. A random sample of 34 utility (gas and electric) stocks such as Boston Edison, Wisconsin Energy, and Texas Utilities was studied for x_2, profit as a percentage of stockholder equity. The result was $\bar{x}_2 = 10.1$ (Source: *Fortune 500*, Vol. 135, No. 8). Assume $\sigma_1 = 4.1$ and $\sigma_2 = 2.7$.

(a) Let μ_1 represent the population mean profit as a percentage of stockholder equity for retail stocks, and let μ_2 represent the population mean profit as a percentage of stockholder equity for utility stocks. Find a 95% confidence interval for $\mu_1 - \mu_2$.

(b) Examine the confidence interval and explain what it means in the context of this problem. Does the interval consist of numbers that are all positive? all negative? of different signs? At the 95% level of confidence, does it appear that the profit as a percentage of stockholder equity for retail stocks is higher than that for utility stocks?

(c) Test the claim that the profit as a percentage of stockholder equity for retail stocks is higher than that for utility stocks. Use $\alpha = 0.01$.

5. *Wildlife: Wolves* A random sample of 18 adult male wolves from the Canadian Northwest Territories gave an average weight $\bar{x}_1 = 98$ lb with estimated sample standard deviation $s_1 = 6.5$ lb. Another sample of 24 adult male wolves from Alaska gave an average weight $\bar{x}_2 = 90$ lb with estimated sample standard deviation $s_2 = 7.3$ lb (Source: *The Wolf*, by L. D. Mech, University of Minnesota Press).

(a) Let μ_1 represent the population mean weight of adult male wolves from the Northwest Territories, and let μ_2 represent the population mean weight of adult male wolves from Alaska. Find a 75% confidence interval for $\mu_1 - \mu_2$.

(b) Examine the confidence interval and explain what it means in the context of this problem. Does the interval consist of numbers that are all positive? all negative? of different signs? At the 75% level of confidence, does it appear that the average weight of adult male wolves from the Northwest Territories is greater than that of the Alaska wolves?

(c) Test the claim that the average weight of adult male wolves from the Northwest Territories is different from that of Alaska wolves. Use $\alpha = 0.01$.

6. *Wildlife: Wolves* A random sample of 17 wolf litters in Ontario, Canada, gave an average of $\bar{x}_1 = 4.9$ wolf pups per litter with estimated sample standard deviation $s_1 = 1.0$. Another random sample of 6 wolf litters in Finland gave an average of $\bar{x}_2 = 2.8$ wolf pups per litter with sample standard deviation $s_2 = 1.2$ (see source for Problem 5).

(a) Find an 85% confidence interval for $\mu_1 - \mu_2$, the difference in population mean litter size between Ontario and Finland.

(b) Examine the confidence interval and explain what it means in the context of this problem. Does the interval consist of numbers that are all positive? all negative? of different signs? At the 85% level of confidence, does it appear that the average litter size of wolf pups in Ontario is greater than the average litter size in Finland?

(c) Test the claim that the average litter size of wolf pups in Ontario is greater than the average litter size of wolf pups in Finland. Use $\alpha = 0.01$.

7. *Survey Response: Validity* The book *Survey Responses: An Evaluation of Their Validity*, by E. J. Wentland and K. Smith (Academic Press), includes studies reporting the accuracy of answers to questions from surveys. A study by Locander and others considered the question "Are you a registered voter?" Accuracy of response was confirmed by a check of city voting records. Two methods of survey were used: a face-to-face interview and a telephone interview. A random sample of 93 people were asked the voter registration question face-to-face. Seventy-nine respondents gave accurate answers (as verified by city records). Another random sample of 83 people were asked the same question during a telephone interview. Seventy-four respondents gave accurate answers. Assume that the samples are representative of the general population.

(a) Let p_1 be the population proportion of all people who answer the voter-registration question accurately during a face-to-face interview. Let p_2 be the population proportion of all people who answer the question accurately during a telephone interview. Find a 95% confidence interval for $p_1 - p_2$.

(b) Does the interval contain numbers that are all positive? all negative? mixed? Comment on the meaning of the confidence interval in the context of this problem. At the 95% level, do you detect any difference in the proportion of accurate responses from face-to-face interviews compared with the proportion of accurate responses from telephone interviews?

(c) Test the claim that there is a difference in the proportion of accurate responses from face-to-face interviews compared with telephone interviews. Use $\alpha = 0.05$.

8. *Survey Response: Validity* Locander and others (see reference in Problem 7) also studied the accuracy of responses on questions involving more sensitive material than voter registration. From public records, individuals were identified as having been charged with drunken driving not less than 6 months or more than 12 months from the starting date of the study. Two random samples from this group were studied. In the first sample of 30 individuals, the respondents were asked in a face-to-face interview if they had been charged with drunken driving in the last 12 months. Of these 30 people interviewed face-to-face, 16 answered the question accurately. The second random sample consisted of 46 people who had been charged with drunken driving. During a telephone interview, 25 of these responded accurately to the question asking if they had been charged with drunken driving during the past 12 months. Assume that the samples are representative of all people recently charged with drunken driving.

(a) Let p_1 represent the population proportion of all people with recent charges of drunken driving who respond accurately to a face-to-face interview asking if they have been charged with drunken driving during the past 12 months. Let p_2 represent the population proportion of people who respond accurately to the same question when it is asked in a telephone interview. Find a 90% confidence interval for $p_1 - p_2$.

(b) Does the interval found in part (a) contain numbers that are all positive? all negative? mixed? Comment on the meaning of the confidence interval in the context of this problem. At the 90% level, do you detect any differences in the proportion of accurate responses to the question from face-to-face interviews as compared with the proportion of accurate responses from telephone interviews?

(c) Test the claim that there is a difference in the proportion of accurate responses from face-to-face interviews compared with the proportion of accurate responses from telephone interviews. Use $\alpha = 0.05$.

9. *Marketing: Sporting Goods* A marketing consultant was hired to visit a random sample of five sporting goods stores across the state of California. Each store was part of a large franchise of sporting goods stores. The consultant taught the managers of each store better ways to advertise and display their goods. The net sales for 1 month before and 1 month after the consultant's visit were recorded as follows for each store (in thousands of dollars):

Store	1	2	3	4	5
Before visit	57.1	94.6	49.2	77.4	43.2
After visit	63.5	101.8	57.8	81.2	41.9

Do the data indicate that the average net sales improved? (Use $\alpha = 0.05$.)

10. *Psychology: Creative Thinking* Six sets of identical twins were randomly selected from a population of identical twins. One child was taken at random from each pair to form an experimental group. These children participated in a program designed to promote creative thinking. The other child from each pair was part of the control group that did not participate in the program to promote creative thinking. At the end of the program, a creative problem-solving test was given, with the results shown in the following table:

Twin pair	A	B	C	D	E	F
Experimental group	53	35	12	25	33	47
Control group	39	21	5	18	21	42

Higher scores indicate better performance in creative problem solving. Do the data support the claim that the program of the experimental group promoted creative problem solving? (Use $\alpha = 0.01$.)

DATA HIGHLIGHTS: GROUP PROJECTS

Break into small groups and discuss the following topic. Organize a brief outline in which you summarize the main points of your group discussion.

"Sweets May Not Be Culprit in Hyper Kids" was a *USA Today* (February 3, 1994) headline reporting results of a study that appeared in the *New England Journal of Medicine*. In this study, the subjects were 25 normal preschoolers ages 3 to 5, and 23 kids ages 6 to 10, who had been described as "sensitive to sugar." The kids and their families were put on three different diets for 3 weeks each. One diet was high in sugar, one was low in sugar and contained aspartame, and one was low in sugar and contained saccharin. The diets were all free of other additives, artificial food coloring, preservatives, and chocolate. All food in the households was removed, and then meals were delivered to the families. Researchers gathered information about the kids' behavior from parents, babysitters, and teachers. In addition, researchers tested the kids for memory, concentration, reading, and math skills. The result: "We couldn't find any difference in terms of their behavior or their learning on any of the three diets," said Mark Wolraich, professor of pediatrics at Vanderbilt University Medical Center, who oversaw the project. In another interview, Dr. Wolraich was quoted as saying, "Our study would say there is no evidence sugar has an adverse effect on children's behavior."

(a) This research involved comparisons of several means, not just two. However, let us take a simplified view of the problem and consider the difference in behavior when children consumed the diet with high sugar compared with their behavior when they consumed the diet with aspartame and low sugar. List some variables that might be measured to reflect the behavior of the children.

(b) Let's assume that the general null hypothesis was that there is no difference in children's behavior when they have a diet high in sugar. Was the evidence sufficient to allow the researchers to reject the null hypothesis and conclude that there are differences in children's behavior when they have a diet high in sugar? When we cannot reject H_0, have we *proved* that H_0 is true? In your own words, paraphrase the comments made by Dr. Wolraich.

LINKING CONCEPTS: WRITING PROJECTS

Discuss the following topic in class or review it on your own. Then write a brief but complete essay in which you summarize the main points. Please include formulas and graphs as appropriate.

Is there a relationship between confidence intervals and two-tailed hypothesis tests? The answer is yes. Let c be the level of confidence used to construct a confidence interval from sample data. Let α be the level of significance for a two-tailed hypothesis test. The following statement applies to hypothesis tests of the mean:

For a two-tailed hypothesis test with level of significance α and null hypothesis H_0: $\mu = k$, we *reject* H_0 whenever k falls *outside* the $c = 1 - \alpha$ confidence interval for μ based on the sample data. When k falls within the $c = 1 - \alpha$ confidence interval, we do not reject H_0.

For a one-tailed hypothesis test with level of significance α and null hypothesis H_0: $\mu = k$, we *reject* H_0 whenever k falls *outside* the $c = 1 - 2\alpha$ confidence interval for μ based on the sample data. When k falls within the $c = 1 - 2\alpha$ confidence interval, we do not reject H_0.

A corresponding relationship between confidence intervals and two-tailed hypothesis tests is also valid for other parameters, such as p, $\mu_1 - \mu_2$, and $p_1 - p_2$.

(a) Consider the hypotheses H_0: $\mu_1 - \mu_2 = 0$ and H_1: $\mu_1 - \mu_2 \neq 0$. Suppose a 95% confidence interval for $\mu_1 - \mu_2$ contains only positive numbers. Should you reject the null hypothesis when $\alpha = 0.05$? Why or why not?

(b) Consider the hypotheses H_0: $p_1 - p_2 = 0$ and H_1: $p_1 - p_2 > 0$. Suppose a 98% confidence interval for $p_1 - p_2$ contains only positive numbers. Should you reject or fail to reject H_0 at the $\alpha = 0.01$ level of significance?

USING TECHNOLOGY

Application
Paired Difference Test

Suppose a random sample of eight nurses' schedules were changed from the day shift to a rotating shift of some night work and some day work. For each of these nurses, the information shown in the table below was recorded about level of tension and anxiety at the end of a day shift and also at the end of a night shift. Higher numbers indicate higher levels of tension.

Data for Nurses Working Both Shifts

Nurse	1	2	3	4	5	6	7	8
Day shift (B)	1.5	3	2	3	2	2	1	2
Night shift (A)	3.5	2	4	4	3.5	2	3.5	3

(a) Explain why the sample data for the day shift cannot be thought of as independent of the sample data for the night shift.
(b) Let us say that A is the random variable representing tension levels of night nurses and B is the random variable representing tension levels of day nurses. If we wanted to test the claim that nurses have a higher level of tension after the night shift, what would we use for the null hypothesis? What would we use for the alternate hypothesis? Choose the appropriate hypotheses and enter your choices on the computer. Use a 2% level of significance. Shall we accept or reject the claim that after a night shift, nurses express more feelings of tension, on average, than they do after a day shift?
(c) What is the smallest level of significance at which these data will allow us to accept the claim that after a night shift, nurses express more feelings of tension?

Technology Hints: Paired Difference Test

TI-84Plus/TI-83Plus/TI-nspire (with TI-84Plus keypad)

Enter the first number of each data pair in the list L_1 and the corresponding value in the second list L_2. Create the list L_3 by subtracting L_2 from L_1. List L_3 contains the differences d. Our next step is to conduct a t test on the differences in L_3. Press **STAT**, select **TESTS**, and choose **Option 2:T-Test.** Use **Inpt:Data.** The value for μ_0 is 0 because we are testing the hypothesis $H_0: \mu_0 = 0$. Indicate that the data are in list L_3 with **Freq: 1.** Select the appropriate symbol ($\neq, <, >$) for the alternate hypothesis, and then choose **Calculate**, which gives the value of the sample test statistic along with its t value and P-value, or **Draw**, which shows the area corresponding to the P-value of the sample test statistic and gives the t value and P-value of the sample test statistic.

Excel 2010

Enter the data in two columns. Then use the ribbon choices **Data Tab ➤ Data Analysis ➤ t-Test: Paired Two-Sample for Means.** The output contains the mean and variance for each data column, the degrees of freedom for the test, the sample t statistic, the P-value for a one-tailed test, the P-value for a two-tailed test, and the critical values t_0 for the specified level of significance for both a one-tailed test and a two-tailed test. Another way to conduct a paired difference t test is to press **Insert Function** (f_x) and select **Statistical** for the category and then **T.TEST** for the function. In the dialogue box, use **1** (paired) for the type of test.

Minitab

Enter the data in two columns. Then use the menu selections **Stats ➤ Basic Statistics ➤ Paired-t**. The output includes an option for a confidence interval. Other items include the number of data pairs n, the mean, the standard deviation, and the standard error of the mean, for each data set as well as for the differences. The t value of the sample test statistic is given along with the P-value. There is also an option to generate a histogram, dotplot, or boxplot for the differences.

SPSS

Enter the data in two columns in the Data View window. In the Variable View window, name and label the two columns. Be sure the data type is numeric. Use the commands **Analyze ➤ Compare Means ➤ Paired-Samples T Test**. In the dialogue box, select Options to set the confidence level for a confidence interval. Results for a two-tailed test with $H_0: \mu_d = 0$ and $H_1: \mu_d \neq 0$ are given, with test statistic, standard deviation of the paired differences, standard error, t value of the test statistic, degrees of freedom, and P-value of the test statistic. In the output, **Sig(2-tailed)** is the P-value of the test statistic for a two-tailed test. A confidence interval for the mean of the paired differences is also given, as well as the mean, standard deviation, and standard error for each column of data.

Other Applications
Technology Hints: Inferences for $\mu_1 - \mu_2$

TI-84Plus/TI-83Plus/TI-*n*spire (with TI-84Plus keypad)

You have the option of using raw data entered into two separate lists or summary statistics. Press the **STAT** key, select the **TESTS** option, and then use option **3:2-SampZTest** or option **4:2-SampTTest**. The output gives the z or t value of the sample test statistic and the P-value for the test. For confidence intervals, use option **9:2-SampZInt** or option **0:2-SampTTest** under the **TESTS** option.

Excel 2010

Enter the data in two columns. Then use the ribbon selection ➤ **Data Tab ➤ Data Analysis ➤ z-Test Two Sample for Means** if you know σ_1 and σ_2. Otherwise, use **t-Test: Two Sample Assuming Unequal Variances**. The output provides the mean and variance for each variable, the z or t value of the sample test statistic, the P-values for a one-tailed test and a two-tailed test, and the critical z_0 or t_0 values for a one-tailed test and a two-tailed test. Another way to conduct a paired difference of means test is to press **Insert Function** (fx) and select **Statistical** for the category and then **T.TEST** for the function. In the dialogue box for type of test, use **2** for the t test with Satterthwaite's approximation for degrees of freedom or **3** for tests with pooled standard deviation.

Minitab

When testing the difference of means from independent samples, Minitab always uses the t distribution. The P-value of the sample test statistic will be slightly larger than the P-value generated using the normal distribution. Enter the data in two columns. Then use the menu selections ➤ **Stat** ➤ **Basic Statistics** ➤ **2-Sample** t. Do not check equal variances. The output gives an option for a confidence interval as well as for dotplots or boxplots of the two variables. The output displays the mean, standard deviation, and standard error of the mean for each of the variables. The t value of the sample test statistic and the P-value are based on the Student's t distribution.

SPSS

Enter all the data in one column. In an adjacent column, enter a grouping variable with two values to separate the cases into two independent samples. For instance, use the value 1 to indicate that a data value is from the first sample, and the value 2 to indicate that a data value is from the second sample. Then use the menu choices **Analyze** ➤ **Compare Means** ➤ **Independent-Samples T Test.** In the dialogue box, move the column containing the data into the Test Variable box and the column containing the grouping variables into the Grouping Variable box. Then press Define Groups and enter the grouping values, such as 1 and 2. A confidence interval can be set by using the Options button. The output gives the mean, standard deviation, and standard error for each sample. Test results show the t value, degrees of freedom, and P-value for a two-tailed test for both the case of equal variances and the case of unequal variances. Confidence intervals are also included in the output.

11

Business Wire/Getty Images Publicity/Getty Images

Photo Researchers/Alamy

Make everything as simple as possible, but no simpler.

—ALBERT EINSTEIN

Today Professor Einstein's theories are well accepted as part of modern physics. However, this was not always the case. When Einstein first discovered relativity and the bending of light in large gravitational fields, his theories were held suspect. Physicists studied starlight as it passed near our sun (a large gravitational field). They discovered that starlight did not travel in a straight line. Instead it bent very slightly, exactly as Einstein had predicted!

Many studies in natural science, medicine, social science, economics, and business require testing one established theory against another new theory. Statistical tests you will learn in this chapter are of great value in such work.

For online student resources, visit the Brase/Brase, *Understanding Basic Statistics,* 6th edition website at **http://www.cengagebrain.com.**

ADDITIONAL TOPICS USING INFERENCE

PREVIEW QUESTIONS

How do you decide if random variables are dependent or independent?
(SECTION 11.1)

How do you decide if different populations share the same proportions of specified characteristics? (SECTION 11.1)

How do you decide if two distributions are not only dependent, but actually the same distribution? (SECTION 11.2)

How do you construct tests for σ? (SECTION 11.3)

How do you test a correlation coefficient? (SECTION 11.4)

What is the standard error of estimate? How do you compute it, and where is it used? (SECTION 11.4)

How do you compute confidence intervals for a least-squares prediction? (SECTION 11.4)

The slope of the least-squares line represents rate of growth. How can you determine if the rate of growth is statistically significant? (SECTION 11.4)

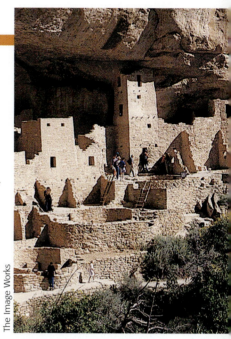

Mesa Verde National Park

FOCUS PROBLEM

Archaeology in Bandelier National Monument

Archaeologists at Washington State University did an extensive summer excavation at Burnt Mesa Pueblo in Bandelier National Monument. Their work is published in the book *Bandelier Archaeological Excavation Project: Summer 1990 Excavations at Burnt Mesa Pueblo and Casa del Rito*, edited by T. A. Kohler.

One question the archaeologists asked was: Is the raw material used by prehistoric Indians for stone tool manufacture independent of the archaeological excavation site? Two different excavation sites at Burnt Mesa Pueblo gave the information in the table below.

Use a chi-square test with 5% level of significance to test the claim that raw material used for construction of stone tools and excavation site are independent. (See Problem 17 of Section 11.1.)

Stone Tool Construction Material, Burnt Mesa Pueblo

Material	Site A	Site B	Row Total
Basalt	731	584	1315
Obsidian	102	93	195
Pedernal chert	510	525	1035
Other	85	94	179
Column Total	1428	1296	2724

Archaeological excavation site

PART I: INFERENCES USING THE CHI-SQUARE DISTRIBUTION

Overview of the Chi-Square Distribution

So far, we have used several probability distributions for hypothesis testing and confidence intervals, with the most frequently used being the normal distribution and the Student's t distribution. In this chapter, we will use another probability distribution, namely, the chi-square distribution (where *chi* is pronounced like the first two letters in the word *kite*). In Part I, we will see applications of the chi-square distribution.

Chi is a Greek letter denoted by the symbol χ, so chi-square is denoted by the symbol χ^2. Because the distribution is of chi-*square* values, the χ^2 values begin at 0 and then are all positive. The graph of the χ^2 distribution is not symmetrical, and like the Student's t distribution, it depends on the number of degrees of freedom. Figure 11-1 shows the χ^2 distribution for several degrees of freedom (*d.f.*).

As the degrees of freedom increase, the graph of the chi-square distribution becomes more bell-like and begins to look more and more symmetrical.

> The **mode (high point)** of a chi-square distribution with n degrees of freedom occurs over $n - 2$ (for $n \geq 3$).

Table 5 of the Appendix shows critical values of chi-square distributions for which a designated area falls to the *right* of the critical value. Table 11-1 gives an excerpt from Table 5. Notice that the row headers are degrees of freedom, and the column headers are areas in the *right* tail of the distribution. For instance, according to the table, for a χ^2 distribution with 3 degrees of freedom, the area occurring to the *right* of $\chi^2 = 0.072$ is 0.995. For a χ^2 distribution with 4 degrees of freedom, the area falling to the *right* of $\chi^2 = 13.28$ is 0.010.

In the next three sections, we will see how to apply the chi-square distribution to different applications.

FIGURE 11-1

The χ^2 Distribution

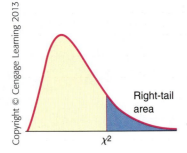

Right-tail area

χ^2

TABLE 11-1   Excerpt from Table 5 of the Appendix: The χ^2 Distribution

	Area of the Right Tail					
d.f.	0.995	0.990	0.975	...	0.010	0.005
⋮	⋮	⋮	⋮		⋮	⋮
3	0.072	0.115	0.216		11.34	12.84
4	0.207	0.297	0.484		13.28	14.86

Copyright © Cengage Learning 2013

SECTION 11.1

Chi-Square: Tests of Independence and of Homogeneity

FOCUS POINTS

- Set up a test to investigate independence of random variables.
- Use contingency tables to compute the sample χ^2 statistic.
- Find or estimate the P-value of the sample χ^2 statistic and complete the test.
- Conduct a test of homogeneity of populations.

Innovative Machines Incorporated has developed two new letter arrangements for computer keyboards. The company wishes to see if there is any relationship between the arrangement of letters on the keyboard and the number of hours it takes a new typing student to learn to type at 20 words per minute. Or, from another point of view, is the time it takes a student to learn to type *independent* of the arrangement of the letters on a keyboard?

To answer questions of this type, we test the hypotheses

Hypotheses

> H_0: Keyboard arrangement and learning times *are independent.*
>
> H_1: Keyboard arrangement and learning times *are not independent.*

Test of independence

Chi-square distribution

In problems of this sort, we are testing the *independence* of two factors. The probability distribution we use to make the decision is the *chi-square distribution*. Recall from the overview of the chi-square distribution that *chi* is pronounced like the first two letters of the word *kite* and is a Greek letter denoted by the symbol χ. Thus chi-square is denoted by χ^2.

Innovative Machines' first task is to gather data. Suppose the company took a random sample of 300 beginning typing students and randomly assigned them to learn to type on one of three keyboards. The learning times for this sample are shown in Table 11-2. These learning times are the *observed frequencies* O.

Observed frequency of a cell, *O*
Contingency table

Table 11-2 is called a *contingency table*. The *shaded boxes* that contain observed frequencies are called *cells*. The row and column totals are not considered to be cells. This contingency table is of size 3×3 (read "three-by-three") because there are three rows of cells and three columns. When giving the size of a contingency table, we always list the number of *rows first*.

> To determine the **size** of a contingency table, count the number of rows containing data and the number of columns containing data. The size is
>
> Number of rows $\times$ Number of columns
>
> where the symbol "$\times$" is read "by." The number of rows is always given first.

TABLE 11-2   **Keyboard versus Time to Learn to Type at 20 wpm**

Keyboard	21–40 h	41–60 h	61–80 h	Row Total
A	#1 25	#2 30	#3 25	80
B	#4 30	#5 71	#6 19	120
Standard	#7 35	#8 49	#9 16	100
Column Total	90	150	60	300 Sample size

GUIDED EXERCISE 1 | *Size of contingency table*

Give the sizes of the contingency tables in Figures 11-2(a) and (b). Also, count the number of cells in each table. (Remember, each pink shaded box is a cell.)

(a) **FIGURE 11-2(a)** Contingency Table

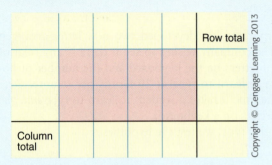

Row total

Column total

Copyright © Cengage Learning 2013

(b) **FIGURE 11-2(b)** Contingency Table

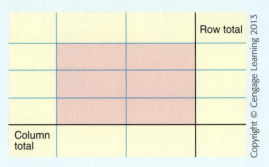

Row total

Column total

Copyright © Cengage Learning 2013

(a) ⇨ There are two rows and four columns, so this is a 2 × 4 table. There are eight cells.

(b) ⇨ Here we have three rows and two columns, so this is a 3 × 2 table with six cells.

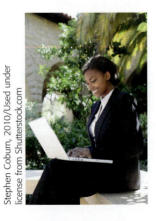

Stephen Coburn, 2010/Used under license from Shutterstock.com

Expected frequency of a cell, E

We are testing the null hypothesis that the keyboard arrangement and the time it takes a student to learn to type are *independent*. We use this hypothesis to determine the *expected frequency* of each cell.

For instance, to compute the expected frequency of cell 1 in Table 11-2, we observe that cell 1 consists of all the students in the sample who learned to type on keyboard A and who mastered the skill at the 20-words-per-minute level in 21 to 40 hours. By the assumption (null hypothesis) that the two events are independent, we use the multiplication law to obtain the probability that a student is in cell 1.

$$P(\text{cell 1}) = P(\text{keyboard A } and \text{ skill in 21–40 h})$$

$$= P(\text{keyboard A}) \cdot P(\text{skill in 21–40 h})$$

Because there are 300 students in the sample and 80 used keyboard A,

$$P(\text{keyboard A}) = \frac{80}{300}$$

Also, 90 of the 300 students learned to type in 21–40 hours, so

$$P(\text{skill in 21–40 h}) = \frac{90}{300}$$

Using these two probabilities and the assumption of independence,

$$P(\text{keyboard A } and \text{ skill in 21–40 h}) = \frac{80}{300} \cdot \frac{90}{300}$$

Finally, because there are 300 students in the sample, we have the *expected frequency E* for cell 1.

$$E = P(\text{student in cell 1}) \cdot (\text{no. of students in sample})$$

$$= \frac{80}{300} \cdot \frac{90}{300} \cdot 300 = \frac{80 \cdot 90}{300} = 24$$

We can repeat this process for each cell. However, the last step yields an easier formula for the expected frequency E.

Formula for expected frequency *E*

$$E = \frac{(\text{Row total})(\text{Column total})}{\text{Sample size}}$$

Note: If the expected value is not a whole number, do *not* round it to the nearest whole number.

Let's use this formula in Example 1 to find the expected frequency for cell 2.

EXAMPLE 1   EXPECTED FREQUENCY

Find the expected frequency for cell 2 of Table 11-2.

SOLUTION: Cell 2 is in row 1 and column 2. The *row total* is 80, and the *column total* is 150. The size of the sample is still 300.

$$E = \frac{(\text{Row total})(\text{Column total})}{\text{Sample size}}$$
$$= \frac{(80)(150)}{300} = 40$$

GUIDED EXERCISE 2 | **Expected frequency**

Table 11-3 contains the *observed frequencies O* and *expected frequencies E* for the contingency table giving keyboard arrangement and number of hours it takes a student to learn to type at 20 words per minute. Fill in the missing expected frequencies.

TABLE 11-3 Complete Contingency Table of Keyboard Arrangement and Time to Learn to Type

Keyboard	21–40 h	41–60 h	61–80 h	Row Total
A	#1 $O = 25$ $E = 24$	#2 $O = 30$ $E = 40$	#3 $O = 25$ $E = \_\_$	80
B	#4 $O = 30$ $E = 36$	#5 $O = 71$ $E = \_\_$	#6 $O = 19$ $E = \_\_$	120
Standard	#7 $O = 35$ $E = \_\_$	#8 $O = 49$ $E = 50$	#9 $O = 16$ $E = 20$	100
Column Total	90	150	60	300 Sample Size

For cell 3, we have
$$E = \frac{(80)(60)}{300} = 16$$

For cell 5, we have
$$E = \frac{(120)(150)}{300} = 60$$

For cell 6, we have
$$E = \frac{(120)(60)}{300} = 24$$

For cell 7, we have
$$E = \frac{(100)(90)}{300} = 30$$

Computing the sample test statistic χ^2

Now we are ready to compute the sample statistic χ^2 for the typing students. The χ^2 value is a measure of the sum of the differences between observed frequency O and expected frequency E in each cell. These differences are listed in Table 11-4.

As you can see, if we sum the differences between the observed frequencies and the expected frequencies of the cells, we get the value zero. This total certainly does not reflect the fact that there were differences between the observed and expected frequencies. To obtain a measure whose sum does reflect the magnitude of the differences, we square the differences and work with the quantities $(O - E)^2$. But instead of using the terms $(O - E)^2$, we use the values $(O - E)^2/E$.

TABLE 11-4	Differences between Observed and Expected Frequencies		
Cell	Observed O	Expected E	Difference $(O - E)$
1	25	24	1
2	30	40	−10
3	25	16	9
4	30	36	−6
5	71	60	11
6	19	24	−5
7	35	30	5
8	49	50	−1
9	16	20	−4
			$\Sigma(O - E) = 0$

Copyright © Cengage Learning 2013

We use this expression because a small difference between the observed and expected frequency is not nearly as important when the expected frequency is large as it is when the expected frequency is small. For instance, for both cells 1 and 8, the squared difference $(O - E)^2$ is 1. However, this difference is more meaningful in cell 1, where the expected frequency is 24, than it is in cell 8, where the expected frequency is 50. When we divide the quantity $(O - E)^2$ by E, we take the size of the difference with respect to the size of the expected value. We use the sum of these values to form the sample statistic χ^2:

Sample test statistic, χ^2

$$\chi^2 = \Sigma \frac{(O - E)^2}{E}$$

where the sum is over all cells in the contingency table.

COMMENT If you look up the word *irony* in a dictionary, you will find one of its meanings is described as "the difference between actual (or observed) results and expected results." Because irony is so prevalent in much of our human experience, it is not surprising that statisticians have incorporated a related chi-square distribution into their work.

GUIDED EXERCISE 3 | **Sample χ^2**

(a) Complete Table 11-5.

The last two rows of Table 11-5 are

TABLE 11-5 Data of Table 11-4

Cell	O	E	$O - E$	$(O - E)^2$	$(O - E)^2/E$
1	25	24	1	1	0.04
2	30	40	−10	100	2.50
3	25	16	9	81	5.06
4	30	36	−6	36	1.00
5	71	60	11	121	2.02
6	19	24	−5	25	1.04
7	35	30	5	25	0.83
8	49	50	___	___	___
9	16	20	___	___	___

$$\Sigma \frac{(O - E)^2}{E} = \text{\_\_\_\_}$$

Cell	O	E	$O - E$	$(O - E)^2$	$(O - E)^2/E$
8	49	50	−1	1	0.02
9	16	20	−4	16	0.80

Copyright © Cengage Learning 2013

$$\Sigma \frac{(O - E)^2}{E} = \text{total of last column} = 13.31$$

Copyright © Cengage Learning 2013

(b) Compute the statistic χ^2 for this sample.

Since $\chi^2 = \Sigma \dfrac{(O - E)^2}{E}$, then $\chi^2 = 13.31$.

Notice that when the observed frequency and the expected frequency are very close, the quantity $(O - E)^2$ is close to zero, and so the statistic χ^2 is near zero. As the difference increases, the statistic χ^2 also increases. To determine how large the sample statistic can be before we must reject the null hypothesis of independence, we find the *P*-value of the statistic in the chi-square distribution, Table 5 of the Appendix, and compare it to the specified level of significance α. The *P*-value depends on the number of degrees of freedom. To test independence, the degrees of freedom *d.f.* are determined by the following formula:

Degrees of freedom for test of independence

Degrees of freedom = (Number of rows − 1) · (Number of columns − 1)

or $\qquad\qquad d.f. = (R - 1)(C - 1)$

where R = number of cell rows

C = number of cell columns

Degrees of freedom for test of independence

GUIDED EXERCISE 4 | **Degrees of freedom**

Determine the number of degrees of freedom in the example of keyboard arrangements (see Table 11-2). Recall that the contingency table had three rows and three columns.

➡ $d.f. = (R - 1)(C - 1)$
$= (3 - 1)(3 - 1) = (2)(2) = 4$

Finding the P-value for tests of independence

To test the hypothesis that the letter arrangement on a keyboard and the time it takes to learn to type at 20 words per minute are independent at the $\alpha = 0.05$ level of significance, we estimate the *P*-value shown in Figure 11-3 on the next page for the sample test statistic $\chi^2 = 13.31$ (calculated in Guided Exercise 3). We then compare the *P*-value to the specified level of significance α.

For tests of independence, we always use a *right-tailed* test on the chi-square distribution. This is because we are testing to see if the χ^2 measure of the difference between the observed and expected frequencies is too large to be due to chance alone.

In Guided Exercise 4, we found that the degrees of freedom for the example of keyboard arrangements is 4. From Table 5 of the Appendix, in the row headed by *d.f.* = 4, we see that the sample $\chi^2 = 13.31$ falls between the entries 13.28 and 14.86.

FIGURE 11-3

P-value

Right-Tail Area	0.010	0.005
d.f. = 4	13.28	14.86
		↑
		Sample χ^2 = 13.31

The corresponding *P*-value falls between 0.005 and 0.010. From technology, we get *P*-value ≈ 0.0098.

Since the *P*-value is less than the level of significance $\alpha = 0.05$, we reject the null hypothesis of independence and conclude that keyboard arrangement and learning time are *not* independent.

Tests of independence for two statistical variables involve a number of steps. A summary of the procedure follows.

PROCEDURE

HOW TO TEST FOR INDEPENDENCE OF TWO STATISTICAL VARIABLES

Setup

Construct a contingency table in which the rows represent one statistical variable and the columns represent the other. Obtain a random sample of observations, which are assigned to the cells described by the rows and columns. These assignments are called the **observed values O** from the sample.

Procedure

1. Set the level of significance α and use the hypotheses

   H_0: The variables are independent.

   H_1: The variables are not independent.

2. For each cell, compute the **expected frequency E** (do not round, but give as a decimal number).

$$E = \frac{(\text{Row total})(\text{Column total})}{\text{Sample size}}$$

Requirements

You need a sample size large enough so that, for each cell, $E \geq 5$. Now each cell has two numbers, the observed frequency O from the sample and the expected frequency E.

Next compute the sample *chi-square test statistic*

$$\chi^2 = \Sigma \frac{(O - E)^2}{E} \text{ with degrees of freedom } d.f. = (R - 1)(C - 1)$$

where the sum is over all cells in the contingency table and

   R = number of rows in contingency table

   C = number of columns in contingency table

3. Use the chi-square distribution (Table 5 of the Appendix) and a *right-tailed test* to find (or estimate) the *P-value* corresponding to the test statistic.

4. *Conclude* the test. If *P*-value $\leq \alpha$, then reject H_0. If *P*-value $> \alpha$, then do not reject H_0.

5. *Interpret your conclusion* in the context of the application.

GUIDED EXERCISE 5 | **Testing for independence**

Super Vending Machines Company is to install soda pop machines in elementary schools and high schools. The market analysts wish to know if flavor preference and school level are independent. A random sample of 200 students was taken. Their school level and soda pop preferences are given in Table 11-6. Is independence indicated at the $\alpha = 0.01$ level of significance?

STEP 1: State the null and alternate hypotheses.

➡ H_0: School level and soda pop preference are independent.

H_1: School level and soda pop preference are not independent.

STEP 2:

(a) Complete the contingency Table 11-6 by filling in the required expected frequencies.

➡ The expected frequency

for cell 5 is $\dfrac{(40)(80)}{200} = 16$

for cell 6 is $\dfrac{(40)(120)}{200} = 24$

for cell 7 is $\dfrac{(20)(80)}{200} = 8$

for cell 8 is $\dfrac{(20)(120)}{200} = 12$

TABLE 11-6 School Level and Soda Pop Preference

Soda Pop	High School	Elementary School	Row Total
Kula Kola	$O = 33$ #1 $E = 36$	$O = 57$ #2 $E = 54$	90
Mountain Mist	$O = 30$ #3 $E = 20$	$O = 20$ #4 $E = 30$	50
Jungle Grape	$O = 5$ #5 $E = \underline{\quad}$	$O = 35$ #6 $E = \underline{\quad}$	40
Diet Pop	$O = 12$ #7 $E = \underline{\quad}$	$O = 8$ #8 $E = \underline{\quad}$	20
Column Total	80	120	200 Sample Size

Copyright © Cengage Learning 2013

Note: In this example, the expected frequencies are all whole numbers. If the expected frequency has a decimal part, such as 8.45, do *not* round the value to the nearest whole number; rather, give the expected frequency as the decimal number.

(b) Fill in Table 11-7 and use the table to find the sample statistic χ^2.

➡ The last three rows of Table 11-7 should read as follows:

TABLE 11-7 Computational Table for χ^2

Cell	O	E	O − E	$(O - E)^2$	$(O - E)^2/E$
1	33	36	−3	9	0.25
2	57	54	3	9	0.17
3	30	20	10	100	5.00
4	20	30	−10	100	3.33
5	5	16	−11	121	7.56
6	35	24	11	—	—
7	12	8	—	—	—
8	8	12	—	—	—

Copyright © Cengage Learning 2013

Cell	O	E	O − E	$(O - E)^2$	$(O - E)^2/E$
6	35	24	11	121	5.04
7	12	8	4	16	2.00
8	8	12	−4	16	1.33

Copyright © Cengage Learning 2013

χ^2 = total of last column

$$= \Sigma \frac{(O - E)^2}{E} = 24.68$$

(c) What is the size of the contingency table? Use the number of rows and the number of columns to determine the degrees of freedom.

➡ The contingency table is of size 4 × 2. Since there are four rows and two columns,

$$d.f. = (4 - 1)(2 - 1) = 3$$

Continued

GUIDED EXERCISE 5 *continued*

STEP 3: Use Table 5 of the Appendix to estimate the P-value of the sample statistic $\chi^2 = 24.68$ with $d.f. = 3$.

Right-Tail Area	0.005
$d.f. = 3$	12.84

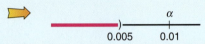

As the χ^2 values increase, the area to the right decreases, so

P-value < 0.005

STEP 4: Conclude the test by comparing the P-value of the sample statistic χ^2 to the level of significance $\alpha = 0.01$.

Since the P-value is less than α, we reject the null hypothesis of independence. Technology gives P-value ≈ 0.00002.

STEP 5: *Interpret* the test result in the context of the application.

At the 1% level of significance, we conclude that school level and soda pop preference are dependent.

TECH NOTES   The TI-84Plus/TI-83Plus/TI-*n*spire calculators, Excel, and Minitab all support chi-square tests of independence. In each case, the observed data are entered in the format of the contingency table.

TI-84Plus/TI-83Plus/TI-*n*spire (with TI-84Plus keypad) Enter the observed data into a matrix. Set the dimension of matrix **[B]** to match that of the matrix of observed values. Expected values will be placed in matrix **[B]**. Press **STAT, TESTS,** and select option **C:χ^2-Test.** The output gives the sample χ^2 with the P-value.

Excel 2010 Enter the table of observed values. Use the formulas of this section to compute the expected values. Enter the corresponding table of expected values. Finally, use **Insert Function** (f_x) ➤ **Statistical** ➤ **CHISQ.TEST.** Excel returns the P-value of the sample χ^2 value.

Minitab Enter the contingency table of observed values. Use the menu selection **Stat** ➤ **Tables** ➤ **Chi-Square Test.** The output shows the contingency table with expected values and the sample χ^2 with P-value.

Tests of Homogeneity

We've seen how to use contingency tables and the chi-square distribution to test for independence of two random variables. The same process enables us to determine whether several populations share the same proportions of distinct categories. Such a test is called a *test of homogeneity.*

According to the dictionary, among the definitions of the word *homogeneous* are "of the same structure" and "composed of similar parts." In statistical jargon, this translates as a test of homogeneity to see if two or more populations share specified characteristics in the same proportions.

Test of homogeneity

> A **test of homogeneity** tests the claim that *different populations* share the *same proportions* of specified characteristics.

The computational processes for conducting tests of independence and tests of homogeneity are the same. However, there are two main differences in the initial setup of the two types of tests, namely, the sampling method and the hypotheses.

Tests of Independence Compared to Tests of Homogeneity

1. **Sampling method**

   For tests of independence, we use one random sample and observe how the sample members are distributed among distinct categories.

   For tests of homogeneity, we take random samples from each different population and see how members of each population are distributed over distinct categories.

2. **Hypotheses**

   For tests of independence,

   H_0: The variables are independent.

   H_1: The variables are not independent.

   For tests of homogeneity,

   H_0: Each population shares respective characteristics in the same proportion.

   H_1: Some populations have different proportions of respective characteristics.

EXAMPLE 2   TEST OF HOMOGENEITY

Image Source/Getty Images

Pets—who can resist a cute kitten or puppy? Tim is doing a research project involving pet preferences among students at his college. He took random samples of 300 female and 250 male students. Each sample member responded to the survey question "If you could own only one pet, what kind would you choose?" The possible responses were: dog, cat, other pet, no pet. The results of the study follow.

Pet Preference

Gender	Dog	Cat	Other Pet	No Pet
Female	120	132	18	30
Male	135	70	20	25

Does the same proportion of males as females prefer each type of pet? Use a 1% level of significance.

We'll answer this question in several steps.

(a) First make a cluster bar graph showing the percentages of females and the percentages of males favoring each category of pet. From the graph, does it appear that the proportions are the same for males and females?

   SOLUTION: The cluster graph shown in Figure 11-4 was created using Minitab. Looking at the graph, it appears that there are differences in the proportions of females and males preferring each type of pet. However, let's conduct a statistical test to verify our visual impression.

FIGURE 11-4

Pet Preference by Gender

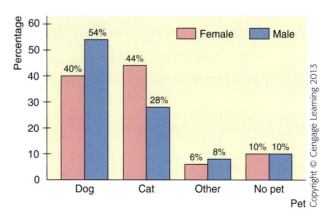

Copyright © Cengage Learning 2013

(b) Is it appropriate to use a test of homogeneity?

SOLUTION: Yes, since there are separate random samples for each designated population, male and female. We also are interested in whether each population shares the same proportion of members favoring each category of pet.

(c) State the hypotheses and conclude the test by using the Minitab printout.

H_0: The proportions of females and males naming each pet preference are the same.

H_1: The proportions of females and males naming each pet preference are not the same.

```
Chi-Square Test: Dog, Cat, Other, No Pet

Expected counts are printed below observed counts
Chi-Square contributions are printed below expected counts

              Dog        Cat      Other     No Pet      Total
1             120        132         18         30        300
           139.09     110.18      20.73      30.00
            2.620      4.320      0.359      0.000

2             135         70         20         25        250
           115.91      91.82      17.27      25.00
            3.144      5.185      0.431      0.000

Total         255        202         38         55        550

Chi-Sq =  16.059,  DF =  3,  P-Value  =  0.001
```

Since the P-value is less than α, we reject H_0 at the 1% level of significance.

(d) *Interpret* the results.

SOLUTION: It appears from the sample data that male and female students at Tim's college have different preferences when it comes to selecting a pet.

PROCEDURE

HOW TO TEST FOR HOMOGENEITY OF POPULATIONS

Setup

Obtain random samples from each of the populations. For each population, determine the number of members that share a distinct specified characteristic. Make a contingency table with the different populations as the rows (or columns) and the characteristics as the columns (or rows). The values recorded in the cells of the table are the **observed values O** taken from the samples.

Procedure

1. Set the level of significance and use the hypotheses.

    H_0: The proportion of each population sharing specified characteristics is the same for all populations.

    H_1: The proportion of each population sharing specified characteristics is not the same for all populations.

2. Follow steps 2–5 of the procedure used to test for independence.

It is important to observe that when we reject the null hypothesis in a test of homogeneity, we don't know which proportions differ among the populations. We know only that the populations differ in some of the proportions sharing a characteristic.

VIEWPOINT | Loyalty! Going, Going, Gone!

Was there a time in the past when people worked for the same company all their lives, regularly purchased the same brand names, always voted for candidates from the same political party, and loyally cheered for the same sports team? One way to look at this question is to consider tests of statistical independence. Is customer loyalty independent of company profits? Can a company maintain its productivity independent of loyal workers? Can politicians do whatever they please independent of the voters back home? Americans may be ready to act on a pent-up desire to restore a sense of loyalty in their lives. For more information, see American Demographics, *Vol. 19, No. 9.*

SECTION 11.1 PROBLEMS

 aplia™   Access end-of-section exercises at www.aplia.com

SECTION 11.2 | # Chi-Square: Goodness of Fit

FOCUS POINTS

- Set up a test to investigate how well a sample distribution fits a given distribution.
- Use observed and expected frequencies to compute the sample χ^2 statistic.
- Find or estimate the *P*-value and complete the test.

Last year, the labor union bargaining agents listed five categories and asked each employee to mark the *one* most important to her or him. The categories and corresponding percentages of favorable responses are shown in Table 11-8 on the next page. The bargaining agents need to determine if the *current* distribution of responses "fits" last year's distribution or if it is different.

In questions of this type, we are asking whether a population follows a specified distribution. In other words, we are testing the hypotheses

Hypotheses

> H_0: The population fits the given distribution.
>
> H_1: The population has a different distribution.

We use the chi-square distribution to test "goodness-of-fit" hypotheses.

TABLE 11-8   Bargaining Categories (Last Year)

Category	Percentage of Favorable Responses
Vacation time	4%
Salary	65%
Safety regulations	13%
Health and retirement benefits	12%
Overtime policy and pay	6%

Copyright © Cengage Learning 2013

Computing sample χ^2

Just as with tests of independence, we compute the sample statistic:

$$\chi^2 = \Sigma \frac{(O - E)^2}{E} \text{ with degrees of freedom} = k - 1$$

where E = expected frequency

O = observed frequency

$\frac{(O - E)^2}{E}$ is summed for each category in the distribution

k = number of categories in the distribution

Next we use the chi-square distribution table (Table 5 of the Appendix) to estimate the P-value of the sample χ^2 statistic. Finally, we compare the P-value to the level of significance α and conclude the test.

Goodness-of-fit test

In the case of a *goodness-of-fit test*, we use the null hypothesis to compute the expected values for the categories. Let's look at the bargaining category problem to see how this is done.

In the bargaining category problem, the two hypotheses are

H_0: The present distribution of responses is the same as last year's.

H_1: The present distribution of responses is different.

The null hypothesis tells us that the *expected frequencies* of the present response distribution should follow the percentages indicated in last year's survey. To test this hypothesis, a random sample of 500 employees was taken. If the null hypothesis is true, then there should be 4%, or 20 responses, out of the 500 rating vacation time as the most important bargaining issue. Table 11-9 gives the other expected values and all the information necessary to compute the sample statistic χ^2. We see that the sample statistic is

$$\chi^2 = \Sigma \frac{(O - E)^2}{E} = 14.15$$

Steve Skjold/PhotoEdit

TABLE 11-9   Observed and Expected Frequencies for Bargaining Categories

Category	O	E	$(O - E)^2$	$(O - E)^2/E$
Vacation time	30	4% of 500 = 20	100	5.00
Salary	290	65% of 500 = 325	1225	3.77
Safety	70	13% of 500 = 65	25	0.38
Health and retirement	70	12% of 500 = 60	100	1.67
Overtime	40	6% of 500 = 30	100	3.33
	$\Sigma O = 500$	$\Sigma E = 500$		$\Sigma \frac{(O - E)^2}{E} = 14.15$

Copyright © Cengage Learning 2013

Larger values of the sample statistic χ^2 indicate greater differences between the proposed distribution and the distribution followed by the sample. The larger the χ^2 statistic, the stronger the evidence to reject the null hypothesis that the population distribution fits the given distribution. Consequently, goodness-of-fit tests are always *right-tailed* tests.

Type of test

> For *goodness-of-fit tests*, we use a *right-tailed* test on the chi-square distribution. This is because we are testing to see if the χ^2 measure of the difference between the observed and expected frequencies is too large to be due to chance alone.

To test the hypothesis that the present distribution of responses to bargaining categories is the same as last year's, we use the chi-square distribution (Table 5 of the Appendix) to estimate the *P*-value of the sample statistic $\chi^2 = 14.15$. To estimate the *P*-value, we need to know the number of *degrees of freedom*. In the case of a goodness-of-fit test, the degrees of freedom are found by the following formula:

Degrees of freedom

> **Degrees of freedom for goodness-of-fit test**
>
> $$d.f. = k - 1$$
>
> where k = number of categories

Notice that when we compute the expected values E, we must use the null hypothesis to compute all but the last one. To compute the last one, we can subtract the previous expected values from the sample size. For instance, for the bargaining issues, we could have found the number of responses for overtime policy by adding the other expected values and subtracting that sum from the sample size 500. We would again get an expected value of 30 responses. The degrees of freedom, then, is the number of E values that *must* be computed by using the null hypothesis.

For the bargaining issues, we have

$$d.f. = 5 - 1 = 4$$

where $k = 5$ is the number of categories.

We now have the tools necessary to use Table 5 of the Appendix to estimate the *P*-value of $\chi^2 = 14.15$. Figure 11-5 shows the *P*-value. In Table 5, we use the row headed by $d.f. = 4$. We see that $\chi^2 = 14.15$ falls between the entries 13.28 and 14.86. Therefore, the *P*-value falls between the corresponding right-tail areas 0.005 and 0.010.

P-value

To test the hypothesis that the distribution of responses to bargaining issues is the same as last year's at the 1% level of significance, we compare the *P*-value of the statistic to $\alpha = 0.01$.

We see that the *P*-value is less than α, so we reject the null hypothesis that the distribution of responses to bargaining issues is the same as last year's. At the 1% level of significance, we can say that the evidence supports the conclusion that this year's responses to the issues are different from last year's.

Goodness-of-fit tests involve several steps that can be summarized as follows.

FIGURE 11-5

P-value

Sample $\chi^2 = 14.15$

Right-Tail Area	0.010	0.005
$d.f. = 4$	13.28	14.86
		Sample $\chi^2 = 14.15$

PROCEDURE

> ## HOW TO TEST FOR GOODNESS OF FIT
>
> ### Setup
>
> First, each member of a population needs to be classified into exactly one of several different categories. Next, you need a specific (theoretical) distribution that assigns a fixed probability (or percentage) that a member of the population will fall into one of the categories. You then need a random sample size n from the population. Let O represent the *observed number* of data from the sample that fall into each category. Let E represent the *expected number* of data from the sample that, in theory, would fall into each category.
>
> > O = observed frequency count of a category using sample data
> >
> > E = expected frequency of a category
> >
> > = (sample size n)(probability assigned to category)
>
> ### Requirements
>
> The sample size n should be large enough so that $E \geq 5$ in each category.
>
> ### Procedure
>
> 1. Set the *level of significance* α and use the *hypotheses*
>
> H_0: The population fits the specified distribution of categories.
>
> H_1: The population has a different distribution.
>
> 2. For each category, compute $(O - E)^2/E$, then compute the sample *test statistic*
>
> $$\chi^2 = \Sigma \frac{(O - E)^2}{E} \quad \text{with } d.f. = k - 1$$
>
>   where the sum is taken over all categories and k = number of categories.
>
> 3. Use the chi-square distribution (Table 5 of the Appendix) and a *right-tailed test* to find (or estimate) the *P-value* corresponding to the sample test statistic.
>
> 4. *Conclude* the test. If *P*-value $\leq \alpha$, then reject H_0. If *P*-value $> \alpha$, then do not reject H_0.
>
> 5. *Interpret your conclusion* in the context of the application.

One important application of goodness-of-fit tests is genetics theory. Such an application is shown in Guided Exercise 6.

GUIDED EXERCISE 6 | ## Goodness-of-fit test

According to genetics theory, red–green colorblindness in humans is a recessive sex-linked characteristic. In this case, the gene is carried on the X chromosome only. We will denote an X chromosome with the colorblindness gene by X_c and one without the gene by X_n. Women have two X chromosomes, and they will be red–green colorblind only if both chromosomes have the gene, designated $X_c X_c$. A woman can have normal vision but still carry the colorblindness gene if only one of the chromosomes has the gene, designated $X_c X_n$. A man carries an X and a Y chromosome; if the X chromosome carries the colorblindness gene ($X_c Y$), the man is colorblind.

According to genetics theory, if a man with normal vision ($X_n Y$) and a woman carrier ($X_c X_n$) have a child, the probabilities that the child will have red–green colorblindness, will have normal vision and not carry the gene, or will have normal vision and carry the gene are given by the *equally likely* events in Table 11-10.

Continued

$P(\text{child has normal vision and is not a carrier}) = P(X_n Y) + P(X_n X_n) = \dfrac{1}{2}$

$P(\text{child has normal vision and is a carrier}) = P(X_c X_n) = \dfrac{1}{4}$

$P(\text{child is red-green colorblind}) = P(X_c Y) = \dfrac{1}{4}$

TABLE 11-10
Red–Green Colorblindness

	Father	
Mother	X_n	Y
X_c	$X_c X_n$	$X_c Y$
X_n	$X_n X_n$	$X_n Y$

Copyright © Cengage Learning 2013

To test this genetics theory, Genetics Labs took a random sample of 200 children whose mothers were carriers of the colorblindness gene and whose fathers had normal vision. The results are shown in Table 11-11. We wish to test the hypothesis that the population follows the distribution predicted by the genetics theory (see Table 11-10). Use a 1% level of significance.

(a) State the null and alternate hypotheses. What is α?

➡ H_0: The population fits the distribution predicted by genetics theory.

H_1: The population does not fit the distribution predicted by genetics theory.

$\alpha = 0.01$.

(b) Fill in the rest of Table 11-11 and use the table to compute the sample statistic χ^2.

TABLE 11-11 Colorblindness Sample

Event	O	E	$(O - E)^2$	$(O - E)^2/E$
Red–green colorblindness	35	50	225	4.50
Normal vision, noncarrier	105	____	____	____
Normal vision, carrier	60	____	____	____

Copyright © Cengage Learning 2013

➡ **TABLE 11-12 Completion of Table 11-11**

Event	O	E	$(O - E)^2$	$(O - E)^2/E$
Red–green colorblindness	35	50	225	4.50
Normal vision, noncarrier	105	100	25	0.25
Normal vision, carrier	60	50	100	2.00

Copyright © Cengage Learning 2013

The sample statistic is $\chi^2 = \Sigma \dfrac{(O - E)^2}{E} = 6.75$.

(c) There are $k = 3$ categories listed in Table 11-11. Use this information to compute the degrees of freedom.

➡ $d.f. = k - 1$
$= 3 - 1 = 2$

(d) Find the P-value for $\chi^2 = 6.75$.

➡ Using Table 5 of the Appendix and the fact that goodness-of-fit tests are right-tailed tests, we see that

Right-Tail Area	0.050	0.025
$d.f. = 2$	5.99	7.38
		↑ Sample $\chi^2 = 6.75$

Copyright © Cengage Learning 2013

$0.025 < P\text{-value} < 0.050$

(e) Conclude the test for $\alpha = 0.01$.

➡ For $\alpha = 0.01$, we have

```
      α
 ─────┼────────(──────────────)──
    0.01     0.025        0.050
```

Since P-value $> \alpha$, do not reject H_0.

(f) *Interpret* the conclusion in the context of the application.

➡ At the 1% level of significance, there is insufficient evidence to conclude that the population follows a distribution different from that predicted by genetics theory.

What description would you use for marathon runners? How about age distribution? Body weight? Length of stride? Heart rate? Blood pressure? What countries do these runners come from? What are their best running times? Make your own estimated distribution for these variables, and then consider a goodness-of-fit test for your distribution compared with available data. For more information on marathon runners, visit this text's site at **http://www.cengagebrain.com** *and find links to the Honolulu marathon site and to the* Runner's World *site.*

SECTION 11.2
PROBLEMS

Access end-of-section exercises at www.aplia.com

SECTION 11.3

Testing a Single Variance or Standard Deviation

FOCUS POINTS

- Set up a test for a single variance σ^2.
- Compute the sample χ^2 statistic.
- Use the χ^2 distribution to estimate a *P*-value and conclude the test.

Testing σ^2

Many problems arise that require us to make decisions about variability. In this section, we will test hypotheses about the variance (or standard deviation) of a population. It is customary to talk about variance instead of standard deviation because our techniques employ the sample variance rather than the standard deviation. Of course, the standard deviation is just the square root of the variance, so any discussion about variance is easily converted to a similar discussion about standard deviation.

Test of variance

Let us consider a specific example in which we might wish to test a hypothesis about the variance. Almost everyone has had to wait in line. In a grocery store, bank, post office, or registration center, there are usually several checkout or service areas. Frequently, each service area has its own independent line. However, many businesses and government offices are adopting a "single-line" procedure.

In a single-line procedure, there is only one waiting line for everyone. As any service area becomes available, the next person in line gets served. The old independent-lines procedure has a line at each service center. An incoming customer simply picks the shortest line and hopes it will move quickly. In either procedure, the number of clerks and the rate at which they work is the same, so the average waiting time is the *same*. What is the advantage of the single-line procedure? The difference is in the *attitudes* of people who wait in the lines. A lengthy waiting line will be more acceptable if the variability of waiting times is smaller, even though the average waiting time is the same. When the variability is small, the inconvenience of waiting (although it might not be reduced) does become more predictable. This means impatience is reduced and people are happier.

To test the hypothesis that variability is less in a single-line process, we use the chi-square distribution. The next theorem tells us how to use the sample and population variance to compute values of χ^2.

THEOREM 11.1   If we have a *normal* population with variance σ^2 and a random sample of n measurements is taken from this population with sample variance s^2, then

$$\chi^2 = \frac{(n-1)s^2}{\sigma^2}$$

has a chi-square distribution with degrees of freedom $d.f. = n - 1$.

Recall that the chi-square distribution is *not* symmetrical and that there are different chi-square distributions for different degrees of freedom. Table 5 of the Appendix gives chi-square values for which the area α is to the *right* of the given chi-square value.

Table 11-13 summarizes the techniques for using the chi-square distribution (Table 5 of the Appendix) to find P-values for a right-tailed test, a left-tailed test,

TABLE 11-13	*P*-values for Chi-Square Distribution Table (Table 5 of the Appendix)

(a) Two-tailed test

Remember that the *P*-value is the probability of getting a test statistic as extreme as, or more extreme than, the test statistic computed from the sample. For a two-tailed test, we need to account for corresponding equal areas in both the upper and lower tails. This means that in each tail, we have an area of *P*-value/2. The total *P*-value is then

$$P\text{-value} = 2\left(\frac{P\text{-value}}{2}\right)$$

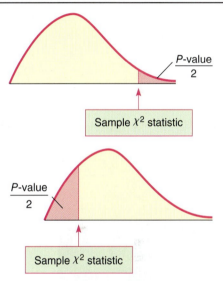

Be sure to choose the area in the appropriate tail (left or right) so that $\dfrac{P\text{-value}}{2} \le 0.5$.

(b) Right-tailed test

Since the chi-square table gives right-tail probabilities, you can use the table directly to find or estimate the *P*-value.

(c) Left-tailed test

Since the chi-square table gives right-tail probabilities, you first find or estimate the quantity $1 - (P\text{-value})$ of the right tail. Then subtract from 1 to get the *P*-value of the left-tail.

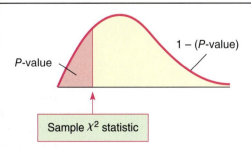

and a two-tailed test. Example 3 demonstrates the technique of finding P-values for a left-tailed test, Example 4 demonstrates the technique for a two-tailed test, and Guided Exercise 7 uses a right-tailed test.

Now let's use Theorem 11.1 and our knowledge of the chi-square distribution to determine if a single-line procedure has less variance of waiting times than independent lines.

EXAMPLE 3 TESTING THE VARIANCE (LEFT-TAILED TEST)

For years, a large discount store has used independent lines to check out customers. Historically the standard deviation of waiting times is 7 minutes. The manager tried a new single-line procedure. A random sample of 25 customers using the single-line procedure was monitored, and it was found that the standard deviation for waiting times was only $s = 5$ minutes. Assume the waiting times are normally distributed. Use $\alpha = 0.05$ to test the claim that the variance in waiting times is reduced for the single-line method.

Establish H_0 and H_1

SOLUTION: As a null hypothesis, we assume that the variance of waiting times is the same as that of the former independent-lines procedure. The alternate hypothesis is that the variance for the single-line procedure is less than that for the independent-lines procedure. If we let σ be the standard deviation of waiting times for the single-line procedure, then σ^2 is the variance, and we have

Sample χ^2 value and degrees of freedom

$$H_0: \sigma^2 = 49 \qquad H_1: \sigma^2 < 49 \qquad (\text{use } 7^2 = 49)$$

P-value

We use the chi-square distribution to test the hypotheses. Assuming that the waiting times are normally distributed, we compute our observed value of χ^2 by using Theorem 11.1, with $n = 25$.

$$s = 5 \quad \text{so} \quad s^2 = 25 \quad (\text{observed from sample})$$

$$\sigma = 7 \quad \text{so} \quad \sigma^2 = 49 \quad (\text{from } H_0: \sigma^2 = 49)$$

$$\chi^2 = \frac{(n-1)s^2}{\sigma^2} = \frac{(25-1)25}{49} \approx 12.24$$

$$d.f. = n - 1 = 25 - 1 = 24$$

Next we estimate the P-value for $\chi^2 = 12.24$. Since we have a left-tailed test, the P-value is the area of the chi-square distribution that lies to the *left* of $\chi^2 = 12.24$, as shown in Figure 11-6.

To estimate the P-value on the left, we consider the fact that the area of the right tail is between 0.975 and 0.990. To find an estimate for the area of the left tail, we *subtract* each right-tail endpoint from 1. The P-value (area of the left tail) is in the interval

$$1 - 0.990 < P\text{-value of left tail} < 1 - 0.975$$

$$0.010 < P\text{-value} < 0.025$$

Test conclusion

To conclude the test, we compare the P-value to the level of significance $\alpha = 0.05$.

Checkout lines

Mark Richards/PhotoEdit

Since the P-value is less than α, we reject H_0.

Interpretation At the 5% level of significance, we conclude that the variance of waiting times for a single line is less than the variance of waiting times for multiple lines.

FIGURE 11-6

P-value

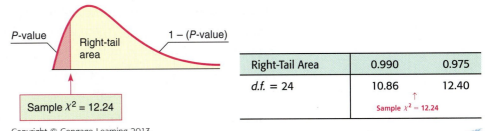

Copyright © Cengage Learning 2013

The steps used in Example 3 for testing the variance σ^2 are summarized as follows:

PROCEDURE

How to test σ^2

Requirements

You first need to know that a random variable x has a normal distribution. In testing σ^2, the normal assumption must be strictly observed (whereas in testing means, we can say "normal" or "approximately normal"). Next you need a random sample (size $n \geq 2$) of values from the x distribution for which you compute the sample variance s^2.

Procedure

1. In the context of the problem, state the *null hypothesis* H_0 and the *alternate hypothesis* H_1, and set the *level of significance* α.

2. Use the value of σ^2 given in the null hypothesis H_0, the sample variance s^2, and the sample size n to compute the sample *test statistic*

    $$\chi^2 = \frac{(n-1)s^2}{\sigma^2} \quad \text{with degrees of freedom } d.f. = n - 1$$

3. Use a chi-square distribution and the type of test to find or estimate the *P-value*. Use the procedures shown in Table 11-13 and Table 5 of the Appendix.

4. *Conclude* the test. If *P*-value $\leq \alpha$, then reject H_0. If *P*-value $> \alpha$, then do not reject H_0.

5. *Interpret your conclusion* in the context of the application.

EXAMPLE 4

Testing the variance (two-tailed test)

Let x be a random variable that represents weight loss (in pounds) after following a certain diet for 6 months. After extensive study, it is found that x has a normal distribution with $\sigma = 5.7$ pounds. A new modification of the diet has been implemented. A random sample of $n = 21$ people use the modified diet for 6 months. For these people, the sample standard deviation of weight loss is $s = 4.1$ pounds. Does this result indicate that the variance of weight loss for the modified diet is different (either way) from the variance of weight loss for the original diet? Use $\alpha = 0.01$.

(a) What is the level of significance? State the null and alternate hypotheses.

SOLUTION: We are using $\alpha = 0.01$. The standard deviation of weight loss for the original diet is $\sigma = 5.7$ pounds, so the variance is $\sigma^2 = 32.49$. The null hypothesis is that the weight loss variance for the modified diet is the same as that for the original diet. The alternate hypothesis is that the variance is different.

$$H_0: \sigma^2 = 32.49 \quad H_1: \sigma^2 \neq 32.49$$

(b) Compute the sample χ^2 statistic and the degrees of freedom.

SOLUTION: Using sample size $n = 21$, sample standard deviation $s = 4.1$ pounds, and $\sigma^2 = 32.49$ from the null hypothesis, we have

$$\chi^2 = \frac{(n-1)s^2}{\sigma^2} = \frac{(21-1)4.1^2}{32.49} \approx 10.35$$

with degrees of freedom $d.f. = n - 1 = 21 - 1 = 20$.

(c) Use the chi-square distribution (Table 5 of the Appendix) to estimate the P-value.

FIGURE 11-7

P-value

Right-Tail Area	0.975	0.950
$d.f. = 20$	8.59	10.85
		Sample $\chi^2 \approx 10.35$

Copyright © Cengage Learning 2013

SOLUTION: For a *two-tailed* test, the area beyond the sample χ^2 represents *half* the total P-value, or $(P\text{-value})/2$. Figure 11-7 shows this region, which is to the left of $\chi^2 \approx 10.35$. However, Table 5 of the Appendix gives the areas in the *right tail*. We use Table 5 to find the area in the right tail and then subtract from 1 to find the corresponding area in the left tail.

From the table, we see that the right-tail area falls in the interval between 0.950 and 0.975. Subtracting each endpoint of the interval from 1 gives us an interval containing $(P\text{-value})/2$. Multiplying by 2 gives an interval for the P-value.

$$1 - 0.975 < \frac{P\text{-value}}{2} < 1 - 0.950 \quad \text{Subtract right-tail-area endpoints from 1.}$$

$$0.025 < \frac{P\text{-value}}{2} < 0.050$$

$$0.05 < P\text{-value} < 0.10 \qquad \text{Multiply each part by 2.}$$

(d) Conclude the test.

SOLUTION: The P-value is greater than $\alpha = 0.01$, so we do not reject H_0.

(e) *Interpret* the conclusion in the context of the application.

SOLUTION: At the 1% level of significance, there is insufficient evidence to conclude that the variance of weight loss using the modified diet is different from the variance of weight loss using the original diet.

GUIDED EXERCISE 7 | Testing the variance (right-tailed test)

Certain industrial machines require overhaul when wear on their parts introduces too much variability to pass inspection. A government official is visiting a dentist's office to inspect the operation of an X-ray machine. If the machine emits too little radiation, clear photographs cannot be obtained. However, too much radiation can be harmful to the patient. Government regulations specify an average emission of 60 millirads with standard deviation σ of 12 millirads, and the machine has been set for these readings. After examining the machine, the inspector is satisfied that the average emission is still 60 millirads. However, there is wear on certain mechanical parts. To test variability, the inspector takes a random sample of 30 X-ray emissions and finds the sample standard deviation to be $s = 15$ millirads. Does this support the claim that the variance is too high (i.e., the machine should be overhauled)? Use a 1% level of significance.

The radiation emissions follow a normal distribution. Let σ be the (population) standard deviation of emissions (in millirads) of the machine in its present condition.

(a) What is α? State H_0 and H_1.

⟹ $\alpha = 0.01$. Government regulations specify that $\sigma = 12$. This means that the variance $\sigma^2 = 144$. We are to test the claim that the variance is higher than government specifications allow.

H_0: $\sigma^2 = 144$ and H_1: $\sigma^2 > 144$

(b) Compute the sample statistic χ^2 and corresponding degrees of freedom.

⟹ Using $n = 30$, $s = 15$, and $\sigma^2 = 144$ from H_0,

$$\chi^2 = \frac{(n-1)s^2}{\sigma^2} = \frac{(30-1)15^2}{144} \approx 45.3$$

Degrees of freedom $d.f. = n - 1 = 30 - 1 = 29$

(c) Estimate the P-value for the sample $\chi^2 = 45.3$ with $d.f. = 29$.

⟹ Since this is a *right-tailed* test, we look up P-values directly in the chi-square table (Table 5 of the Appendix).

FIGURE 11-8 *P-value*

Sample $\chi^2 = 45.3$

Right-Tail Area	0.050	0.025
$d.f. = 29$	42.56	45.72

Sample $\chi^2 \approx 45.3$

Copyright © Cengage Learning 2013

$0.025 < $ P-value $ < 0.050$

(d) Conclude the test.

⟹ The P-value for $\chi^2 = 45.3$ is greater than $\alpha = 0.01$.

```
     α
  ———+————————(—————————)———
   0.01      0.025      0.050
```

Fail to reject H_0.

(e) *Interpret* the conclusion in the context of the application.

⟹ At the 1% level of significance, there is insufficient evidence to conclude that the variance of the radiation emitted by the machine is greater than that specified by government regulations. The evidence does not indicate that an adjustment is necessary at this time.

VIEWPOINT | Adoption—A Good Choice!

Cuckoos are birds that are known to lay their eggs in the nests of other (host) birds. The host birds then hatch the eggs and adopt the cuckoo chicks as their own. Birds such as the meadow pipit, tree pipit, hedge sparrow, robin, and wren all have played host to cuckoo eggs and adopted their chicks. L. H. C. Tippett (1902–1985) was a pioneer in the field of statistical quality control who collected data on cuckoo eggs found in the nests of other birds. For more information and data from Tippett's study, visit the site for this text at **http://www.cengagebrain.com** *and find a link to DASL, the Carnegie Mellon University Data and Story Library. Find Biology under Data Subjects, and then select the Cuckoo Egg Length Data file.*

PART II: INFERENCES RELATING TO LINEAR REGRESSION

SECTION 11.4

Inferences for Correlation and Regression

FOCUS POINTS

- Test the correlation coefficient ρ.
- Use sample data to compute the standard error of estimate S_e.
- Test the slope β of the least-squares line.
- Find a confidence interval for the value of y predicted for a specified value of x.

Learn more, earn more! We have probably all heard this platitude. The question is whether or not there is some truth in the statement. Do college graduates have an improved chance at a better income? Is there a trend in the general population to support the "learn more, earn more" statement?

Consider the following variables: x = percentage of the population 25 or older with at least 4 years of college and y = percentage *growth* in per capita income over the past 7 years. A random sample of six communities in Ohio gave the information (based on *The Rating Guide to Life in America's Small Cities*, by G. S. Thomas) shown in Table 11-14.

If we use what we learned in Sections 4.1 and 4.2, we can compute the correlation coefficient r and the least-squares line $\hat{y} = a + bx$ using the data of Table 11-14. However, r is only a *sample* correlation coefficient and $\hat{y} = a + bx$ is only a "sample-based" least-squares line. What if we used *all* possible data pairs (x, y) from *all* U.S. cities, not just the six towns in Ohio? If we accomplished this seemingly impossible task, we would have the *population* of all (x, y) pairs.

Population correlation coefficient, ρ

From this population of (x, y) pairs, we could (in theory) compute the *population correlation coefficient*, which we call ρ (Greek letter rho, pronounced like "row"). We could also compute the least-squares line for the entire population, which we denote as $y = \alpha + \beta x$ using more Greek letters, α (alpha) and β (beta).

Sample Statistic		Population Parameter
r	$\rightarrow$	ρ
a	$\rightarrow$	α
b	$\rightarrow$	β
$\hat{y} = a + bx$	$\rightarrow$	$y = \alpha + \beta x$

Requirements for statistical inference

Requirements for inferences concerning linear regression

To make inferences regarding the population correlation coefficient ρ and the slope β of the population least-squares line, we need to be sure that

(a) The set (x, y) of ordered pairs is a *random sample* from the population of all possible such (x, y) pairs.

(b) For each fixed value of x, the y values have a normal distribution. All of the y distributions have the same variance, and, for a given x value, the distribution of y values has a mean that lies on the least-squares line. We also assume that for a fixed *y*, each *x* has its own normal distribution. In most cases, the results are still accurate if the distributions are simply mound-shaped and symmetrical, and the *y* variances are approximately equal.

We assume these conditions are met for all inferences presented in this section.

TABLE 11-14	Education and Income Growth Percentages					
x	9.9	11.4	8.1	14.7	8.5	12.6
y	37.1	43.0	33.4	47.1	26.5	40.2

Copyright © Cengage Learning 2013

Testing the Correlation Coefficient

The first topic we want to study is the statistical significance of the sample correlation coefficient *r*. To do this, we construct a statistical test of *ρ*, the population correlation coefficient. The test will be based on the following theorem:

THEOREM 11.2   Let *r* be the sample correlation coefficient computed using data pairs (x, y). We use the null hypothesis

H_0: *x* and *y* have no linear correlation, so $\rho = 0$

The alternate hypothesis may be

$H_1: \rho > 0$ or $H_1: \rho < 0$ or $H_1: \rho \neq 0$

The conversion of *r* to a Student's *t* distribution is

$$t = \frac{r\sqrt{n-2}}{\sqrt{1-r^2}} \quad \text{with } d.f. = n - 2$$

where *n* is the number of sample data pairs (x, y) $(n \geq 3)$.

PROCEDURE

Problem 13 discusses how sample size might affect the significance of *r*.

Serial correlation describes the extent to which the result of one time period of a time series is related to the result in the next period. Problems 15–17 at the end of this section discuss serial correlation.

HOW TO TEST THE POPULATION CORRELATION COEFFICIENT *ρ*

1. Use the *null hypothesis* H_0: $\rho = 0$. In the context of the application, state the *alternate hypothesis* $(\rho > 0$ or $\rho < 0$ or $\rho \neq 0)$ and set the *level of significance* α.

2. Obtain a random sample of $n \geq 3$ data pairs (x, y) and compute the sample *test statistic* where *r* is the sample correlation coefficient and *n* is the sample size

$$t = \frac{r\sqrt{n-2}}{\sqrt{1-r^2}} \quad \text{with degrees of freedom } d.f. = n - 2$$

3. Use a Student's *t* distribution and the type of test, one-tailed or two-tailed, to find (or estimate) the *P-value* corresponding to the test statistic.

4. *Conclude* the test. If *P-value* $\leq$ α, then reject H_0. If *P-value* $>$ α, then do not reject H_0.

5. *Interpret your conclusion* in the context of the application.

EXAMPLE 5   TESTING *ρ*

Let's return to our data from Ohio regarding the percentage of the population with at least 4 years of college and the percentage of growth in per capita income (Table 11-14). We'll develop a test for the population correlation coefficient *ρ*.

TABLE 11-15	Excerpt from Student's *t* Distribution Table	
✓one-tail area	0.010	0.005
two-tail area	0.020	0.010
d.f. = 4	3.747	4.604

↑
Sample *t* = 3.84

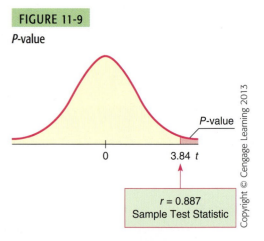

FIGURE 11-9

P-value

P-value

0   3.84 *t*

r = 0.887
Sample Test Statistic

SOLUTION: First, we compute the sample correlation coefficient *r*. Using a calculator, statistical software, or a "by-hand" calculation from Section 4.1, we find

$$r \approx 0.887$$

Now we test the correlation coefficient ρ. Remember that *x* represents percentage college graduates and *y* represents percentage salary increases in the general population. We suspect the population correlation is positive, $\rho > 0$. Let's use a 1% level of significance:

$H_0: \rho = 0$ (no linear correlation)

$H_1: \rho > 0$ (positive linear correlation)

Convert the sample test statistic *r* = 0.887 to *t* using *n* = 6.

$$t = \frac{r\sqrt{n-2}}{\sqrt{1-r^2}} = \frac{0.887\sqrt{6-2}}{\sqrt{1-0.887^2}} \approx 3.84 \quad \text{with } d.f. = n - 2 = 6 - 2 = 4$$

The *P*-value for the sample test statistic *t* = 3.84 is shown in Figure 11-9. Since we have a right-tailed test, we use the one-tail area in the Student's *t* distribution (Table 4 of the Appendix).

From Table 11-15, we see that

$$0.005 < P\text{-value} < 0.010$$

Since the interval containing the *P*-value is less than the level of significance α = 0.01, we reject H_0 and conclude that the population correlation coefficient between *x* and *y* is positive. Technology gives *P*-value ≈ 0.0092.

α

——(━━━━━)——
0.005   0.010

Interpretation Although we have shown that *x* and *y* are positively correlated, we have not shown that an increase in education *causes* an increase in earnings.

Standard Error of Estimate

Sometimes a scatter diagram clearly indicates the existence of a linear relationship between *x* and *y*, but it can happen that the points are widely scattered about the least-squares line. We need a method (besides just looking) for measuring the spread of a set of points about the least-squares line. There are three common methods of measuring the spread. One method uses the *standard error of estimate*. The others are the *coefficient of correlation* and the *coefficient of determination*.

For the standard error of estimate, we use a measure of spread that is in some ways like the standard deviation of measurements of a single variable. Let

$$\hat{y} = a + bx$$

be the predicted value of y from the least-squares line. Then $y - \hat{y}$ is the difference between the y value of the *data point* (x, y) shown on the scatter diagram (Figure 11-10) and the $\hat{y}$ value of the point on the *least-squares line* with the same x value.

FIGURE 11-10

The Distance between Points (x, y) and $(x, \hat{y})$

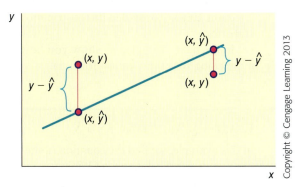

Copyright © Cengage Learning 2013

Residual

The quantity $y - \hat{y}$ is known as the *residual*. To avoid the difficulty of having some positive and some negative values, we square the quantity $(y - \hat{y})$. Then we sum the squares and, for technical reasons, divide this sum by $n - 2$. Finally, we take the square root to obtain the *standard error of estimate*, denoted by S_e.

Standard error of estimate, S_e

$$\text{Standard error of estimate} = S_e = \sqrt{\frac{\Sigma(y - \hat{y})^2}{n - 2}} \qquad (1)$$

where $\hat{y} = a + bx$ and $n \geq 3$.

Note: To compute the standard error of estimate, we require that there be at least three points on the scatter diagram. If we had only two points, the line would be a perfect fit, since two points determine a line. In such a case, there would be no need to compute S_e.

The nearer the scatter points lie to the least-squares line, the smaller S_e will be. In fact, if $S_e = 0$, it follows that each $y - \hat{y}$ is also zero. This means that all the scatter points lie *on* the least-squares line if $S_e = 0$. The larger S_e becomes, the more scattered the points are.

The formula for the standard error of estimate is reminiscent of the formula for the standard deviation. It, too, is a measure of dispersion. However, the standard deviation involves differences of data values from a mean, whereas the standard error of estimate involves the differences between experimental and predicted y values for a given x (i.e., $y - \hat{y}$).

The actual computation of S_e using Equation (1) is quite long because the formula requires us to use the least-squares-line equation to compute a predicted value $\hat{y}$ for *each* x value in the data pairs. There is a computational formula that we strongly recommend you use. However, as with all the computation formulas, be careful about rounding. This formula is sensitive to rounding, and you should carry as many digits as seems reasonable for your problem. Answers will vary, depending on rounding used. We give the formula here and follow it with an example of its use.

PROCEDURE

HOW TO FIND THE STANDARD ERROR OF ESTIMATE S_e

1. Obtain a random sample of $n \geq 3$ data pairs (x, y).

2. Use the procedures of Section 4.2 to find a and b from the sample least-squares line $\hat{y} = a + bx$.

Computation formula for S_e

3. The standard error of estimate is

$$S_e = \sqrt{\frac{\Sigma y^2 - a\Sigma y - b\Sigma xy}{n - 2}} \qquad (2)$$

With a considerable amount of algebra, Equations (1) and (2) can be shown to be mathematically equivalent. Equation (1) shows the strong similarity between the standard error of estimate and the standard deviation. Equation (2) is a shortcut calculation formula because it involves few subtractions. The sums Σx, Σy, Σx^2, Σy^2, and Σxy are provided directly on most calculators that support two-variable statistics.

In the next example, we show you how to compute the standard error of estimate using the computation formula.

EXAMPLE 6   LEAST-SQUARES LINE AND S_e

In this example, we find the equation of the least-squares line and the standard error of estimate for the data from Ohio regarding x = percentage of the population with at least 4 years of college and y = percentage growth in per capita income.

x	9.9	11.4	8.1	14.7	8.5	12.6
y	37.1	43.0	33.4	47.1	26.5	40.2

Copyright © Cengage Learning 2013

SOLUTION: To find the equation of the least-squares line $\hat{y} = a + bx$ and the value of S_e, we note that $n = 6$ and compute the following sums:

$$\Sigma x = 65.2; \ \Sigma y = 227.3; \ \Sigma x^2 = 740.68; \ \Sigma y^2 = 8877.67;$$
$$\Sigma xy = 2552.17$$

Using formulas (3) and (4) from Chapter 4, we have

$$b = \frac{n\Sigma xy - (\Sigma x)(\Sigma y)}{n\Sigma x^2 - (\Sigma x)^2} = \frac{6(2552.17) - (65.2)(227.3)}{6(740.68) - (65.2)^2} \approx 2.554$$

$$\bar{x} = \frac{\Sigma x}{n} = \frac{65.2}{6} \approx 10.87 \quad \text{and} \quad \bar{y} = \frac{\Sigma y}{n} = \frac{227.3}{6} \approx 37.88$$

$$a = \bar{y} - b\bar{x} \approx 37.88 - 2.554(10.87) \approx 10.12$$

The equation of the least-squares line is

$$\hat{y} \approx 10.12 + 2.554x$$

Note: The values of a and b for the least-squares line are readily available on calculators supporting two-variable statistics.

$$S_e = \sqrt{\frac{\Sigma y^2 - a\Sigma y - b\Sigma xy}{n - 2}}$$

$$\approx \sqrt{\frac{8877.67 - 10.12(227.3) - 2.554(2552.17)}{6 - 2}} \approx \sqrt{\frac{59.152}{4}} \approx 3.8$$

TECH NOTES   Although many calculators that support two-variable statistics and linear regression do not provide the value of the standard error of estimate S_e directly, they do provide the sums required for the calculation of S_e. The TI-84Plus/TI-83Plus/TI-*n*spire, Excel, and Minitab all provide the value of S_e.

TI-84Plus/TI-83Plus/TI-*n*spire (with TI-84Plus keypad) The value for S_e is given as s under STAT, TEST, option E: LinRegTTest.

Excel 2010 Use **Insert Function** $\boxed{f_x}$, select **Statistical**, and choose the function **STEYX**.

Minitab Use the menu choices **Stat ➤ Regression ➤ Regression**. The value for S_e is given as s in the display.

Inferences about the Slope β

Population slope of the least-squares line, β

Recall that $\hat{y} = a + bx$ is the sample-based least-squares line and $y = \alpha + \beta x$ is the population least-squares line computed (in theory) from the population of all (x, y) data pairs. In many real-world applications, the *population slope β* is very important because β measures the rate at which y changes per unit change in x. Our next topic is to develop statistical tests for β. Our work is based on the following theorem:

THEOREM 11.3 Let b be the slope of the sample least-squares line $\hat{y} = a + bx$ computed from a random sample of $n \geq 3$ data pairs (x, y). Let β be the slope of the population least-squares line $y = \alpha + \beta x$, which is in theory computed from the population of all (x, y) data pairs. Let S_e be the standard error of estimate computed from the sample. Then

$$t = \frac{b - \beta}{S_e \Big/ \sqrt{\Sigma x^2 - \dfrac{1}{n}(\Sigma x)^2}}$$

has a Student's t distribution with degrees of freedom $d.f. = n - 2$.

COMMENT The expression $S_e \Big/ \sqrt{\Sigma x^2 - \dfrac{1}{n}(\Sigma x)^2}$ is called the *standard error* for b.

Using this theorem, we can construct a procedure for statistical tests of β.

PROCEDURE

HOW TO TEST β

Setup

Obtain a random sample of $n \geq 3$ data pairs (x, y). Use the procedure of Section 4.2 to find b, the slope of the sample least-squares line. Use Equation (2) of this section to find S_e, the standard error of estimate.

Procedure for a statistical test of β

1. Use the *null hypothesis* H_0: $\beta = 0$. Use an *alternate hypothesis* H_1 appropriate to your application ($\beta > 0$ or $\beta < 0$ or $\beta \neq 0$). Set the level of significance α.

2. Use the null hypothesis H_0: $\beta = 0$ and the values of S_e, n, Σx, Σx^2, and b to compute the *sample test statistic*

$$t = \frac{b}{S_e}\sqrt{\Sigma x^2 - \frac{1}{n}(\Sigma x)^2} \quad \text{with } d.f. = n - 2$$

3. Use a Student's t distribution and the type of test, one-tailed or two-tailed, to find (or estimate) the *P-value* corresponding to the test statistic.

4. *Conclude* the test. If $P\text{-value} \leq \alpha$, then reject H_0. If $P\text{-value} > \alpha$, then do not reject H_0.

5. *Interpret your conclusion* in the context of the application.

EXAMPLE 7   TESTING β

Returning to the sample data regarding x = percentage of the population 25 or older with at least 4 years of college and y = percentage growth in per capita income over the past 7 years (data in Table 11-14), test the claim that the slope β of the least-squares line is positive. Use $\alpha = 0.01$.

SOLUTION: In Example 6, we found the equation of the least-squares line and the standard error of estimate S_e to be

$$\hat{y} \approx 10.12 + 2.554x \qquad \text{and} \qquad S_e \approx 3.8$$

We also found the sums $\Sigma x^2 = 740.68$ and $\Sigma x = 65.2$. Note that the number of data pairs is $n = 6$.

To test the claim that β is positive, use the hypotheses

$$H_0: \beta = 0 \qquad \text{and} \qquad H_1: \beta > 0$$

Convert the sample test statistic $b = 2.554$ to t:

$$t = \frac{b}{S_e}\sqrt{\Sigma x^2 - \frac{1}{n}(\Sigma x)^2} \approx \frac{2.554}{3.8}\sqrt{740.68 - \frac{1}{6}(65.2)^2} \approx 3.81$$

with $d.f. = n - 2 = 6 - 2 = 4$.

To conclude the test, we estimate the P-value of the sample test statistic. Scanning Table 4 of the Appendix, we see that in the row with $d.f. = 4$, the sample t value 3.81 falls between the entries 3.747 and 4.604. These entries correspond to the one-tail area between 0.010 and 0.005. Therefore,

$$0.005 < P\text{-value} < 0.010$$

Since P-value $\leq \alpha$ of 0.01, we reject H_0 and conclude that the population slope of the least-squares line is positive.

Interpretation This means that at the 1% level of significance, the evidence indicates that the percentage growth in per capita income increases as the percentage of the population 25 or older with at least 4 years of college increases.

Examples 5–7 utilize the data of Table 11-14 regarding education levels and income growth. It is interesting to note that for the same data, the t value of the sample correlation coefficient r and the t value of the sample slope b are essentially the same. This is not an accident! Recall from Section 4.2, formula (6), that

$$b = r\left(\frac{s_y}{s_x}\right) \qquad \text{where } s_y \text{ and } s_x \text{ are the standard deviations of the } y \text{ and } x \text{ values of the data pairs, respectively.}$$

Using this relationship and some algebra, it can be shown that:

t values of r and b are equal

> For tests of the correlation coefficient r and the slope b of the least-squares line, both b and r convert to the same t value, with $d.f. = n - 2$, where n is the number of data pairs.

In the next Guided Exercise, we take advantage of this fact.

Problem 14 discusses the fact that for the same data, the Student's t value for r equals that for b.

COMPUTATION HINTS
(a) Most two-variable calculators provide the value of the correlation coefficient r. Find the t value for r. The t value for b is the same.
(b) Most computer-based statistical packages provide the t value for b. The t value for r is the same.

GUIDED EXERCISE 8 | ## Inference for ρ and β

How fast do puppies grow? That depends on the puppy. How about male wolf pups in the Helsinki Zoo (Finland)? Let x = age in weeks and y = weight in kilograms for a random sample of male wolf pups. The following data are based on the article "Studies of the Wolf in Finland *Canis lupus L*" (*Ann. Zool. Fenn.* 2: 215–259) by E. Pulliainen, University of Helsinki.

x	8	10	14	20	28	40	45
y	7	13	17	23	30	34	35

$\Sigma x = 165; \ \Sigma y = 159; \ \Sigma x^2 = 5169; \ \Sigma y^2 = 4317; \ \Sigma xy = 4659$

(a) Verify the following values:

$r \approx 0.959; \ b \approx 0.712$

 Find the values for r and b on your calculator or use the formulas of Sections 4.1 and 4.2.

(b) Use a 1% level of significance to test the claim that $\rho > 0$.

Convert the sample correlation coefficient $r \approx 0.959$ to a t value.

$$t = \frac{r\sqrt{n-2}}{\sqrt{1-r^2}} \approx \frac{0.959\sqrt{5}}{\sqrt{1-0.959^2}} \approx 7.57$$

$H_0: \rho = 0$; $H_1: \rho > 0$; $d.f. = 5$. Using the Student's t distribution, Table 4 of the Appendix, we see that the P-value corresponding to the sample t value of 7.57 is less than 0.0005.

(c) *Interpret* the results.

Since the P-value is less than α of 0.01, we reject H_0 and conclude that there is a positive correlation between age and weight.

(d) Use a 1% level of significance to test the claim that $\beta > 0$.

Convert $b \approx 0.712$ to a t value. Since the t value for b equals the t value for r, we use the result of part (b) and conclude that $t \approx 7.57$.

$H_0: \beta = 0$; $H_1: \beta > 0$; $d.f. = 5$. As in part (b), the P-value corresponding to the sample t value of 7.56 is less than 0.0005.

(e) *Interpret* the results.

We reject H_0 at the 1% level of significance and conclude that the slope of the least-squares line is positive.

TECH NOTES The t value corresponding to the correlation coefficient r is the same as the t value corresponding to b, the slope of the least-squares line. Consequently, the two tests $H_0: \rho = 0$ and $H_0: \beta = 0$ (with similar corresponding alternate hypotheses) have the same conclusions. The TI-84Plus/TI-83Plus/TI-*n*spire uses this fact explicitly. Minitab and Excel show the two-tailed P-value for the slope b of the least-squares line. Excel also shows confidence intervals for β. The displays show data from Guided Exercise 8 regarding the age and weight of wolf pups. The displays are for a two-tailed test.

TI-84Plus/TI-83Plus/TI-*n*spire (with TI-84Plus keypad) Under **STAT**, select **TEST** and use option **E:LinRegTTest**.

```
LinRegTTest
 y=a+bx
 β≠0 and ρ≠0
 t=7.5632
 p=6.4075E⁻4
 df=5.0000
↓a=5.9317
```

```
LinRegTTest
 y=a+bx
 β≠0 and ρ≠0
↑b=.7120
 s=3.3676
 r²=.9196
 r=.9590
```

Note that the value of S_e is given as s.

Excel 2010 Use the ribbon selection **Data tab ➤ Data Analysis ➤ Regression.**

Regression Statistics	
Multiple R	0.958966516
R Square	0.919616778
Adjusted R Square	0.903540133
Standard Error	3.367628886
Observations	7

Value of S_e

	Coefficients	Standard Error	t Stat	P-value	Lower 95%	Upper 95%
Intercept	5.931681179	2.558126184	2.318760198	0.068158803	-0.644180779	12.50754314
X Variable 1	0.711989283	0.094138596	7.563202697	0.000640746	0.469998714	0.953979853

b for two-tailed test

Minitab Use the menu selection **Stat ➤ Regression ➤ Regression.** The value of S_e is S; P is the P-value of a two-tailed test. For a one-tailed test, divide the P-value by 2.

```
Regression Analysis
The regression equation is
y = 5.93 + 0.712 x
Predictor        Coef        StDev          T          P
Constant        5.932       2.558       2.32      0.068
x               0.71199     0.09414     7.56      0.001
S  = 3.368      R-Sq = 92.0%      R-Sq(adj) = 90.4%
```

Confidence Intervals for *y*

The least-squares line gives us a predicted value $\hat{y}$ for a specified x value. However, we used sample data to get the equation of the line. The line derived from the population of all data pairs is likely to have a slightly different slope, which we designate by the symbol β for population slope, and a slightly different y intercept, which we designate by the symbol α for population intercept. In addition, there is some random error ϵ, so the true y value is

$$y = \alpha + \beta x + \epsilon$$

Because of the random variable ϵ, for each x value there is a corresponding distribution of y values. The methods of linear regression were developed so that the distribution of y values for a given x would be centered on the population regression line. Furthermore, the distributions of y values corresponding to each x value all have the same standard deviation, estimated by the standard error of estimate S_e.

Using all this background, the theory tells us that for a specific x, a *c confidence interval for y* is given by the next procedure.

c confidence interval for y

PROCEDURE

How to find a confidence interval for a predicted y from the least-squares line

1. Obtain a random sample of $n \geq 3$ data pairs (x, y).

2. Use the procedure of Section 4.2 to find $\hat{y} = a + bx$. You also need to find $\bar{x}$ from the sample data and the standard error of estimate S_e using Equation (2) of this section.

3. The c confidence interval for y for a specified value of x is

$$\hat{y} - E < y < \hat{y} + E$$

where

$$E = t_c S_e \sqrt{1 + \frac{1}{n} + \frac{n(x - \bar{x})^2}{n\Sigma x^2 - (\Sigma x)^2}}$$

$\hat{y} = a + bx$ is the predicted value of y from the least-squares line for a *specified* x value

c = confidence level $(0 < c < 1)$

t_c = critical value from Student's t distribution for c confidence level using $d.f. = n - 2$

S_e = standard error of estimate

n = number of data pairs $(n \geq 3)$

The formulas involved in the computation of a c confidence interval look complicated. However, they involve quantities we have already computed or values we can easily look up in tables. The next example illustrates this point.

EXAMPLE 8   CONFIDENCE INTERVAL FOR PREDICTION

June and Jim are partners in the chemistry lab. Their assignment is to determine how much copper sulfate ($CuSO_4$) will dissolve in water at 10, 20, 30, 40, 50, 60, and 70°C. Their lab results are shown in the table, where y is the weight in grams of copper sulfate that will dissolve in 100 g of water at x°C.

x	10	20	30	40	50	60	70
y	17	21	25	28	33	40	49

Find the equation of the least-squares line and find a 95% confidence interval for the amount of copper sulfate that will dissolve in 100 g of water at 45°C.

SOLUTION: Using a calculator with appropriate formulas, we find

$$\Sigma x = 280; \ \Sigma y = 213; \ \Sigma x^2 = 14{,}000; \ \Sigma y^2 = 7229; \ \Sigma xy = 9940$$
$$\bar{x} = 40; \ \bar{y} \approx 30.429; \ b \approx 0.50714; \ a \approx 10.143; \ S_e \approx 2.35$$

Next we use the equation of the least-squares line to find $\hat{y}$ for $x = 45$°C.

$$\hat{y} = a + bx$$
$$\hat{y} \approx 10.14 + 0.51x$$
$$\hat{y} \approx 10.14 + 0.51(45) \quad \text{Use 45 in place of } x.$$
$$\hat{y} \approx 33$$

A 95% confidence interval is then

$$\hat{y} - E < y < \hat{y} + E$$
$$33 - E < y < 33 + E$$

where $E = t_c S_e \sqrt{1 + \dfrac{1}{n} + \dfrac{n(x - \bar{x})^2}{n\Sigma x^2 - (\Sigma x)^2}}$.

Using $n - 2 = 7 - 2 = 5$ degrees of freedom, we find from Table 4 of the Appendix that $t_{0.95} = 2.571$.

$$E \approx (2.571)(2.35)\sqrt{1 + \dfrac{1}{7} + \dfrac{7(45 - 40)^2}{7(14,000) - (280)^2}}$$

$$\approx (2.571)(2.35)\sqrt{1.15179} \approx 6.5$$

A 95% confidence interval for y is

$$33 - 6.5 \le y \le 33 + 6.5$$
$$26.5 \le y \le 39.5$$

Interpretation This means that we are 95% sure that the interval between 26.5 g and 39.5 g is one that contains the predicted amount of copper sulfate that will dissolve in 100 g of water at 45°C. The interval is fairly wide but would decrease with more sample data.

GUIDED EXERCISE 9 | **Confidence interval for prediction**

Let's use the data of Example 8 to compute a 95% confidence interval for $y =$ amount of copper sulfate that will dissolve at $x = 15$°C.

(a) From Example 8, we have

$$\hat{y} \approx 10.14 + 0.51x$$

Evaluate $\hat{y}$ for $x = 15$.

⟹ $\hat{y} \approx 10.14 + 0.51x$
     $\approx 10.14 + 0.51(15)$
     ≈ 17.8

(b) The bound E on the error of estimate is

$$E = t_c S_e \sqrt{1 + \dfrac{1}{n} + \dfrac{n(x - \bar{x})^2}{n\Sigma x^2 - (\Sigma x)^2}}$$

From Example 8, we know that $S_e \approx 2.35$, $\Sigma x = 280$, $\Sigma x^2 = 14,000$, $\bar{x} = 40$, and $n = 7$. Find $t_{0.95}$ and compute E.

⟹ $t_{0.95} = 2.571$ for $d.f. = n - 2 = 5$

$$E \approx (2.571)(2.35)\sqrt{1 + \dfrac{1}{7} + \dfrac{7(15 - 40)^2}{7(14,000) - (280)^2}}$$

$$\approx (2.571)(2.35)\sqrt{1.366071} \approx 7.1$$

(c) Find a 95% confidence interval for y.

$$\hat{y} - E \le y \le \hat{y} + E$$

⟹   The confidence interval is

$$17.8 - 7.1 \le y \le 17.8 + 7.1$$
$$10.7 \le y \le 24.9$$

As we compare the results of Guided Exercise 9 and Example 8, we notice that the 95% confidence interval of y values for $x = 15$°C is 7.1 units above and below the least-squares line, while the 95% confidence interval of y values for $x = 45$°C is only 6.5 units above and below the least-squares line. This comparison reflects the general property that confidence intervals for y are narrower the nearer we are to the mean $\bar{x}$ of the x values. As we move near the extremes of the

FIGURE 11-11

95% Confidence Band for Predicted Values $\hat{y}$

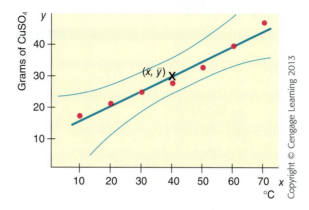

x distribution, the confidence intervals for y become wider. This is another reason that we should not try to use the least-squares line to predict y values for x values beyond the data extremes of the sample x distribution.

Confidence prediction band

If we were to compute a 95% confidence interval for all x values in the range of the sample x values, the *confidence prediction band* would curve away from the least-squares line, as shown in Figure 11-11.

TECH NOTES   Minitab provides confidence intervals for predictions. Use the menu selection **Stat ➤ Regression ➤ Regression**. Under **Options**, enter the observed x value and set the confidence level. In the output, the confidence interval for predictions is designated by %PI.

VIEWPOINT | Hawaii Island Hopping!

*Suppose you want to go camping in Hawaii. Yes! Hawaii has both state and federal parks where you can enjoy camping on the beach or in the mountains. However, you will probably need to rent a car to get to the different campgrounds. How much will the car rental cost? That depends on the islands you visit. For car rental data and regression statistics that you can compute regarding costs on different Hawaiian islands, visit this textbook's site at **http://www.cengagebrain.com** and find the link to Hawaiian Islands.*

**SECTION 11.4
PROBLEMS**

aplia™   Access end-of-section exercises at www.aplia.com

Chapter Review

In this chapter, we introduced applications of the chi-square probability distribution and inferences for correlation and linear regression. In Part I, we looked at

- Properties of the chi-square distribution
- Tests of independence or homogeneity
- Tests of goodness-of-fit
- Tests of variance σ^2

In Part II, we returned to the linear regression model and linear correlation as introduced in Chapter 4. The inferences discussed were

- Testing the correlation coefficient ρ
- Testing the slope β of the least-squares line
- Confidence intervals for predicted values based on a specific value for x and the least-squares line

Procedure displays within Chapter 11 show methods and assumptions for all tests and confidence intervals discussed.

VIEWPOINT | Movies and Money!

Young adults are the movie industry's best customers. However, going to the movies is expensive, which may explain why attendance rates increase with household income. Using what you have learned in this chapter, you can create appropriate chi-square tests to determine how good a fit exists between national percentage rates of attendance by household income and attendance rates in your demographic area. For more information and national data, see American Demographics *(Vol. 18, No. 12).*

CHAPTER REVIEW PROBLEMS

Part I: Inferences Using the Chi-Square Distribution

1. *Statistical Literacy* Of the following random variables, which have only non-negative values: z, t, chi-square?

2. *Statistical Literacy* Of the following probability distributions, which are always symmetrical: normal, Student's t, chi-square?

3. *Critical Thinking* Suppose you took random samples from three distinct age groups. Through a survey, you determined how many respondents from each age group preferred to get news from TV, newspapers, the Internet, or another source (respondents could select only one mode). What type of test would be appropriate to determine if there is sufficient statistical evidence to claim that the proportions of each age group preferring the different modes of obtaining news are not the same? Select from tests of independence, homogeneity, and goodness-of-fit.

Before you solve Problems 4–8, first classify the problem as one of the following:

> Chi-square test of independence or homogeneity
> Chi-square goodness-of-fit
> Chi-square for testing σ^2 or σ

Then, in each of the problems when a test is to be performed, do the following:

(i) Give the value of the level of significance. State the null and alternate hypotheses.
(ii) Find the sample test statistic.
(iii) Find or estimate the P-value of the sample test statistic.
(iv) Conclude the test.
(v) *Interpret* the conclusion in the context of the application.

4. *Education: Exams* Professor Fair believes that extra time does not improve grades on exams. He randomly divided a group of 300 students into two groups and gave them all the same test. One group had exactly 1 hour in which to finish the test, and the other group could stay as long as desired. The results are shown in the following table. Test at the 0.01 level of significance that time to complete a test and test results are independent.

Time	A	B	C	F	Row Total
1 h	23	42	65	12	142
Unlimited	17	48	85	8	158
Column Total	40	90	150	20	300

5. *Tires: Blowouts* A consumer agency is investigating the blowout pressures of Soap Stone tires. A Soap Stone tire is said to blow out when it separates from the wheel rim due to impact forces usually caused by hitting a rock or a pothole in the road. A random sample of 30 Soap Stone tires were inflated to the recommended pressure, and then forces measured in foot-pounds were applied to each tire (1 foot-pound is the force of 1 pound dropped from a height of 1 foot). The customer complaint is that some Soap Stone tires blow out under small-impact forces, while other tires seem to be well made and don't have this fault. For the 30 test tires, the sample standard deviation of blowout forces was 1353 foot-pounds.

Soap Stone claims its tires will blow out at an average pressure of 20,000 foot-pounds, with a standard deviation of 1020 foot-pounds. The average blowout force is not in question, but the variability of blowout forces is in question. Using a 0.01 level of significance, test the claim that the variance of blowout pressures is more than Soap Stone claims it is.

6. *Packaging: Corn Flakes* A machine that puts corn flakes into boxes is adjusted to put an average of 15 ounces into each box, with standard deviation of 0.25 ounce. If a random sample of 12 boxes gave a sample standard deviation of 0.38 ounce, do these data support the claim that the variance has increased and the machine needs to be brought back into adjustment? (Use a 0.01 level of significance.)

7. *Teacher Ratings: Grades* Professor Stone complains that student teacher ratings depend on the grade the student receives. In other words, according to Professor Stone, a teacher who gives good grades gets good ratings, and a teacher who gives bad grades gets bad ratings. To test this claim, the Student Assembly took a random sample of 300 teacher ratings on which the student's grade for the course also was indicated. The results are given in the following table. Test the hypothesis that teacher ratings and student grades are independent at the 0.01 level of significance.

Rating	A	B	C	F (or withdrawal)	Row Total
Excellent	14	18	15	3	50
Average	25	35	75	15	150
Poor	21	27	40	12	100
Column Total	60	80	130	30	300

Copyright © Cengage Learning 2013

8. *Sociology: Age Distribution* A sociologist is studying the age of the population in Blue Valley. Ten years ago, the population was such that 20% were under 20 years old, 15% were in the 20- to 35-year-old bracket, 30% were between 36 and 50, 25% were between 51 and 65, and 10% were over 65. A study done this year used a random sample of 210 residents. This sample showed

Under 20	20–35	36–50	51–65	Over 65
26	27	69	68	20

At the 0.01 level of significance, has the age distribution of the population of Blue Valley changed?

Part II: Inferences Relating to Linear Regression

9. *Statistical Literacy* What does it mean to say that the sample correlation coefficient r is significant?

10. *Physiology: Children* The following problem is based on information taken from the pediatrics section of *The Merck Manual* (a commonly used reference in medical schools and nursing programs). Let x be the body weight of a child (in kilograms), and let y be the metabolic rate of the child (in 100 kcal/24 h).

x	3.0	5.0	9.0	11.0	15.0	17.0	19.0	21.0
y	1.4	2.7	5.0	6.0	7.1	7.8	8.3	8.8

$\Sigma x = 100; \Sigma y = 47.1; \Sigma x^2 = 1552; \Sigma y^2 = 327.83; \Sigma xy = 710.3$

(a) Verify that $S_e \approx 0.518$.
(b) Verify that $r \approx 0.984$. Test that ρ is positive. Use $\alpha = 0.01$.
(c) Verify that $b \approx 0.402$. Test that β is positive. Use $\alpha = 0.01$.
(d) Verify that $\hat{y} \approx 0.856 + 0.402x$. Find a 90% confidence interval for the predicted metabolic rate for a child weighing 10 kg.

11. *Baseball: Batting Averages and Home Runs* In baseball, is there a linear correlation between batting average and home run percentage? Let x represent the batting average of a professional baseball player. Let y represent the home run percentage (number of home runs per 100 times at bat). A random sample of $n = 7$ professional baseball players gave the following information (Reference: *The Baseball Encyclopedia*, Macmillan Publishing Company):

x	0.243	0.259	0.286	0.263	0.268	0.339	0.299
y	1.4	3.6	5.5	3.8	3.5	7.3	5.0

$\Sigma x = 1.957$; $\Sigma y = 30.1$; $\Sigma x^2 \approx 0.553$; $\Sigma y^2 = 150.15$; $\Sigma xy \approx 8.753$

(a) Verify that $S_e \approx 0.647$.
(b) Verify that $r \approx 0.948$. Test that ρ is positive. Use $\alpha = 0.01$.
(c) Verify that $b \approx 55.166$. Test that β is positive. Use $\alpha = 0.01$.
(d) Verify that $\hat{y} \approx -11.123 + 55.166x$. Find a 90% confidence interval for the predicted home run percentage for a player with a batting average of 0.310.

DATA HIGHLIGHTS: GROUP PROJECTS

Break into small groups and discuss the following topic. Organize a brief outline in which you summarize the main points of your group discussion.

The *Statistical Abstract of the United States* reported information about the percentage of arrests of all drunk drivers according to age group. In the following table, the entry 3.7 in the first row means that in the entire United States, about 3.7% of all people arrested for drunk driving were in the age group 16–17 years. The Freemont County Sheriff's Office obtained data about the number of drunk drivers arrested in each age group over the past several years. In the following table, the entry 8 in the first row means that eight people in the age group 16–17 years were arrested for drunk driving in Freemont County.

Distribution of Drunk-Driving Arrests by Age

Age	National Percentage	Number in Freemont County
16–17	3.7	8
18–24	18.9	35
25–29	12.9	23
30–34	10.3	19
35–39	8.5	12
40–44	7.9	14
45–49	8.0	16
50–54	7.9	13
55–59	6.8	10
60–64	5.7	9
65 and over	9.4	15
	100%	174

Use a chi-square test with 5% level of significance to test the claim that the age distribution of drunk drivers arrested in Freemont County is the same as the national age distribution of drunk drivers arrested.

(a) State the null and alternate hypotheses.
(b) Find the value of the chi-square test statistic from the sample.
(c) Find the degrees of freedom and the *P*-value of the test statistic.
(d) Decide whether you should reject or fail to reject the null hypothesis.
(e) Interpret your conclusion in the context of the problem.
(f) How could you gather data and conduct a similar test for the city or county in which you live? Explain.

LINKING CONCEPTS: WRITING PROJECTS

Discuss each of the following topics in class or review the topics on your own. Then write a brief but complete essay in which you summarize the main points. Please include formulas and graphs as appropriate.

Consider the results of a study conducted by the makers of Advil. A random sample of patients were assigned to one of two groups: One received ibuprofen and the other received a placebo. Then patients were asked if they experienced stomach upset (yes or no). The results of the study follow:

Group	Stomach Upset		Row Total
	Yes	No	
Ibuprofen	8	664	672
Placebo	6	645	651
Column total	14	1309	1323

Copyright © Cengage Learning 2013

1. In Chapter 5 on probability (Section 5.2), we considered contingency tables and the probabilities of events described by different combinations of the rows and columns. Use these techniques to compute the probability that a patient had stomach upset *given* that the patient was in the ibuprofen group. Compute the probability that a patient had stomach upset *given* that the patient was in the placebo group.

2. Are the proportions of patients having stomach upset in the two different treatment groups the same? We treated questions of this type in Section 10.3, where we applied the normal distribution to test the difference of proportions. Use such a test to determine whether or not the proportions are different at the 5% level of significance.

3. Now consider the 2 × 2 contingency table. We can apply techniques of tests of independence to see if the proportions of patients experiencing stomach upset in the two treatment groups are the same or not. When we apply these techniques to tests of proportions, we assume that the sample size for each treatment group is *assigned* ahead of time and is not itself a random variable. In such a case, we apply a test of homogeneity. Perform such a test.

4. Compare the results of Problems 2 and 3. Do the test conclusions agree? When testing two proportions, you can use either the normal distribution test of Section 10.3 or a chi-square test. Can you think of any advantages of one choice over the other?

USING TECHNOLOGY

Tests of Independence

A study involving people who process food gave the following information about work shift and number of sick days:

Shift	Number of Sick Days				
	0	1	2	3	4 or more
Day	134	44	24	10	61
Aft/ev	90	39	23	18	99
Night	107	37	21	20	82
Rotating	56	20	14	17	92

Source: United States Department of Health, Education, and Welfare, NIOSH Technical Report. Tasto, Colligan, et al. *Health Consequences of Shift Work*. Washington: GPO, 1978, 29. (*Note:* This table was adapted from Table 7 on page 25 of the source.)

Use a 1% level of significance to test the null hypothesis that work shift and number of sick days are independent against the alternate hypothesis that they are not independent. What is the sample chi-square value? What is the test conclusion?

Technology Hints

TI-84Plus/TI-83Plus/TI-*n*spire (with TI-84Plus keypad)

Press the **Matrix** key and enter the observed values in a matrix. Name the matrix [A]. Then enter dimension matrix [B] with the same dimensions. Matrix [B] will hold the expected values. Press the **STAT** key and under the **TESTS** menu, select option **C:χ^2-Test. Calculate** will give the sample χ^2 value and its P value. **Draw** will show the sample test statistic on the χ^2 distribution with the P-value shaded.

Excel 2010

There are no built-in commands for executing a chi-square test. The *Technology Guide* shows a multistep process for conducting a chi-square test. The Tech Note of Section 11.1 gives a brief outline of the process.

Minitab

In Minitab, enter data for tests of independence in columns. Be sure to enter data only and not row or column totals. Then use the menu choices **Stat ➤ Tables ➤ Chi-Square Test.**

SPSS

SPSS utilizes raw data to make a contingency table and then provides an option for a chi-square test of independence. For example, rather than entering the data for number of sick days versus shift as a table, in SPSS you enter all 1009 cases, where the shift variable shows the row category (day, aft/eve, night, rotating) of the entry and the sick-days variable shows the corresponding column category (number of sick days). Specifically, you would enter 134 cases with shift variable "day" and sick-day variable "0," etc. Then use the menu choices **Analyze ➤ Descriptive Statistics ➤ Crosstabs....** In the dialogue box, move the row variable and the column variable to the designated boxes. Select the Statistics option and check Chi-square. To see the expected values, select the Cells option and check Expected. The output includes a contingency table, the sample χ^2 test statistic, degrees of freedom, and the P-value of the test statistic.

Cumulative Review Problems

CHAPTERS 10–11

For all hypothesis tests, please provide the following information:

(i) What is the level of significance? State the null and alternate hypotheses.

(ii) *Check Requirements* What sampling distribution will you use? What assumptions are you making? What is the value of the sample test statistic?

(iii) Find (or estimate) the *P*-value. Sketch the sampling distribution and show the area corresponding to the *P*-value.

(iv) Based on your answers in parts (i)–(iii), will you reject or fail to reject the null hypothesis? Are the data statistically significant at level α?

(v) *Interpret* your conclusion in the context of the application.

1. *Inferences for Linear Regression* Plate tectonics and the spread of the ocean floor are very important in modern studies of earthquakes and earth science in general. A random sample of islands in the Indian Ocean gave the following information, where x = age of a volcanic island in the Indian Ocean (units in 10^6 years) and y = distance of the island from the center of the midoceanic ridge (units in 100 km).

x	120	83	60	50	35	30	20	17
y	30	16	15.5	14.5	22	18	12	0

Source: King, Cuchlaine A. M. *Physical Geography.* Oxford: Basil Blackwell, 1980, pp. 77–86 and 196–206. Reprinted by permission of the publisher.

$\Sigma x = 415$; $\Sigma y = 128$; $\Sigma x^2 = 30{,}203$; $\Sigma y^2 = 2558.5$; $\Sigma xy = 8133$

(a) Verify that the standard error of estimate $S_e \approx 6.50$.

(b) Verify that the sample correlation coefficient $r \approx 0.709$. Use a 5% level of significance to test that ρ is positive.

Photo Researchers

(c) Verify that the equation of the least-squares line is $\hat{y} \approx 7.07 + 0.172x$. Use a 5% level of significance to test that β is positive.

(d) Given that the age of a volcanic island is 55 (units in 10^6 years), predict how far the island is from the center of the midoceanic ridge (units in 100 km). Find an 85% confidence interval for the prediction.

2. *Goodness of Fit* A recent national study gave the following information about declared majors of undergraduate college students: humanities, 15%; science, 23%; business, 29%; education, 11%; social science, 10%; all other fields, 12% (Reference: *Statistical Abstract of the United States*). A random sample of 215 college students in Colorado gave the following counts regarding college majors: humanities, 24; science, 54; business, 70; education, 16; social science, 30; all other fields, 21. Does this information indicate that the distribution of college majors for Colorado students is different from the national distribution? Use $\alpha = 0.01$.

3. *Test of Independence: Agriculture* Three types of fertilizer were used on 132 identical plots of maize. Each plot was harvested and the yield (in kg) was recorded (Reference: Caribbean Agricultural Research and Development Institute).

Yield (kg)	Type of Fertilizer			Row Total
	I	II	III	
0–2.9	12	10	15	37
3.0–5.9	18	21	11	50
6.0–8.9	16	19	10	45
Column total	46	50	36	132

Use a 5% level of significance to test the hypothesis that type of fertilizer and yield of maize are independent. Interpret the results.

4. *Testing Variances: Iris* Random samples of two species of iris gave the following petal lengths (in cm) (Reference: R. A. Fisher, *Annals of Eugenics*, Vol. 7):

x_1: *Iris virginica*	5.1 5.9 4.5 4.9 5.7 4.8 5.8 6.4 5.6 5.9
x_2: *Iris versicolor*	4.5 4.8 4.7 5.0 3.8 5.1 4.4 4.2

Use a 5% level of significance to test the claim that the population standard deviation of x_1 is larger than 0.55.

5. *Testing Paired Differences* Phosphorus is a chemical that is found in many household cleaning products. Unfortunately, phosphorus also finds its way into surface water, where it can harm fish, plants, and other wildlife. Two methods of phosphorus reduction are being studied. At a random sample of 7 locations, both methods were used and the total phosphorus reduction (mg/L) was recorded (Reference: EPA report 832-R-93-005).

Site	1	2	3	4	5	6	7
Method I	0.013	0.030	0.015	0.055	0.007	0.002	0.010
Method II	0.014	0.058	0.017	0.039	0.017	0.001	0.013

Do these data indicate a difference (either way) in the average reduction of phosphorus between the two methods? Use $\alpha = 0.05$.

6. *Testing and Estimating $\mu_1 - \mu_2$, σ_1 and σ_2 Unknown* In the airline business, "on-time" flight arrival is important for connecting flights and general customer satisfaction. Is there a difference between summer and winter average on-time flight arrivals? Let x_1 be a random variable that represents percentage of on-time arrivals at major airports in the summer. Let x_2 be a random variable that represents percentage of on-time arrivals at major airports in the winter. A random sample of $n_1 = 16$ major airports showed that $\bar{x}_1 = 74.8\%$, with $s_1 = 5.2\%$. A random sample of $n_2 = 18$ major airports showed that $\bar{x}_2 = 70.1\%$, with $s_2 = 8.6\%$ (Reference: *Statistical Abstract of the United States*).

(a) Does this information indicate a difference (either way) in the population mean percentage of on-time arrivals for summer compared to winter? Use $\alpha = 0.05$.

(b) Find an 95% confidence interval for $\mu_1 - \mu_2$.

(c) What assumptions about the original populations have you made for the methods used?

7. **Testing and Estimating a Difference of Proportions**
$p_1 - p_2$ How often do you go out dancing? This question was asked by a professional survey group on behalf of the National Arts Survey. A random sample of $n_1 = 95$ single men showed that $r_1 = 23$ went out dancing occasionally. Another random sample of $n_2 = 92$ single women showed that $r_2 = 19$ went out dancing occasionally.

Corbis

(a) Do these data indicate that the proportion of single men who go out dancing occasionally is higher than the proportion of single women? Use a 5% level of significance. List the assumptions you made in solving this problem. Do you think these assumptions are realistic?

(b) Compute a 90% confidence interval for the population difference of proportions $p_1 - p_2$ of single men and single women who occasionally go out dancing.

APPENDIX: TABLES

1. Random Numbers
2. Binomial Probability Distribution $C_{n,r}p^r q^{n-r}$
3. Areas of a Standard Normal Distribution
4. Critical Values for Student's t Distribution
5. The χ^2 Distribution

TABLE 1 Random Numbers

92630	78240	19267	95457	53497	23894	37708	79862	76471	66418
79445	78735	71549	44843	26104	67318	00701	34986	66751	99723
59654	71966	27386	50004	05358	94031	29281	18544	52429	06080
31524	49587	76612	39789	13537	48086	59483	60680	84675	53014
06348	76938	90379	51392	55887	71015	09209	79157	24440	30244
28703	51709	94456	48396	73780	06436	86641	69239	57662	80181
68108	89266	94730	95761	75023	48464	65544	96583	18911	16391
99938	90704	93621	66330	33393	95261	95349	51769	91616	33238
91543	73196	34449	63513	83834	99411	58826	40456	69268	48562
42103	02781	73920	56297	72678	12249	25270	36678	21313	75767
17138	27584	25296	28387	51350	61664	37893	05363	44143	42677
28297	14280	54524	21618	95320	38174	60579	08089	94999	78460
09331	56712	51333	06289	75345	08811	82711	57392	25252	30333
31295	04204	93712	51287	05754	79396	87399	51773	33075	97061
36146	15560	27592	42089	99281	59640	15221	96079	09961	05371
29553	18432	13630	05529	02791	81017	49027	79031	50912	09399
23501	22642	63081	08191	89420	67800	55137	54707	32945	64522
57888	85846	67967	07835	11314	01545	48535	17142	08552	67457
55336	71264	88472	04334	63919	36394	11196	92470	70543	29776
10087	10072	55980	64688	68239	20461	89381	93809	00796	95945
34101	81277	66090	88872	37818	72142	67140	50785	21380	16703
53362	44940	60430	22834	14130	96593	23298	56203	92671	15925
82975	66158	84731	19436	55790	69229	28661	13675	99318	76873
54827	84673	22898	08094	14326	87038	42892	21127	30712	48489
25464	59098	27436	89421	80754	89924	19097	67737	80368	08795
67609	60214	41475	84950	40133	02546	09570	45682	50165	15609
44921	70924	61295	51137	47596	86735	35561	76649	18217	63446
33170	30972	98130	95828	49786	13301	36081	80761	33985	68621
84687	85445	06208	17654	51333	02878	35010	67578	61574	20749
71886	56450	36567	09395	96951	35507	17555	35212	69106	01679
00475	02224	74722	14721	40215	21351	08596	45625	83981	63748
25993	38881	68361	59560	41274	69742	40703	37993	03435	18873
92882	53178	99195	93803	56985	53089	15305	50522	55900	43026
25138	26810	07093	15677	60688	04410	24505	37890	67186	62829
84631	71882	12991	83028	82484	90339	91950	74579	03539	90122

TABLE 1 *continued*

34003	92326	12793	61453	48121	74271	28363	66561	75220	35908
53775	45749	05734	86169	42762	70175	97310	73894	88606	19994
59316	97885	72807	54966	60859	11932	35265	71601	55577	67715
20479	66557	50705	26999	09854	52591	14063	30214	19890	19292
86180	84931	25455	26044	02227	52015	21820	50599	51671	65411
21451	68001	72710	40261	61281	13172	63819	48970	51732	54113
98062	68375	80089	24135	72355	95428	11808	29740	81644	86610
01788	64429	14430	94575	75153	94576	61393	96192	03227	32258
62465	04841	43272	68702	01274	05437	22953	18946	99053	41690
94324	31089	84159	92933	99989	89500	91586	02802	69471	68274
05797	43984	21575	09908	70221	19791	51578	36432	33494	79888
10395	14289	52185	09721	25789	38562	54794	04897	59012	89251
35177	56986	25549	59730	64718	52630	31100	62384	49483	11409
25633	89619	75882	98256	02126	72099	57183	55887	09320	73463
16464	48280	94254	45777	45150	68865	11382	11782	22695	41988

Source: Reprinted from *A Million Random Digits with 100,000 Normal Deviates* by the Rand Corporation
(New York: The Free Press, 1955). Copyright 1955 and 1983 by the Rand Corporation. Used by permission.

TABLE 2 Binomial Probability Distribution $C_{n,r} p^r q^{n-r}$

This table shows the probability of r successes in n independent trials, each with probability of success p.

n	r	.01	.05	.10	.15	.20	.25	.30	.35	.40	.45	.50	.55	.60	.65	.70	.75	.80	.85	.90	.95
2	0	.980	.902	.810	.723	.640	.563	.490	.423	.360	.303	.250	.203	.160	.123	.090	.063	.040	.023	.010	.002
	1	.020	.095	.180	.255	.320	.375	.420	.455	.480	.495	.500	.495	.480	.455	.420	.375	.320	.255	.180	.095
	2	.000	.002	.010	.023	.040	.063	.090	.123	.160	.203	.250	.303	.360	.423	.490	.563	.640	.723	.810	.902
3	0	.970	.857	.729	.614	.512	.422	.343	.275	.216	.166	.125	.091	.064	.043	.027	.016	.008	.003	.001	.000
	1	.029	.135	.243	.325	.384	.422	.441	.444	.432	.408	.375	.334	.288	.239	.189	.141	.096	.057	.027	.007
	2	.000	.007	.027	.057	.096	.141	.189	.239	.288	.334	.375	.408	.432	.444	.441	.422	.384	.325	.243	.135
	3	.000	.000	.001	.003	.008	.016	.027	.043	.064	.091	.125	.166	.216	.275	.343	.422	.512	.614	.729	.857
4	0	.961	.815	.656	.522	.410	.316	.240	.179	.130	.092	.062	.041	.026	.015	.008	.004	.002	.001	.000	.000
	1	.039	.171	.292	.368	.410	.422	.412	.384	.346	.300	.250	.200	.154	.112	.076	.047	.026	.011	.004	.000
	2	.001	.014	.049	.098	.154	.211	.265	.311	.346	.368	.375	.368	.346	.311	.265	.211	.154	.098	.049	.014
	3	.000	.000	.004	.011	.026	.047	.076	.112	.154	.200	.250	.300	.346	.384	.412	.422	.410	.368	.292	.171
	4	.000	.000	.000	.001	.002	.004	.008	.015	.026	.041	.062	.092	.130	.179	.240	.316	.410	.522	.656	.815
5	0	.951	.774	.590	.444	.328	.237	.168	.116	.078	.050	.031	.019	.010	.005	.002	.001	.000	.000	.000	.000
	1	.048	.204	.328	.392	.410	.396	.360	.312	.259	.206	.156	.113	.077	.049	.028	.015	.006	.002	.000	.000
	2	.001	.021	.073	.138	.205	.264	.309	.336	.346	.337	.312	.276	.230	.181	.132	.088	.051	.024	.008	.001
	3	.000	.001	.008	.024	.051	.088	.132	.181	.230	.276	.312	.337	.346	.336	.309	.264	.205	.138	.073	.021
	4	.000	.000	.000	.002	.006	.015	.028	.049	.077	.113	.156	.206	.259	.312	.360	.396	.410	.392	.328	.204
	5	.000	.000	.000	.000	.000	.001	.002	.005	.010	.019	.031	.050	.078	.116	.168	.237	.328	.444	.590	.774
6	0	.941	.735	.531	.377	.262	.178	.118	.075	.047	.028	.016	.008	.004	.002	.001	.000	.000	.000	.000	.000
	1	.057	.232	.354	.399	.393	.356	.303	.244	.187	.136	.094	.061	.037	.020	.010	.004	.002	.000	.000	.000
	2	.001	.031	.098	.176	.246	.297	.324	.328	.311	.278	.234	.186	.138	.095	.060	.033	.015	.006	.001	.000
	3	.000	.002	.015	.042	.082	.132	.185	.236	.276	.303	.312	.303	.276	.236	.185	.132	.082	.042	.015	.002
	4	.000	.000	.001	.006	.015	.033	.060	.095	.138	.186	.234	.278	.311	.328	.324	.297	.246	.176	.098	.031
	5	.000	.000	.000	.000	.002	.004	.010	.020	.037	.061	.094	.136	.187	.244	.303	.356	.393	.399	.354	.232
	6	.000	.000	.000	.000	.000	.000	.001	.002	.004	.008	.016	.028	.047	.075	.118	.178	.262	.377	.531	.735
7	0	.932	.698	.478	.321	.210	.133	.082	.049	.028	.015	.008	.004	.002	.001	.000	.000	.000	.000	.000	.000
	1	.066	.257	.372	.396	.367	.311	.247	.185	.131	.087	.055	.032	.017	.008	.004	.001	.000	.000	.000	.000
	2	.002	.041	.124	.210	.275	.311	.318	.299	.261	.214	.164	.117	.077	.047	.025	.012	.004	.001	.000	.000
	3	.000	.004	.023	.062	.115	.173	.227	.268	.290	.292	.273	.239	.194	.144	.097	.058	.029	.011	.003	.000
	4	.000	.000	.003	.011	.029	.058	.097	.144	.194	.239	.273	.292	.290	.268	.227	.173	.115	.062	.023	.004
	5	.000	.000	.000	.001	.004	.012	.025	.047	.077	.117	.164	.214	.261	.299	.318	.311	.275	.210	.124	.041
	6	.000	.000	.000	.000	.000	.001	.004	.008	.017	.032	.055	.087	.131	.185	.247	.311	.367	.396	.372	.257
	7	.000	.000	.000	.000	.000	.000	.000	.001	.002	.004	.008	.015	.028	.049	.082	.133	.210	.321	.478	.698

TABLE 2 *continued*

										P										
n	r	.01	.05	.10	.15	.20	.25	.30	.35	.40	.45	.50	.55	.60	.65	.70	.75	.80	.85	.90	.95
8	0	.923	.663	.430	.272	.168	.100	.058	.032	.017	.008	.004	.002	.001	.000	.000	.000	.000	.000	.000	.000
	1	.075	.279	.383	.385	.336	.267	.198	.137	.090	.055	.031	.016	.008	.003	.001	.000	.000	.000	.000	.000
	2	.003	.051	.149	.238	.294	.311	.296	.259	.209	.157	.109	.070	.041	.022	.010	.004	.001	.000	.000	.000
	3	.000	.005	.033	.084	.147	.208	.254	.279	.279	.257	.219	.172	.124	.081	.047	.023	.009	.003	.000	.000
	4	.000	.000	.005	.018	.046	.087	.136	.188	.232	.263	.273	.263	.232	.188	.136	.087	.046	.018	.005	.000
	5	.000	.000	.000	.003	.009	.023	.047	.081	.124	.172	.219	.257	.279	.279	.254	.208	.147	.084	.033	.005
	6	.000	.000	.000	.000	.001	.004	.010	.022	.041	.070	.109	.157	.209	.259	.296	.311	.294	.238	.149	.051
	7	.000	.000	.000	.000	.000	.000	.001	.003	.008	.016	.031	.055	.090	.137	.198	.267	.336	.385	.383	.279
	8	.000	.000	.000	.000	.000	.000	.000	.000	.001	.002	.004	.008	.017	.032	.058	.100	.168	.272	.430	.663
9	0	.914	.630	.387	.232	.134	.075	.040	.021	.010	.005	.002	.001	.000	.000	.000	.000	.000	.000	.000	.000
	1	.083	.299	.387	.368	.302	.225	.156	.100	.060	.034	.018	.008	.004	.001	.000	.000	.000	.000	.000	.000
	2	.003	.063	.172	.260	.302	.300	.267	.216	.161	.111	.070	.041	.021	.010	.004	.001	.000	.000	.000	.000
	3	.000	.008	.045	.107	.176	.234	.267	.272	.251	.212	.164	.116	.074	.042	.021	.009	.003	.001	.000	.000
	4	.000	.001	.007	.028	.066	.117	.172	.219	.251	.260	.246	.213	.167	.118	.074	.039	.017	.005	.001	.000
	5	.000	.000	.001	.005	.017	.039	.074	.118	.167	.213	.246	.260	.251	.219	.172	.117	.066	.028	.007	.001
	6	.000	.000	.000	.001	.003	.009	.021	.042	.074	.116	.164	.212	.251	.272	.267	.234	.176	.107	.045	.008
	7	.000	.000	.000	.000	.000	.001	.004	.010	.021	.041	.070	.111	.161	.216	.267	.300	.302	.260	.172	.063
	8	.000	.000	.000	.000	.000	.000	.000	.001	.004	.008	.018	.034	.060	.100	.156	.225	.302	.368	.387	.299
	9	.000	.000	.000	.000	.000	.000	.000	.000	.000	.001	.002	.005	.010	.021	.040	.075	.134	.232	.387	.630
10	0	.904	.599	.349	.197	.107	.056	.028	.014	.006	.003	.001	.000	.000	.000	.000	.000	.000	.000	.000	.000
	1	.091	.315	.387	.347	.268	.188	.121	.072	.040	.021	.010	.004	.002	.000	.000	.000	.000	.000	.000	.000
	2	.004	.075	.194	.276	.302	.282	.233	.176	.121	.076	.044	.023	.011	.004	.001	.000	.000	.000	.000	.000
	3	.000	.010	.057	.130	.201	.250	.267	.252	.215	.166	.117	.075	.042	.021	.009	.003	.001	.000	.000	.000
	4	.000	.001	.011	.040	.088	.146	.200	.238	.251	.238	.205	.160	.111	.069	.037	.016	.006	.001	.000	.000
	5	.000	.000	.001	.008	.026	.058	.103	.154	.201	.234	.246	.234	.201	.154	.103	.058	.026	.008	.001	.000
	6	.000	.000	.000	.001	.006	.016	.037	.069	.111	.160	.205	.238	.251	.238	.200	.146	.088	.040	.011	.001
	7	.000	.000	.000	.000	.001	.003	.009	.021	.042	.075	.117	.166	.215	.252	.267	.250	.201	.130	.057	.010
	8	.000	.000	.000	.000	.000	.000	.001	.004	.011	.023	.044	.076	.121	.176	.233	.282	.302	.276	.194	.075
	9	.000	.000	.000	.000	.000	.000	.000	.000	.002	.004	.010	.021	.040	.072	.121	.188	.268	.347	.387	.315
	10	.000	.000	.000	.000	.000	.000	.000	.000	.000	.001	.001	.003	.006	.014	.028	.056	.107	.197	.349	.599
11	0	.895	.569	.314	.167	.086	.042	.020	.009	.004	.001	.000	.000	.000	.000	.000	.000	.000	.000	.000	.000
	1	.099	.329	.384	.325	.236	.155	.093	.052	.027	.013	.005	.002	.001	.000	.000	.000	.000	.000	.000	.000
	2	.005	.087	.213	.287	.295	.258	.200	.140	.089	.051	.027	.013	.005	.002	.001	.000	.000	.000	.000	.000

TABLE 2 *continued*

n	r	.01	.05	.10	.15	.20	.25	.30	.35	.40	.45	.50	.55	.60	.65	.70	.75	.80	.85	.90	.95
												p									
11	3	.000	.014	.071	.152	.221	.258	.257	.225	.177	.126	.081	.046	.023	.010	.004	.001	.000	.000	.000	.000
	4	.000	.001	.016	.054	.111	.172	.220	.243	.236	.206	.161	.113	.070	.038	.017	.006	.002	.000	.000	.000
	5	.000	.000	.002	.013	.039	.080	.132	.183	.221	.236	.226	.193	.147	.099	.057	.027	.010	.002	.000	.000
	6	.000	.000	.000	.002	.010	.027	.057	.099	.147	.193	.226	.236	.221	.183	.132	.080	.039	.013	.002	.000
	7	.000	.000	.000	.000	.002	.006	.017	.038	.070	.113	.161	.206	.236	.243	.220	.172	.111	.054	.016	.001
	8	.000	.000	.000	.000	.000	.001	.004	.010	.023	.046	.081	.126	.177	.225	.257	.258	.221	.152	.071	.014
	9	.000	.000	.000	.000	.000	.000	.001	.002	.005	.013	.027	.051	.089	.140	.200	.258	.295	.287	.213	.087
	10	.000	.000	.000	.000	.000	.000	.000	.000	.001	.002	.005	.013	.027	.052	.093	.155	.236	.325	.384	.329
	11	.000	.000	.000	.000	.000	.000	.000	.000	.000	.000	.000	.001	.004	.009	.020	.042	.086	.167	.314	.569
12	0	.886	.540	.282	.142	.069	.032	.014	.006	.002	.001	.000	.000	.000	.000	.000	.000	.000	.000	.000	.000
	1	.107	.341	.377	.301	.206	.127	.071	.037	.017	.008	.003	.001	.000	.000	.000	.000	.000	.000	.000	.000
	2	.006	.099	.230	.292	.283	.232	.168	.109	.064	.034	.016	.007	.002	.001	.000	.000	.000	.000	.000	.000
	3	.000	.017	.085	.172	.236	.258	.240	.195	.142	.092	.054	.028	.012	.005	.001	.000	.000	.000	.000	.000
	4	.000	.002	.021	.068	.133	.194	.231	.237	.213	.170	.121	.076	.042	.020	.008	.002	.001	.000	.000	.000
	5	.000	.000	.004	.019	.053	.103	.158	.204	.227	.223	.193	.149	.101	.059	.029	.011	.003	.001	.000	.000
	6	.000	.000	.000	.004	.016	.040	.079	.128	.177	.212	.226	.212	.177	.128	.079	.040	.016	.004	.000	.000
	7	.000	.000	.000	.001	.003	.011	.029	.059	.101	.149	.193	.223	.227	.204	.158	.103	.053	.019	.004	.000
	8	.000	.000	.000	.000	.001	.002	.008	.020	.042	.076	.121	.170	.213	.237	.231	.194	.133	.068	.021	.002
	9	.000	.000	.000	.000	.000	.000	.001	.005	.012	.028	.054	.092	.142	.195	.240	.258	.236	.172	.085	.017
	10	.000	.000	.000	.000	.000	.000	.000	.001	.002	.007	.016	.034	.064	.109	.168	.232	.283	.292	.230	.099
	11	.000	.000	.000	.000	.000	.000	.000	.000	.000	.001	.003	.008	.017	.037	.071	.127	.206	.301	.377	.341
	12	.000	.000	.000	.000	.000	.000	.000	.000	.000	.000	.000	.001	.002	.006	.014	.032	.069	.142	.282	.540
15	0	.860	.463	.206	.087	.035	.013	.005	.002	.000	.000	.000	.000	.000	.000	.000	.000	.000	.000	.000	.000
	1	.130	.366	.343	.231	.132	.067	.031	.013	.005	.002	.000	.000	.000	.000	.000	.000	.000	.000	.000	.000
	2	.009	.135	.267	.286	.231	.156	.092	.048	.022	.009	.003	.001	.000	.000	.000	.000	.000	.000	.000	.000
	3	.000	.031	.129	.218	.250	.225	.170	.111	.063	.032	.014	.005	.002	.000	.000	.000	.000	.000	.000	.000
	4	.000	.005	.043	.116	.188	.225	.219	.179	.127	.078	.042	.019	.007	.002	.001	.000	.000	.000	.000	.000
	5	.000	.001	.010	.045	.103	.165	.206	.212	.186	.140	.092	.051	.024	.010	.003	.001	.000	.000	.000	.000
	6	.000	.000	.002	.013	.043	.092	.147	.191	.207	.191	.153	.105	.061	.030	.012	.003	.001	.000	.000	.000
	7	.000	.000	.000	.003	.014	.039	.081	.132	.177	.201	.196	.165	.118	.071	.035	.013	.003	.001	.000	.000
	8	.000	.000	.000	.001	.003	.013	.035	.071	.118	.165	.196	.201	.177	.132	.081	.039	.014	.003	.000	.000
	9	.000	.000	.000	.000	.001	.003	.012	.030	.061	.105	.153	.191	.207	.191	.147	.092	.043	.013	.002	.000
	10	.000	.000	.000	.000	.000	.001	.003	.010	.024	.051	.092	.140	.186	.212	.206	.165	.103	.045	.010	.001

TABLE 2 *continued*

n	r	.01	.05	.10	.15	.20	.25	.30	.35	.40	.45	.50	.55	.60	.65	.70	.75	.80	.85	.90	.95
15	11	.000	.000	.000	.000	.000	.000	.001	.002	.007	.019	.042	.078	.127	.179	.219	.225	.188	.116	.043	.005
	12	.000	.000	.000	.000	.000	.000	.000	.000	.002	.005	.014	.032	.063	.111	.170	.225	.250	.218	.129	.031
	13	.000	.000	.000	.000	.000	.000	.000	.000	.000	.001	.003	.009	.022	.048	.092	.156	.231	.286	.267	.135
	14	.000	.000	.000	.000	.000	.000	.000	.000	.000	.000	.000	.002	.005	.013	.031	.067	.132	.231	.343	.366
	15	.000	.000	.000	.000	.000	.000	.000	.000	.000	.000	.000	.000	.000	.002	.005	.013	.035	.087	.206	.463
16	0	.851	.440	.185	.074	.028	.010	.003	.001	.000	.000	.000	.000	.000	.000	.000	.000	.000	.000	.000	.000
	1	.138	.371	.329	.210	.113	.053	.023	.009	.003	.001	.000	.000	.000	.000	.000	.000	.000	.000	.000	.000
	2	.010	.146	.275	.277	.211	.134	.073	.035	.015	.006	.002	.001	.000	.000	.000	.000	.000	.000	.000	.000
	3	.000	.036	.142	.229	.246	.208	.146	.089	.047	.022	.009	.003	.001	.000	.000	.000	.000	.000	.000	.000
	4	.000	.006	.051	.131	.200	.225	.204	.155	.101	.057	.028	.011	.004	.001	.000	.000	.000	.000	.000	.000
	5	.000	.001	.014	.056	.120	.180	.210	.201	.162	.112	.067	.034	.014	.005	.001	.000	.000	.000	.000	.000
	6	.000	.000	.003	.018	.055	.110	.165	.198	.198	.168	.122	.075	.039	.017	.006	.001	.000	.000	.000	.000
	7	.000	.000	.000	.005	.020	.052	.101	.152	.189	.197	.175	.132	.084	.044	.019	.006	.001	.000	.000	.000
	8	.000	.000	.000	.001	.006	.020	.049	.092	.142	.181	.196	.181	.142	.092	.049	.020	.006	.001	.000	.000
	9	.000	.000	.000	.000	.001	.006	.019	.044	.084	.132	.175	.197	.189	.152	.101	.052	.020	.005	.000	.000
	10	.000	.000	.000	.000	.000	.001	.006	.017	.039	.075	.122	.168	.198	.198	.165	.110	.055	.018	.003	.000
	11	.000	.000	.000	.000	.000	.000	.001	.005	.014	.034	.067	.112	.162	.201	.210	.180	.120	.056	.014	.001
	12	.000	.000	.000	.000	.000	.000	.000	.001	.004	.011	.028	.057	.101	.155	.204	.225	.200	.131	.051	.006
	13	.000	.000	.000	.000	.000	.000	.000	.000	.001	.003	.009	.022	.047	.089	.146	.208	.246	.229	.142	.036
	14	.000	.000	.000	.000	.000	.000	.000	.000	.000	.001	.002	.006	.015	.035	.073	.134	.211	.277	.275	.146
	15	.000	.000	.000	.000	.000	.000	.000	.000	.000	.000	.000	.001	.003	.009	.023	.053	.113	.210	.329	.371
	16	.000	.000	.000	.000	.000	.000	.000	.000	.000	.000	.000	.000	.000	.001	.003	.010	.028	.074	.185	.440
20	0	.818	.358	.122	.039	.012	.003	.001	.000	.000	.000	.000	.000	.000	.000	.000	.000	.000	.000	.000	.000
	1	.165	.377	.270	.137	.058	.021	.007	.002	.000	.000	.000	.000	.000	.000	.000	.000	.000	.000	.000	.000
	2	.016	.189	.285	.229	.137	.067	.028	.010	.003	.001	.000	.000	.000	.000	.000	.000	.000	.000	.000	.000
	3	.001	.060	.190	.243	.205	.134	.072	.032	.012	.004	.001	.000	.000	.000	.000	.000	.000	.000	.000	.000
	4	.000	.013	.090	.182	.218	.190	.130	.074	.035	.014	.005	.001	.000	.000	.000	.000	.000	.000	.000	.000
	5	.000	.002	.032	.103	.175	.202	.179	.127	.075	.036	.015	.005	.001	.000	.000	.000	.000	.000	.000	.000
	6	.000	.000	.009	.045	.109	.169	.192	.171	.124	.075	.037	.015	.005	.001	.000	.000	.000	.000	.000	.000
	7	.000	.000	.002	.016	.055	.112	.164	.184	.166	.122	.074	.037	.015	.005	.001	.000	.000	.000	.000	.000
	8	.000	.000	.000	.005	.022	.061	.114	.161	.180	.162	.120	.073	.035	.014	.004	.001	.000	.000	.000	.000
	9	.000	.000	.000	.001	.007	.027	.065	.116	.160	.177	.160	.119	.071	.034	.012	.003	.000	.000	.000	.000

p

TABLE 2 *continued*

n	r	.01	.05	.10	.15	.20	.25	.30	.35	.40	.45	.50	.55	.60	.65	.70	.75	.80	.85	.90	.95
20	10	.000	.000	.000	.000	.002	.010	.031	.069	.117	.159	.176	.159	.117	.069	.031	.010	.002	.000	.000	.000
	11	.000	.000	.000	.000	.000	.003	.012	.034	.071	.119	.160	.177	.160	.116	.065	.027	.007	.001	.000	.000
	12	.000	.000	.000	.000	.000	.001	.004	.014	.035	.073	.120	.162	.180	.161	.114	.061	.022	.005	.000	.000
	13	.000	.000	.000	.000	.000	.000	.001	.005	.015	.037	.074	.122	.166	.184	.164	.112	.055	.016	.002	.000
	14	.000	.000	.000	.000	.000	.000	.000	.001	.005	.015	.037	.075	.124	.171	.192	.169	.109	.045	.009	.000
	15	.000	.000	.000	.000	.000	.000	.000	.000	.001	.005	.015	.036	.075	.127	.179	.202	.175	.103	.032	.002
	16	.000	.000	.000	.000	.000	.000	.000	.000	.000	.001	.005	.014	.035	.074	.130	.190	.218	.182	.090	.013
	17	.000	.000	.000	.000	.000	.000	.000	.000	.000	.000	.001	.004	.012	.032	.072	.134	.205	.243	.190	.060
	18	.000	.000	.000	.000	.000	.000	.000	.000	.000	.000	.000	.001	.003	.010	.028	.067	.137	.229	.285	.189
	19	.000	.000	.000	.000	.000	.000	.000	.000	.000	.000	.000	.000	.000	.002	.007	.021	.058	.137	.270	.377
	20	.000	.000	.000	.000	.000	.000	.000	.000	.000	.000	.000	.000	.000	.000	.001	.003	.012	.039	.122	.358

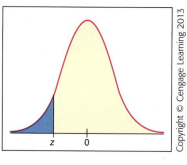

The table entry for z is the area to the left of z.

TABLE 3   Areas of a Standard Normal Distribution

(a) Table of Areas to the Left of z

z	.00	.01	.02	.03	.04	.05	.06	.07	.08	.09
−3.4	.0003	.0003	.0003	.0003	.0003	.0003	.0003	.0003	.0003	.0002
−3.3	.0005	.0005	.0005	.0004	.0004	.0004	.0004	.0004	.0004	.0003
−3.2	.0007	.0007	.0006	.0006	.0006	.0006	.0006	.0005	.0005	.0005
−3.1	.0010	.0009	.0009	.0009	.0008	.0008	.0008	.0008	.0007	.0007
−3.0	.0013	.0013	.0013	.0012	.0012	.0011	.0011	.0011	.0010	.0010
−2.9	.0019	.0018	.0018	.0017	.0016	.0016	.0015	.0015	.0014	.0014
−2.8	.0026	.0025	.0024	.0023	.0023	.0022	.0021	.0021	.0020	.0019
−2.7	.0035	.0034	.0033	.0032	.0031	.0030	.0029	.0028	.0027	.0026
−2.6	.0047	.0045	.0044	.0043	.0041	.0040	.0039	.0038	.0037	.0036
−2.5	.0062	.0060	.0059	.0057	.0055	.0054	.0052	.0051	.0049	.0048
−2.4	.0082	.0080	.0078	.0075	.0073	.0071	.0069	.0068	.0066	.0064
−2.3	.0107	.0104	.0102	.0099	.0096	.0094	.0091	.0089	.0087	.0084
−2.2	.0139	.0136	.0132	.0129	.0125	.0122	.0119	.0116	.0113	.0110
−2.1	.0179	.0174	.0170	.0166	.0162	.0158	.0154	.0150	.0146	.0143
−2.0	.0228	.0222	.0217	.0212	.0207	.0202	.0197	.0192	.0188	.0183
−1.9	.0287	.0281	.0274	.0268	.0262	.0256	.0250	.0244	.0239	.0233
−1.8	.0359	.0351	.0344	.0336	.0329	.0322	.0314	.0307	.0301	.0294
−1.7	.0446	.0436	.0427	.0418	.0409	.0401	.0392	.0384	.0375	.0367
−1.6	.0548	.0537	.0526	.0516	.0505	.0495	.0485	.0475	.0465	.0455
−1.5	.0668	.0655	.0643	.0630	.0618	.0606	.0594	.0582	.0571	.0559
−1.4	.0808	.0793	.0778	.0764	.0749	.0735	.0721	.0708	.0694	.0681
−1.3	.0968	.0951	.0934	.0918	.0901	.0885	.0869	.0853	.0838	.0823
−1.2	.1151	.1131	.1112	.1093	.1075	.1056	.1038	.1020	.1003	.0985
−1.1	.1357	.1335	.1314	.1292	.1271	.1251	.1230	.1210	.1190	.1170
−1.0	.1587	.1562	.1539	.1515	.1492	.1469	.1446	.1423	.1401	.1379
−0.9	.1841	.1814	.1788	.1762	.1736	.1711	.1685	.1660	.1635	.1611
−0.8	.2119	.2090	.2061	.2033	.2005	.1977	.1949	.1922	.1894	.1867
−0.7	.2420	.2389	.2358	.2327	.2296	.2266	.2236	.2206	.2177	.2148
−0.6	.2743	.2709	.2676	.2643	.2611	.2578	.2546	.2514	.2483	.2451
−0.5	.3085	.3050	.3015	.2981	.2946	.2912	.2877	.2843	.2810	.2776
−0.4	.3446	.3409	.3372	.3336	.3300	.3264	.3228	.3192	.3156	.3121
−0.3	.3821	.3783	.3745	.3707	.3669	.3632	.3594	.3557	.3520	.3483
−0.2	.4207	.4168	.4129	.4090	.4052	.4013	.3974	.3936	.3897	.3859
−0.1	.4602	.4562	.4522	.4483	.4443	.4404	.4364	.4325	.4286	.4247
−0.0	.5000	.4960	.4920	.4880	.4840	.4801	.4761	.4721	.4681	.4641

For values of z less than −3.49, use 0.000 to approximate the area.

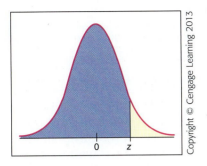

The table entry for z is the area to the left of z.

z	.00	.01	.02	.03	.04	.05	.06	.07	.08	.09
0.0	.5000	.5040	.5080	.5120	.5160	.5199	.5239	.5279	.5319	.5359
0.1	.5398	.5438	.5478	.5517	.5557	.5596	.5636	.5675	.5714	.5753
0.2	.5793	.5832	.5871	.5910	.5948	.5987	.6026	.6064	.6103	.6141
0.3	.6179	.6217	.6255	.6293	.6331	.6368	.6406	.6443	.6480	.6517
0.4	.6554	.6591	.6628	.6664	.6700	.6736	.6772	.6808	.6844	.6879
0.5	.6915	.6950	.6985	.7019	.7054	.7088	.7123	.7157	.7190	.7224
0.6	.7257	.7291	.7324	.7357	.7389	.7422	.7454	.7486	.7517	.7549
0.7	.7580	.7611	.7642	.7673	.7704	.7734	.7764	.7794	.7823	.7852
0.8	.7881	.7910	.7939	.7967	.7995	.8023	.8051	.8078	.8106	.8133
0.9	.8159	.8186	.8212	.8238	.8264	.8289	.8315	.8340	.8365	.8389
1.0	.8413	.8438	.8461	.8485	.8508	.8531	.8554	.8577	.8599	.8621
1.1	.8643	.8665	.8686	.8708	.8729	.8749	.8770	.8790	.8810	.8830
1.2	.8849	.8869	.8888	.8907	.8925	.8944	.8962	.8980	.8997	.9015
1.3	.9032	.9049	.9066	.9082	.9099	.9115	.9131	.9147	.9162	.9177
1.4	.9192	.9207	.9222	.9236	.9251	.9265	.9279	.9292	.9306	.9319
1.5	.9332	.9345	.9357	.9370	.9382	.9394	.9406	.9418	.9429	.9441
1.6	.9452	.9463	.9474	.9484	.9495	.9505	.9515	.9525	.9535	.9545
1.7	.9554	.9564	.9573	.9582	.9591	.9599	.9608	.9616	.9625	.9633
1.8	.9641	.9649	.9656	.9664	.9671	.9678	.9686	.9693	.9699	.9706
1.9	.9713	.9719	.9726	.9732	.9738	.9744	.9750	.9756	.9761	.9767
2.0	.9772	.9778	.9783	.9788	.9793	.9798	.9803	.9808	.9812	.9817
2.1	.9821	.9826	.9830	.9834	.9838	.9842	.9846	.9850	.9854	.9857
2.2	.9861	.9864	.9868	.9871	.9875	.9878	.9881	.9884	.9887	.9890
2.3	.9893	.9896	.9898	.9901	.9904	.9906	.9909	.9911	.9913	.9916
2.4	.9918	.9920	.9922	.9925	.9927	.9929	.9931	.9932	.9934	.9936
2.5	.9938	.9940	.9941	.9943	.9945	.9946	.9948	.9949	.9951	.9952
2.6	.9953	.9955	.9956	.9957	.9959	.9960	.9961	.9962	.9963	.9964
2.7	.9965	.9966	.9967	.9968	.9969	.9970	.9971	.9972	.9973	.9974
2.8	.9974	.9975	.9976	.9977	.9977	.9978	.9979	.9979	.9980	.9981
2.9	.9981	.9982	.9982	.9983	.9984	.9984	.9985	.9985	.9986	.9986
3.0	.9987	.9987	.9987	.9988	.9988	.9989	.9989	.9989	.9990	.9990
3.1	.9990	.9991	.9991	.9991	.9992	.9992	.9992	.9992	.9993	.9993
3.2	.9993	.9993	.9994	.9994	.9994	.9994	.9994	.9995	.9995	.9995
3.3	.9995	.9995	.9995	.9996	.9996	.9996	.9996	.9996	.9996	.9997
3.4	.9997	.9997	.9997	.9997	.9997	.9997	.9997	.9997	.9997	.9998

TABLE 3(a) *continued*

For values of z greater than 3.49, use 1.000 to approximate the area.
Copyright © Cengage Learning 2013

TABLE 3 *continued*

(b) Confidence Intervals, Critical Values z_c

Level of Confidence c	Critical Value z_c
0.70, or 70%	1.04
0.75, or 75%	1.15
0.80, or 80%	1.28
0.85, or 85%	1.44
0.90, or 90%	1.645
0.95, or 95%	1.96
0.98, or 98%	2.33
0.99, or 99%	2.58

Copyright © Cengage Learning 2013

TABLE 3 *continued*

(c) Hypothesis Testing, Critical Values z_0

Level of Significance	$\alpha = 0.05$	$\alpha = 0.01$
Critical value z_0 for a left-tailed test	−1.645	−2.33
Critical value z_0 for a right-tailed test	1.645	2.33
Critical values $\pm z_0$ for a two-tailed test	±1.96	±2.58

Copyright © Cengage Learning 2013

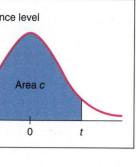

c is a confidence level

Area c

−t   0   t

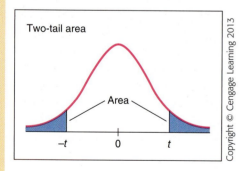

One-tail area

Right-tail area

0   t

Left-tail area

−t   0

Two-tail area

Area

−t   0   t

TABLE 4	Critical Values for Student's *t* Distribution								
one-tail area	0.250	0.125	0.100	0.075	0.050	0.025	0.010	0.005	0.0005
two-tail area	0.500	0.250	0.200	0.150	0.100	0.050	0.020	0.010	0.0010
d.f. \ c	0.500	0.750	0.800	0.850	0.900	0.950	0.980	0.990	0.999
1	1.000	2.414	3.078	4.165	6.314	12.706	31.821	63.657	636.619
2	0.816	1.604	1.886	2.282	2.920	4.303	6.965	9.925	31.599
3	0.765	1.423	1.638	1.924	2.353	3.182	4.541	5.841	12.924
4	0.741	1.344	1.533	1.778	2.132	2.776	3.747	4.604	8.610
5	0.727	1.301	1.476	1.699	2.015	2.571	3.365	4.032	6.869
6	0.718	1.273	1.440	1.650	1.943	2.447	3.143	3.707	5.959
7	0.711	1.254	1.415	1.617	1.895	2.365	2.998	3.499	5.408
8	0.706	1.240	1.397	1.592	1.860	2.306	2.896	3.355	5.041
9	0.703	1.230	1.383	1.574	1.833	2.262	2.821	3.250	4.781
10	0.700	1.221	1.372	1.559	1.812	2.228	2.764	3.169	4.587
11	0.697	1.214	1.363	1.548	1.796	2.201	2.718	3.106	4.437
12	0.695	1.209	1.356	1.538	1.782	2.179	2.681	3.055	4.318
13	0.694	1.204	1.350	1.530	1.771	2.160	2.650	3.012	4.221
14	0.692	1.200	1.345	1.523	1.761	2.145	2.624	2.977	4.140
15	0.691	1.197	1.341	1.517	1.753	2.131	2.602	2.947	4.073
16	0.690	1.194	1.337	1.512	1.746	2.120	2.583	2.921	4.015
17	0.689	1.191	1.333	1.508	1.740	2.110	2.567	2.898	3.965
18	0.688	1.189	1.330	1.504	1.734	2.101	2.552	2.878	3.922
19	0.688	1.187	1.328	1.500	1.729	2.093	2.539	2.861	3.883
20	0.687	1.185	1.325	1.497	1.725	2.086	2.528	2.845	3.850
21	0.686	1.183	1.323	1.494	1.721	2.080	2.518	2.831	3.819
22	0.686	1.182	1.321	1.492	1.717	2.074	2.508	2.819	3.792
23	0.685	1.180	1.319	1.489	1.714	2.069	2.500	2.807	3.768
24	0.685	1.179	1.318	1.487	1.711	2.064	2.492	2.797	3.745
25	0.684	1.198	1.316	1.485	1.708	2.060	2.485	2.787	3.725
26	0.684	1.177	1.315	1.483	1.706	2.056	2.479	2.779	3.707
27	0.684	1.176	1.314	1.482	1.703	2.052	2.473	2.771	3.690
28	0.683	1.175	1.313	1.480	1.701	2.048	2.467	2.763	3.674
29	0.683	1.174	1.311	1.479	1.699	2.045	2.462	2.756	3.659
30	0.683	1.173	1.310	1.477	1.697	2.042	2.457	2.750	3.646
35	0.682	1.170	1.306	1.472	1.690	2.030	2.438	2.724	3.591
40	0.681	1.167	1.303	1.468	1.684	2.021	2.423	2.704	3.551
45	0.680	1.165	1.301	1.465	1.679	2.014	2.412	2.690	3.520
50	0.679	1.164	1.299	1.462	1.676	2.009	2.403	2.678	3.496
60	0.679	1.162	1.296	1.458	1.671	2.000	2.390	2.660	3.460
70	0.678	1.160	1.294	1.456	1.667	1.994	2.381	2.648	3.435
80	0.678	1.159	1.292	1.453	1.664	1.990	2.374	2.639	3.416
100	0.677	1.157	1.290	1.451	1.660	1.984	2.364	2.626	3.390
500	0.675	1.152	1.283	1.442	1.648	1.965	2.334	2.586	3.310
1000	0.675	1.151	1.282	1.441	1.646	1.962	2.330	2.581	3.300
∞	0.674	1.150	1.282	1.440	1.645	1.960	2.326	2.576	3.291

For degrees of freedom *d.f.* not in the table, use the closest *d.f.* that is *smaller*.

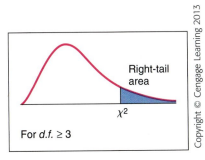

Right-tail area

χ^2

For d.f. ≥ 3

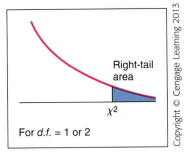

Right-tail area

χ^2

For d.f. = 1 or 2

TABLE 5 The χ^2 Distribution

d.f.	Right-Tail Area									
	.995	.990	.975	.950	.900	.100	.050	.025	.010	.005
1	0.0^4393	0.0^3157	0.0^3982	0.0^2393	0.0158	2.71	3.84	5.02	6.63	7.88
2	0.0100	0.0201	0.0506	0.103	0.211	4.61	5.99	7.38	9.21	10.60
3	0.072	0.115	0.216	0.352	0.584	6.25	7.81	9.35	11.34	12.84
4	0.207	0.297	0.484	0.711	1.064	7.78	9.49	11.14	13.28	14.86
5	0.412	0.554	0.831	1.145	1.61	9.24	11.07	12.83	15.09	16.75
6	0.676	0.872	1.24	1.64	2.20	10.64	12.59	14.45	16.81	18.55
7	0.989	1.24	1.69	2.17	2.83	12.02	14.07	16.01	18.48	20.28
8	1.34	1.65	2.18	2.73	3.49	13.36	15.51	17.53	20.09	21.96
9	1.73	2.09	2.70	3.33	4.17	14.68	16.92	19.02	21.67	23.59
10	2.16	2.56	3.25	3.94	4.87	15.99	18.31	20.48	23.21	25.19
11	2.60	3.05	3.82	4.57	5.58	17.28	19.68	21.92	24.72	26.76
12	3.07	3.57	4.40	5.23	6.30	18.55	21.03	23.34	26.22	28.30
13	3.57	4.11	5.01	5.89	7.04	19.81	22.36	24.74	27.69	29.82
14	4.07	4.66	5.63	6.57	7.79	21.06	23.68	26.12	29.14	31.32
15	4.60	5.23	6.26	7.26	8.55	22.31	25.00	27.49	30.58	32.80
16	5.14	5.81	6.91	7.96	9.31	23.54	26.30	28.85	32.00	34.27
17	5.70	6.41	7.56	8.67	10.09	24.77	27.59	30.19	33.41	35.72
18	6.26	7.01	8.23	9.39	10.86	25.99	28.87	31.53	34.81	37.16
19	6.84	7.63	8.91	10.12	11.65	27.20	30.14	32.85	36.19	38.58
20	7.43	8.26	8.59	10.85	12.44	28.41	31.41	34.17	37.57	40.00
21	8.03	8.90	10.28	11.59	13.24	29.62	32.67	35.48	38.93	41.40
22	8.64	9.54	10.98	12.34	14.04	30.81	33.92	36.78	40.29	42.80
23	9.26	10.20	11.69	13.09	14.85	32.01	35.17	38.08	41.64	44.18
24	9.89	10.86	12.40	13.85	15.66	33.20	36.42	39.36	42.98	45.56
25	10.52	11.52	13.12	14.61	16.47	34.38	37.65	40.65	44.31	46.93
26	11.16	12.20	13.84	15.38	17.29	35.56	38.89	41.92	45.64	48.29
27	11.81	12.88	14.57	16.15	18.11	36.74	40.11	43.19	46.96	49.64
28	12.46	13.56	15.31	16.93	18.94	37.92	41.34	44.46	48.28	50.99
29	13.21	14.26	16.05	17.71	19.77	39.09	42.56	45.72	49.59	52.34
30	13.79	14.95	16.79	18.49	20.60	40.26	43.77	46.98	50.89	53.67
40	20.71	22.16	24.43	26.51	29.05	51.80	55.76	59.34	63.69	66.77
50	27.99	29.71	32.36	34.76	37.69	63.17	67.50	71.42	76.15	79.49
60	35.53	37.48	40.48	43.19	46.46	74.40	79.08	83.30	88.38	91.95
70	43.28	45.44	48.76	51.74	55.33	85.53	90.53	95.02	100.4	104.2
80	51.17	53.54	57.15	60.39	64.28	96.58	101.9	106.6	112.3	116.3
90	59.20	61.75	65.65	69.13	73.29	107.6	113.1	118.1	124.1	128.3
100	67.33	70.06	74.22	77.93	82.36	118.5	124.3	129.6	135.8	140.2

Source: From H. L. Herter, *Biometrika,* June 1964. Printed by permission of the Biometrika Trustees.

0.0^4393 means that there are 4 0's between the decimal and 3; $0.0^4393 = 0.0000393$.

ANSWERS AND KEY STEPS TO ODD-NUMBERED PROBLEMS

CHAPTER 1

Chapter 1 Review Problems

1. Because of the requirement that each number appear only once in any row, column, or box, it would be very inefficient to use a random-number table to select the numbers. It's better to simply look at existing numbers, list possibilities that meet the requirement, and eliminate numbers that don't work.

3. (a) Stratified. (b) Students on your campus with work-study jobs. (c) Hours scheduled; quantitative; ratio. (d) Rating of applicability of work experience to future employment; qualitative; ordinal. (e) Statistic. (f) 60%; the people choosing not to respond may have some characteristics, such as not working many hours, that would bias the study. (g) No. The sample frame is restricted to one campus.

5. Assign digits so that 3 out of the 10 digits 0 through 9 correspond to the answer "Yes" and 7 of the digits correspond to the answer "No." One assignment is digits 0, 1, and 2 correspond to "Yes," while digits 3, 4, 5, 6, 7, 8, and 9 correspond to "No." Starting with line 1, block 1 of Table 1, this assignment of digits gives the sequence No, Yes, No, No, Yes, No, No.

7. (a) Observational study. (b) Experiment.

9. Possible directions on survey questions: Give height in inches, give age as of last birthday, give GPA to one decimal place, and so forth. Think about the types of responses you wish to have on each question.

11. (a) Experiment, since a treatment is imposed on one colony. (b) The control group receives normal daylight–darkness conditions. The treatment group has light 24 hours per day. (c) The number of fireflies living at the end of 72 hours. (d) Ratio.

CHAPTER 2

Chapter 2 Review Problems

1. (a) Bar graph, Pareto chart, pie chart. (b) All.

3. Any large gaps between bars or stems with leaves at the beginning or end of the data set might indicate that the extreme data values are outliers.

5. (a) Yes, with lines used instead of bars. However, because of the perspective nature of the drawing, the lengths of the bars do not represent the mileages. Thus,

the scale for each bar changes. (b) Yes. The scale does not change, and the viewer is not distracted by the graphic of the highway.

7. Problems with Tax Returns

9. (a) Class width = 11.

Class Limits	Class Boundaries	Midpoint	Frequency	Relative Frequency	Cumulative Frequency
69–79	68.5–79.5	74	2	0.03	2
80–90	79.5–90.5	85	3	0.05	5
91–101	90.5–101.5	96	8	0.13	13
102–112	101.5–112.5	107	19	0.32	32
113–123	112.5–123.5	118	22	0.37	54
124–134	123.5–134.5	129	3	0.05	57
135–145	134.5–145.5	140	3	0.05	60

(b, c) Trunk Circumference (mm)—Histogram and Relative-Frequency Histogram

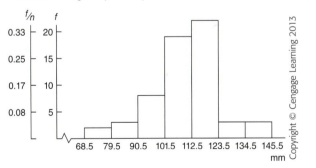

(d) Skewed slightly left.
(e) Trunk Circumference (mm)—Ogive

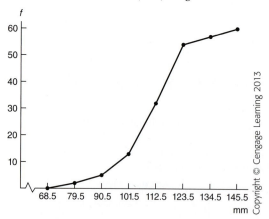

11. (a) The 1240s. (b) 75. (c) 1203–1212. Little if any repairs or new construction.

CHAPTER 3

Chapter 3 Review Problems

1. (a) Variance and standard deviation. (b) Box-and-whisker plot.
3. (a) For both data sets, mean = 20 and range = 24. (b) The C1 distribution seems more symmetrical because the mean and median are equal, and the median is in the center of the interquartile range. In the C2 distribution, the mean is less than the median. (c) The C1 distribution has a larger interquartile range that is symmetrical around the median. The C2 distribution has a very compressed interquartile range, with the median equal to Q_3.
5. (a) Low = 31; Q_1 = 40; median = 45; Q_3 = 52.5; high = 68; IQR = 12.5.
Percentage of Democratic Vote by County

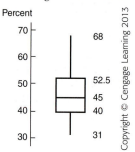

(b) Class width = 8.

Class	Midpoint	f
31–38	34.5	11
39–46	42.5	24
47–54	50.5	15
55–62	58.5	7
63–70	66.5	3

$\bar{x} \approx 46.1$; $s \approx 8.64$; 28.82 to 63.38.
(c) $\bar{x} \approx 46.15$; $s \approx 8.63$.

7. Mean weight = 156.25 pounds.
9. (a) No. (b) $34,206 to $68,206. (c) $10,875.
11. $\Sigma w = 16$, $\Sigma wx = 121$, average = 7.56.

CUMULATIVE REVIEW PROBLEMS

Chapters 1–3

1. (a) Median, percentile. (b) Mean, variance, standard deviation.
2. (a) Gap between first bar and rest of bars or between last bar and rest of bars. (b) Large gap between data on far left or far right side and rest of data. (c) Several empty stems after stem including lowest values or before stem including highest values. (d) Data beyond fences placed at $Q_1 - 1.5(IQR)$ and $Q_3 + 1.5(IQR)$.
3. (a) Same. (b) Set B has a higher mean. (c) Set B has a higher standard deviation. (d) Set B has a much longer whisker beyond Q_3.
4. (a) Set A, because 86 is the relatively higher score since a larger percentage of scores fall below it. (b) Set B, because 86 is more standard deviations above the mean.
5. Assign consecutive numbers to all the wells in the study region. Then use a random-number table, computer, or calculator to select 102 values that are less than or equal to the highest number assigned to a well in the study region. The sample consists of the wells with numbers corresponding to those selected.
6. Ratio.
7. 7 | 0 represents a pH level of 7.0

7	0 0 0 0 0 0 0 0 1 1 1 1 1 1 1 1 1 1
7	2 2 2 2 2 2 2 2 2 2 3 3 3 3 3 3 3 3 3 3
7	4 4 4 4 4 4 4 4 4 5 5 5 5 5 5 5 5
7	6 6 6 6 6 6 6 6 6 7 7 7 7 7 7
7	8 8 8 8 8 9 9 9 9 9
8	0 1 1 1 1 1 1 1
8	2 2 2 2 2 2 2
8	4 5
8	6 7
8	8 8

8. Clear the decimals. Then the highest value is 88 and the lowest is 70. The class width for the whole numbers is 4. For the actual data, the class width is 0.4.

Class Limits	Class Boundaries	Midpoint	Frequency	Relative Frequency
7.0–7.3	6.95–7.35	7.15	39	0.38
7.4–7.7	7.35–7.75	7.55	32	0.31
7.8–8.1	7.75–8.15	7.95	18	0.18
8.2–8.5	8.15–8.55	8.35	9	0.09
8.6–8.9	8.55–8.95	8.75	4	0.04

Levels of pH in West Texas Wells—Histogram and Relative-Frequency Histogram

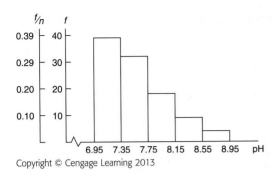

9. Levels of pH in West Texas Wells—Ogive

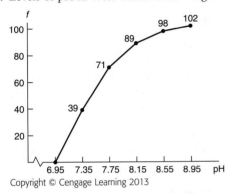

10. Range = 1.8; $\bar{x} \approx 7.58$; median = 7.5; mode = 7.3.
11. (a) Use a calculator or computer. (b) $s^2 \approx 0.20$; $s \approx 0.45$; $CV \approx 5.9\%$.
12. 6.68 to 8.48.
13. Levels of pH in West Texas Wells

IQR = 0.7.
14. Skewed right. Lower values are more common.
15. 89%; 50%.
16. No, there are no gaps in the plot, but only 6 out of 102, or about 6%, have pH levels at or above 8.4. Eight wells are neutral.
17. Half the wells have pH levels between 7.2 and 7.9. The data are skewed toward the high values, with the upper half of the pH levels spread out more than the lower half. The upper half ranges between 7.5 and 8.8, while the lower half is clustered between 7 and 7.5.
18. The report should emphasize the relatively low mean, median, and mode, and the fact that half the wells have a pH level less than 7.5. The data are clustered at the low end of the range.

Chapter 4 Review Problems

1. r will be close to 0.
3. Results are more reliable for interpolation.
5. (a) Age and Mortality Rate for Bighorn Sheep

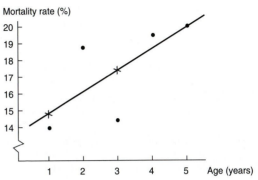

(b) $\bar{x} = 3$; $\bar{y} \approx 17.38$; $b \approx 1.27$; $\hat{y} \approx 13.57 + 1.27x$.
(c) $r \approx 0.685$; $r^2 \approx 0.469$; 46.9% explained.
7. (a) Weight of One-Year-Old versus Weight of Adult

(b) $\bar{x} \approx 21.43$; $\bar{y} \approx 126.79$; $b \approx 1.285$; $\hat{y} \approx 99.25 + 1.285x$. (c) $r \approx 0.468$; $r^2 \approx 0.219$; 21.9% explained. (d) 124.95 pounds.
9. (a) Weight of Mail versus Number of Employees Required

(b) $\bar{x} \approx 16.38$; $\bar{y} \approx 10.13$; $b \approx 0.554$;
$\hat{y} \approx 1.051 + 0.554x$.   (c) $r \approx 0.913$; $r^2 \approx 0.833$;
83.3% explained.   (d) 9.36.

CHAPTER 5

Chapter 5 Review Problems

1. (a) The individual does not own a cell phone.   (b) The individual owns a cell phone as well as a laptop computer.   (c) The individual owns either a cell phone or a laptop computer, and maybe both.   (d) The individual owns a cell phone, given he or she owns a laptop computer.   (e) The individual owns a laptop computer, given he or she owns a cell phone.
3. For independent events A and B, $P(A) = P(A|B)$.
5. (a) P(offer job 1 *and* offer job 2) = 0.56. The probability of getting offers for both jobs is less than the probability of getting each individual job offer.
   (b) P(offer job 1 *or* offer job 2) = 0.94. The probability of getting at least one of the job offers is greater than the probability of getting each individual job offer. It seems worthwhile to apply for both jobs since the probability is high of getting at least one offer.
7. (a) No. You need to know that the events are independent or you need to know the value of $P(A|B)$ or $P(B|A)$.   (b) Yes. For independent events, $P(A \text{ and } B) = P(A) \cdot P(B)$.
9. P(asked) = 24%; P(received | asked) = 45%; P(asked *and* received) = (0.24)(0.45) = 10.8%.
11. (a) Drop a fixed number of tacks and count how many land flat side down. Then form the ratio of the number landing flat side down to the total number dropped.   (b) Up, down.   (c) P(up) = 160/500 = 0.32; P(down) = 340/500 = 0.68.
13. (a)

Outcomes x	2	3	4	5	6
$P(x)$	0.028	0.056	0.083	0.111	0.139

   (b)

x	7	8	9	10	11	12
$P(x)$	0.167	0.139	0.111	0.083	0.056	0.028

15. $C_{8,2} = [8!/(2!6!)] = (8 \cdot 7)/2 = 28$.
17. $4 \cdot 4 \cdot 4 \cdot 4 \cdot 4 = 1024$ choices; P(all correct) = $1/1024 \approx 0.00098$.
19. $10 \cdot 10 \cdot 10 = 1000$.

CHAPTER 6

Chapter 6 Review Problems

1. A description of all distinct possible values of a random variable x, with a probability assignment $P(x)$ for each value or range of values. $0 \le P(x) \le 1$ and $\Sigma P(x) = 1$.
3. (a) Yes. $\mu = 2$ and $\sigma \approx 1.3$. Numbers of successes above 5.25 are unusual.   (b) No. It would be unusual to get more than five questions correct.

5. (a) 38; 11.6.
   (b) Duration of Leases in Months

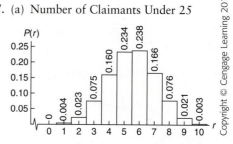

7. (a) Number of Claimants Under 25

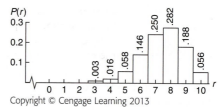

   (b) $P(r \ge 6) = 0.504$.   (c) $\mu = 5.5$; $\sigma \approx 1.57$.
9. (a) 0.039.   (b) 0.403.   (c) 8.
11. (a) Number of Good Grapefruit

   (b) 0.244; 0.999.   (c) 7.5.   (d) 1.37.
13. $P(r \le 2) = 0.000$ (to three digits). The data seem to indicate that the percent favoring the increase in fees is less than 85%.

CUMULATIVE REVIEW PROBLEMS

1. The specified ranges of readings are disjoint and cover all possible readings.
2. Essay.
3. Yes; the events constitute the entire sample space.
4. (a) 0.85.   (b) 0.70.   (c) 0.70.   (d) 0.30.   (e) 0.15.
   (f) 0.75.   (g) 0.30.   (h) 0.05.
5.

x	$P(x)$
5	0.25
15	0.45
25	0.15
35	0.10
45	0.05

   $\mu \approx 17.5$; $\sigma \approx 10.9$.
6. (a) $p = 0.10$.   (b) $\mu = 1.2$; $\sigma \approx 1.04$.   (c) 0.718.
   (d) 0.889.

7. (a) Blood Glucose Level

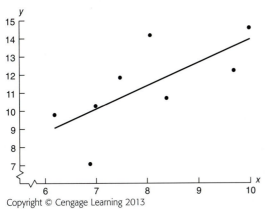

(b) $\hat{y} \approx 1.135 + 1.279x$.   (c) $r \approx 0.700$; $r^2 \approx 0.490$; 49% of the variance in y is explained by the model and the variance in x.   (d) 12.65.

CHAPTER 7

Chapter 7 Review Problems

1. Normal probability distributions are distributions of continuous random variables. They are symmetrical about the mean and bell-shaped. Most of the data fall within 3 standard deviations of the mean. The mean and median are the same.
3. Normal; $\mu_{\bar{x}} = 10$; $\sigma_{\bar{x}} = 1$.
5. The points lie close to a straight line.
7. $\sigma_{\bar{x}} = \sigma/\sqrt{n}$.
9. (a) A normal distribution.   (b) The mean μ of the x distribution.   (c) $\sigma/\sqrt{n}$, where σ is the standard deviation of the x distribution.   (d) They will both be approximately normal with the same mean, but the standard deviations will be $\sigma/\sqrt{50}$ and $\sigma/\sqrt{100}$, respectively.
11. (a) 0.9821.   (b) 0.3156.   (c) 0.2977.
13. 1.645.
15. (a) 0.8665.   (b) 0.7330.
17. (a) 0.0166.   (b) 0.9750.
19. (a) 0.9772.   (b) 17.3 hours.
21. $P(98 \le \bar{x} \le 102) = P(-1.33 \le z \le 1.33) = 0.8164$.
23. (a) Yes, np and nq both exceed 5.   (b) $\mu_{\hat{p}} = 0.4$; $\sigma_{\hat{p}} = 0.1$.

CHAPTER 8

Chapter 8 Review Problems

1. See text.
3. (a) No, the probability that μ is in the interval is either 0 or 1.   (b) Yes, 99% confidence intervals are constructed in such a way that 99% of all such confidence intervals based on random samples of the designated size will contain μ.
5. Interval for a mean; 176.91 to 180.49.
7. Interval for a mean.   (a) Use a calculator.   (b) 64.1 cm to 84.3 cm.

9. Interval for a proportion; 0.50 to 0.54.
11. Interval for a proportion.   (a) $\hat{p} = 0.4072$.   (b) 0.333 to 0.482.
13. (a) $P(A_1 < \mu_1 < B_1 \text{ and } A_2 < \mu_2 < B_2) = (0.80)(0.80) = 0.64$. The complement of the event $A_1 < \mu_1 < B_1$ and $A_2 < \mu_2 < B_2$ is that either μ_1 is not in the first interval or μ_2 is not in the second interval, or both. Thus, P(at least one interval fails) $= 1 - P(A_1 < \mu_1 < B_1 \text{ and } A_2 < \mu_2 < B_2) = 1 - 0.64 = 0.36$.   (b) Suppose $P(A_1 < \mu_1 < B_1) = c$ and $P(A_2 < \mu_2 < B_2) = c$. If we want the probability that both hold to be 90%, and if x_1 and x_2 are independent, then $P(A_1 < \mu_1 < B_1 \text{ and } A_2 < \mu_2 < B_2) = 0.90$ means $P(A_1 < \mu_1 < B_1) \cdot P(A_2 < \mu_2 < B_2) = 0.90$, so $c^2 = 0.90$, or $c = 0.9487$.   (c) In order to have a high probability of success for the whole project, the probability that each component will perform as specified must be significantly higher.

CHAPTER 9

Chapter 9 Review Problems

1. Look at the original x distribution. If it is normal or $n \ge 30$, and σ is known, use the standard normal distribution. If the x distribution is mound-shaped or $n \ge 30$, and σ is unknown, use the Student's t distribution. The $d.f.$ is $n - 1$.
3. A larger sample size increases the $|z|$ or $|t|$ value of the sample test statistic.
5. Single mean.   (a) $\alpha = 0.05$; H_0: $\mu = 11.1$; H_1: $\mu \ne 11.1$.   (b) Standard normal; $z = -3.00$.   (c) P-value = 0.0026; on standard normal curve, shade areas to the right of 3.00 and to the left of -3.00.   (d) P-value of 0.0026 $\le$ 0.05 for α; reject H_0.   (e) At the 5% level of significance, the evidence is sufficient to say that the miles driven per vehicle in Chicago are different from the national average.
7. Single mean.   (a) $\alpha = 0.01$; H_0: $\mu = 0.8$; H_1: $\mu > 0.8$.   (b) Student's t, $d.f. = 8$; $t \approx 4.390$.   (c) $0.0005 < P$-value < 0.005; on t graph, shade area to the right of 4.390. From TI-84, P-value ≈ 0.0012.   (d) P-value interval ≤ 0.01 for α; reject H_0.   (e) At the 1% level of significance, the evidence is sufficient to say that the Toylot claim of 0.8 A is too low.
9. Single proportion.   (a) $\alpha = 0.01$; H_0: $p = 0.60$; H_1: $p < 0.60$.   (b) Standard normal; $z = -3.01$.   (c) P-value = 0.0013; on standard normal curve, shade area to the left of -3.01.   (d) P-value of 0.0013 ≤ 0.01 for α; reject H_0.   (e) At the 1% level of significance, the evidence is sufficient to show that the mortality rate has dropped.
11. Single mean.   (a) $\alpha = 0.05$; H_0: $\mu = 7$ oz; H_1: $\mu \ne 7$ oz.   (b) Student's t, $d.f. = 7$; $t \approx 1.697$.   (c) $0.100 < P$-value < 0.150; on t graph, shade areas to the right of 1.697 and to the left of -1.697. From TI-84, P-value ≈ 0.1335.   (d) P-value interval > 0.05 for α; do not reject H_0.   (e) At the 5% level of significance, the evidence is insufficient to show that the population mean amount of coffee per cup is different from 7 oz.

CUMULATIVE REVIEW PROBLEMS

1. $\mu = 0$; $\sigma = 1$.
2. 95%.
3. Essay based on material from Chapter 7 and Section 1.2.
4. (a) $\sigma \approx 1.7$. (b) 0.1314. (c) 0.1075.
5. (a) Yes. Both np and nq are greater than 5.
   (b) $\mu \approx 73.5$; $\sigma \approx 5.6$. (c) $P(r \geq 65) \approx P(x \geq 64.5) \approx P(z \geq -1.61) \approx 0.9463$.
6. (a) Because of the large sample size, the central limit theorem describes the $\bar{x}$ distribution (approximately).
   (b) $P(\bar{x} \leq 6820) = P(z \leq -2.75) = 0.0030$. (c) The probability that the average white blood cell count for 50 healthy adults is as low as or lower than 6820 is very small, 0.0030. Based on this result, it would be reasonable to gather additional facts.
7. (a) Yes, np and nq both exceed 5. (b) $\mu_{\hat{p}} = 0.45$; $\sigma_{\hat{p}} \approx 0.09$.
8. (a) i. $\alpha = 0.01$; H_0: $\mu = 2.0$ μg/L; H_1: $\mu > 2.0$ μg/L.
   ii. Standard normal; $z = 2.53$.
   iii. P-value ≈ 0.0057; on standard normal curve, shade area to the right of 2.53.
   iv. P-value of $0.0057 \leq 0.01$ for α; reject H_0.
   v. At the 1% level of significance, the evidence is sufficient to say that the population mean discharge level of lead is higher.
   (b) 2.13 μg/L to 2.99 μg/L. (c) $n = 48$.
9. (a) Use rounded results to compute t in part (b).
   (b) i. $\alpha = 0.05$; H_0: $\mu = 10\%$; H_1: $\mu > 10\%$.
   ii. Student's t, $d.f. = 11$; $t \approx 1.248$.
   iii. $0.100 < P$-value < 0.125; on t graph, shade area to the right of 1.248. From TI-84, P-value ≈ 0.1190.
   iv. P-value interval > 0.05 for α; fail to reject H_0.
   v. At the 5% level of significance, the evidence does not indicate that the patient is asymptomatic.
   (c) 9.27% to 11.71%.
10. (a) i. $\alpha = 0.05$; H_0: $p = 0.10$; H_1: $p \neq 0.10$; yes, $np > 5$ and $nq > 5$; necessary to use normal approximation to the binomial.
    ii. Standard normal; $\hat{p} \approx 0.147$; $z = 1.29$.
    iii. P-value $= 2P(z > 1.29) \approx 0.1970$; on standard normal curve, shade areas to the right of 1.29 and to the left of -1.29.
    iv. P-value of $0.1970 > 0.05$ for α; fail to reject H_0.
    v. At the 5% level of significance, the data do not indicate any difference from the national average for the population proportion of crime victims.
    (b) 0.063 to 0.231. (c) From sample, $p \approx \hat{p} \approx 0.147$; $n = 193$.
11. (a) Essay. (b) Outline of study.
12. Essay.

3. Difference of means.
   (a) Use a calculator. (b) $d.f. \approx 71$; to use Table 4, round down to $d.f. \approx 70$; $E \approx 0.83$; interval from -0.06 to 1.6. (c) Because the interval contains both positive and negative values, we cannot conclude at the 95% confidence level that there is any difference in soil water content in the two fields. (d) Student's t distribution, because σ_1 and σ_2 are unknown. Both samples are large, so no assumptions about the original distributions are needed. (e) (i) $\alpha = 0.01$; H_0: $\mu_1 = \mu_2$; H_1: $\mu_1 > \mu_2$.
   (ii) Student's t; $d.f. = 71$ (use $d.f. = 70$ in Table 4 of the Appendix); $t \approx 1.841$. (iii) $0.025 < P$-value < 0.050; on t graph, shade area to the right of 1.841. From TI-84, $d.f. \approx 140.50$; P-value ≈ 0.034. (iv) P-value > 0.01 for α; fail to reject H_0. (v) At the 1% level of significance, the evidence does not show that the population mean soil water content of the first field is higher than that of the second.
5. Difference of means.
   (a) $d.f. \approx 17$; $E \approx 2.5$; interval from 5.5 to 10.5 lb.
   (b) Yes. The interval contains values that are all positive. At the 75% level of confidence, it appears that the average weight of adult male wolves from the Northwest Territories is greater. (c) (i) $\alpha = 0.01$; H_0: $\mu_1 = \mu_2$; H_1: $\mu_1 \neq \mu_2$. (ii) Student's t; $d.f. = 17$; $t \approx 3.743$. (iii) $0.001 < P$-value < 0.010; on t graph, shade areas to the right of 3.743 and to the left of -3.743. (iv) P-value ≤ 0.01 for α; reject H_0. (v) At the 1% level of significance, the evidence is sufficient to conclude that the average weight of adult male wolves in the Northwest Territories is different from that of Alaska wolves.
7. Difference of proportions.
   (a) $\hat{p}_1 = 0.8495$; $\hat{p}_2 = 0.8916$; -0.1409 to 0.0567.
   (b) No. The interval contains both negative and positive numbers. We do not detect a difference in the proportions at the 95% confidence level. (c) (i) $\alpha = 0.05$; H_0: $p_1 = p_2$; H_1: $p_1 \neq p_2$. (ii) Normal distribution; $z \approx -0.83$. (iii) P-value $= 2(0.2033) = 0.4066$; on the normal graph, shade areas to the left of -0.83 and to the right of 0.83. (iv) P-value of $0.4066 > \alpha = 0.05$; fail to reject H_0. (v) At the 5% level of significance, there is insufficient evidence to conclude that the proportion of accurate responses from face-to-face interviews differs from the proportion for telephone interviews.
9. (i) $\alpha = 0.05$; H_0: $\mu_d = 0$; H_1: $\mu_d < 0$. (ii) Student's t, $d.f. = 4$; $\bar{d} \approx -4.94$; $t = -2.832$. (iii) $0.010 < P$-value < 0.025; on t graph, shade area to the left of -2.832. From TI-84, P-value ≈ 0.0236. (iv) P-value ≤ 0.05 for α; reject H_0. (v) At the 5% level of significance, there is sufficient evidence to claim that the population average net sales improved.

CHAPTER 10

Chapter 10 Review Problems

1. Two random samples are independent if the selection of sample data from one population is completely unrelated to the selection of sample data from the other population.

CHAPTER 11

Chapter 11 Review

1. Chi-square.
3. Test of homogeneity.
5. Chi-square test of σ^2. (i) $\alpha = 0.01$; H_0: $\sigma^2 =$

1,040,400; H_1: $\sigma^2 > 1,040,400$. (ii) $\chi^2 \approx 51.03$; $d.f. = 29$. (iii) $0.005 < P\text{-value} < 0.010$. (iv) Reject H_0. (v) At the 1% level of significance, there is sufficient evidence to conclude that the variance is greater than claimed.

7. Chi-square test of independence. (i) $\alpha = 0.01$; H_0: Student grade and teacher rating are independent; H_1: Student grade and teacher rating are not independent. (ii) $\chi^2 \approx 9.80$; $d.f. = 6$. (iii) $0.100 < P\text{-value} < 0.900$. From TI-84, $P\text{-value} \approx 0.1337$. (iv) Do not reject H_0. (v) At the 1% level of significance, there is insufficient evidence to claim that student grade and teacher rating are not independent.

9. We reject the null hypothesis that ρ is zero.

11. (a) Use a calculator. (b) (i) $\alpha = 0.01$; H_0: $\rho = 0$; H_1: $\rho > 0$. (ii) $t \approx 6.665$; $d.f. = 5$. (iii) $0.0005 < P\text{-value} < 0.005$. From TI-84, $P\text{-value} \approx 0.0006$. (iv) Reject H_0. (v) At the 1% level of significance, there is evidence that ρ is positive. (c) (i) $\alpha = 0.01$; H_0: $\beta = 0$; H_1: $\beta > 0$. (ii) $t \approx 6.665$; $d.f. = 5$. (iii) $0.0005 < P\text{-value} < 0.005$. From TI-84, $P\text{-value} \approx 0.0006$. (iv) Reject H_0. (v) At the 1% level of significance, there is evidence that β is positive. (d) $\hat{y} \approx 5.98$; $E \approx 1.48$; 4.498 to 7.458.

CUMULATIVE REVIEW PROBLEMS
CHAPTERS 10–11

1. (a) Use a calculator. (b) (i) $\alpha = 0.05$; H_0: $\rho = 0$; H_1: $\rho > 0$. (ii) Student's t, $d.f. = 6$; $t \approx 2.466$. (iii) $0.010 < P\text{-value} < 0.025$. On the t graph, shade area to the right of 2.466. From TI-84, $P\text{-value} \approx 0.0244$. (iv) $P\text{-value}$ interval $< \alpha = 0.05$; reject H_0. (v) At the 5% level of significance, there is evidence of a positive correlation between age of a volcanic island in the Indian Ocean and distance of the island from the center of the mid-oceanic ridge. (c) (i) $\alpha = 0.05$; H_0: $\beta = 0$; H_1: $\beta > 0$. (ii) Student's t, $d.f. = 6$; $t \approx 2.466$. (iii) $0.010 < P\text{-value} < 0.025$. On the t graph, shade area to the right of 2.466. From TI-84, $P\text{-value} \approx 0.0244$. (iv) $P\text{-value}$ interval $< \alpha = 0.05$; reject H_0. (v) At the 5% level of significance, there is evidence of a positive slope for the regression line for age of a volcanic island in the Indian Ocean and distance of the island from the center of the midoceanic ridge. (d) $\hat{y} \approx 16.5$ (units in 100 km); $d.f. = 6$; $t_{0.85} = 1.650$; $E \approx 11.4$; 5.1 to 27.9 (units in 100 km).

2. (i) $\alpha = 0.01$; H_0: Colorado major distribution is the same as national distribution; H_1: Colorado major distri-bution is different from national distribution. (ii) Chi-square; $d.f. = 5$; $\chi^2 \approx 10.196$. (iii) $0.050 < P\text{-value} < 0.100$; shade area to the right of 10.196. (iv) $P\text{-value}$ interval $> \alpha = 0.01$; fail to reject H_0. (v) At the 1% level of significance, the evidence is insufficient to claim that the distribution of college majors selected by Colorado students is different from the national distribution of college majors.

3. (i) $\alpha = 0.05$; H_0: Yield and fertilizer type are indepen-dent; H_1: Yield and fertilizer type are not indepen-dent. (ii) Chi-square distribution $\chi^2 \approx 5.005$; $d.f. = 4$. (iii) $0.100 < P\text{-value} < 0.900$. From TI-84, $P\text{-value} \approx 0.2868$. (iv) Do not reject H_0. (v) At the 5% level of significance, the evidence is insufficient to conclude that fertilizer type and yield are not independent.

4. (i) $\alpha = 0.05$; H_0: $\sigma = 0.55$; H_1: $\sigma > 0.55$. (ii) Chi-square distribution $s \approx 0.602$; $d.f. = 9$; $\chi^2 \approx 10.78$. (iii) $0.100 < P\text{-value} < 0.900$. (iv) Do not reject H_0. (v) At the 5% level of significance, there is insuffi-cient evidence to conclude that the standard deviation of petal lengths is greater than 0.55.

5. (i) $\alpha = 0.05$; H_0: $\mu_d = 0$; H_1: $\mu_d \neq 0$. (ii) Student's t, $d.f. = 6$; $\bar{d} \approx -0.0039$; $t \approx -0.771$. (iii) $0.250 < P\text{-value} < 0.500$; on t graph, shade areas to the right of 0.771 and to the left of -0.771. From TI-84, $P\text{-value} \approx 0.4699$. (iv) $P\text{-value}$ interval > 0.05 for α; fail to reject H_0. (v) At the 5% level of significance, the evidence does not show a population mean difference in phosphorus reduction between the two methods.

6. (a) (i) $\alpha = 0.05$; H_0: $\mu_1 = \mu_2$; H_1: $\mu_1 \neq \mu_2$. (ii) Stu-dent's t, $d.f. = 15$; $t \approx 1.952$. (iii) $0.050 < P\text{-value} < 0.100$; on t graph, shade areas to the right of 1.952 and to the left of -0.1952. From TI-84, $P\text{-value} \approx 0.0609$. (iv) $P\text{-value}$ interval > 0.05 for α; fail to reject H_0. (v) At the 5% level of significance, the evidence does not show any difference in the population mean pro-portion of on-time arrivals in summer versus winter. (b) -0.43% to 9.835%. (c) x_1 and x_2 distributions are approximately normal (mound-shaped and symmetrical).

7. (a) (i) $\alpha = 0.05$; H_0: $p_1 = p_2$; H_1: $p_1 > p_2$. (ii) Stan-dard normal; $\hat{p}_1 \approx 0.242$; $\hat{p}_2 \approx 0.207$; $\bar{p} \approx 0.2246$; $z \approx 0.58$. (iii) $P\text{-value} \approx 0.2810$; on standard normal curve, shade area to the right of 0.58. (iv) $P\text{-value}$ interval > 0.05 for α; fail to reject H_0. (v) At the 5% level of significance, the evidence does not indicate that the pop-ulation proportion of single men who go out dancing oc-casionally differs from the proportion of single women. Since $n_1\bar{p}$, $n_1\bar{q}$, $n_2\bar{p}$, and $n_2\bar{q}$ are all greater than 5, the normal approximation to the binomial is justified. (b) -0.065 to 0.139.

INDEX